Scramjet Propulsion

Aerospace Series

Scramjet Propulsion: A Practical Introduction
Dora Musielak

Aircraft System Classifications: A Handbook of Characteristics and Design Guidelines
Allan Seabridge and Mohammad Radaei

UAS Integration into Civil Airspace: Policy, Regulations and Strategy
Douglas M. Marshall

Introduction to UAV Systems, Fifth Edition
Paul G. Fahlstrom, Thomas J. Gleason, Mohammad H. Sadraey

Introduction to Flight Testing
James W. Gregory, Tianshu Liu

Foundations of Space Dynamics
Ashish Tewari

Essentials of Supersonic Commercial Aircraft Conceptual Design
Egbert Torenbeek

Design of Unmanned Aerial Systems
Mohammad H. Sadraey

Future Propulsion Systems and Energy Sources in Sustainable Aviation
Saeed Farokhi

Flight Dynamics and Control of Aero and Space Vehicles
Rama K. Yedavalli

Design and Development of Aircraft Systems, 3rd Edition
Allan Seabridge, Ian Moir

Helicopter Flight Dynamics: Including a Treatment of Tiltrotor Aircraft, 3rd Edition
Gareth D. Padfield CEng, PhD, FRAeS

Space Flight Dynamics, 2nd Edition
Craig A. Kluever

Performance of the Jet Transport Airplane: Analysis Methods, Flight Operations, and Regulations
Trevor M. Young

Small Unmanned Fixed-wing Aircraft Design: A Practical Approach
Andrew J. Keane, András Sóbester, James P. Scanlan

Advanced UAV Aerodynamics, Flight Stability and Control: Novel Concepts, Theory and Applications
Pascual Marqués, Andrea Da Ronch

Differential Game Theory with Application to Missiles and Autonomous Systems Guidance
Farhan A. Faruqi

Introduction to Nonlinear Aeroelasticity
Grigorios Dimitriadis

Introduction to Aerospace Engineering with a Flight Test Perspective
Stephen Corda

Aircraft Control Allocation
Wayne Durham, Kenneth A. Bordignon, Roger Beck

Remotely Piloted Aircraft Systems: A Human Systems Integration Perspective
Nancy J. Cooke, Leah J. Rowe, Winston Bennett Jr., DeForest Q. Joralmon

Theory and Practice of Aircraft Performance
Ajoy Kumar Kundu, Mark A. Price, David Riordan

Adaptive Aeroservoelastic Control
Ashish Tewari

The Global Airline Industry, 2nd Edition
Peter Belobaba, Amedeo Odoni, Cynthia Barnhart, Christos Kassapoglou

Introduction to Aircraft Aeroelasticity and Loads, 2nd Edition
Jan R.Wright, Jonathan Edward Cooper

Theoretical and Computational Aerodynamics
Tapan K. Sengupta

Aircraft Aerodynamic Design: Geometry and Optimization
András Sóbester, Alexander I J Forrester

Stability and Control of Aircraft Systems: Introduction to Classical Feedback Control
Roy Langton

Aerospace Propulsion
T.W. Lee

Civil Avionics Systems, 2nd Edition
Ian Moir, Allan Seabridge, Malcolm Jukes

Aircraft Flight Dynamics and Control
Wayne Durham

Modelling and Managing Airport Performance
Konstantinos Zografos, Giovanni Andreatta, Amedeo Odoni

Advanced Aircraft Design: Conceptual Design, Analysis and Optimization of Subsonic Civil Airplanes
Egbert Torenbeek

Design and Analysis of Composite Structures: With Applications to Aerospace Structures, 2nd Edition
Christos Kassapoglou

Aircraft Systems Integration of Air-Launched Weapons
Keith A. Rigby

Understanding Aerodynamics: Arguing from the Real Physics
Doug McLean

Aircraft Design: A Systems Engineering Approach
Mohammad H. Sadraey

Theory of Lift: Introductory Computational Aerodynamics in MATLAB/Octave
G.D. McBain

Sense and Avoid in UAS: Research and Applications
Plamen Angelov

Morphing Aerospace Vehicles and Structures
John Valasek

Spacecraft Systems Engineering, 4th Edition
Peter Fortescue, Graham Swinerd, John Stark

Unmanned Aircraft Systems: UAVS Design, Development and Deployment
Reg Austin

Visit www.wiley.com to view more titles in the Aerospace Series.

Scramjet Propulsion

A Practical Introduction

Dora Musielak
University of Texas at Arlington

WILEY

This edition first published 2023

Registered Office(s)
John Wiley & Sons, Inc., 111 River Street, Hoboken, NJ 07030, USA
John Wiley & Sons Ltd, The Atrium, Southern Gate, Chichester, West Sussex, PO19 8SQ, UK

Editorial Office
9600 Garsington Road, Oxford, OX4 2DQ, UK

For details of our global editorial offices, customer services, and more information about Wiley products visit us at www.wiley.com.

Wiley also publishes its books in a variety of electronic formats and by print-on-demand. Some content that appears in standard print versions of this book may not be available in other formats.

Library of Congress Cataloging-in-Publication Data
Names: Musielak, Dora, author.
Title: Scramjet propulsion : a practical introduction / Dora Musielak.
Description: Hoboken, NJ : Wiley, 2023.
Identifiers: LCCN 2022027389 (print) | LCCN 2022027390 (ebook) | ISBN 9781119640608 (hardback) | ISBN 9781119640592 (adobe pdf) | ISBN 9781119640639 (epub)
Subjects: LCSH: Airplanes–Jet propulsion. | Airplanes–Scramjet engines.
Classification: LCC TL709 .M87 2023 (print) | LCC TL709 (ebook) | DDC 629.134/353–dc23/eng/20220812
LC record available at https://lccn.loc.gov/2022027389
LC ebook record available at https://lccn.loc.gov/2022027390

Cover image: © NASA
Cover design by Wiley

Set in 9.5/12.5pt STIXTwoText by Straive, Pondicherry, India
Printed and bound by CPI Group (UK) Ltd, Croydon, CR0 4YY

C9781119640608_021122

Contents

Preface

Flying at hypersonic speeds will revolutionize travel across the globe. One could fly from Texas and be in Paris within two hours, for example, or the flight time from Frankfurt to Sydney could be reduced from 22 hours to five. To make such progress in flight travel requires an innovative, technologically advanced air-breathing propulsion system.

The scramjet engine can power such hypersonic cruise aircraft, and it can also integrate with combined cycle propulsion required for innovative reusable spaceplanes and launch vehicles to place people and satellites in Earth orbit. The requirements to operate efficiently and reliably over such a wide range of flight velocity conditions and altitudes make hypersonic air-breathing propulsion a rather challenging field of study. Powered hypersonic flight received renewed interest after the successful demonstration flight tests carried out in the United States and in Australia in the 2000s. The NASA X-43A, AFRL/DARPA X-51A, and U.S. AFRL/Australia DST HIFiRE programs proved that both hydrogen and hydrocarbon fueled scramjet engines can effectively propel a vehicle to hypersonic speed. New programs are now funded to expand the scramjet capability thus demonstrated, focusing on combined cycle engines, maturing the technologies required for sustained air-breathing hypersonic flight.

Scramjet Propulsion: A Practical Introduction is the outcome of a professional development short course that I offered in the autumn of 2018, sponsored by the American Institute of Aeronautics and Astronautics (AIAA). The main objective of that online course was to provide an up-to-date exposition of technical topics related to the scramjet engine, focusing on the R&D that thrust the development of critical technologies. The course attracted students and professionals working in the field, as well as newcomers from many regions of the world.

This book is by no means a comprehensive, descriptive survey of the entire subject of hypersonic air-breathing propulsion. Just as the short course, this book is a primer to get you ready to tackle hypersonic air-breathing propulsion, and I hope you use it as a foundations text or a springboard to more advanced study. I focus on some important physical phenomena required to understand, to first order, the behavior of air-breathing engines, and the technologies that are needed to advance scramjet engine and vehicle development.

Scramjet Propulsion: A Practical Introduction provides a personal perception of hypersonic air-breathing propulsion R&D efforts based on my experience teaching, reviewing, and contributing to research studies and publications. This book is intended as a guide to be of practical value to students, engineers, and professionals engaged in programs that support R&D work related to the field. I tried to provide a general overview of what is feasible based on scramjets already developed rather than present the mathematical derivations underlying the engineering principles found in textbooks. This work, therefore, can be easily read by someone who does not have previous experience in propulsion engineering.

The main objective of this work is to provide an overview of the technologies required for the development and maturing of scramjets, including combined cycle engines to power hypersonic cruise aircraft and transatmospheric reusable spaceplanes. A brief perusal of the contents shows that each chapter is formulated to meet this objective. The important topics of hypersonic vehicle design, structural analysis, flight trajectories, and control systems are omitted. However, the material covered provides the basis for their study.

Chapter 1 introduces the concept of hypersonic flight and describes high-speed air-breathing propulsion, highlighting the scramjet engine. After a summary of important R&D programs, brief and pointed summaries of the history of hypersonic air-breathing propulsion are given along with the current status. Chapter 2 provides technical background, and it introduces the notation, defines important concepts, reviews fundamental formulas, and gives a qualitative glimpse of the vehicle-integrated propulsion system.

Chapter 3 is devoted to aerothermodynamics of vehicle-integrated scramjets, reviewing viscous flow phenomena and boundary-layer transition. The design and performance of the scramjet main components (inlet, combustor, and nozzle) are treated in Chapters 4, 5, and 8, respectively. The presentation is not what a design engineer would consider rigorous, but the material in these chapters conveys the essence of the scramjet flowpath design, and it also provides some ideas on the conditions and limitations associated with the performance relationships under study. Chapter 6 addresses endothermic fuels, while Chapter 7 gives a phenomenological description of dual-mode combustion, an important topic for scramjet engines that will operate in the low hypersonic flight regime.

Chapter 9 contains information on high-temperature materials, structures, and thermal management systems. Chapter 10 reviews combined cycle propulsion, including formulations for ideal cycle analysis of turbojets and ramjets. Chapter 11 focuses on ground testing, since sophisticated analysis and computational fluid dynamic (CFD) codes and methodology are used to obtain solutions to hypersonic air-breathing propulsion flowfields, Chapter 12 presents results obtained using computer codes of varying degrees of rigor, emphasizing their scope and contribution to the development of scramjet engines. An overview of flight testing is presented in Chapter 13, including the methods pursued, the experience gained during past scramjet demonstration flights, lessons learned, and infrastructure for safe testing of a scramjet-powered vehicle. The book concludes with an outlook of what the future may bring. *Powering the Future of Transcontinental Flight and Access to Space* encapsulates the current status of air-breathing propulsion R&D programs pursuing this future.

Most high-speed propulsion designs and experimental data are classified ITAR or proprietary and are therefore outside the scope of this book. To compensate for this, the material I include is based on knowledge and research published in books, technical reports, public briefings, conference papers, and journal articles. I made every effort to cite sources and provide full references at the end of each chapter so that readers may obtain further details on the addressed topics. All images used to illustrate hypersonic vehicle concepts have been cleared for public release and most are available through the Internet.

In my own study, I benefited greatly from three excellent textbooks: *Hypersonic Airbreathing Propulsion* by William H. Heiser and David T. Pratt, *Hypersonic Aerothermodynamics* by John J. Bertin, and *Hypersonic and High Temperature Gas Dynamics* by John D. Anderson, Jr. These texts are the golden standard for learning the subject matter, and I recommend the reader acquire them to complement this book.

I sincerely hope that this work will serve as a source of information and technical insight for the many students, engineers, and program managers involved in the exciting study, R&D, and ultimate application of scramjets for hypersonic flight.

Acknowledgment

First and foremost, I express my deepest appreciation to NASA for the breathtaking discoveries that always inspire me, for the knowledge I acquired from its distinguished researchers, and for the prestigious research fellowships it bestowed on me. This book is only possible thanks to the hypersonic propulsion pioneers and dedicated researchers, brilliant engineers, and scientists all over the world (too many to name individually) who have built the body of work synthetized in the following chapters. Hence, I acknowledge some of them in the citations to recognize their superb contributions. In particular, I wish to express my gratitude to Tom Drozda (NASA LaRC); Phil Drummond (NASA LaRC); Christer Fureby (Lund University, Sweden); Nick Gibbons (The Centre for Hypersonics, University of Queensland); Peyman Givi (University of Pittsburgh); Antonella Ingenito (University of Rome "La Sapienza"); Suppandipillai Jeyakumar (India); Ivan Bermejo-Moreno (University of Southern California); Sebastian Karl (DLR, Germany); Tobias Langener (ESA-ESTEC, The Netherlands); Mary Jo Long-Davis (NASA GRC); Christian Messe (University of Stuttgart, Germany, now at Berkely Lab); Michael Smart (Hypersonix, Australia); and Axel Vincent-Rondennier (ONERA, France). Of course, this book may not do justice to their technical efforts, but citing some of their work is my way to acknowledge them.

At Wiley, I thank Editor Lauren Poplawski for her enthusiasm and support for this project and Sarah Lemore, Associate Managing Editor, who kept me on track and ensured the manuscript met editorial standards. To the entire Wiley production team, especially Isabella Proietti, Sindujaabirami (Abi) Ravichandiran, and Ramya Vengaiyan a sincere thank you for your excellent work. I also wish to recognize the staff at the University of Texas at Arlington (UTA) Library for their helpfulness and willingness to find special publications required for my research. I also wish to thank the following organizations for permission to use images from their hypersonic propulsion programs: NASA, Hypersonix (Australia), and the Research and Technology Organization, North Atlantic Treaty Organization (NATO RTO).

To my family, words cannot express enough my gratitude for their steadfast love and encouragement: to my intelligent daughters Dasi and Lauren, thank you for being so wise, fearless, and independent, and to my scholarly husband Zdzislaw, thank you for standing by me always, cheering and loving me when spacetime seemed to suddenly darken. With you all in my life, my universe is limitless and splendorous.

Dora Musielak

1

Introduction to Hypersonic Air-Breathing Propulsion

We begin a study of hypersonic air-breathing propulsion systems, engines that take the oxidizer from the surrounding atmosphere and propel vehicles to sustained speeds greatly in excess of the local speed of sound, higher than Mach 5. Hypersonic flight with air-breathing propulsion is pursued for its potential to realize cost-effective access to space and high-speed cruise. Applications include civil transports, scramjet-powered missiles to fly in the Mach 6–8 range, both tactical and strategic, single-stage space planes, and multiple-stage-to-orbit vehicle configurations. Renewed interest in hypersonic sustained flight has increased research and development activities and made substantial advances in required technologies. The remarkable performance improvements promised by high-speed air-breathing propulsion were brought within our reach by the recent development of technologies related to scramjet engines and by their demonstration in flight. Figure 1.1 depicts an artistic view of NASA's X-43A aircraft that flew at Mach 7 and 10 to demonstrate the viability of hydrogen-fueled scramjet propulsion.

Hypersonic air-breathing propulsion is based on ramjets and scramjets, the simplest jet engines to propel a vehicle to hypersonic speeds within the atmosphere. These engines have no internal moving parts, as they do not require turbomachinery (mechanical compressor/turbine) to process the ingested ambient air.

The ramjet engine has three main components: an inlet, a combustion chamber, and a nozzle. The dynamic action of the freestream air is used to produce the compression in the inlet as the vehicle flies at high speed. This action is referred to as the ram effect. The higher the velocity of the incoming air, the greater the pressure rise. The fundamental principle underlying ram compression in the ramjet inlet lies in converting the kinetic energy of the air into pressure. The compressed air then enters the combustion zone where it is mixed with the fuel and burned. The hot, high-pressure gas flow then accelerates back to a supersonic exit speed through the nozzle to develop thrust.

The most distinctive feature of the ramjet is that combustion of fuel with air takes place after the flow has been slowed internally to subsonic speeds. Moreover, the air flow is compressed in several steps, including passing through one or more oblique shock waves generated by the forebody of the vehicle or of the diffuser, deceleration of the supersonic flow in a convergent duct, transforming the supersonic flow into subsonic flow through a normal shock wave system, and further decelerating the subsonic flow in a divergent duct. Ramjets

Scramjet Propulsion: A Practical Introduction, First Edition. Dora Musielak.

Figure 1.1 Artistic rendition of scramjet-powered hypersonic cruiser. Source: NASA.

are suitable for applications where the flight Mach number is in the range 3–5 and are used mainly for supersonic flight.

When the flight Mach number exceeds about 5, deceleration of the ingested airflow to subsonic conditions would cause it to reach unacceptable high temperatures. To extend the flight regime above Mach 5, the scramjet was conceived. In this type of ramjet, the hypersonic inlet airflow is diffused only to supersonic speed prior to mixing it with fuel in the combustor. Hence, the combustion process takes place at locally supersonic conditions. The engine operating in this mode becomes a *scramjet*, an acronym standing for "supersonic combustion ramjet," a name used to emphasize that the combustion of fuel and air must occur in a supersonic flowfield relative to the engine. Scramjets in fully supersonic combustion mode begin to produce thrust flying at speeds of at least Mach 4 and would operate as long as there is sufficient air to pass and process through its inlet; the theoretical maximum operational speed for scramjets is unknown, but it could effectively reach about Mach 12.

There is a wide range of speed and altitude over which air-breathing propulsion is capable of higher specific impulse (I_{sp}) than is rocket propulsion. The I_{sp} parameter indicates how much thrust the engine produces per every unit mass of propellant (fuel plus oxidizer) it uses per second. Since the air-breathing engine does not need to carry oxidizer on board, its specific impulse is much higher than that of the rocket. Scramjets are therefore the most efficient air-breathing engines, that is, with the highest (fuel)-specific impulse, at flight Mach numbers above 5. To capitalize on such advantage, much effort has been devoted to developing hypersonic air-breathing propulsion (HAP) systems to achieve hypersonic flight within the Earth's atmosphere. One such HAP concept is the dual-mode scramjet (DMSJ), an engine that operates both as ramjet (subsonic combustion) and as scramjet (supersonic combustion) in order to propel a vehicle in a flight Mach number ranging from 3.5 to 12 (the upper limit in scramjet operational Mach number is still unknown). However, due to its minimum functional speed, scramjets require acceleration by other means in order to become operational for takeoff.

For some military applications, air-breathing hypersonic vehicles can be air or ground-launched attached to a rocket motor that will accelerate the craft to the take-off speed of the scramjet. Other applications require to integrate the scramjet engine with a low-speed propulsion system (e.g. turbojet, turbofan) in order to provide the capability of propel a vehicle from the runway all the way to its maximum hypersonic speed.

Powered by scramjet engines, hypersonic vehicles scoop the oxygen required for fuel combustion from the atmosphere, and this reduces tankage requirements and airframe mass. In fact, for missile propulsion, the ramjet is very competitive with the rocket because it is simple in construction and has greater range for the same propellant weight. These characteristics are particularly attractive for military applications where simplicity and low initial cost are essential features of devices that must function on demand and never return. Moreover, hypersonic vehicles propelled by air-breathing propulsion promise affordable and rapid access to space and hypersonic cruise. Scramjet propulsion flight demonstrator programs (e.g. X-43A, X-51A, HIFiRE) have already proven that HAP vehicles are technically feasible. However, more flights and flight-test programs are required to demonstrate sustained cruise and acceleration to establish the DMSJ engine as a viable and mature hypersonic air-breathing propulsion system.

Moving at hypersonic speeds, a vehicle will naturally generate a massive amount of heat that must be properly managed. The vehicle and its integrated propulsion system must be fabricated with advanced materials designed to withstand those high temperatures, materials with high strength, and high toughness. Hypersonic vehicles travel very fast, getting hot enough to melt most traditional metals, so engineers are developing new material formulations for hypersonic craft to survive such harsh environment.

This book intends to provide the technical background to describe the fundamental characteristics of high-speed air-breathing engines, focusing on the technologies that are being developed to advance the DMSJ to power future hypersonic flight.

1.1 Hypersonic Flow and Hypersonic Flight

For air-breathing propulsion, hypersonic flight is interpreted to mean flight speeds V_0 higher than five times the speed of sound, that is,

$$M_0 \equiv \frac{V_0}{a_0} > 5$$

where M_0 denotes a vehicle's flight Mach number and a_0 is the local speed of sound.

For the analysis of hypersonic air-breathing propulsion, we can define hypersonic flow as the regime where the calorically perfect gas model for air becomes invalid. For calorically perfect gas or temperatures less than 400 K, the specific heats c_p and c_v are constant. As the air temperature increases, in the range of temperature 400 K $< T <$ 1700 K air behaves as a thermally perfect gas, the value of the specific heats is function of temperature; and thus the specific heat ratio ($\gamma = c_p/c_v$) is also a function of temperature.

At temperatures above 1700 K (3000 °R), the equilibrium of specific heat (c_p) of air depends strongly upon both temperature and pressure because chemical reactions have become important. Hence, γ reaches a value of 1.286 when the formation of nitric oxide

(NO) begins. When nitrogen is released during combustion, it combines with oxygen atoms to create NO, which then combines with oxygen to create nitrogen dioxide (NO_2). At temperatures above 1700 K, chemical reaction and dissociation become very complicated and cannot be treated with a simple gas model.

We can also consider the value of the freestream total or stagnation temperature that would cause real gas effects to occur. Let us consider the total to static temperature ratio,

$$\frac{T_{t0}}{T_0} = 1 + \frac{\gamma - 1}{2} M_0^2 \tag{1.1}$$

where T_{t0} is the freestream total temperature, M_0 is the flight Mach number, and the subscript 0 denotes the undisturbed freestream flow conditions far ahead of the vehicle as seen from the reference frame of the vehicle. When we substitute a representative value of the freestream air static temperature, $T_0 = 227$ K with $\gamma = 1.4$, we find that at a Mach number $M_0 = 6$, the stagnation temperature is $T_{t0} = 1861.4$ K, a value which exceeds the limit temperature for thermal equilibrium flow. At temperatures above 1700 K, the thermally perfect model for air is no longer valid because at this condition the formation of NO begins.

From (1.1), it is clear that for hypersonic flow, we have

$$\frac{\gamma - 1}{2} M_0^2 \gg 1$$

Formally, the basic assumption for all hypersonic flow theories is $M \gg 1$. This definition implies that the internal thermodynamic energy of the freestream fluid particles is small compared with the kinetic energy of the freestream for hypersonic flows. Some textbooks also define hypersonic flow as that with Mach numbers at which supersonic linear theory fails. However, these definitions do not provide a quantitative description of the boundary between supersonic and hypersonic flows.

When a vehicle moves through the atmosphere, at supersonic or hypersonic speed, a shock curved layer forms upstream of the vehicle. Known as a bow shock, this is a curved propagating disturbance shock wave characterized by an abrupt, nearly discontinuous, change in pressure, temperature, and density, and entropy increase. In hypersonic flight, as the freestream velocity becomes very large while its freestream thermodynamic state remains fixed, this produces extremely high temperatures in the shock layer. As the freestream Mach number increases, the shock waves approach or hug the bounding surface of a body more and more closely, and the resulting thin shock layer increases the wall heating of the hypersonic vehicle. For a slender aircraft, hypersonic effects are very clear when it flies at high Mach numbers, as the bow shock gets closer to the body and aerodynamic heating dominates the physics of its flight.

From aerodynamics perspective, we define hypersonic flow when the following flow phenomena become progressively more important as Mach number increases:

- High Temperature Effects → aerodynamic heating increases proportionally with M_0^2.
- Low Density Flow → aerodynamic equations (Euler and Navier–Stokes equations) break down.
- Thin Shock Layers → shock layer is composed of the flowfield between the oblique shock wave and the vehicle. As the Mach number increases, the shock wave gets closer to the vehicle.

- Entropy Layer → entropy increase becomes greater as shock strength increases.
- Viscous Interaction → increased flow temperature (due to friction heat) near body surface causes boundary layer to become thicker as speed increases, resulting in high drag.

Each factor plays a huge role in the design and operation of practical hypersonic vehicles.

Hypersonic flight regime is characterized by flow velocities that exceed five times the local speed of sound ($M_0 > 5$) and where shock layers are thin and viscous drag and heating loads are very high. Inside the air-breathing engine flowpath, the term hypersonic refers to regions of high stagnation temperatures at which chemical reactions become important and simple models of gas behavior break down.

1.2 Chemical Propulsion Systems

A propulsion system is a sophisticated engineering device capable of generating a thrust force to propel a vehicle in a particular direction or to accelerate it in flight. All methods to produce a thrust force for propelling a vehicle are based on one principle: the time rate of change of momentum of a fluid accelerated by the system under consideration. The fluid may be a propellant stored in the vehicle and carried along during its flight, as in the case of rocket propulsion, or it can be a mixture of a fuel and an oxidizer that is taken from the surrounding environment, such as in the air-breathing propulsion systems we study herein. In both cases, the fluid is chemically reacted in a combustion chamber, and the hot products of combustion are expelled at high velocity through a propulsion nozzle. Hence, we refer to chemical propulsion systems as those in which chemical energy resulting from a chemical reaction of a fluid and an oxidizer cause breaking or forming of chemical bonds with the resulting gas achieving a high temperature, i.e. thermal energy is released by the fuel during combustion.

Chemical propulsion can be divided into two major groups: air-breathing and rocket engines, as illustrated in Figure 1.2. Air-breathing propulsion can be further divided into gas turbine jet engines (turbojet, turbofan) and non-rotor jet engines (pulsejet, ramjet, and scramjet), all based on the open air-standard power cycle. Jet propulsion is a specialized field that deals with systems in which the generation of thrust is achieved by direct expansion of a gas (the working fluid) used by the engine.

Air-breathing jet engines operate on an open air-standard power cycle known as Brayton cycle. That is, the working fluid undergoes a series of thermodynamic processes (compression, energy addition, and expansion) arranged in a manner that the processed gas can produce thrust. The Brayton cycle represents the ideal conversion of thermal energy to mechanical energy.

The low-pressure air from the surrounding atmosphere enters the engine and is compressed to a higher pressure. The compressed air is then heated by injecting fuel into it, allowing the air stream and the fuel to mix together, ignite, and burn at almost constant pressure in a combustion chamber. The high-pressure hot gaseous products of combustion

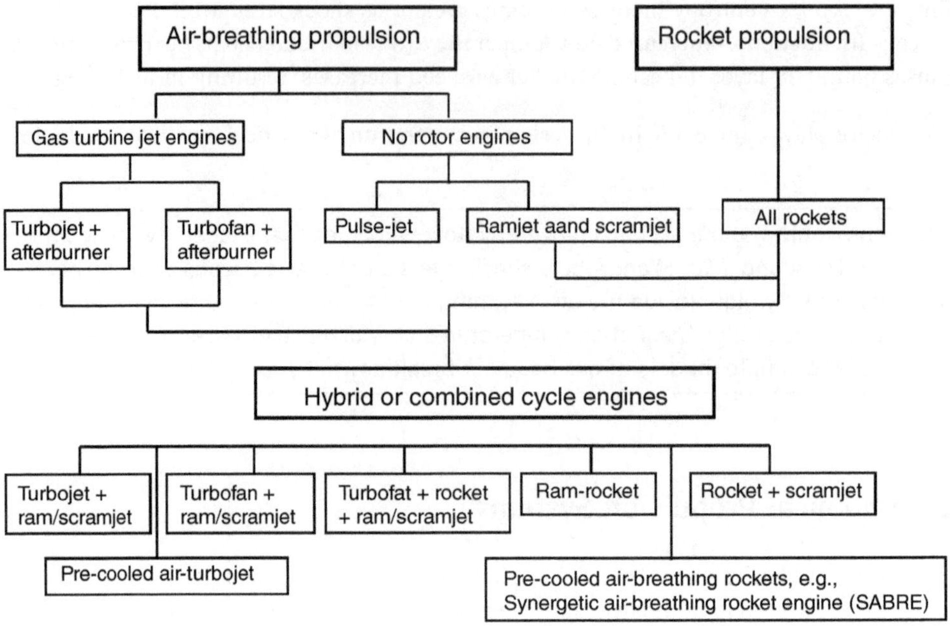

Figure 1.2 Classification of chemical propulsion.

accelerate and are expanded in the nozzle and finally exhausts into the atmosphere. A reaction force or thrust is generated by the flow because its high temperature has more velocity and momentum leaving than it did entering the engine. This reaction force is known as the internal or uninstalled thrust.

1.2.1 Turbojets

For flight in the range of $0 < M_0 < 3$, the jet engine requires a mechanical compressor to increase the pressure of the incoming atmospheric air before fuel is added and combusted, particularly in the low end of the speed range. In the case of a turbojet (Figure 1.3), the ingested air is compressed mechanically by an axial compressor, and the hot gases of combustion are first expanded in a turbine attached to the compressor by a common driveshaft. The expansion process is limited so that the work necessary to drive the compressor is supplied only by the turbine, and the rest of the expansion occurs in the exhaust nozzle.

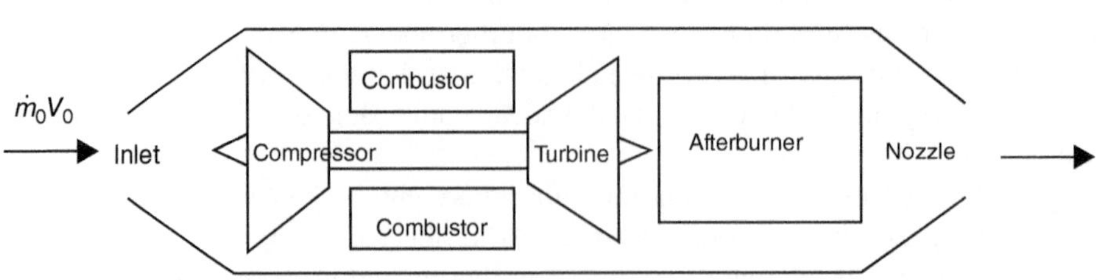

Figure 1.3 Representation of turbojet with afterburner.

Atmospheric air taken in through the turbojet inlet is compressed up to 3–12 times its original pressure in the axial compressor. Fuel is added to the air and burned in a combustion chamber to raise the temperature of the fluid mixture to about 593–704 °C (1100–1300 °F). The resulting hot gas is passed through a turbine, which drives the compressor. If the turbine and compressor are efficient enough, the gas pressure at the turbine discharge will be nearly twice the atmospheric pressure, and this excess pressure is sent to the nozzle to produce a high-velocity stream of hot gas which produces the thrust force. Substantial increases in thrust can be obtained by employing an afterburner. This is a second combustion chamber positioned after the turbine and before the nozzle (see Figure 1.3). The afterburner increases the temperature of the gas ahead of the exhaust nozzle. The result of this increase in temperature is an increase of about 40% in thrust at takeoff, and a much larger percentage at high speeds once the vehicle is in the air.

For supersonic flight in the Mach number range of $2 < M_0 < 5$, the ram pressure produced by the inlet in slowing down the incoming air to subsonic speeds is high enough that there is no longer need for an air compressor, and therefore no need for a turbine. As a consequence, the turbomachinery is unnecessary and the engine design configuration is simplified. The resulting propulsion system is called ramjet.

The effect of Mach number on total or stagnation pressure of the freestream, denoted by p_{t0}, becomes immediately clear through the relation

$$p_{t0} = p_0 \left(1 + \frac{\gamma - 1}{2} M_0^2\right)^{\gamma/(\gamma - 1)} \tag{1.2}$$

where p_0 is the static pressure of the ambient air.

When the Mach number is $M_0 = 4$, Eq. (1.2) yields a ram pressure ratio $p_{t0}/p_0 = 151.8$, and from Eq. (1.1), we obtain a total to static temperature ratio $T_{t0}/T_0 = 4.2$; together these two ratios give air conditions more than adequate for operating an engine without turbomachinery. For this flight regime, the ramjet is the air-breathing propulsion of choice. The total pressure to static pressure ratio (p_{t0}/p_0) of the freestream is often referred to as *ram pressure ratio*. The natural air compression brought on by the aircraft's high speed is known as ram pressure.

1.2.2 Ramjets

Ramjets are the air-breathing engines of choice for supersonic flight in the Mach number range 3–5. Conceptually, a ramjet is the simplest air-breathing jet engine, consisting of three main components: an inlet diffuser, a combustor or burner, and a thrust nozzle. Figure 1.4 depicts a conventional ramjet of circular cross-section that uses a spike-like inlet diffuser.

The incoming atmospheric air flowing at supersonic speed is first compressed at the nose of the vehicle by going through the oblique shock anchored by the central spike and then after a terminal normal shock at the cowl lip, the air flows into the interior of the engine. In the diffuser, most of the air's kinetic energy is converted into a pressure raise. After compression by the inlet shocks, the air has slowed considerably, reaching subsonic velocity. In the subsonic diffuser, the air flow gains an additional increase in pressure before it reaches the combustion chamber.

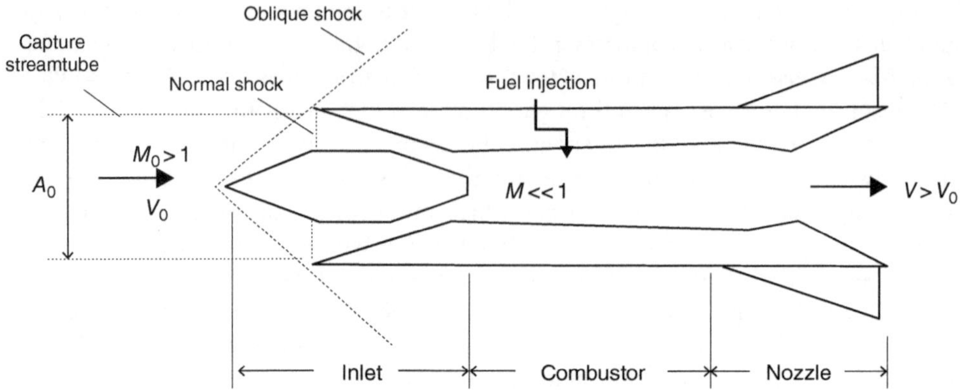

Figure 1.4 Schematic of a conventional ramjet engine.

The combustor section in the ramjet is a chamber similar to an afterburner. In the combustor fuel is injected and mixed with the compressed air and ignited. Because the combustor's cross-section is relatively small, the airflow velocity is much higher in the ramjet combustor than in a turbojet burner. Therefore, ramjets require flame holders in order to stabilize the flame and prevent blowout. The hot combustion gas then expands, passing through a convergent–divergent nozzle before being expelled at high velocity to produce thrust. In principle, a ramjet can operate at flight Mach numbers as low as M_0~0.5 or as high as M_0~5.0. However, the specific thrust would be very low at these velocities.

At hypersonic flight velocities $M_0 > 5.0$, the temperature increase that would result from decelerating the airflow to subsonic speed in the ramjet inlet becomes so large that adding thermal energy to the flow in the combustor would not yield a significant temperature rise, mainly because the normal products of combustion would be highly dissociated. If the air temperature would raise further in the inlet (by decelerating the high velocity air to subsonic values), it would cause the total temperature change in the combustor to be reduced to eventually zero. This is the reason why the ramjet engine is not effective for operation at flight Mach numbers above 5. Since the ramjet lacks turbomachinery, it requires boosting to some minimal airspeed before it can sustain combustion and generate thrust greater than its own drag.

Ramjet: A propulsion system that utilizes the natural "ram effect" to compress the ingested freestream air and in which fuel/air combustion occurs at subsonic speeds. The ramjet can operate effectively in the flight Mach range of $1.0 < M_0 < 6.0$.

1.2.3 Scramjets

For hypersonic flight speeds, $M_0 > 5$, the temperature increase accompanying the ram compression to subsonic speeds is so high that little or no additional heat can be added in the combustor by burning fuel. The only alternative is to use the inlet to slow the flow down

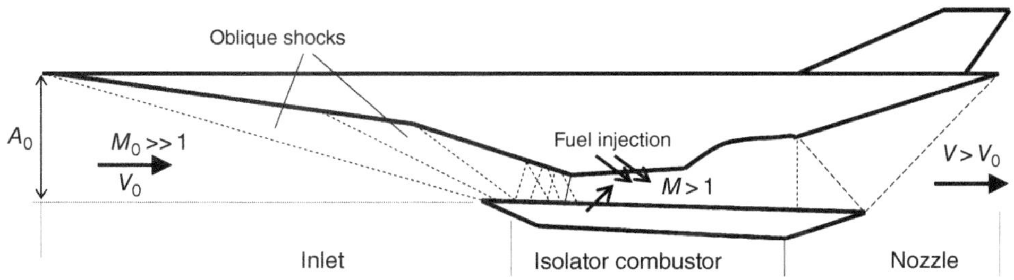

Figure 1.5 Schematic of vehicle-integrated scramjet engine.

from the flight speed to some lower supersonic Mach number, thereby not increasing the temperature too much. However, then fuel must be added to a supersonic stream, mixed, and combusted. Such a supersonic combustion ramjet is known as a scramjet (Figure 1.5).

The scramjet is capable of operating in the hypersonic regime $M_0 > 5$. The scramjet overcomes the limitation outlined above by partially compressing and decelerating the incoming air but still keeping it at supersonic velocity throughout the engine. Hence, the inlet flow must be diffused only to the compressed state required by combustion, remaining at supersonic speed prior to entering the combustor. The scramjet engine is in fact a ramjet engine in which the combustion process takes place at supersonic conditions.

A scramjet engine (Figure 1.5) consists of four basic components; an inlet, an isolator that is used to "isolate" the supersonic inlet flow from the back pressure growth in the combustor, a combustor, and a nozzle aft of the combustor. In the scramjet inlet, the freestream air is first partially decelerated and compressed from the freestream condition (point 0) to the combustor entry condition by means of a combination of isentropic compression and oblique shock waves. This compression process must be sufficient to provide a large enough static temperature ratio across the inlet (in the range of 6–8) for optimum thermodynamic cycle efficiency, and to produce high enough values of pressure and temperature at the combustor entrance to support complete and stable combustion at supersonic flow conditions. The airflow temperature and that of the combustion products cannot be allowed to rise above the range in which the combustion of the fuel with air can be completed.

Careful design of the supersonic combustion duct is required to avoid thermal choking. The hot gases are then accelerated further through the expansion process from the combustor exit condition to the freestream ambient condition. Since the accelerating gases are supersonic, the exhaust nozzle of the scramjet is of diverging geometry. In fact, some of the expansion and acceleration occur outside the scramjet confining flowpath by using the afterbody of the vehicle as a free expanding surface.

Scramjets are mechanically simple but aerodynamically more complex than any other jet engine. Figure 1.5 depicts a scramjet engine that is closely integrated with the vehicle. In such configuration, the bottom of vehicle becomes an integral part of the high-speed air-breathing propulsion system. To capture the required airflow for maximum propulsion, it is necessary to use entire forebody underneath the vehicle as a compression surface. Since a streamtube of air enormously larger than the physical opening of the engine forms upstream of the vehicle, a long forebody is needed to increase the capture area A_0. In this manner, the vehicle's forebody serves as a portion of the external inlet, and the afterbody completes the nozzle expansion process. In the configuration of Figure 1.5, the engine flowpath includes the vehicle external lower surfaces and the engine cowl internal surfaces.

The inlet of the scramjet achieves air compression both externally and internally through oblique shocks. The series of oblique shock reflections inside the inlet create a shock train, which is contained in the isolator. The isolator is a duct between the inlet proper and the entrance to the combustor that serves to prepare the air flow for mixing with the fuel and supersonic combustion process. This duct acts to isolate the two main engine components. The air flow that emerges from the isolator is supersonic as it enters the combustor, requiring that it has a Mach number approximately a third of the flight Mach number.

Combustion at supersonic speeds is rather challenging. Consider this, if the air enters the combustor at a speed of 999 m/s (~Mach 3), it would transverse a meter length in about 1 ms. Achieving efficient fuel injection, mixing, and chemical reaction in such short timescales require ingenious and complex design schemes for the combustor and the fuel injection systems. We shall examine these issues in Chapter 5.

The hot gases generated in the combustor of the scramjet require a nozzle exit area even larger than the original freestream capture area in order to expand effectively and produce thrust. Hence, the entire afterbody underneath the vehicle is used as a free expansion surface, as illustrated in Figure 1.5.

There are two aspects of ram compression that are unfavorable to thrust production. These are (i) the inlet total pressure recovery that exponentially deteriorates with flight Mach number, and (ii) the rising gas temperature in the inlet that reduces the total temperature difference across the combustor to zero.

Decelerating a supersonic flow in the inlet of the ram/scramjet is accompanied by shock waves that produce a total pressure loss much greater than, and in addition to, the wall friction loss. Hence, the overall pressure ratio of the inlet is the product of the ram pressure ratio and the diffuser pressure ratio. Because of shocks, only a portion of the ram total pressure can be recovered. A well-designed inlet for a scramjet is required to minimize the sources of irreversibilities. We will address this topic in subsequent chapters.

Moreover, the deceleration process in the inlet produces a high temperature rise, and at some limiting flight speed, the temperature will approach the limit set by the wall materials and cooling methods available for the engine. Thus, when the temperature increase due to deceleration reaches the limit, the propulsion system is no longer effective.

Finally, since ram/scramjets cannot generate thrust for take-off, a vehicle needs some other propulsion device to accelerate these engines to the supersonic speeds required to start generating thrust. That is a reason why ramjets are paired with a rocket booster and often used to propel air launched missiles. Previous scramjet flight demonstration tests required that the hypersonic vehicles be air-lifted by a carrier aircraft and then boosted to the altitude test conditions using a rocket booster. After separating from the launch aircraft and booster, the scramjet performed the fueled propulsion test in free flight.

> ***Scramjet***: Supersonic combustion ramjet engine that can operate effectively in the Mach flight range $5.0 < M_0 < 15$. The pure scramjet is the true hypersonic air-breathing engine.

Dual-mode Scramjet: Air-breathing engine that can operate as a ramjet and as a scramjet, depending on whether it combusts its fuel at subsonic or supersonic conditions, transitioning to each engine mode as needed. It can propel a vehicle in the flight regime $3.0 < M_0 < 15$.

The idea of the DMSJ evolved from the conflicting requirements of the hypersonic engine to have high cycle thermal efficiency (that depends on having a high compression pressure ratio), and prevent or minimize dissociation of the working fluid (which can occur if the static compression temperature of the gas exceeds a known value). In other words, the static temperature at the beginning of combustion cannot be increased indefinitely, but must be limited. If the temperature becomes too large, adding thermal energy to the flow in the combustor (via fuel injection) would not yield a significant temperature rise, mainly because the normal products of combustion would be highly dissociated and would cause the total temperature change in the combustor to be reduced to zero. Moreover, at some limiting flight velocity, the compression temperature would approach the limit set by the wall materials and cooling methods available for the propulsion system.

Thus, to prevent or control those detrimental effects, the DMSJ must be designed so that the combustion process occurs at subsonic flow conditions (ramjet) for flight Mach numbers less than about 5, and be supersonic (scramjet) for hypersonic flight. The dual-mode combustor must transition seamlessly from subsonic combustion to supersonic combustion, and vice versa, as required by the flight speed.

To determine the maximum allowable temperatures in a scramjet requires a combination of elaborate computations and experienced judgment because these temperatures depend on many interrelated variables, including flight altitude and Mach number, inlet losses, fuel type, fuel/air ratio, and burner and exhaust system geometry. However, we shall carry out analysis based on fundamental thermo-aerodynamic equations to determine the values of compression temperature that are acceptable. The limit on compression temperature should also help us predict the values of Mach number that the compressed air must have as it enters the combustor.

1.2.4 Combined Cycle Propulsion

A DMSJ capable of producing thrust in the Mach 3–12 range can be combined with other engine cycles to develop efficient concepts with sufficient performance and operability to power hypersonic aircraft and spaceplanes. Many studies have considered the DMSJ as the core of combined-cycle propulsion systems for single-stage-to-orbit (SSTO) vehicles by providing most of the orbital ascent energy. Such concepts are also relevant to two-stage-to-orbit (TSTO) systems with an air breathing first stage (see Chapter 10).

An engine representative of combined air-breathing cycle is the turboramjet, an engine that can operate both as a turbojet and ramjet engine, intended for high supersonic speed flight. There are many other hybrid or combined cycle propulsion concepts that adapt the different types of engines into another propulsion cycle that yields performance characteristics needed to span a hypersonic flight regime. For example, the air turborocket (ATR) takes elements of an air-breathing jet engine and combines them with a rocket. One can also conceive a propulsion system composed of a turbojet, a scramjet, and a rocket combined in such a manner that the resulting engine can power a launch vehicle to orbital velocities.

Turbine-based combined cycle (TBCC) propulsion concepts integrate a gas turbine engine (turbojet or turbofan) and a DMSJ in a dual-flowpath configuration. Rocket-based combined cycle (RBCC) propulsion concepts integrate rocket thrusters with a DMSJ flowpath.

Precooled air-breathing propulsion utilize methods to reduce the high temperature of air entering the inlet system. The precooled air can then be compressed to combustion chamber operating pressures with reduced power requirements. This approach can be used in turbo engines, or in air-breathing rockets (e.g. SABRE).

The concept of the precooled engine is very attractive. According to Eq. (1.1), at Mach 5 air, the stagnation temperature is six times the ambient static temperature. For $T_0 = 223.5$ K (at 27 km altitude), this is $T_{t0} = 1341$ K, a high temperature that makes it impractical to compress the air mechanically due to large power required, and because the resulting compressor delivery air temperature would be too high for the combustor.

The air-breathing rocket concept is designed with an inlet to take the oxidizer for combustion from the surrounding atmosphere rather than carrying it onboard as traditional rockets do. One example of such air-breathing rocket is the Synergetic Air-Breathing Rocket Engine (SABRE) now being developed by Reaction Engines Limited (REL), a private company based in the United Kingdom. SABRE operates in both air-breathing and rocket modes, relying on precooling technology to rapidly cool the incoming air (from ~1000 °C to ambient), enabling SABRE to operate at higher speeds than existing conventional air-breathing engines. This propulsion system is scalable so that it can power hypersonic cruise aircraft and the Earth-to-orbit spaceplane Skylon.

We should note that other forms of energy – electrical, nuclear, plasma, and so on – can be and are used for deep space propulsion, and we can conceive new engineering ideas using those forms of energy to augment chemical propulsion. However, the subject matter in this book is confined to the study of hypersonic air-breathing engines (DMSJs).

1.3 Classes of Hypersonic Vehicles

Vehicles that fly at hypersonic speeds include rocket launchers that deliver payloads to Earth orbits, spacecraft designed to return from space intact, and aircraft to cruise through the atmosphere. For military applications, potential mission areas include long-range cruise missiles, flexible high-altitude atmospheric interceptors, and responsive hypersonic aircraft for global payload delivery. Hypersonic flight offers superior speed and range and rapid response to time-critical targets, making possible global fast-reaction reconnaissance missions. While comparable to flight time of ballistic missile, hypersonic vehicle offers advantage of flexible recall and en route re-direction.

Hypersonic flight that occurs strictly within the Earth's atmosphere is called endo-atmospheric, and if it occurs outside the atmosphere, it becomes exo-atmospheric flight. An endo-atmospheric vehicle is one that remains within the Earth's atmosphere (at altitudes below 100 km). An exo-atmospheric vehicle is a space access vehicle with speeds up to Mach

25 (orbital). Hypersonic vehicles include winged and glide re-entry vehicles and can be classified as powered and unpowered gliders. Powered vehicles can use rocket propulsion, air-breathing propulsion, or a combination of the two distinct propulsion cycle systems. In a boost glide system, a rocket accelerates its payload (a weapon or a missile) to high speeds above the atmosphere. The payload then separates from the rocket and glides unpowered to its destination.

Hypersonic air-breathing propulsion vehicles can be divided into two classes: accelerators and cruisers. The accelerator vehicle is one where the dominant design characteristic is low drag per unit inlet capture area (D/A), since the cross-section attributes a large percentage to propulsion. The cruiser aircraft is characterized by its high lift-to-drag ratio (L/D) since thin/flat fuselages with high fineness ratios are needed to reduce drag as the aircraft move through the atmosphere.

The fuel is a design-influencing parameter for shaping the hypersonic vehicle. Hydrogen-fueled vehicles must be very volumetric efficient to contain the low-density fuel. On the other hand, for hydrocarbon-fueled vehicles, the concern is loading because of high-density fuel; they tend to be more like waveriders, vehicles which, at their design point, fly with an attached shock wave all along the leading edge. A conventional hypersonic vehicle usually flies with the shock wave detached from the leading edge. Propulsion–vehicle integration places tight volume constraints on vehicle packaging requirements as this restricts available space for fuel tanks.

Hypersonic vehicles powered by air-breathing propulsion are considered for a variety of missions such as long-range civil transports, first stage launch vehicles, and a variety of military applications such as air-launch missiles, combat, and reconnaissance airplanes. Three vehicle types are summarized in Table 1.1, where we also indicated representative characteristics, including the type of fuel for each application, where LH_2 denotes liquid hydrogen, and LHC denotes liquid hydrocarbon fuels. The expected flight regimes of air-breathing aircraft are illustrated in Figure 1.6. One can quickly estimate that a trip across the United States from coast to coast could take 39 minutes by flying at Mach 5, whereas a Mach 6 military missile could fly a downrange distance of 2220 km in 25 minutes.

Table 1.1 Representative air-breathing hypersonic vehicles.

Mission	Flight Mach number M_0	Propulsion system	Flow path geometry	Fuel	Flight duration	Vehicle length, ft (m)
Hypersonic cruise	0–5 0–8	Turbo-ramjet Turbo-dual-mode scramjet	Variable geometry, actively cooled	LHC LH_2	1–3 hours	100–200 (30–61)
Tactical missile	5–8	Scramjet	Fixed geometry, passively cooled	LHC	10–30 minutes	5–16 (1.5–5)
Trans-atmospheric vehicle	0–20	TBCC RBCC	Variable geometry, actively cooled	LH_2	20–60 minutes	100–200 (30–61)

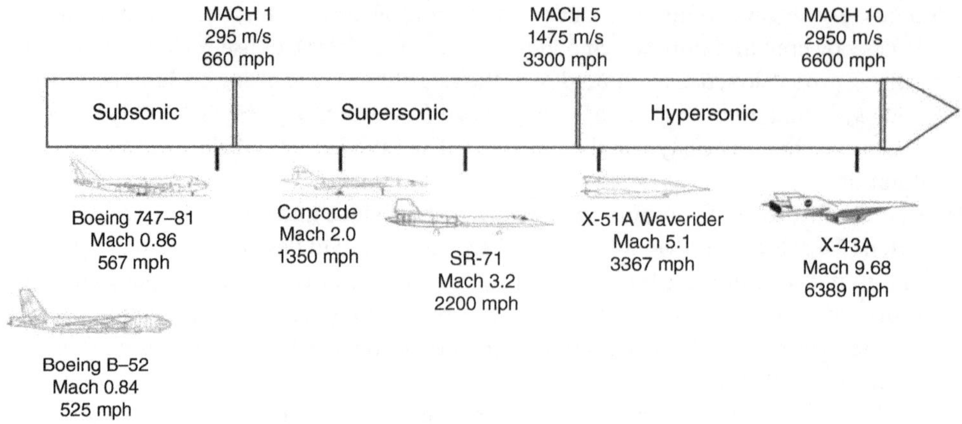

Figure 1.6 Flight regimes for air-breathing propulsion.

A hypersonic air-breathing vehicle must follow a flight trajectory within the atmosphere, maintaining structural heating and load limits while delivering enough airflow to its propulsion system to ensure an optimum level of net thrust for acceleration. The optimal flight altitude of a stratospheric aircraft depends on the maximum Mach number and the required air dynamic pressure. To ensure these operational flight limits, hypersonic vehicles are designed to operate within a narrow range of dynamic pressure, which should be between 23 and 95 kPa (500–2000 psf; see Chapter 2). Figure 1.6 depicts the nominal velocities of representative vehicles flying within Earth's atmosphere. The speed of sound is a function of temperature and decreases with altitude. On a standard day at sea-level conditions, the speed of sound is about 340 m/s, or 761 mph, or 1100 ft/s, or 1224.7 km/h. Hence, the velocity values given in Figure 1.6 are approximated, assuming sonic speed is 295 m/s or 660 mph.

> The fastest supersonic air-breathing crewed aircraft, the SR-71, flew at slightly more than Mach 3.2. In 2004, the autonomous X-43A lifting-body, scramjet-propelled hypersonic aircraft more than tripled the maximum speed of the jet-powered SR-71.

Many different scramjets have been conceived, each of which was designed for a particular application. The engine configuration and choice of fuel vary, depending on the mission requirements and thus the vehicle design and its integration with the propulsion system may be completely different.

Depending on the application, a scramjet design may require rectangular or circular combustor geometries. The engine may incorporate a variety of inlet concepts, including axisymmetric, planar two-dimensional and highly three-dimensional designs. For example, for a Mach 6 cruise missile, a round scramjet geometry with an inward-turning inlet may be more appropriate, while a Mach 10 cruise lifting-body global reach aircraft may require a planar configuration with a tightly asymmetric engine/airframe integration.

The choice of fuel depends on the mission requirements and the Mach number. Military applications may opt for hydrocarbon fuels; vehicles for access to space are more likely to use hydrogen, while civil transports can use either type of fuel. The limit cruise Mach number with hydrocarbon fuel is expected to be approximately 8, as this speed is considered the extent to which a dual-mode ramjet/scramjet can be cooled with endothermic fuels for optimum performance contraction ratios. For higher Mach number cruise, liquid hydrogen (LH_2) has more cooling capacity and provides considerably more range than LHC for the same cruise Mach number.

The focus of scramjet R&D is mainly concentrated in the following applications, which require ram/scramjet propulsion development in the given Mach number regimes:

- The first-stage vehicle of a TSTO system could utilize a hydrogen-fueled dual-mode scramjet operating at $3.0 < M_0 < 12$ and combined with either a turbine or rocket-based accelerator.
- A cruise aircraft could be propelled by an LH_2- or LHC-fueled dual-mode scramjet for flight at $3.0 < M_0 < 7$, or use a simple LHC ramjet for $3.0 < M_0 < 5$; either option would be combined with a turbine engine for the low-speed regime of its mission.
- An air-launched long-range cruise missile may be propelled by an LHC fuel ramjet for $3.0 < M_0 < 5$, boosted via a solid rocket to takeover speed.

These and many other possible applications allow for variations in flowpath and vehicle geometry to meet design and mission requirements such as weight, surface area, volume, simplicity, and affordability for more practical sustained hypersonic flight. Table 1.1 summarizes the overall characteristics of representative hypersonic vehicles. As indicated, a turboramjet is the propulsion of choice for a hypersonic Mach 5 cruiser. However, for an air-launched, rocket boosted Mach 6 tactical missile, a pure scramjet engine is a better option.

1.4 Scramjet Engine–Vehicle Integration

After many studies conducted in the 1960s, researchers determined that, for fast aircraft flight speeds (M_0~10), an air-breathing propulsion system would be too large to mount under the wings of the aircraft. Most importantly, axisymmetric, strut-mounted scramjet designs were found to incur very severe shock interaction heating and aerodynamic drag. Hence, the engine had to be closely integrated with the airframe to avoid or minimize such problems.

To reduce installation effects, the National Aeronautics and Space Administration (NASA) Hyper-X airframe-integrated scramjet was designed to be a flying engine, or a slice of the original hypersonic research engine (HRE), an axisymmetric, podded propulsion system. The vehicle forebody replaced the podded engine's spike inlet, and the vehicle afterbody replaced the plug nozzle. In subsequent designs, the scramjet had to be highly integrated with the vehicle; that is, the propulsion system would become the entire undersurface of the vehicle with the forebody providing initial compression of the engine inlet flow, and the afterbody acting

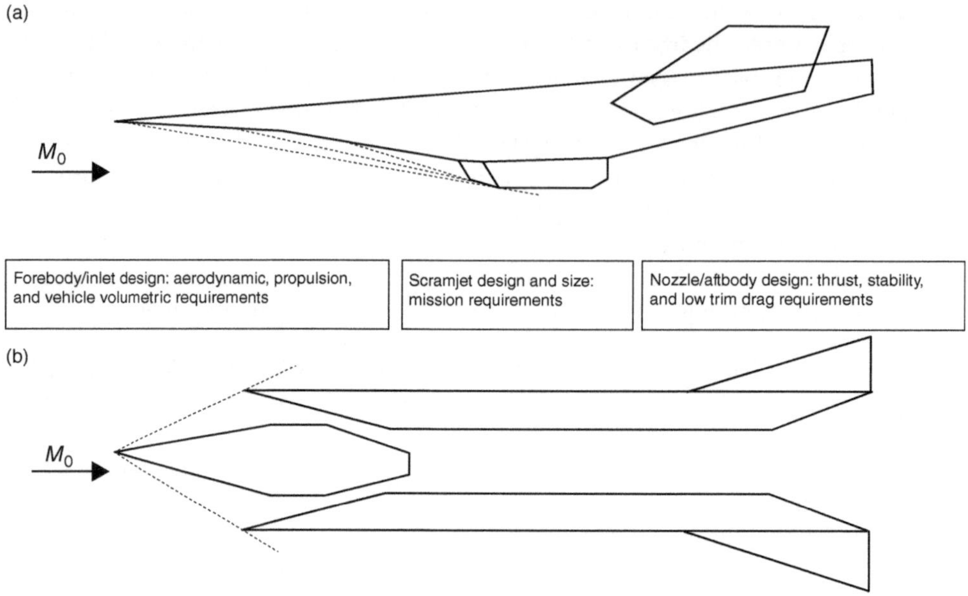

Figure 1.7 Design features for efficient scramjet engine/airframe integration (a) asymmetric; (b) axisymmetric.

as the one-sided nozzle expanding the engine exhaust flow. The design features of this integrated system are illustrated in Figure 1.7a for the asymmetric design configuration, where the bottom of the vehicle may be regarded as the engine, and the top as the airplane.

Hypersonic missiles can be designed in typical axisymmetric configuration. As shown in Figure 1.7b, the airflow is captured near the front of the vehicle, transported internally through ducts, and exhausted axially at the aft end. This type of airframe–propulsion integration requires careful separation of thrust and drag forces during the development process, but the engine performance and missile control are relatively independent. Whether asymmetric or axisymmetric, a well-designed, highly integrated vehicle must provide both maximum inlet capture area and maximum nozzle expansion area while maintaining minimum cowl drag.

The scramjet engine–airframe integration process represents the first design opportunity to maximize the performance of hypersonic air-breathing vehicles. For example, the scramjet inlet must be designed within the forebody compression field to obtain maximum engine performance under all flight conditions. At the same time, the scramjet nozzle design is primarily governed by thrust and stability requirements. Thus, the location of the scramjet, orientation of the thrust vector, and the resulting trim penalties must be examined across the entire flight envelope. The asymmetric airframe design offers a considerably higher level of performance but requires considerably more attention to vehicle–engine integration issues than the axisymmetric design.

Ultimately, the optimum aerodynamic integration design of an air-breathing hypersonic vehicle also requires careful integration of engine and airframe structures, materials, cooling, controls, and all other required vehicle-propulsion subsystems. We will study some of these issues in subsequent chapters.

1.5 Chemical Propulsion Performance Comparison

All chemical propulsion systems (rockets and airbreathers) can be compared using a unifying figure of merit, namely their specific impulse, I_{sp}. This parameter determines the ability of a propulsion system to produce thrust with the least amount of fuel or propellant consumption. For an air-breathing engine, its specific impulse is given by this formula:

$$I_{sp,a} = \frac{F}{g_0 \dot{m}_f} \tag{1.3}$$

where F is the thrust and g_0 is the gravitational acceleration on the surface of the Earth, that is 9.807 m/s^2 or 32.2 ft/s^2. The dimension of $g_0 \dot{m}_f$ in the denominator of Eq. (1.3) is the weight flow rate of the fuel based on Earth's gravity, or force per unit time. Consequently, the dimension of specific impulse is "Force/Force/second" that simplifies to just the "second."

For a rocket, $I_{sp} = F/g_0 \dot{m}_p$, where $g_0 \dot{m}_p$ is the propellant weight flow rate, i.e. the sum of oxidizer and fuel weight flow rates. This means that both substances (oxidizer plus fuel) contribute to the "expenditure" in the rocket to produce thrust, and as such the oxidizer flow rate $\dot{m}_O$ needs to be accounted for as well in the calculation of specific impulse, that is

$$I_{sp,r} = \frac{F}{g_0 \dot{m}_p} = \frac{F}{g_0 \left(\dot{m}_f + \dot{m}_O \right)} \tag{1.4}$$

Figure 1.8 shows representative values for I_{sp} of ram/scramjet and rocket propulsion systems as a function of flight Mach number. First note that the specific impulse of air-breathing engines depends on Mach number. However, for rockets their I_{sp} is independent of Mach number, and the highest value (~460 seconds) corresponds to liquid propellant rocket engines. This I_{sp} is an order of magnitude lower that the specific impulse of air-breathing engines operating at Mach numbers up to 10.

Although the scramjet performance shown in this plot is theoretical, the trends depicted in Figure 1.8 are significative. Later in the book, we will address the following facts about air-breathing propulsion systems:

- At low velocities, turbojet engines produce the highest thrust with the least amount of fuel consumption. For the regime of flight Mach numbers $0 < M_0 < 3$, the I_{sp} range between $\sim$2000 and 3500 seconds.
- The I_{sp} of turbojets drops with flight Mach number. Near $M_0 = 3$, a conventional ramjet begins to outperform the turbojet.
- The scramjet is a better propulsion choice when the ramjet engine performance deteriorates (near $M_0 = 5$).

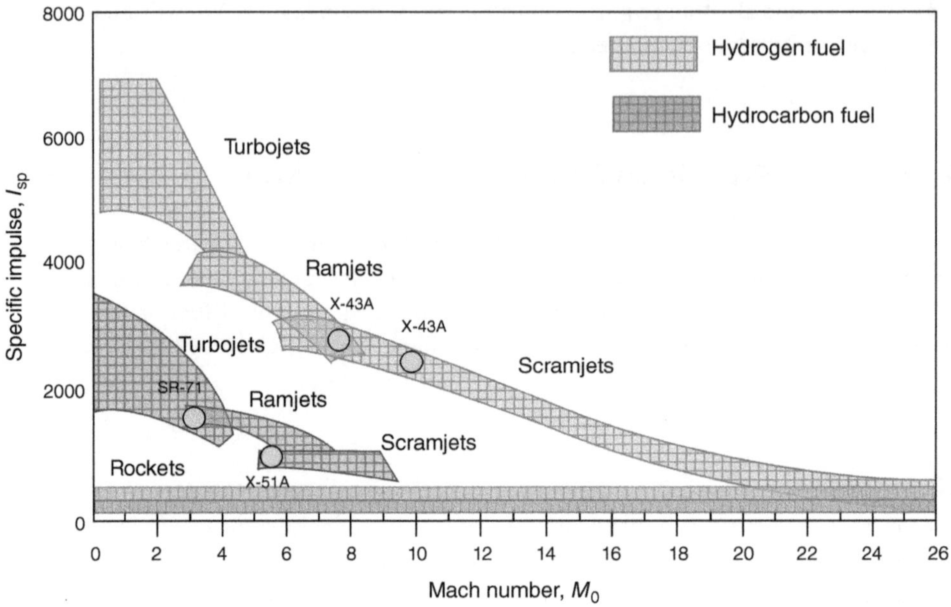

Figure 1.8 Representative specific impulse variation with Flight Mach number for air-breathing and rocket propulsion systems. Source: Image adapted from plot by NASA and USAF.

- The specific impulse for all air-breathing engines is much higher when hydrogen is used as fuel in comparison to hydrocarbon fuels. The use of a more energetic fuel leads to higher I_{sp}, since a smaller (nearly two and half times lower), consumption of fuel is needed to achieve the same temperature in the combustor.
- Superimposed on this performance map are the estimated I_{sp} of the scramjet engines flown to date: the X-43A fueled by liquid hydrogen, and the X-51A and HIFiRE, both fueled by hydrocarbons (see Section 1.7). Although the exact amount of thrust they generated is classified, these scramjets outperformed rocket propulsion at the flight Mach numbers at which they were flight tested.
- Rockets consume more propellants than air breathers. The specific fuel consumption for a typical cryogenic fuel rocket engine is 222 g/(kN s) or 7.83 lbm/(lbf h), a rather high value.

It is clear that no one propulsion system is optimum over the entire flight Mach number range of interest, from takeoff to orbital speeds. Hence, for sustained high-speed cruise, when the velocity exceeds the capability of conventional turbojet propulsion, a DMSJ would outperform rocket propulsion. Although the upper operational Mach number of scramjets is unknown, scramjet performance may approach that of a chemical rocket at flight Mach numbers M_0~12.

For space launch vehicles, the performance advantage of air-breathing propulsion over rocket propulsion is derived from the increase in performance efficiency, which results from the reduced propellant mass required. By taking a fraction of the oxidizer directly from the atmosphere, the propellant required for a mission is vastly reduced and could enable horizontal takeoff for space access. Horizontal takeoff will allow lower thrust loading thereby reducing overall engine weight. Among the propulsion technologies considered for future launch vehicles that include DMSJs are RBCC engines and TBCC engines, and the air-breathing rocket with supercooling. These concepts will be described in Chapter 10.

1.6 Hypersonic Air-Breathing Propulsion Historical Overview

The concept of the scramjet engine dates back to the 1950s. The potential for this type of high-speed propulsion grew as several countries in the world began studies and experiments, building scramjet models for research and development.

At the onset of the "space race," the United States (US) and the Soviet Union established hypersonics programs and built facilities for ground tests. For example, the US Air Force (USAF) started studies on hypersonic aerodynamics for both ballistic and lifting vehicles. Since the 1960s, the USAF tested hypersonic devices as intercontinental ballistic missiles, reentry and launch vehicles. In this period, "X-series" vehicles were built and tested as the United States aimed to develop a vehicle capable of reaching hypersonic speed. The experimental X-series program lasted a few decades and developed various prototypes of research aircrafts.

Initially, the challenge was to reach transonic and hypersonic flight with a piloted aircraft. The famous North American X-15 was a rocket-powered aircraft able to reach Mach 6.7. The X-15 was the first trans-atmospheric aircraft to test the high thermal load of hypersonic flight, showing the challenges to overcome the aerodynamic heating effects with the technologies available at that time.

An excellent history of hypersonics research in the United States is found in NASA History Series, "Facing the Heat Barrier: A History of Hypersonics" by Heppenheimer (2006). It includes the early work in Germany on the V-2 rocket in the 1940s, the development of Mach 3 aircraft such as the SR-71, and the technology development during the NASP Program.

1.6.1 Development Efforts in the United States

In 1964, NASA initiated the Hypersonic Research Engine (HRE) Project with the initial aim to develop a scramjet engine for flight testing. The goals of HRE were to design, develop, and construct a high-performance hypersonic research ram/scramjet engine for flight tests over the speed range of Mach 4–8. The three phases of the HRE project included project definition, research engine development, and flight test using the X-15A-2 research airplane, which was modified to carry hydrogen fuel for the HRE.

Because the X-15 program was canceled in 1968, the project goal of an engine flight test was eliminated. Thus, the focus of HRE became ground tests of full-scale engine models. Two axisymmetric full-scale engine models, having 18-in.-diameter cowls, were fabricated and tested: a structural model, and a combustion/propulsion model. Although never flight tested, the HRE generated a vast amount of experience with scramjet propulsion, especially during the ground-test programs. A brief historical review of the HRE project, with important features, typical data results, and lessons learned, was published by Andrews and Mackley (1994).

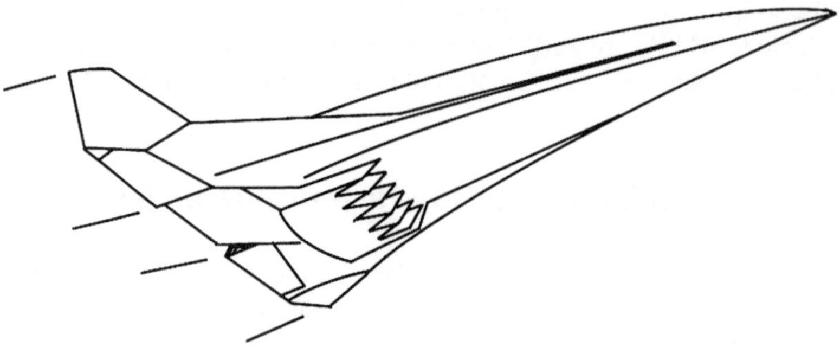

Figure 1.9 Conceptual rendering of NASP aerospace plane flying through the atmosphere on its ascent to low-Earth orbit. Source: NASA.

The interest for hypersonic flight with air-breathing propulsion was reignited in the 1980s. From 1986 through 1993, the US invested in air-breathing hypersonics, focusing on the National Aero-Space Plane (NASP), a joint Air Force-NASA program. The piloted Rockwell X-30 was the advanced technology demonstrator vehicle concept, part of the plan to develop a SSTO spaceplane, and a hypersonic passenger cruiser capable of two-hour flights from America to Asia. A major goal of the NASP program was to test the capability of hypersonic air-breathing propulsion. Preliminary research suggested a maximum speed of Mach 8 for the scramjet engine. Figure 1.9 is an illustrative depiction of the X-30 vehicle in flight.

To develop the NASP experimental vehicles, engineers faced enormous technical challenges. Two issues in particular made the X-30 seem impossible: the required combined cycle propulsion that would take the vehicle from take-off and efficiently produce thrust up to Mach 25 to reach low Earth orbit, and developing the high-temperature materials capable of surviving the harsh aerothermodynamic environments during ascent and re-entry. There were, of course, many other technical issues brought about by the very restrictive structural weight fraction limits, the reusability requirement, and many others. The political challenges proved to be even greater. In 1994, the NASP program was de-scoped to a flight test program named the Hypersonic Systems Technology Program, HySTP. Soon, however, the HySTP was cancelled. This was a major setback to the hypersonics propulsion community.

After the Air Force's withdrawal from NASP, the NASA Langley Research Center (LaRC) opted to develop a smaller unmanned hydrogen-fueled hypersonic X-plane (see Section 1.9).

1.6.2 Development Programs around the World

In parallel to the American efforts to develop a spaceplane, some European nations and the Soviet Union conceived their own hypersonics programs in response to the X-30 project. In 1986, the Soviet Union began development of the Tupelov Tu-2000, an experimental crewed concept intended to test technologies for a SSTO spaceplane. The Tu-2000 would be propelled by a system consisting of seven turbojets, one scramjet, and two rockets, all fueled by liquid hydrogen. Like NASP, the Tu-2000 program was terminated in the 1990s.

1.6.2.1 The German Sänger TSTO

In 1987, the German Ministry for Research and Technology (BMFT) initiated a Hypersonics Technology Program (HTP) oriented toward developing the Sänger I, a two-stage fully reusable, space transportation system. The piloted first or lower stage was a blended wing/body vehicle conceived for horizontal takeoff and landing on conventional runways. This first stage was known as the European Hypersonic Transport Vehicle (EHTV) because it had commonality with a Mach 6.8 hypersonic passenger aircraft.

The first stage of the Sänger I vehicle would carry the second stage attached on top, take off horizontally from a runway, and climb to an altitude of 30 km using ramjet engines. The second stage would then detach and accelerate to orbital speeds and altitudes using its LOX/LH2 rocket engine. Later, other concepts included a hypersonic passenger airliner, and an enlarged version of the two-stage launch vehicle, named Sänger II, intended for deploying payloads and astronauts via the conceptual Horus (Horizontal Upper Stage) spaceplane. These ideas drew the support of the German Aerospace Center (DLR), leading to further detailed studies being conducted as a part of the German national-level hypersonic study.

The Sänger Space Transportation System was defined in the first phase of the German HTP, and studies began to define the lower and upper stages.

German engineers identified the most critical key technologies for the realization of a Sänger transportation system, including propulsion, aerothermodynamics, materials and structures, and subsystems. Consequently, in the last phase of the HTP, a Technology Development and Verification Plan (TDVP) was established. This plan comprised all efforts needed for ground and inflight experimentation to demonstrate the technological readiness level of these most critical technologies, including the air-breathing propulsion system of the first stage of the Sänger system. Although the baseline propulsion system was a combined cycle turbo-ram concept, the trade-offs with regard to the optimum stage separation Mach-number led also to the consideration of a dual-mode ram-scramjet propulsion system.

1.6.2.2 The Russian Kholod Project

In the 1970s, the Kholod project was carried out in Russia. To study ram-scramjet propulsion this project developed a flight vehicle designed to test a hypersonic propulsion system in the harsh environmental conditions that are difficult or impossible to reproduce on the ground. The Russian Central Institute of Aviation Motors (CIAM) designed and built a two-dimensional scramjet model with a three-shock inlet and divergent combustor. This scramjet was tested in the CIAM free-jet test facility in freestream at Mach numbers 5 and 6. In the 1980s, the first successful flight tests of the cold scramjet model were carried out.

The scramjet was flown atop the SA5 surface-to-air rocket-powered missile. The experimental flight support unit was known as the Hypersonic Flying Laboratory (HFL), "Kholod." The HFL included the scramjet engine and propellant, engine control, engine cooling, instrumentation, and telemetry systems. Figure 1.10 shows a schematic of the complete HFL.

In the 1990s, CIAM designed and flight tested a Mach 6 dual mode, hydrogen-fueled scramjet engine. The first flight test of the small-scale scramjet took place in November 1991. The second flight test was carried out in November 1992. The trajectory segment over which the scramjet operated lasted for 23 seconds. Researchers reported stable combustion at hydrogen injection both from the subsonic and supersonic section of the fuel manifold at

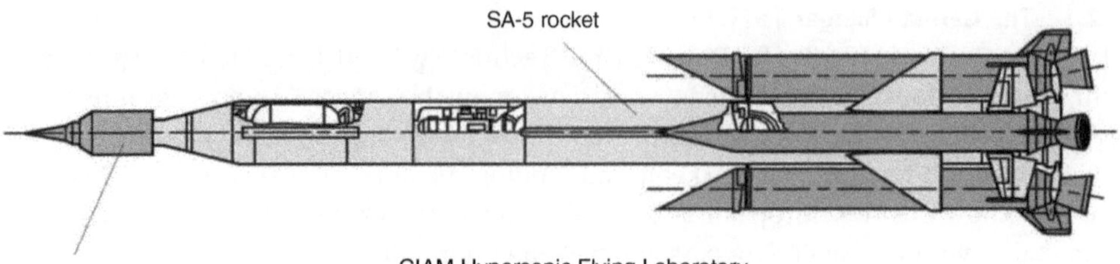

Figure 1.10 Russia's CIAM Hypersonic Flying Laboratory. Source: Roudakov et al. (1998)/NASA/Public Domain.

flight Mach number 5.3. In 1995, CIAM teamed with France for the third flight attempt, which unfortunately was unsuccessful. The scramjet engine failed to operate due to onboard systems problem (Sabel'nikov and Penzin 2000).

In addition to their HFL, Russia built a strong program for hypersonic technology and established a cost-effective way of developing these technologies with a complete launch system and flight test infrastructure. Valuable flight and ground test data with identical full-scale engines were obtained to hypersonic velocities.

1.6.2.3 France

The French National Aerospace Research Center ONERA (*Office National d'Etudes et Recherches Aérospatiales*) and MBDA (Matra, BAe Dynamics and Alenia), the European multinational developer and manufacturer of missiles, have led a large Research and Technology effort to acquire detailed knowledge on high-speed air-breathing propulsion, and to develop corresponding technologies. For operational, civilian or military, application of the high-speed air-breathing propulsion, French researchers focused on two key points: develop technologies for a low-weight propulsion system with robust fuel-cooled structure for the combustor, and develop the capability to predict with a reasonable accuracy and to optimize the aero-propulsive balance (or generalized thrust-minus-drag).

In 1992, the French National Research and Technology Program for Advanced Hypersonic Propulsion (PREPHA) was established. PREPHA (*Programme de Recherche et Technologie sur la Propulsion Hypersonic Avancée*) focus was on the study and ground testing of a hydrogen-fueled scramjet. After the end of PREPHA, Matra, BAe Dynamics and Alenia (MBDA) and ONERA continued further work to preserve the intellectual and material investment and to improve mastery of hypersonic air-breathing propulsion. ONERA and DLR led the Joint Airbreathing Propulsion for Hypersonic Application Research (JAPHAR) program to study a hydrogen-fueled dual mode ramjet for the Mach number range from 4 to 8, defining a methodology for ground and flight performance demonstration. MBDA and EADS-Launch Vehicles (EADS-LV) pursued innovative technology for fuel-cooled composite material structures. MBDA and ONERA led the Promethee R&D military program to improve knowledge on hydrocarbon-fueled dual mode ramjet for Mach 8 missile application. Steelant (2010) provides an excellent overview of European hypersonic projects.

In January 2003, MBDA and ONERA began a flight test program named LEA (Falempin and Serre 2011). The nonrecoverable LEA experimental vehicle designed to operate in the

Mach range 4 to 8 was planned to be air-launched by rocket booster, and after booster separation and stabilization, the small-scale vehicle would fly autonomously during 20-30 seconds. It is unclear whether LEA was satisfactorily flight tested. Today, France continues the technology development effort on different aspects that contribute to ensure the performance and thermal and mechanical strength of the supersonic combustion chamber. This includes (i) variable geometry needed to optimize the performance on the overall flight Mach number range; (ii) endothermic fuel used as coolant for combustion chamber structure; and (iii) design of the fuel-cooled structure itself.

1.6.2.4 Japan

In Japan, research activities related to the scramjet date back to the late 1970s. The activities during that period were limited to fundamental, laboratory-scale experiments on supersonic combustion at universities. The decade after the early 1980s was relatively less active in this field except for the experimental work at the Kakuda Research Center of the National Aerospace Laboratory (NAL-KRC), which was stimulated by previous studies.

In the past decade, the Japan Aerospace Exploration Agency (JAXA) has been promoting R&D to establish technologies for a Mach 5, 300 passenger class hypersonic aircraft that can cross the Pacific Ocean in two hours. With the main focus on a hypersonic turbojet engine that can operate continuously from takeoff to Mach 5, research and development in Japan focus on the new engine and a heat-resistant structure.

1.7 Scramjet Flight Demonstration Programs

After more than 40 years of ground-based scramjet research, the United States and Russia moved hypersonic air-breathing technology to the flight environment, which is the last stage preceding prototype development. In the 1990s, Russia flew several axisymmetric scramjet models in the HFL Kholod, as described earlier. In 1998, CIAM teamed with NASA to conduct the fourth successful flight test of a DMSJ at Mach 6.5. Valuable subsonic and supersonic combustion data were obtained from Mach 3.5 to greater than Mach 6.4. Preliminary flight test results were reported by Roudakov et al. (1998).

In the mid-1990s, NASA initiated the joint LaRC and DFRC Hyper-X Program to advance hypersonic air-breathing propulsion and related technologies from the laboratory to flight demonstration. The Hyper-X hypersonic research program aimed to demonstrate scramjet engine technologies that promise to increase payload capacity – or reduce vehicle size for the same payload – for future hypersonic aircraft and reusable space launch vehicles. In 2001, the Hyper-X Flight program demonstrated the viability of the hydrogen-fueled scramjet engine to power a hypersonic flight research vehicle (designated X-43A) to hypersonic speeds up to Mach 10. The X-43A was intended as the first in a series of vehicles envisioned to validate in flight air-breathing propulsion systems, operating from the ground to Mach 15 and beyond. After the X-43A, the X-43C Project was conceived to demonstrate Mach 5 to Mach 7 acceleration with a vehicle powered by a fuel-cooled scramjet engine using hydrocarbon fuel. It would use technology from the NASA Hyper-X Program and the USAF HyTech Program. Unfortunately, the X-43C project was not funded.

Prior to the NASA Hyper-X scramjet flight demonstration, the Australian Centre for Hypersonics at the University of Queensland carried out an innovative in-flight supersonic combustion experiment known as the HyShot. On 30 July 2002, the team launched a small combustor test article, lofted by a two-stage booster into the upper atmosphere over Australia's Woomera test range. HyShot demonstrated five seconds of hypersonic combustion at Mach 7.6 as it plunged toward Earth, capturing important data on scramjet combustion performance. The Australian HyShot program was only a component (combustor) flight test.

In 1994, the Defense Advanced Research Projects Agency (DARPA) sponsored the Hypersonic Collaborative Australia/US Experiment (HyCAUSE) that incorporated ground tests, and CFD analysis of two-dimensional and inward-turning scramjet engine concepts. It also included development of a Mach 10, hydrogen-fueled flight vehicle to be tested over a range of dynamic pressures. After conducting studies and wind tunnel experiments at the University of Queensland, a three-dimensional scramjet geometry was selected for flight testing. The flight test was conducted on June 2007. A two-stage booster was used to launch the vehicle to follow a ballistic trajectory, aiming to place the scramjet engine at the Mach 10 condition with dynamic pressures between 500 and 1600 psf during reentry (Walker et al. 2008). Unfortunately, an anomaly occurred during the reorientation maneuver, and the HyCAUSE experiment was unsuccessful.

Between May 2010 and 2015, the USAF partnered with DARPA, Pratt & Whitney Rocketdyne, and Boeing and established the Scramjet Engine Demonstrator-WaveRider (SED-WR) Program. The main goal was to conduct flight tests to demonstrate hypersonic air-breathing propulsion with a hydrocarbon-fueled scramjet, aiming to accelerate the vehicle through several Mach numbers. The SED-WR Program built a scaled vehicle known as the X-51A Waverider, to demonstrate scramjet operation in flight, starting from the NASA's X-43C test vehicle design.

The USAF also teamed with NASA and the Australian Defence Science and Technology Organisation (DSTO) to carry out the Hypersonic International Flight Research Experimentation (HIFiRE) Program. HIFiRE was conceived to study basic hypersonic phenomena through flight experimentation. In the following paragraphs, we review some aspects of the three hypersonic flight programs.

1.7.1 NASA Hyper-X Flight Program (X-43A Research Vehicle)

The Hyper-X Flight Program was a seven-year ground and flight test program that explored alternatives to rocket power for space access vehicles. Hyper-X began in late 1995 as a joint effort of NASA Langley and Dryden Research Centers with industry partnership. The main goal was to demonstrate the in-flight performance of a hydrogen-fueled, airframe-integrated scramjet at flight Mach numbers of 5, 7, and 10. The project was subsequently redirected to focus only on Mach 7 and 10 flight tests, with ground engine research continuing at Mach 5. The Mach 7 flight data were meant to allow direct comparison with ground test results for the same engine and flowpath combination, whereas the Mach 10 flight test results would provide 5–10 seconds of data at that flight Mach number, or thousands of times the duration available from ground test results (Harsha et al. 2005). The Hyper-X program goal was to verify and demonstrate experimental techniques, computational methods and analytical design tools required to advance hypersonic, hydrogen-fueled, scramjet-

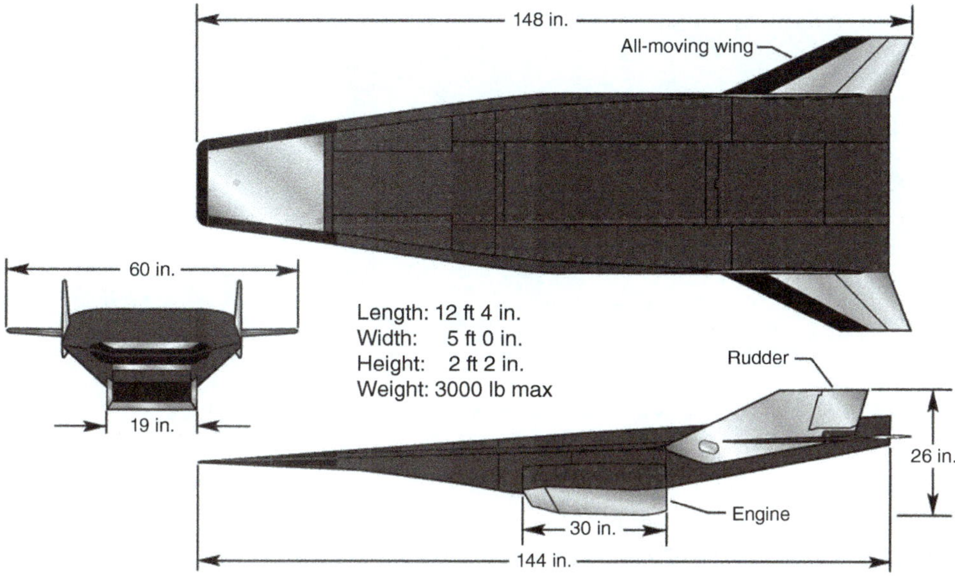

Figure 1.11 The dimensions of the NASA Hyper-X research vehicle (X-43A). Source: From Redifer et al. (2007)/NASA/Public Domain.

powered aircraft. To keep the costs down, researchers conceived a small-scale hypersonic vehicle to be air-launched with the NASA B-52B airplane and boosted to take-off speed by an off-the-shelf rocket booster.

With a length of 3.759 m (12 ft 4 in), the 1361 kg (3000 lbm) research vehicle designated X-43A (Figure 1.11) is a subscale design based on an existing Mach 10 cruise, global-reach mission configuration, and the extensive NASP database. Since the X-43A could not be photographically scaled from the 61 m (200 ft) long Mach 10 cruise vehicle, the minimum subscale size for flight testing was determined to be about 12 ft. Three vehicles were constructed. The third had the hydrogen-fueled scramjet engine, horizontal control surfaces, tails, rudders, and carbon–carbon leading edges modified to handle thermal loads for the Mach 10 mission. The vehicle design incorporated an inlet cowl with a hinged moveable door for engine close off and inlet starting. The X-43A vehicle was designed to achieve only a few seconds of powered flight at a single point design condition with a heat-sink engine.

The Hyper-X flight stack consisted of the Pegasus solid propellant rocket booster and the subscale X-43A vehicle. The stack was mounted under the wing of the B-52B aircraft with the X-43A attached to the booster through a load carrying adaptor (Figure 1.12). The B-52B carried the stack to an altitude of about 9.144 km (30 kft), and then the booster fired to take the hypersonic vehicle to the proper flight altitude. The first flight attempted in June 2001 was unsuccessful, due to the launch rocket booster going out of control early in the flight. However, two subsequent X-43A flight tests were conducted as planned.

On 27 March 2004, the Mach 7 vehicle was boosted to approximately 28.96 km (95 kft) altitude. The X-43A separated from the booster and the scramjet engine ignited. The fully autonomous vehicle flew preprogrammed over the Western Test Range off the California coast. Test data were transmitted to aircraft and ground stations. This second hypersonic flight test at an altitude of 95 kft (980 psf dynamic pressure) demonstrated acceleration

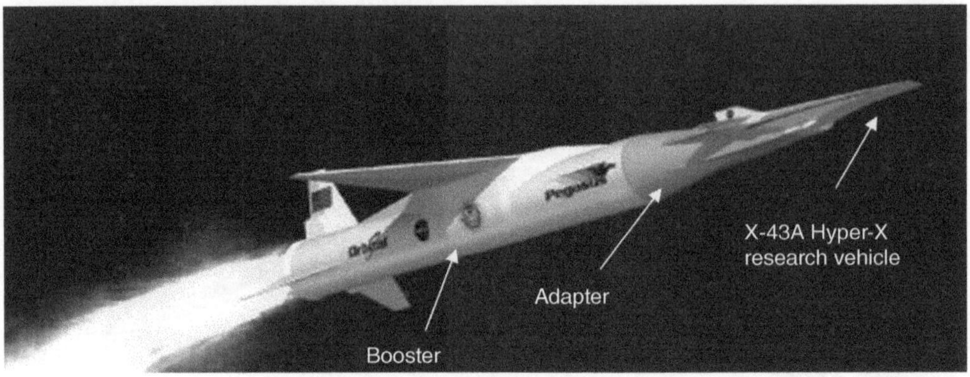

Figure 1.12 NASA X-43A vehicle attached to the booster rocket. Source: NASA.

of the scramjet-propelled X-43A vehicle during climbing flight at Mach 6.83. The vehicle accelerated under its own power for 24 km (15 miles) in 11 seconds (Voland et al. 2005).

On 16 November 2004, the X-43A achieved the highest hypersonic speed with the hydrogen-fueled scramjet flying at an altitude of 33.53 km (110 kft). The booster delivered the stack to the stage separation point at slightly lower Mach number and dynamic pressure (930 psf) than expected. The stage separation system performed smoothly, accomplishing the first known successful nonsymmetric, high-dynamic pressure, and high Mach number stage separation. The vehicle cruised at Mach 9.68, at 110 kft altitude, covering more than 32 km (20 miles) in 10.5 seconds of scramjet powered flight, making the X-43A the world's first autonomous scramjet-propelled aircraft. The X-43A/Hyper-X program provided the first free flight data on scramjet propulsion, demonstrating that the predictive design tools were accurate. Measured scramjet thrust (classified value) matched predictions to within better than two percent.

More details on the nose-to-tail CFD solutions for the actual flight condition are given in Chapter 12. These solutions demonstrate the significant increase in computational throughput, which permit full 3D solutions for the entire X-43A vehicle, which were not possible at the earlier stages of the Hyper-X Program.

1.7.2 Air Force Scramjet Engine Demonstrator-WaveRider Program (X-51A)

After the successful flight demonstrations of the NASA X-43A scramjet vehicle, the most significant activity in the United States was the Air Force SED-WR Program. The joint Air Force Research Laboratory (AFRL)/Defense Advanced Research Projects Agency (DARPA) X-51A Scramjet Engine Demonstrator-WaveRider (SED) vehicle was the flight demonstration of AFRL's Hypersonic Technology (HyTech) program. The HyTech program had several successful engine ground tests, but true scramjet viability was demonstrated with the X-51A SED in a series of 4 flight tests that began in 2010. With the X-51A waverider, the program demonstrated a practical scramjet engine burning JP-7 jet fuel at flight speeds between Mach 6.0 and 7.0+. At the same time, it aimed to extend the scramjet burn time to 300 seconds or more – a full 5 minutes of hypersonic flight.

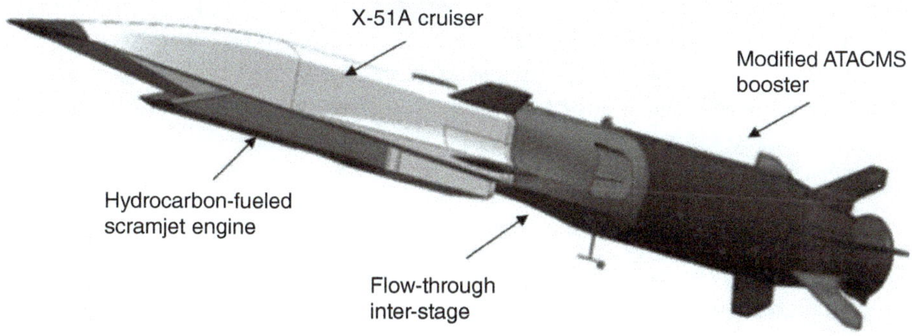

Figure 1.13 USAF X-51A Stack. Source: From Mutzman et al. (2007)/American Institute of Aeronautics and Astronautics.

Hydrocarbons fuels are of interest because they are easier to handle than hydrogen and are most suited for missile applications. In 2005, the Propulsion Directorate of the USAF began working with Pratt & (P&W) Rocketdyne and the Boeing Company to design the X-51A flight vehicle based on technology developed under AFRL/PR's Hydrocarbon Scramjet Engine Technology (HySET) Program. Although NASA had cancelled the X-43C project, the JP-7-fueled scramjet engine had been conceived, and thus it became the basis for the design of the Pratt & Whitney Rocketdyne SJX61 scramjet, which Boeing integrated into the expendable X-51A waverider cruise vehicle (Figure 1.13). Differences between the X-43C and X-51A engines include a variable geometry versus fixed inlet, flowpath width, and bolted corner construction versus welded for the X-51A, but the scramjet engines are similar and some pieces of the fuel system are exactly the same (Hank et al. 2008).

The goals of the program were to acquire ground and flight data on an actively cooled, self-controlled operating scramjet engine. This data was intended to develop and fine-tune the "rules and tools," or the understanding of the governing physical phenomena and computational design tools used for scramjet design. An important objective was to demonstrate viability of an endothermically fueled scramjet in flight. The most important outcome was for the scramjet to produce greater thrust than drag, as only then it could be proven the viability of a free-flying, scramjet powered vehicle.

As shown in Figure 1.13, the shape of the 4.27 m (14 ft) long X-51A vehicle is different from that of the X-43A vehicle, as it was designed to ride on the shock wave to reduce drag – hence the name "waverider." Its contoured, shrouded nose hides the scramjet inlet, ensuring proper airflow into the engine, and its fuselage is boxier than the sleek surfboard design of the X-43A. However, both vehicles share some design similarities. For example, the X-51A's nose cap was made of tungsten to ensure it survived the surface temperature it would experience, which was predicted to be 816 °C (1500 °F). The fins and engine were made of Inconel. The X-51A fuselage structure was made of aluminum and, due to low melting point, the structure was covered with special Boeing materials developed for use on the Space Shuttle Orbiter. These included BRI16 tiles around the engine and inlet, lightweight sprayable ablative materials on the upper surfaces, and a high-density honeycomb around the nozzle.

The SED-WR Program planned to execute four flight tests of the SED-WR vehicle. Each waverider was to be carried aloft by a B-52 launch aircraft to an altitude of about 10.67 km

(35 000 ft) and released. For the flight test, the X-51A vehicle was attached to a flow-through interstage, a section to connect with the rocket booster, as depicted in Figure 1.13. The interstage included the B-52 attachment and served as aerodynamic fuel preheating prior to lighting the scramjet proper. During the boost phase, air flew through the scramjet engine and out a duct in the interstage. This was done to warm the engine and the fuel circulating through it, thermally cracking the JP-7 fuel to ready it for regenerative cooling of the vehicle.

Initially boosted by an Army ATACMS solid rocket, the X-51A waverider was designed to take over at approximately Mach 4.5 and then accelerate on scramjet power to a preset flight speed. The first flight of the X-51A on 26 May 2010 made history, demonstrating scramjet propulsion burning JP-7 fuel for more than two minutes, with the scramjet accelerating the waverider to Mach 4.87, nearly 3400 miles/h. The fuel-cooled scramjet performed as planned transmitting normal telemetry for 143 seconds, then observing a decrease in thrust and acceleration for another 30 seconds. An anomaly then resulted in a loss of telemetry before fuel setting for higher Mach number, and the flight test was terminated; the vehicle was destroyed by flight controllers on command.

The second flight occurred on 13 June 2011. The rocket booster took the X-51A vehicle to the proper condition, and after separation, the scramjet propelled the waverider initially burning ethylene. When the engine attempted to transition to JP-7 fuel operation, it experienced an inlet unstart. An attempt to restart and orient the vehicle to optimize the engine start condition was unsuccessful. The waverider continued in a controlled flight orientation until it descended into the ocean within the test range.

The third flight on 14 August 2012 experienced a fin failure at boost, which was unrelated to hypersonic propulsion technology. Although the X-51A and its rocket booster safely separated from the B-52 aircraft, after 16 seconds under the rocket booster, a fault was identified with one of the control fins in the vehicle. Once the X-51 separated from the rocket booster, approximately 15 seconds later, the cruiser was not able to maintain control due to the faulty control fin.

The final flight of the X-51A occurred on 1 May 2013 and was the most successful in terms of meeting all the flight demonstration test objectives. The waverider flew more than 426 km (230 nautical miles) in just over six minutes, reaching a peak speed of Mach 5.1. This was the longest (240 seconds) of four X-51A test flights and the longest air-breathing hypersonic flight on JP-7 fuel.

The X-51A is considered the world's fastest jet powered aircraft burning hydrocarbon fuels. The more than nine minutes of flight data collected from the X-51A program represents an unprecedented achievement, proving the viability of air-breathing, endothermic hydrocarbon-fueled scramjet propulsion in flight.

1.7.3 Australia–US HIFiRE Program

The HIFiRE program is a collaborative effort between the Australian Defense Science and Technology Office (DSTO) and the US Air Force Research Laboratory (AFRL). The overarching technical objective is to explore fundamental hypersonic flow physics phenomena through focused flight experiments, to validate ground test techniques and verify numerical prediction tools (Dolvin 2009).

A major driver for the HIFiRE program was the need to conduct flight experiments faster and at lower cost than has traditionally been achieved (e.g. air-lifted, rocket boosted). Hence, the program chose to use sounding rockets to boost experimental payloads to hypersonic test conditions, which would be at lower cost but incurred greater technical risk in performing the flight experiments. The HIFiRE opted to test in tree launch sites: Woomera Test Range (WRC) in Australia; Andoya Rocket Range in Norway, and the Pacific Missile Range Facility (PMRF) on the island of Kauai, Hawaii.

Ten flights were conceived, with each project addressing one or more technical challenges in propulsion, aerodynamics, propulsion–airframe integration, aerothermodynamics, flight control, high-temperature materials and structures, thermal management, and/or instrumentation and sensors. The successful flight tests performed to date include HIFiRE-1, a flight experiment that provided a valuable database pertaining to boundary layer transition over a 7° half-angle, circular cone model from supersonic to hypersonic Mach numbers, and a range of Reynolds numbers and angles of incidence. HIFiRE-2 was a flight to develop an alternative test technique for acquiring high enthalpy scramjet flight test data, allowing exploration of accelerating hydrocarbon-fueled scramjet performance and dual mode transition up to and beyond Mach 8 flight.

The primary objectives of HIFiRE-2 were to (i) demonstrate scramjet mode transition end evaluate engine performance and operability; (ii) achieve combustion performance at Mach 8 flight conditions using a hydrocarbon fuel; and (iii) demonstrate a scramjet flight test approach that provides a variable Mach number flight corridor at nearly constant dynamic pressure. The primary flight experiment used an inward turning, two-dimensional, hydrocarbon-fueled scramjet, as depicted in Figure 1.14. The shroud was deployed prior to the flight experiment to expose the scramjet engine.

The successful HIFiRE-2 flight mission from PMRF was carried out on 01 May 2012. Following a suppressed trajectory via delayed stage ignition and gravity turn, the scramjet accelerated under rocket boost from Mach 5.4 to 8+ at 72 kPa dynamic pressure. This was the first-time scramjet mode transition from subsonic to supersonic combustion was demonstrated with a scramjet-powered flight vehicle using unguided sounding rocket techniques.

The HIFiRE 5B flight experiment studied a three-dimensional inlet geometry, which consists of a ramp with sidewalls and gradually merges in an elliptic isolator structure. The

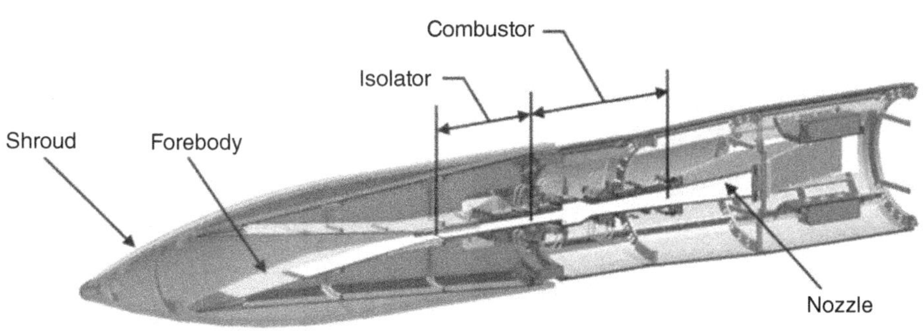

Figure 1.14 HIFiRE-2 hydrocarbon-fueled scramjet. Source: From Cabell et al. (2011)/NASA/Public Domain.

vehicle reached Mach 7.5. The last propulsion test was the HIFiRE-7, launched from Andoya, Norway on 31 March 2015. The scramjet was boosted to a Mach 7 re-entry. As it entered the atmosphere at the correct orientation, the scramjet engine started. However, telemetry was lost when the vehicle was at an altitude of 63 km (before fuel was turned on). The telemetry failure was traced to overheating of electronics in the telemetry power system.

The HIFiRE test technology was used to test partially complete scramjet flowpaths and articles that remain attached to the second-stage booster (Dolvin 2009). Following a ballistic flight trajectory, the scramjet experiment was conducted upon re-entry to the atmosphere at very high flight path angles. The HIFiRE 8 vehicle was intended to cruise at Mach 7 under scramjet power for 30 seconds at approximately zero flight path angle after separation from the booster (Glass et al. 2013).

1.8 New Hypersonic Air-Breathing Propulsion Programs

A diverse set of conceptual designs are now considered for both commercial and military applications. The hypersonic cruise aircraft being developed are summarized in Table 1.2.

In the United States, the Hermeus company is now advancing a Mach 5 aircraft (Figure 1.15). At this speed – over 3000 miles/h – flight times from New York to London will be 90 minutes rather than seven hours. In November 2021, Hermeus unveiled a full-scale prototype of their first aircraft identified as Quarterhorse. The hypersonic aircraft will be propelled by a TBCC ramjet designed around a flight-proven GE J85-21 turbojet. Known for robustness and reliability, the J85 engine has been used in multiple aircraft over the years, including the Northrop F-5 aircraft. Engineers at Hermeus are modifying the turbojet to become part of the TBCC propulsion configuration.

In 2013, Lockheed Martin announced plans to develop a hypersonic unmanned aerial vehicle (UAV) intended for intelligence, surveillance, and reconnaissance. Identified as the SR-72, this aircraft would succeed the retired Lockheed SR-71 Blackbird. The SR-72 airplane would fly at Mach 6 (4000 mph; 6400 km/h). To attain such speeds, Lockheed Martin was collaborating with Aerojet Rocketdyne to develop an air-breathing hypersonic

Table 1.2 Hypersonic cruise vehicles under development.

Aircraft concept	Mach number
Hermeus (Quarterhorse prototype)	5.0
Boeing Cruiser	5.0
Japan JAXA Cruiser	5.0
Reaction Engines LAPCAT A2	5.2
Lockheed Martin SR-72	6.0
Stratolaunch Talon-A	6.0
StratoFly MR3	8.0

Figure 1.15 Artist depiction of Hermeus Mach 5 aircraft. Source: Hermeus Corp.

propulsion system with the ability to accelerate from standstill to Mach 6.0 using the same engine. In 2018, the company pushed back on plans for advancing the SR-72, stating that when technology is matured, it could enable development of a reusable vehicle.

In 2015, the Boeing Company submitted a patent for the Small Launch Vehicle, a four-stage air-breathing space transportation system, which was planed for development in cooperation with Scaled Composites. In June 2016, Australia announced a similar project called SPARTAN, a rocket–scramjet–rocket system intended for dedicated launch capability. Building on the heritage of Sänger I horizontal take-off and landing (HOTOL) concept, the British Company REL is developing a SSTO space transportation system called Skylon, which is powered by an air-breathing rocket called SABRE. In the air-breathing mode, SABRE will pass the atmospheric air through a heat exchanger to be cooled down to 120 K, before it flows into a conventional turbine in the unique rocket engine.

Europe has added a substantial contribution to air-breathing hypersonics. MBDA France and ONERA, for example, have conducted notable research and development to improve knowledge on high-speed air-breathing propulsion, participating in different programs such as PREPHA, WRR, JAPHAR, and PROMETHEE (Falempin 2004). In 2005, the European Commission funded the Long-Term Advanced Propulsion Concepts and Technologies (LAPCAT) program aimed at examining propulsion concepts for sustained hypersonic flight. The LAPCAT project is composed of a consortium of 12 partners from industry, research institutions, and universities and is coordinated by ESA-ESTEC (Steelant et al. 2015). The ambitious mission goal is to reduce traveling time of long-distance flights, e.g. Brussels to Sydney, to about two to four hours. Advanced propulsion concepts and technologies in the flight regime from Mach 4–8 were considered, including TBCC and RBCC propulsion, in combination with corresponding vehicle system studies.

After the JAPHAR program, ONERA directed scramjet research mainly toward military applications, but still maintain a significant activity on civilian applications. In 2003, ONERA and the Japanese agency JAXA began a common experimental research effort focused on strut injectors for scramjet combustors. Between 2008 and 2013, ONERA participated in the LAPCAT II European program, while maintaining in parallel a continuous combustion modeling and CFD code development activity with internal funding. Scherrer et al. (2016) provide an overview of the most significant research activities at ONERA since 1992.

The European Space Agency established the LAPCAT II program, the follow-up of the previous, co-funded EC-project LAPCAT I. Among the several vehicles studied, only two novel concepts – for Mach 5 and Mach 8 cruise flight – are retained in the new program. The 4-year project, co-funded by the European Commission under the theme of air transportation, involves 16 partners representing six European member states. The LAPCAT-A2 Mach 5 airliner will be propelled by a precooled air-turbo ramjet, and the Mach 8 airplane will be powered by a scramjet engine, both engines fueled by liquid hydrogen (LH_2). Figure 1.16 depicts the conceptual design of a Mach 8 cruise waverider (Murray et al. 2009; Langener et al. 2013).

Designed by Reaction Engines, the LAPCAT-A2 is a 300-passenger aircraft conceived to fly at Mach 5 at an altitude between 25 and 28 km and will have a range of 18 700 km. The conceptual vehicle (Figure 1.17) consists of a slender fuselage with a low aspect ratio delta wing positioned slightly aft of the mid fuselage section. This design configuration is conceived to have good supersonic and subsonic lift-to-drag ratios, and acceptable low speed handling qualities for takeoff and landing. Designers selected a leading edge sweep angle of 55° in order to generate a stable separated vortex at high angle of attack. The liquid hydrogen fuel tanks occupy most of the fuselage volume and are split into two large pressurized tanks on either side of the passenger compartment. Storing the fuel in the fuselage instead of in the wings allows circular cross-section tanks, which minimizes insulation and pressure vessel mass. More details on this unique hypersonic cruise transport are given by Steelant (2010).

Under the Sharp Edge Flight Experiment (SHEFEX), the German Aerospace Center (DLR) pursues new, low cost, and safer design principles for space capsules, hypersonic vehicles, and space planes with re-entry capability and their integration into a complete system. SHEFEX III is a small space plane-like vehicle. It should fly even faster and stay in the

Figure 1.16 LAPCAT-II MR2 Mach 8 waverider concept. Source: From Langener et al. (2013) / ISABE.

Figure 1.17 Artistic depiction of LAPCAT A2 aircraft at takeoff. Source: Reaction Engines.

air for 15 minutes, far longer than the previous two experiments. Its launch was expected in 2021 on a Brazilian VLM rocket. It is unclear if the flight was successfully completed.

For military applications, development of hypersonic cruise missiles and drones is vigorously pursued. In the United States, programs have been funded to develop hypersonic vehicles as a part of an effort to acquire the capability to launch attacks against targets around the world in under an hour. Conventional prompt global strike (CPGS) weapons, including maneuverable stealth missiles, are capable of reaching velocities of up to Mach 6.4.

DARPA is pursuing a scramjet-powered Hypersonic Air-breathing Weapons Concept (HAWC). Sponsored by this program, Raytheon developed a classified hypersonic cruise missile powered by a hydrocarbon-fueled scramjet, and successfully completed its first free flight test in September 2021. This test demonstrated the capability that will make hypersonic cruise missiles a highly effective technology for US warfighters, bringing DARPA one step closer to transitioning HAWC to a program that offers next generation capability to the US military.

The US Air Force is also developing a classified air-launched hypersonic cruise missile called the Hypersonic Attack Cruise Missile (HACM), which seems to be based on research carried out under the HAWC program. Another major development effort in the United States is the AGM-183A Air-Launched Rapid Response Weapon (ARRW), a hypersonic glide vehicle designed to be carried by a B-52 bomber. Russia, India, France, and China are developing competing technologies. Due to the classified nature of military R&D programs, hypersonic cruise missiles are outside the scope of this book, and no further discussion of their design and technologies will be given.

1.9 Hypersonic Air-Breathing Propulsion Critical Technologies

There are a number of issues related to hypersonic air-breathing propulsion that must be fully resolved in order to develop future vehicles capable of sustained hypersonic flight. Many of those issues we will treat in subsequent chapters.

The factors that limit the operational Mach number for the various types of air-breathing engines (Figure 1.8) are different. However, it is sufficient at this point to indicate two very important technical limits. At hypersonic speeds, conversion of the air's kinetic energy to thermal energy can raise the air temperature to the level of dissociation, changing its properties considerably. The temperature can be so high (above about 2500 K) that the combustion conditions lose a large fraction of the available chemical energy to dissociation. In addition, since the air stagnation temperature raises with flight Mach number (Eq. 1.1), above about Mach 6, no conventional material can endure the high stagnation temperatures. Hence, advanced high-temperature materials and sophisticated thermal management are required for sustained hypersonic flight vehicles (see Chapter 9).

The scramjet engine is considered for powering hypersonic cruise aircraft, hypersonic missiles, or as part of combined cycle propulsion for reusable access to space vehicles. As we stated earlier, protection against aerodynamic heating drives hypersonic vehicle design, especially for reusable designs. The requirement to operate efficiently and reliably over such a wide range of flight conditions and altitudes makes hypersonic air-breathing propulsion a rather challenging field of study.

For example, for spaceplane applications, whether SSTO or first stage of a TSTO, a reusable air-breathing vehicle is much more difficult to design and develop. Unlike a rocket launcher that passes nearly vertically and quickly through the atmosphere on its way to orbit, an air-breathing space launcher would take a more leveled trajectory (see representative trajectories in Fig. 3.2). Having a lower thrust-to-weight ratio compared to rockets, air-breathing propulsion needs more time to accelerate. Following a prescribed flight path, the air-breathing vehicle will fly at hypersonic speeds for a long time, so atmospheric friction becomes a problem that must be resolved with advanced high-temperature materials and adequate thermal protection schemes.

The hypersonic aerodynamic environment is extremely harsh and hypersonic vehicles function near the edge of system and technology capability. Aerodynamic heating is a common phenomenon in hypersonic flow due to viscous dissipation. When gas flows over a surface, gas in contact is brought to rest as a result of viscosity. When the velocity near a surface decreases, the temperature increases. Aerodynamic heating increases proportionally with the square of the flight Mach number (see Chapter 3). The thin shock layer characterizing hypersonic flow over the vehicle has drastic consequences. For example, even when a weak shock wave impinges on or interferes with the bow shock wave, it yields complex aerodynamics and very high convective heat transfer rates over the hypersonic vehicle. The temperature rise associated with slowing flow near a surface is very high. This is aggravated by skin friction and shock wave heating. Hence, a huge technical challenge for hypersonic air-breathing propulsion is due to the high stagnation temperature it experiences at high Mach numbers.

The combination of high heating rates experienced by surfaces with small curvature leading edges and the long ascent times will result in large total heat loads. For the air-breathing space vehicle, the most severe heating will occur during ascent to orbit at the stagnation point and wing leading edges. For example, a heat transfer estimate for the NASP at the engine cowl lip on the ascent trajectory was about 6×10^8 W/m^2, which is orders of magnitude greater than the heating rates experienced by the former Space Shuttle Orbiter on its atmospheric entry. Moreover, vehicles propelled by air-breathing engines must fly at altitudes that are low enough that there is sufficient oxygen, i.e. relatively high density, for the propulsion system to operate effectively. However, the convective heat transfer and the drag increase as the density increases, i.e. as the altitude decreases. Thus, the trajectory of an air-breathing hypersonic vehicle must represent a compromise between the propulsion requirements and the heat-transfer/drag requirements for sustained flight.

Another technical challenge for hypersonic air-breathing vehicles is that of attaining stable supersonic combustion in the scramjet in order to produce continuous and sufficient thrust to overcome all drag forces. Engine unstart is a poorly understood phenomenon closely related to the dynamics of fuel injection and combustion that must also be studied in order to find control mechanisms to avoid such engine failure. The following are some of the most critical technologies to support advancing hypersonic air-breathing propulsion:

Structures, Materials, and Thermal Management: Aerodynamic heating caused by high-speed flight will be a significant technical challenge. In addition, heat addition produced by combustion at high velocities and temperatures is another significant factor to take into account. Materials for vehicle/propulsion structures must have excellent properties and be adequate to deal with these and other hypersonic flow phenomena.

High-temperature materials for the scramjet flowpath are expected to withstand internal wall temperatures as high as 2204 °C (4000 °F) over a brief 10-minute flight time. Cooling of scramjet's internal surfaces by fuel or regeneratively cooling combined with convection and radiation is essential. The materials for hypersonic structures must have the strength and properties to deal with these and other hypersonic flow phenomena. Ceramic matrix composites (CMC), due to their high-temperature capabilities, have the potential to provide a passive alternative for at least a portion of the scramjet flowpath, especially for nonreusable vehicles with a relatively short flight time (less than 1 minute). Chapter 9 is devoted to these topics.

Boundary Layers and Heat Control: As the flight velocity increases above Mach 5, the oblique shock waves move closer to the surface of the vehicle, and the flowfield between the shock and the body (the shock layer) becomes very hot. In the extreme case of hypersonic flight, the airflow becomes so hot that the air in the shock layer will dissociate or even ionize. Hence, a vehicle flying at hypersonic speeds must be designed to manage, control, and minimize aerodynamic heating, ensure the integrity of its airframe, and ensure its flight trajectory is controlled. We will discuss these issues in Chapters 3 and 9.

Fuels and Injection Techniques: Injecting a fuel into a supersonic airstream remains a topic of intense research and development. At high flight Mach numbers, combustor residence times are very small, and compressibility effects may suppress mixing. Hence, we require superior fuel injector performance to keep combustor length short, reduce drag, and ensure high propulsion performance. Fuel injection technology requires understanding of the fluid mechanics of mixing of fuel with air to flammable proportions, analysis of the chemistry of exothermic chemical reaction between fuel and air after they are mixed, and comprehensive studies of the aerothermochemistry (the coupling of finite-rate processes of mixing and chemical kinetics of the combustion process with one-dimensional gas dynamics) of the supersonic flow within the scramjet combustor. These topics will be covered in Chapters 5 and 6.

CFD: With limited ability to adequately represent hypersonic flow experimentally, hypersonic computational fluid dynamics (CFD) predictions become even more difficult because substantial experimental data for a variety of flows and flight conditions are not available. However, the fluid dynamics of hypersonic flows is complicated by interaction of boundary layer and shear layer with shock waves, leading to flow separation and instability not amenable to simple analysis. Moreover, high-speed reacting turbulent flows are challenging to simulate fully with Reynolds-Averaged Navier-Stokes Simulation (RANS) techniques. Large Eddy Simulation (LES) allows for modeling of small scales of turbulence while resolving large-scale turbulent structures, but this approach is currently limited to low Reynolds number flows.

Predictive methodologies, including CFD and other analytical tools, play a huge role in the analysis and development of hypersonic air-breathing propulsion (see Chapter 12). A significant element of such analysis is the prediction of integrated vehicle aero-propulsive performance, which includes an integration of aerodynamic and propulsion flowfields. This analysis becomes a huge challenge as hypersonic air-breathing vehicle configurations are characterized by highly integrated propulsion flowpath and airframe systems. Hence, development of this class of vehicle requires a comprehensive assessment of propulsion–airframe flowfield interactions and the integrated aero-propulsive performance of all systems.

The tools of analysis must encompass a wide range of modeling capabilities to capture all of the relevant flow physics of the complete scramjet flowpath as well as the external airframe flowfield. This analysis is normally accomplished using a multilevel approach, increasing in complexity and fidelity as the design is matured. The preliminary analysis phase may employ different engineering tools for the various flowpath components (e.g. inlet, combustor, nozzle), which necessitates the development of force accounting systems appropriate for specific configurations.

To date, CFD has been a valuable tool used to interpret aerodynamic and propulsion ground test data, and it will continue to do so as we progress in the maturation of HAP technologies. In Chapter 12, we will present an overview of the methods used in the analysis and pre-flight development of the scramjet engines flown to date and discuss the state of the art in modeling and simulation of HAP flow phenomena, including a survey of CFD codes with their capabilities and limitations.

1.10 Critical Design Issues

In principle, scramjet propulsion is much more efficient than rocket propulsion because scramjets extract oxygen from the air for combustion to produce thrust. This results in a higher specific impulse and greater range for a hypersonic cruise aircraft or missile, or a greater payload fraction if we integrate scramjets in a combined cycle propulsion for an orbital launch vehicle.

Scramjets are conceptually simpler as they do not need heavy turbomachinery. However, they are extremely challenging to design for optimum performance. For example, efficiently injecting, mixing, and combusting a fuel in a supersonic airflow stream is a formidable process that is coupled with turbulence and other complex flow phenomena present in the scramjet engine. Designing the flowpath is a huge challenge that requires not only to optimize all flow processes but also involves advanced materials and thermal management to protect the internal walls, and ensuring that engine thrust exceeds drag and other losses.

In addition to propulsion technical challenges, air-breathing hypersonic flight poses many other design challenges related to the vehicle itself. Sustained high-speed flight within the atmosphere imposes severe thermal loads on the vehicle's surfaces, especially on leading edges and control surfaces.

Moreover, vehicle drag is a strong function of configuration at hypersonic velocities. Thus, only a highly optimized vehicle shape, designed with a highly integrated propulsion system, can meet the flight and performance requirements. However, efficient hypersonic shapes may not be amenable to stable, controlled flight in the subsonic flight regimes that will be encountered at take-off and landing.

The effectiveness and capability of future hypersonic air-breathing vehicles depends, to a great extent, on the maximum integration of the propulsion system with the vehicle airframe. To capture the required airflow for maximum propulsion, it is necessary to use the entire forebody underneath the vehicle as compression surface, shaping it for high efficiency. A long forebody yields the largest air capture area A_0. The bottom of the vehicle (the windward side) becomes an integral part of the propulsion system. The entire vehicle underside must be carefully designed to satisfy engine performance under all flight conditions. Proper aerodynamic integration of propulsion system with the airframe is required for efficient hypersonic flight. This includes integration of vehicle structures, materials, thermal management, controls, and subsystems.

Hypersonic air-breathing engines must ingest sufficient air mass for propulsion. Achieving supersonic combustion is very difficult because the high speed in the combustor gives very little time for mixing and combustion to take place. Other technical issues related to scramjet combustors will be addressed in Chapters 5 and 7.

Future hypersonic vehicles capable of fulfilling a variety of missions may include cruise missiles, military or commercial transports in the Mach 5–7 regime, long-range cruisers in the Mach 5–10 flight regime, and single- or multiple-stage-to-orbit aerospace planes (Erbland 2004). The military needs to get to a target quicker (global engagement), and the commercial sector desires to take payloads into space cheaper and reliably. Hypersonic technologies are thus required to support the development of high-speed air-breathing propulsion systems whether they will power hypersonic missiles, hypersonic cruise aircraft, or space launchers.

In the following chapters, we will review some of the most important aspects of R&D for scramjet-based propulsion and the state of the required technologies for continued advances.

Questions

1. What type of propulsion system is appropriate for a reusable Mach 5 cruise aircraft?

2. Flight range is a figure of merit for endo-atmospheric vehicles (for given payload at given cruise Mach number). How is the vehicle range impacted by the choice of fuel for its scramjet?

3. What is the most appropriate definition of hypersonic cruise flight?

4. In addition to the critical technologies mentioned in this chapter, what other advances should be made to mature hypersonic air-breathing propulsion systems and make routine hypersonic cruise flight possible?

5. What technological issues must be resolved in order to advance the development of a spaceplane powered by a combined cycle propulsion system that integrates air-breathing propulsion and rockets?

References

Andrews, E.H. and Mackley, E. (1994). NASA's hypersonic research engine project: a review. Technical Memorandum, NASA-TM-107759, Hampton, VA (October 1994).

Cabell, K., Hass, N., Storch, A., and Gruber, M. (2011). HIFiRE direct-connect rig (HDCR) phase I scramjet test results from the NASA Langley arc-heated scramjet test facility. *17th AIAA International Space Planes and Hypersonic Systems and Technologies Conference,* San Francisco, CA (11–14 April 2011).

Dolvin, D.J. (2009). Hypersonic International Flight Research and Experimentation: Technology Development and Flight Certification Strategy. AIAA Paper 2009-7228 presented at 16th AIAA/DLR/DGLR International Space Planes and Hypersonic Systems and Technologies Conference, Bremen, Germany (19–22 October 2009).

Erbland, P.J. (2004). Current and near-term RLV/hypersonic vehicle programs. Presented RTO AVT Lecture Series on "Critical Technologies for Hypersonic Vehicle Development", held at the von Kármán Institute, Rhode-St-Genèse, Belgium (10–14 May 2004), and published in RTO-EN-AVT-116.

Falempin, F. (2004). Propulsion systems for hypersonic flight. RTO AVT Lecture Series on Critical Technologies for Hypersonic Vehicle Development held at the von Kármán Institute, Rhode-St-Genèse, Belgium (10–14 May 2004), and published in RTO-EN-AVT-116-09.

Falempin, F. and Serre, L. (2011). French Flight Testing Program LEA: Status in 2011. AIAA Paper 2011–2200. 17th AIAA International Space Planes and Hypersonic Systems and Technologies Conference, San Francisco, CA (11–14 April 2011).

Glass, D.E., Capriotti, D.P., Thomas, R., et al. (2013). Testing of DLR C/C-SiC for HIFiRE 8 scramjet combustor. *Presented at European Workshop on Thermal Protection Systems and Hot Structures*, Noordwijk, the Netherlands (8 April 2013).

Hank, J.M., Murphy, J.S., and Mutzman, R.C. (2008). The X-51A scramjet engine flight demonstration program. *AIAA 2008-2540, Presented at 15th AIAA International Space Planes and Hypersonic Systems and Technologies Conference,* Dayton, OH (28 April–1 May 2008).

Harsha, P.T., Keel, L.C., Castrogiovanni, A., and Sherrill, R.T. (2005). X-43A vehicle design and manufacture. *13th International Space Planes and Hypersonic Systems and Technology Conference*, Capua, Italy (May 2005).

Heppenheimer, T.A. (2006). Facing the heat barrier: a history of hypersonics. The NASA History Series, 2006. (6 November 2013). history.nasa.gov/sp4232-part1.pdf.

Langener, T., Steelant, J., Karl S., Hannemann, K. (2013). Layout and design verification of a small scale scramjet combustion chamber. *Proceedings of the XXI International Symposium on Air Breathing Engines (ISABE 2013)*, Busan, Korea (9–13 September 2013). ISABE Paper 2013-1655.

Murray, N., Steelant, J., Mack, A. (2009). Conceptual design of a Mach 8 hypersonic cruiser with dorsal engine. *Proceedings of the Sixth European Symposium on Aerothermodynamics for Space Vehicles. ESA SP-659*, Versailles, France (3–6 November 2008). European Space Agency. ISBN 978-92-9221-223-0.

Mutzman, R., Murphy, S., and Hank, J. (2007). The X-51A scramjet engine flight demonstration program. *Presented at the National Institute of Aerospace, NIA Session on US Hypersonic Flight Test Programs*, Hampton, VI (14 September 2007).

Redifer, M., Lin, Y., Bessent C.A., and Barklow, C. (2007). The hyper-X flight systems validation program. NASA/TM-2007-214620 (May 2007). NASA Dryden Flight Research Center Edwards, CA.

Roudakov, A., Semenov, V., and Hicks, J. (1998). Recent flight test results of the joint CIAM-NASA Mach 6.5 scramjet flight program. NASA/TP-1998-206548. National Aeronautics and Space Administration, Dryden Flight Research Center Edwards, CA.

Sabel'nikov, V.A. and Penzin, V.I. (2000). Scramjet research and development in Russia. In: *Scramjet Propulsion. Progress in Astronautics and Aeronautics* (ed. E.T. Curran and S.N.B. Murthy), 223–367. American Institute of Aeronautics and Astronautics. ISBN: 1-56347-322-4.

Scherrer, D., Dessornes, O., Ferrier, M. et al. (2016). Research on supersonic combustion and scramjet combustors at ONERA. *Journal Aerospace Lab* (11): 1–20.

Steelant, J. (2010). Hypersonic Technology Development with EU Co-Funded Projects. Presented in the High Speed Propulsion Engine Design-Integration and Thermal Management Lecture Series, RTO-EN-AVT-185-17, Rhode St Genese, Belgium.

Steelant, J., Varvill, R., Defoort, S. et al. (2015). Achievements obtained for sustained hypersonic flight within the LAPCAT-II project. *AIAA 20th International Space Planes and Hypersonic Systems and Technologies Conferences*, Rhode St Genèse, Belgium.

Voland, R.T., Huebner, L.D., and McClinton, C.R. (2005). X-43A hypersonic vehicle technology development. *Paper IAC-05-D2.6.01 Presented at the 56th International Astronautical Congress*, Fukuoka, Japan (17–21 October 2005).

Walker, S., Rodgers, F., Paull, A., and Van Wie, D.M. (2008). HyCAUSE flight test program. *15th AIAA International Space Planes and Hypersonic Systems and Technologies Conference*, Dayton, Ohio (28 April to 1 May 2008). AIAA Paper 2008-2580.

2

Theoretical Background

This chapter is intended as an overview of basic concepts and formulations pertaining to air-breathing propulsion systems. We begin with a schematic of the NASA X-43A experimental vehicle to use as a reference. The X-43A airframe-integrated scramjet engine uses the vehicle forebody and afterbody as external compression and expansion surfaces, which are continuations of the engine's inlet and nozzle, respectively. The rectangular propulsion flowpath is predominantly two-dimensional, with a short external cowl (76.2 cm long, about 20% of the total length of the vehicle), as depicted in Figure 2.1. For X-43A dimensions, see Figure 1.11.

The flow phenomena found in hypersonic air-breathing propulsion are rather complex and can only be adequately described with data obtained with sophisticated numerical analysis and extensive ground and flight testing. However, this review provides a theoretical foundation that may be helpful to put in perspective the topics addressed in subsequent chapters. The concepts herein will be helpful to answer questions such as *what is the optimum cruise altitude of hypersonic aircraft*? It will help us appreciate why the X-51A Waverider flew at an altitude a few kilometers below the X-43A aircraft. This review also provides the performance metrics for air-breathing propulsion (including ram/scramjets) and cautions us that even in preliminary analysis of hypersonic flow, the calorically perfect gas assumption for air is not appropriate.

2.1 Atmospheric Flight

Whether intended as a hypersonic cruiser or as an orbital spaceplane, the vehicles powered by air-breathing propulsion will fly within the Earth's atmosphere, taking a large amount of ambient air to generate thrust to fly and accelerate. The flight profile of such hypersonic vehicles is defined by a predetermined flight corridor that takes into account its flight and propulsion performance. In order to ensure operational flight limits, hypersonic vehicles are designed to operate within a narrow range of dynamic pressures. An optimized flight trajectory will ensure the vehicle operates below its structural heating and load limits while delivering enough airflow to its propulsion system to ensure an optimum level of net thrust for acceleration and cruise.

Scramjet Propulsion: A Practical Introduction, First Edition. Dora Musielak.

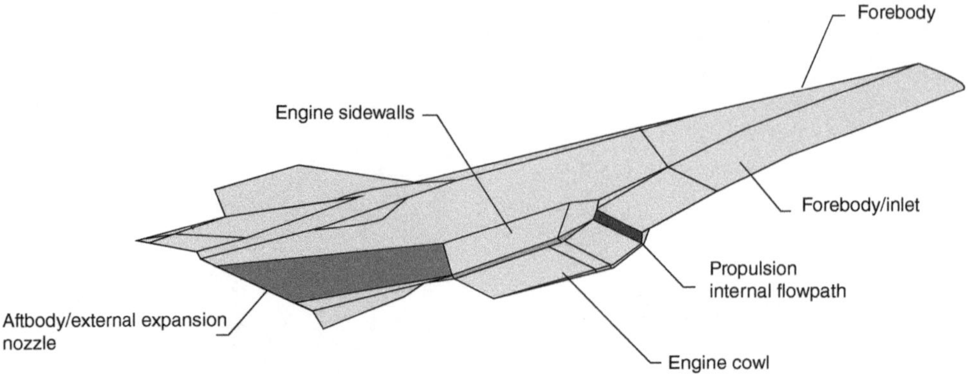

Figure 2.1 Schematic of the NASA X-43A scramjet vehicle. Source: Adapted from Engelund et al. (2000).

2.1.1 Ideal Air-Breathing Propulsion Model: Uninstalled Thrust and Specific Impulse

The study of air-breathing propulsion usually begins with the definition of uninstalled thrust. Viewed as a control volume (CV) submerged in a fluid medium about and through which the fluid flows, the propulsion model incorporates an energy transfer mechanism that increases the kinetic energy of the fluid passing through the system. This mechanism is called the engine and is depicted in Figure 2.2. The idealized air-breathing engine fits the one-dimensional CV model, where the entry and exhaust velocities are assumed to be parallel to the thrust force. In this model, $\dot{m}_0$ is the mass flow rate of the air stream entering the engine through the inlet with cross-sectional area A_0 and V_0 is the freestream velocity of the air coming into the engine (same as the aircraft airspeed) at ambient pressure p_0. The exhaust plane is characterized by the mass flow rate of gas exiting the engine $\dot{m}_{10}$ (including fuel added in the combustor), and V_{10} is the velocity of the hot gas leaving the engine at static pressure p_{10}, which may or may not be equal to p_0. The need to specify engine station numbers is explained in Section 2.4.

Air-breathing engines require to ingest enough airflow in order to generate the thrust F to accelerate the vehicle and to overcome the drag forces on it. According to Newton's laws of motion, the thrust force arises from rearward momentum imparted to a stream of gas (the

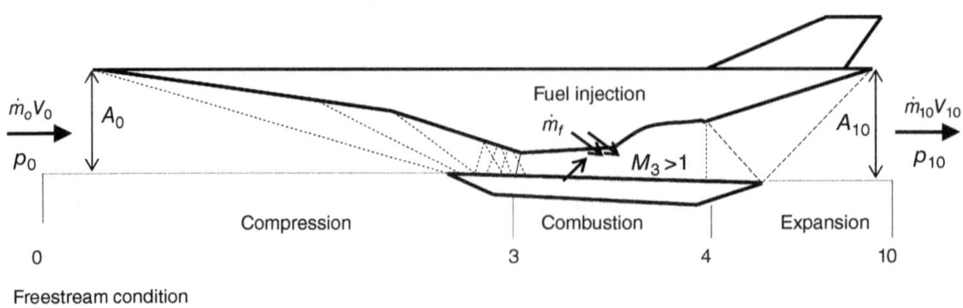

Figure 2.2 Ideal scramjet model with reference station numbers.

reaction mass). Hence, the general thrust equation for the idealized air-breathing jet engine represents the change in momentum flux from the engine inlet to its exhaust. For an engine with a single inlet and single exhaust, the uninstalled thrust is expressed as:

$$\begin{aligned} F &= \dot{m}_{10}V_{10} - \dot{m}_0 V_0 + (p_{10} - p_0)A_{10} = (\dot{m}_0 + \dot{m}_f)V_{10} - \dot{m}_0 V_0 + (p_{10} - p_0)A_{10} \\ &= \dot{m}_0[(1+f)V_{10} - V_0] + (p_{10} - p_0)A_{10} \end{aligned} \tag{2.1}$$

where $\dot{m}_{10}$ is exhaust mass flow comprised the mixture of fuel and air ($\dot{m}_0 + \dot{m}_f$), and F is the internal or uninstalled thrust, which is equal and opposite to the difference in momentum fluxes between the airflows entering and leaving the engine. The last expression written in terms of the fuel-to-air ratio is defined as:

$$f \equiv \frac{\dot{m}_f}{\dot{m}_0} \tag{2.2}$$

The fuel-to-air ratio f is an indicator of the conditions in the combustor. Its upper limit is the value of f that corresponds to complete combustion, known as the stoichiometric fuel-to-air ratio denoted as f_{st}. For scramjets fueled with hydrogen, the value is $f_{st} = 0.0291$, while that for JP-7 fuel is $f_{st} = 0.0694$.

Equation 2.1 includes the three terms that contribute to the uninstalled or net thrust: the first term $\dot{m}_{10}V_{10}$ represents the exhaust momentum through the nozzle, which contributes positively to the propulsion thrust. The second term $\dot{m}_0 V_0$ is due to the engine inlet momentum, denoting an opposing drag force that contributes negatively to the thrust. This term is known as *ram drag*, representing the penalty of bringing air into the engine with a finite momentum, which is inherent in all air-breathing propulsion. The last term $(p_{10} - p_0)$ A_{10} is the pressure area acting on the exhaust nozzle exit plane. This term only contributes to the engine thrust if the static pressure of the exhaust jet p_{10} does not match the static pressure of the ambient p_0. Hence, to maximize the thrust produced, the engine design aims to attain perfect nozzle expansion, a condition where the exhaust flow is perfectly expanded to the external atmospheric pressure, i.e. $p_{10} = p_0$, and ensure that its exhaust speed is greater than the air or flight speed, i.e. $V_{10}/V_0 > 1$.

The ram effect also affects thrust. The ram effect refers to the process when pressure is produced by the dynamic action of the airflow. This happens when the movement of a vehicle relative to the outside atmosphere causes air to be "rammed" into the engine inlet, as it occurs in the ramjet engine. This ram effect increases the airflow into the engine, resulting in higher gross thrust. Ram combines the airspeed increase, the increase in the air pressure, and the airflow into the engine. Therefore, the increase of thrust caused by the ram effect becomes significant as the airspeed increases, and this will compensate for the loss in thrust due to the reduced density at high altitude. That is why subsonic jet-powered aircraft fly at high subsonic speeds and higher altitudes to make use of the ram effect.

A number of factors not explicitly apparent in Eq. (2.1) affect the thrust generated. There are factors related to the engine design, and factors related to the surrounding medium. For example, the terms $\dot{m}_0$, V_{10}, A_{10}, and p_{10} are fixed by the engine, while the terms V_0 and p_0 are fixed by the flight condition. At the same time, the inlet air mass flow rate $\dot{m}_0$, which influences both the momentum thrust and momentum drag, is dependent on several variables, including the flight speed, ambient air temperature and pressure, humidity, flight altitude, and, in the case of a turbo engine, the rotational speed of the compressor.

The specific impulse of an air-breathing propulsion system is defined as the (uninstalled) thrust per unit of fuel flow rate. This parameter represents the ability of a propulsion system to produce thrust with a minimum of fuel expenditure, Eq. (1.3), repeated here for completeness:

$$I_{sp} = \frac{F}{g_0 \dot{m}_f} = \frac{1}{g_0 f} \cdot \frac{F}{\dot{m}_0} \tag{2.3}$$

where g_0 is the acceleration due to gravity at sea level, $g_0 = 9.807\ \text{m/s}^2$.

The specific impulse is the unifying figure of merit for air-breathing engines and rockets. The specific impulse of rocket engines depends on the nozzle exhaust velocity and is constant across the flight regime. For modern liquid propellant rockets, I_{sp}~450 seconds. As we noted in Chapter 1, because scramjet engines have higher specific impulse compared with rocket engines, scramjets are more efficient in the flight regime $5 < M_0 < 12$.

Since the thrust force depends on the engine's ability to ingest, process, and interact with the surrounding atmospheric air, it is important to review the physical characteristics of the environment, as it directly impacts both the aircraft flight characteristics and the performance of its propulsion system.

2.1.2 Earth's Atmosphere

For flight within the atmosphere, it is necessary to know the conditions of the air (temperature, pressure, and density) at the wide range of altitudes that a vehicle and its propulsion system need to operate. This requires a model of the atmosphere that represents either mean conditions or conditions specific to a particular time and place.

Analysis of air-breathing propulsion assumes air is perfect gas, with a chemical composition comprised 76.55% nitrogen + 20.54% oxygen + 1.96% water vapor (by moles). The atmosphere surrounding the Earth is about 480 km (300 miles) thick, but most of the air (about 80%) is found within 16 km (10 miles) over the sea level. There is no exact place where the atmosphere ends; it just gets thinner and thinner, until it merges with outer space. Up to altitudes of approximately 90 km, fluctuating winds and general atmospheric turbulence in all directions keep the air mixed in nearly the same proportions.

The static pressure of air decreases exponentially with increasing geometric altitude *H*. Different models provide the formulation to determine the variation of air properties important for atmospheric flight. Analysis may use quantitative values of air at the different altitudes taken from the U.S. Standard Atmosphere. This standard consists of a series of models that define values for atmospheric temperature, density, pressure, and other properties over a wide range of altitudes. Table 2.1 provides the properties of Earth's atmosphere at selected altitudes over sea level, where the ratios are given with reference to standard, mean conditions at sea level, denoted with the index *SL*.

Although atmospheric air static pressure *p* decreases exponentially with increasing geometric altitude, the static temperature exhibits a much different profile through the layers of the atmosphere. In the troposphere, as the altitude increases from sea level up to an altitude

Table 2.1 Representative properties of the Earth's atmosphere.

Altitude		Pressure ratio	Temperature ratio	Density ratio	Sonic speed ratio	Freestream static conditions
km	kft	p/p_{SL}	T/T_{SL}	ρ/ρ_{SL}	a/a_{SL}	T_0 (K), p_0(N/m^2)
0	0	1.000 0	1.0000	1.000 0	1.0000	288.2, 1.013×10^5
5	16.4	0.533 4	0.8873	0.601 2	0.9420	255.7, 5.403×10^4
10	32.8	0.261 5	0.7748	0.337 6	0.8802	223.3, 2.649×10^4
15	49.2	0.119 5	0.7519	0.159 0	0.8671	216.7, 1.211×10^4
20	65.6	0.054 57	0.7519	0.072 58	0.8671	216.7, 5.528×10^3
25	82.0	0.025 16	0.7689	0.032 72	0.8769	221.6, 2.548×10^3
30	98.4	0.011 81	0.7861	0.015 03	0.8866	226.5, 1.196×10^3
35	114.8	0.005 67	0.8208	0.006 91	0.9060	236.5, 5.743×10^2
40	131.3	0.002 83	0.8688	0.003 26	0.9321	250.4, 2.867×10^2

Ratios are given with reference to standard, mean conditions at sea level:

$p_{SL} = 1.013 \times 10^5\ \mathrm{N/m^2(Pa)} = 2116\ \mathrm{lbf/ft^2}$

$T_{SL} = 288.2\ \mathrm{K} = 518.7\ \mathrm{R}$

$\rho_{SL} = 1.225\ \mathrm{kg/m^3} = 0.076\,47\ \mathrm{lbm/ft^3}$

$a_{SL} = 340.3\ \mathrm{m/s} = 1116\ \mathrm{ft/s}$

Source: Data from U.S. Standard Atmosphere, 1976.

of 11 km (0–36 kft), the air temperature decreases. In the stratosphere, as the altitude increases from 11 km to altitude of 51 km (36–167 kft), the air temperature increases. Hence, the variation in air density will conform to the patterns of pressure and temperature according to the gas law, assuming that the atmosphere is a mixture of perfect gases. In the cruise flight regime $5.0 < M_0 < 12.0$, vehicles will operate at altitudes ranging from 30 to 40 km where the freestream static conditions are $226.5\ \mathrm{K} < T_0 < 250.4\ \mathrm{K}$ and $1.196 \times 10^3\ \mathrm{Pa} < p_0 < 2.867 \times 10^2\ \mathrm{Pa}$. Since the atmosphere is not a quiescent body of gas, the vehicle will encounter wide variations of air properties. Both the scramjet and the airframe designs must consider those air variations and effects on flight.

2.1.3 Dynamic Pressure

Hypersonic vehicles powered by air-breathing propulsion must fly at altitudes which are low enough that there is sufficient oxygen for the engine to operate effectively, i.e. low altitudes of relatively high air density. However, at low altitudes, the convective heat transfer and the drag increase due to the high air density. Hence, the trajectory must represent a compromise between the propulsion requirements and the hypersonic vehicle's heat transfer/drag requirements.

The dynamic pressure represents the increase in a moving fluid's pressure over its static value due to motion. Seen from a moving vehicle, the dynamic pressure, q_0, of the atmospheric air is

$$q_0 = \frac{1}{2}\rho_0 V_0^2 \tag{2.4}$$

where ρ_0 is the air density of the freestream at the flight altitude, and V_0 is the velocity of the vehicle relative to the atmosphere (or vice versa).

In terms of Mach number M_0, the flight velocity is

$$V_0 = a_0 M_0 \tag{2.5}$$

The local speed of sound is

$$a_0 = \sqrt{\gamma_0 R T_0} \tag{2.6}$$

with γ_0 denoting the specific heat ratio for the freestream atmospheric air (see Section 2.2.1). Using the perfect gas law, rewrite the density in terms of static pressure and obtain q_0 as:

$$q_0 = \frac{1}{2}\rho_0 a_0^2 M_0^2 = \frac{p_0}{R_0 T_0} \cdot \gamma_0 R_0 T_0 \cdot \frac{M_0^2}{2} = \frac{1}{2}\gamma_0 p_0 M_0^2 \tag{2.7}$$

We can use Eq. (2.4) or (2.7) to calculate the dynamic pressure of the atmosphere for any combination of altitude and velocity requirement.

Since q_0 is proportional to the air local pressure p_0, it is clear that dynamic pressure is a scaling factor for the forces on the vehicle. For example, if q_0 is too low, the engine may not have enough air to ingest for combustion. Combustion blowout imposes the propulsion cruise limit. If q_0 is too large, the structural forces on the vehicle can be excessive. The dynamic pressure is a scaling factor for the pressures and aerodynamic forces exerted on the vehicle in flight.

When a cruise aircraft is in steady flight, the drag force is equal to the thrust developed by the engine $F_D = F$, where the internal drag is proportional to the air mass ingested by the engine $\dot{m}_0$; the internal thrust is equal and opposite to the difference in momentum fluxes between the airflows entering and leaving the engine, $F \approx \dot{m}_0(V_{10} - V_0)$. At the same time, the lift force is equal to the weight of the vehicle, $F_L = W = mg$. The drag force is the sum of the parasite drag (caused by frictional resistance opposing the airflow over the aircraft), plus the induced drag created by the angle of attack of the air to produce the necessary lift force which supports the weight of the vehicle in flight. The drag F_D and lift F_L are proportional to the dynamic pressure:

$$F_D = AC_D\frac{1}{2}\rho V^2 = AC_D q_0 \tag{2.8}$$

$$F_L = AC_L\frac{1}{2}\rho V^2 = AC_L q_0 \tag{2.9}$$

where A is a reference area such as the wing area or the vehicle's cross-sectional area.

These two expressions imply that (i) if q_0 is too large, the drag force would be great, and thus the required thrust; (ii) if q_0 is too small, the lifting area required for sustained flight would be unreasonably large. Since dynamic pressure imposes a limit that corresponds to

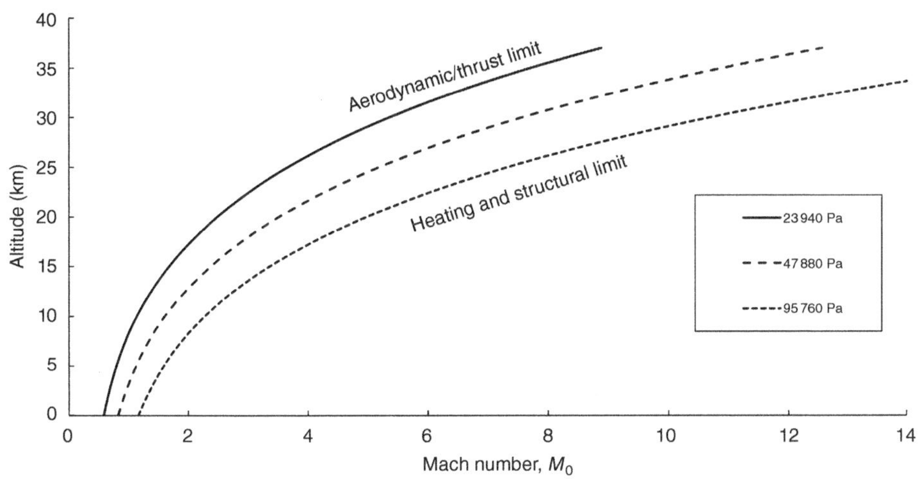

Figure 2.3 Standard day geometric altitude flight trajectory for air-breather vehicles restricted to constant dynamic pressure corridor.

the maximum stagnation pressure which the vehicle structure can bear in the presence of high acceleration loads, this means that sustained (steady) flight at hypersonic speeds must be confined to occur within a restricted corridor of altitudes at each speed (M_0).

Hypersonic vehicles must be designed to fly within a narrow range of dynamic pressures q_0, approximately 23 000–95 000 N/m^2 (500–2000 lbf/ft^2). Above this flight corridor, the vehicle cannot provide enough lift; the lift limit is determined by the maximum level flight altitude at a given velocity. On the other hand, flying below this corridor the surface heating and structural loads on the vehicle would be excessive; the temperature limit is set by the structural thermal limits of the material used in construction of the vehicle. At any given altitude, the maximum velocity attained is temperature-limited by aerodynamic heating effects. At lower altitudes, velocity is limited by aerodynamic force loads rather than by temperature.

The approximate form of the flight corridor is shown in Figure 2.3, where three operational regions are defined in terms of the Mach number and altitude range of each, generated by selecting combinations of q_0 and geometric altitude – p_0 is obtained from US Standard Atmosphere tables – and the flight Mach number is calculated from Eq. (2.7) in the form:

$$M_0 = \left(\frac{2q_0}{\gamma_0 p_0}\right)^{1/2} \tag{2.10}$$

All hypersonic air-breathing vehicles will fly in the stratosphere (low-speed aircraft fly in the troposphere below). According to Figure 2.3, for flight in the range $5.0 < M_0 < 14.0$, hypersonic vehicles will fly in the altitude range 20 km $< H <$ 45 km (65.6–147.6 kft).

The X-43A vehicle flew at Mach 6.83 at an altitude of 29 km (95 kft) where the dynamic pressure was about 50 753 N/m^2 (1060 psf). The X-51A Waverider flew at Mach 5.1 at 21 km (70 kft) altitude. What was the dynamic pressure experienced by the Waverider?

2.1.4 Air Mass Flow Available for Thrust

An air-breathing engine has a much more challenging task than a rocket to propel a vehicle. It must develop a thrust force large enough to overcome atmospheric drag, and for this process the engine must take a huge amount of air from the atmosphere. Since the uninstalled (internal) thrust is proportional to total mass flow rate of air ingested, hypersonic air-breathing engines generate thrust in direct proportion to the rate at which they are able to capture and process the surrounding atmosphere.

Consider a flight at an optimum dynamic pressure q_0. By writing the freestream air density in terms of q_0, the mass flow ingested by the inlet of the engine can be written as:

$$\dot{m}_0 = \rho_0 V_0 A_0 = \frac{2q_0 A_0}{a_0 M_0} \tag{2.11}$$

where A_0 is the inlet capture area, which is the geometric area of the freestream tube that eventually enters the engine inlet.

The capture area can be larger or smaller than the physical opening of the engine inlet, and its size is determined by the entire flow field upstream of the engine face. For a hypersonic vehicle with its scramjet integrated to the airframe, the long forebody compression surfaces are designed to make the freestream capture area A_0 much larger than the physical opening of the engine inlet; A_0 varies slightly with Mach number and angle of attack.

Since a_0 hardly changes over the flight altitude, according to Eq. (2.11) **the freestream mass flow per unit area ($\dot{m}_0/A_0$) primarily varies inversely with Mach number along any trajectory of constant dynamic pressure**. For instance, at Mach 7 aircraft cruising on a flight corridor where dynamic pressure is q_0 = 47 880 N/m^2 would ingest 45.40 kg/s of air per meter square; at a higher dynamic pressure of 60 000 N/m^2, this quantity would be $\dot{m}_0/A_0 \sim 57\,\text{kg/s/m}^2$.

Moreover, as cruise velocity increases, the mass flow decreases, and this will reduce the thrust, as it is clear by rewriting the thrust equation in terms of the dynamic pressure:

$$F = \frac{2q_0 A_0}{a_0 M_0}[(1+f)V_{10} - V_0] + (p_{10} - p_0)A_{10} \tag{2.12}$$

Assuming perfect nozzle expansion, the engine's total uninstalled (internal) thrust is proportional to the total mass flow rate of air ingested, flight velocity, and engine exhaust velocity. Since $\dot{m}_0 = \rho_0 V_0 A_0$, where A_0 is the engine inlet capture area, and ρ_0 is the ambient air density, this means that for high Mach number flight $M_0 > 5$, a scramjet-powered vehicle needs very large capture area A_0 to maximize $\dot{m}_0$. This explains why hypersonic vehicles are designed with long forebody compression surfaces in order to make the freestream capture area A_0 much larger than the physical opening of the engine inlet. Moreover, the exhaust velocity V_{10} must be greater than the inlet air velocity V_0 to yield positive thrust. Above all, hypersonic air-breather vehicles may fly along a trajectory of highest allowable dynamic pressure (without exceeding structural limits) in order to maximize freestream mass airflow per unit area to the engine.

Propelled by a hydrogen-fueled scramjet, the NASA X-43A vehicle flew at M_0 = 6.83 in 2004. Estimate the air mass flow it ingested. For the capture airflow area, assume the width of the rectangular inlet was W = 0.483 m, and the height of captured streamtube was ~0.3W.

Moreover, for air-breathing hypersonic flight, the compression and expansion processes are very critical to the overall propulsion system performance. The fractional velocity change across the engine becomes very small as the flight Mach number increases. This means that a small inefficiency in the nozzle or the inlet could have large consequences for overall scramjet propulsion performance. By determining the fractional velocity change across the scramjet, this issue becomes clear. From Eq. (2.1), the velocity change from free-stream to exhaust is given by:

$$\frac{V_{10}}{V_0} = \left(\frac{1}{1+f}\right)\left(\frac{F}{\dot{m}_0 V_0} + 1\right)$$

Thus, for example, a hydrogen-fueled scramjet, operating at $V_0 = 2950$ m/s with $f = 0.0291$, would have a specific thrust of $F/\dot{m}_0 = 573\ \text{N}\cdot\text{s/kg}$ and a fractional velocity of just $V_{10}/V_0 = 1.20$.

We can also determine the velocity ratio implied by the specific impulse, Eq. (2.3). Write

$$I_{sp} = \frac{1}{g_0 f}\frac{F}{\dot{m}_0} \approx V_0 \frac{(V_{10}/V_0) - 1}{g_0 f} = a_0 M_0 \frac{(V_{10}/V_0) - 1}{g_0 f} \tag{2.13}$$

from where the fractional velocity ratio is

$$\frac{V_{10}}{V_0} - 1 = \frac{f I_{sp} g_0}{a_0 M_0} \tag{2.14}$$

Figure 2.4 shows the fractional velocity as a function of flight Mach number, assuming a constant stoichiometric fuel-to-air ratio in the scramjet: for hydrogen and hydrocarbon fuels, $f = 0.0293$ and $f = 0.0664$, respectively. It is clear that as $M_0 > 8$, the fractional velocities across the air-breathing engine become exceedingly small, regardless of the fuel used.

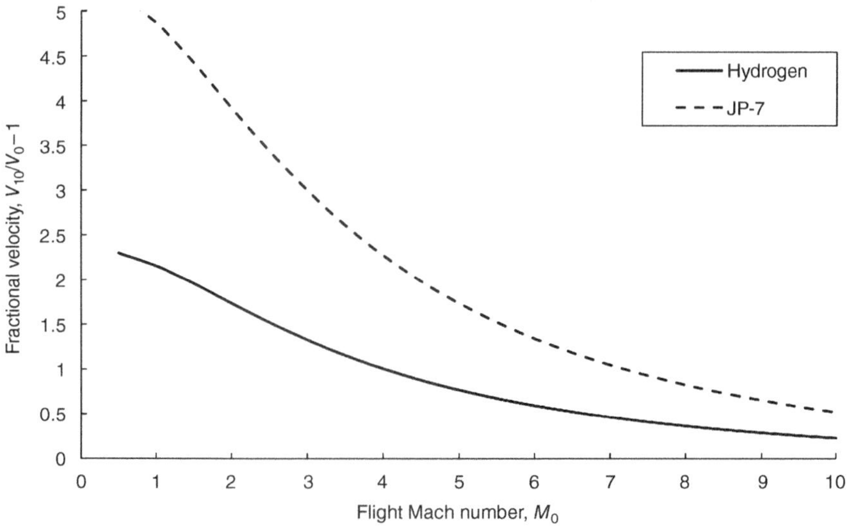

Figure 2.4 Fractional velocity change as function of flight Mach for air-breathing engines.

Hydrogen is considered as the fuel of choice for hypersonic flight because of its very high heating value and its capacity to absorb the thermal loads of high Mach number flight. This was demonstrated with the X-43A vehicle. The scramjet that powered the X-51A Waverider was designed to operate with JP-7 (see Chapter 6 for additional details on the reasons why).

Cruise air-breathing hypersonic vehicles must operate at relatively low altitudes to maintain the mass flow rate for maximum engine thrust. However, flying at low altitudes for extended periods of time, high dynamic pressures and high Reynolds numbers will cause large aerodynamic loads, uncertainty in boundary layer transition, and large total surface heating. These issues are major vehicle design drivers.

2.2 Air Thermodynamic Models

Air is the working fluid for the propulsion system and is also the medium through which hypersonic vehicles must fly. Air is a complex gas mixture when examined at the aerothermodynamic conditions of the engine and of the vehicle in flight, and thus the freestream flow must be characterized, as analysis requires determination of the equilibrium state of air. In this study, the airflow is in thermochemical equilibrium as the density is sufficiently high so that there are sufficient collisions between particles to allow the equilibration of energy transfer between the various modes. For an equilibrium flow, any two thermodynamic properties, e.g. p and T, can be used to uniquely define the state.

2.2.1 Equilibrium Air Chemistry: Perfect Gas Assumption

Air is assumed to obey the ideal or perfect gas equation of state, which is derived from kinetic theory and neglects molecular volume and intermolecular forces. This law is accurate under conditions of relatively low density which correspond to relatively low pressures and/or high temperatures. The equation of state (perfect gas law) is

$$p = \rho RT \tag{2.15}$$

where R is the gas constant (287 N·m/kg·K; 53.3 ft-lbf/lbm-R).

The equilibrium behavior of air also considers the equilibrium of its static enthalpy, specific heat at constant pressure, and the ratio of specific heats. Enthalpy is a thermodynamic quantity equivalent to the total heat content of a system. The specific static enthalpy h is related to the internal energy e of the substance and is defined as:

$$h = e + \frac{p}{\rho} \tag{2.16}$$

Since both internal energy and enthalpy are functions of temperature, these properties define the equilibrium specific heat at constant volume c_v, and the specific heat at constant pressure c_p:

$$c_v \equiv \left(\frac{\partial e}{\partial T}\right)_v ; c_p \equiv \left(\frac{\partial h}{\partial T}\right)_p \tag{2.17}$$

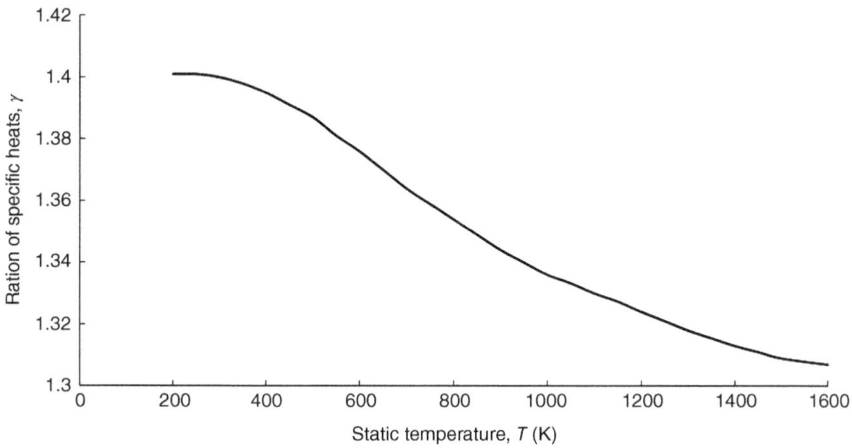

Figure 2.5 Specific heats ratio for air as a function of temperature.

and the ratio of specific heats γ is therefore,

$$\gamma \equiv \frac{c_p}{c_v} \tag{2.18}$$

It can be easily shown that

$$c_p - c_v = R \tag{2.19a}$$

$$\frac{c_p}{R} = \frac{\gamma}{\gamma - 1} \tag{2.19b}$$

Air is assumed to be in perfect chemical and thermodynamic equilibrium such that the behavior of all its constituents are assumed to be perfect gases. Perfect gas means that the equation of state is restricted to a gas which is both calorically and thermally perfect. By *calorically perfect* we mean that the c_p and c_v are constant and so is γ. By *thermally perfect* we mean that the gas obeys the thermal equation of state, Eq. (2.15), and c_p and c_v are functions of temperature, and so $\gamma = \gamma(T)$. Figure 2.5 illustrates the variation of γ as the static temperature changes from room temperature to 1600 K (2880 °R).

In the flight regime $5.0 < M_0 < 11.0$, the air ingested by the scramjet will be compressed by ram compression and shocks in such a manner that the static temperature of the compressed gas in the inlet will be in the range $650\text{ K} < T_3 < 1100\text{ K}$ (see Chapter 4).

High-temperature effects dominate hypersonic aerodynamics and air-breathing propulsion. The kinetic energy of a high-speed, hypersonic flow is dissipated by the influence of friction within the boundary layer. At high air velocity, the extreme viscous dissipation that occurs within boundary layers results in very high temperatures. If the air temperature is high enough, it can excite vibrational energy internally within molecules and cause dissociation and even ionization. The reasons for these changes in the air characteristics are as follows:

1) As its temperature exceeds 400 K, the vibrational energy of the air molecules becomes excited, and this causes the specific heats c_p and c_v to become functions of temperature. As a consequence, the ratio of specific heats, $\gamma = c_p/c_v$, also becomes a function of temperature.

2) As the air temperature approaches 2000 K, oxygen molecules start to dissociate upon collision with other molecules, and the thermodynamic properties of air become a function of both temperature and pressure, and thus $\gamma = f(T, p)$. At such conditions, the thermally perfect gas model breaks down due to changes in gas composition, which changes the molecular weight of the gas mixture. Air at 2000 K has its translational, rotational, and vibrational energy states fully excited. This temperature level describes the combustor or afterburner environment.
3) For air at 1 atm pressure, oxygen dissociation ($O_2 \rightarrow 2O$) begins at about 2000 K, and the molecular oxygen is essentially totally dissociated at 4000 K. At this temperature, nitrogen dissociation ($N_2 \rightarrow 2N$) begins and is essentially totally dissociated at 9000 K. Above a temperature of 9000 K, ions are formed, and air becomes a partially ionized plasma.

As a consequence of these effects in the high hypersonic flight regime, calculating air total or stagnation conditions from static conditions using ideal gas assumption yields significant errors. Therefore, in the study of air-breathing hypersonic propulsion, we must appropriate thermodynamic models to ensure the analysis is valid.

Single constant exponential and multiple-range polynomial curve fits are typically used to determine the variation of γ as a function of temperature. Single constant exponential curve fits fail to accurately model the true thermally perfect behavior of air at temperatures greater than 1600 K. The selection of a suitable curve fit function for c_p is the starting point for the development of thermally perfect compressible flow relations. For the thermally perfect gas model, the following eight-term, fifth-order polynomial expression can be used, where the value of c_p is nondimensionalized by the species gas constant:

$$\frac{c_p}{R} = A_1\left(\frac{1}{T^2}\right) + A_2\left(\frac{1}{T}\right) + A_3 + A_4(T) + A_5(T^2) + A_6(T^3) + A_7(T^4) + A_8(T^5) = \sum_{j=1}^{8} A_j T^{j-3} \tag{2.20}$$

For a given temperature range, A_1–A_8 are the coefficients of the curve fit. This is the functional form for c_p of NASP TM-1107 (1990), which provides values of A_1–A_8 for 15 different gas species. There are similar c_p/R coefficient data for over 200 different gas species for a five-term, fourth-order polynomial curve fit equation (A_3–A_7). Equation (2.14) is also valid with these coefficients provided that A_1, A_2, and A_8 are set equal to zero.

The following models for equilibrium air are useful for preliminary aerothermodynamic analysis:

- **Calorically Perfect Gas**: Assumption of a calorically perfect gas (c_p, c_v, and γ constant) is valid for a relatively narrow range of temperatures (217 K < T < 400 K).
- **Thermally Perfect Gas**: In the temperature range (400 K < T < 1700 K), the molecules vibrational energy mode is excited, and c_p, c_v, and γ become functions of temperature. In this state, we model the gas as a thermally perfect gas. Thermally perfect thus denotes a thermally perfect, calorically imperfect gas.
- **Average c_p Model (ACP)**: In preliminary propulsion analysis, air may be treated as calorically perfect, with average values of $\bar{c}_p$ and $\bar{\gamma}$, which are selected according to the operational temperature range of the various engine components.

2.3 Fundamental Equations

Understanding and predicting the basic performance of propulsion systems requires a closed set of governing equations. These are four basic laws of physics that must be satisfied for the working fluid of every propulsion system: (i) conservation of matter (continuity equation), (ii) conservation of momentum (Newton's second law equation), (iii) conservation of energy (first law of thermodynamics), and (iv) second law of thermodynamics. In general, these laws can be represented mathematically in a form that are amenable to numerical manipulation.

As a first approximation, analysis can use steady, one-dimensional aerothermodynamic analysis to gain insights into propulsion cycles from turbojets to scramjets. In such model, the fluid properties are constant across the flow and vary only in the direction of flow (axial direction). The engine working fluid is considered to be a perfect gas, and the state of the system is described by specifying the values of the thermodynamic properties: pressure, temperature, specific internal energy, and density. Specifying any two independent intensive properties will fix the thermodynamic state of a simple system and, therefore, the values of all other thermodynamic properties of the system.

2.3.1 One-Dimensional Aerothermodynamic Equations

Consider one-dimensional, steady-state flow through a CV where the fluid properties are constant across the flow and depend upon a single spatial coordinate (the axial dimension). The velocity vector V has two components, denoted as u and v. The overall flow through the CV may be in more than one dimension and still be uniform at permeable sections of the control surface normal to the flow direction. Consider the conditions at two equilibrium states of the flow, representing the inlet (i) and exit (e) of the CV, where the intensive properties, such as velocity, density, and temperature, are uniform.

The conservation of mass for the CV is given by:

$$\rho_i u_i A_i + \sum \dot{m}_w = \rho_e u_e A_e \tag{2.21}$$

where $\dot{m}_w$ denotes a flow entering through any other CV boundary.

The conservation of momentum equation has two components, one for the axial or x direction and one for the transverse or y direction:

$$p_i A_i + (\rho_i u_i A_i) u_i + \sum \dot{m}_w u_w + F_{wx} = p_i A_i + (\rho_e u_e A_e) u_e \tag{2.22a}$$

where F_{wx} denotes the total axial force (pressure and viscous) acting on the CV boundary other than the inlet and exit, and u_w is the axial component of velocity of a flow entering through any CV boundary other than the inlet:

$$(\rho_i u_i A_i) v_i + \sum \dot{m}_w v_w + F_{wy} = (\rho_e u_e A_e) v_e \tag{2.22b}$$

where F_{wy} represents the total transverse force (pressure and viscous) acting on the boundary of the CV other than the inlet and exit, and v_w is the transverse component of velocity of a flow entering through any CV boundary other than the inlet.

The conservation of energy equation results from application of the first law of thermodynamics to the CV:

$$(\rho_i u_i A_i)\left(h_i + \frac{u_i^2 + v_i^2}{2}\right) + \sum \dot{m}_w\left(h_w + \frac{u_w^2 + v_w^2}{2}\right) + \dot{Q}_{cv} + \dot{W}_{cv} = (\rho_e u_e A_e)\left(h_e + \frac{u_e^2 + v_e^2}{2}\right) \tag{2.23}$$

where $\dot{W}_{cv}$ is the total rate at which the shaft does work on the CV fluid, and $\dot{Q}_{cv}$ is the total rate at which heat is added to the CV fluid from the surroundings. The former is important in cases where machinery, such as compressors or turbines, are present within the CV, and the latter is important when the flow is hot and the walls must be cooled by removing heat.

The equation of state is needed to determine all intensive properties of the fluid once any two intensive properties are known: $\rho = \rho(p, T)$, $h = h(p, T)$, and $e = e(p, T)$. For a calorically perfect gas, $p = \rho RT$, and

$$dh = c_p dT \tag{2.24a}$$

$$de = c_v dT \tag{2.24b}$$

The differential form of the first law of thermodynamics is known as the Gibbs equation, which relates entropy to the flow properties:

$$T\,ds = dh - dp/\rho$$

where $T\,ds$ expresses reversible heat transfer, dh is the differential change in enthalpy, and dp/ρ denotes reversible work done on an open thermodynamic system. For a calorically perfect gas, this energy equation becomes

$$ds = c_p\frac{dT}{T} - R\frac{dp}{p} \tag{2.25}$$

The speed of sound gives the speed at which a planar pressure disturbance small enough to be isentropic progresses through a gas. For an equilibrium chemically reacting mixture, the speed of sound is given by $a^2 = \partial p/\partial \rho$. For calorically perfect gas, $a^2 = \gamma RT$. Thus, the Mach number is written in the CV as:

$$M^2 = \frac{V^2}{a^2} = \frac{u^2 + v^2}{\gamma RT} \tag{2.26}$$

The above expressions represent a closed set of governing equations for a CV and are useful to obtain the relations for the stagnation state, as shown in the following section.

2.3.2 Stagnation State

The stagnation state is a reference state defined as that thermodynamic state which would exist if a fluid were brought to zero velocity and zero potential. The stagnation state must occur without any energy exchange and without losses. That is, the total state is reached in decelerating a flow to rest reversibly and adiabatically and without any external interactions. From conservation of energy for a steady state 1-D flow in a CV, Eq. (2.23), where

$\dot{m}_w = 0$, and $\dot{W}_{cv} = \dot{Q}_{cv} = 0$, the conditions at two equilibrium states (i) and (e) of the flow are given by the adiabatic energy equation:

$$\left(h_i + \frac{V_i^2}{2}\right) = \left(h_e + \frac{V_e^2}{2}\right) = h_t \tag{2.27}$$

where h_t denotes the *total enthalpy*. This expression is valid as long as the end states are equilibrium states. If equilibrium exists all along the flow, the equilibrium equation is valid continuously and may be written $h + V^2/2 \equiv h_t$. The total enthalpy h_t is a fixed property of constant energy flows (even if the gas is not calorically perfect) and may be evaluated at a place in the flowpath where $V = 0$ and the fluid is in equilibrium.

For a vehicle in flight at velocity V_0, the stagnation or total enthalpy is written in the form:

$$h_{t0} = h_0 + \frac{V_0^2}{2} \tag{2.28}$$

The total enthalpy is very high for hypersonic flight since the velocity can be very large, while the static enthalpy of the atmosphere remains fixed, since the air static temperature changes very little at the altitudes of interest for hypersonic cruise.

For calorically and thermally perfect gas, we can use Eq. (2.24a) to write Eq. (2.27) as:

$$c_p T_i + \frac{V_i^2}{2} = c_p T_e + \frac{V_e^2}{2} = c_p T_t \tag{2.29}$$

where T_t is the total temperature. Using Eqs. (2.19b) and (2.26), this expression can be rewritten as:

$$T_i\left(1 + \frac{\gamma - 1}{2} M_i^2\right) = T_e\left(1 + \frac{\gamma - 1}{2} M_e^2\right) = T_t \tag{2.30}$$

Hence, the *total temperature* is a fixed property of constant energy flows of calorically perfect gases:

$$T_t \equiv T\left(1 + \frac{\gamma - 1}{2} M^2\right) \tag{2.31}$$

It must be clear that the stagnation temperature of a gas must be determined from the exact set of equations of state. For preliminary analysis, Eq. (2.31) can be used over ranges for which mean values of the gas constants can be specified. Again, for a vehicle in flight, the stagnation or total temperature is written in the form:

$$T_{t0} = T_0\left(1 + \frac{\gamma - 1}{2} M_0^2\right) \tag{2.32}$$

where γ denotes the mean value for air for the range of interest.

Assuming that hypersonic cruise occurs at altitudes where the air static temperature is ~238.45 K,and $\gamma = 1.401$, Eq. (2.32) reduces to $T_{t0} = 238.45\,\text{K}\left(1 + 1.2005\,M_0^2\right)$. Note the rapid increase of freestream total temperature with flight Mach number as plotted in Figure 2.6. Since the maximum skin temperature is very near the stagnation temperature of flight, this plot helps us establish the temperature operating limit of hypersonic vehicles

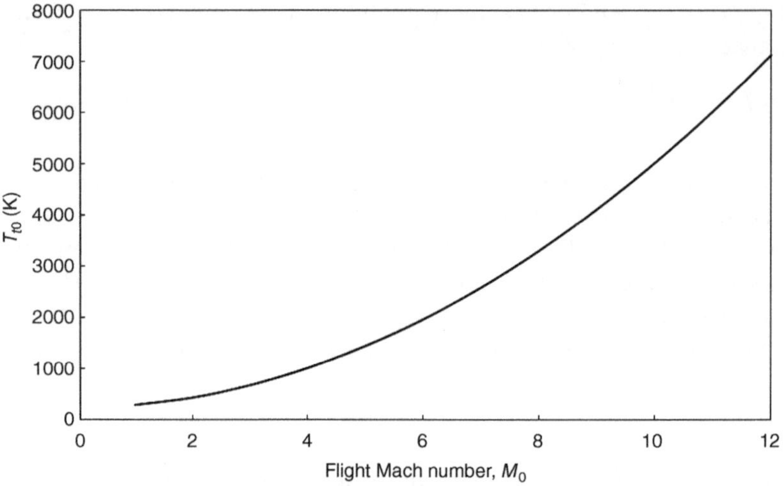

Figure 2.6 Freestream air total temperature variation with flight Mach number at altitudes where T_0 = 238.45 K and γ = 1.401.

from the perspective of materials and the need for wall cooling. For example, at $M_0 = 8.0$, the freestream air has $T_{t0} \cong 3298$ K, a value that may be already unacceptable for conventional materials. In Chapter 9, we shall discuss high-temperature materials and thermal management.

Using the relation $p_t/p = (T_t/T)^{\gamma/(\gamma-1)}$, the *total* or *stagnation pressure* for a vehicle in flight is given as:

$$p_{t0} = p_0\left(1 + \frac{\gamma - 1}{2}M_0^2\right)^{\gamma/(\gamma-1)} \tag{2.33}$$

For subsonic flow where the temperature is less than 500 K, air can be modeled as a perfect gas. In hypersonic flight regime, however, the ideal gas assumption yields significant errors. At M_0~12, one can represent the inviscid portion of flow field by an ideal gas, but in boundary layer flow it requires considering vibrational effects. At $M_0 > 12$, vibrational effects in inviscid flow and dissociation effects in the boundary layer must be considered. However, Eqs. (2.32) and (2.33) are useful to calculate initial values of total pressure and temperature to start a numerical iteration process.

2.4 Thermodynamic Cycle of Air-Breathing Propulsion

Viewed as a thermodynamic cycle, an engine converts chemical energy stored in the fuel into mechanical energy for vehicle thrust. The thermodynamic cycle analysis models a semi-ideal heat engine based on the Brayton cycle. The overall cycle performance of a hypersonic air-breathing propulsion system results from individual component performance and interactions among various components. Preliminary propulsion cycle analysis considers the thermodynamic changes of the working fluid (air and products of combustion) as it flows through the engine. In such analysis, the actual combustion process is replaced with a model of heat addition, typically at constant pressure. In this model, the energy made available by the chemical reactions is represented by the heat that must be removed from the final combustion products in order to return them to the same temperature as the initial reactants (at the same pressure).

Hypersonic flows cannot be accurately described by conventional (equilibrium) thermodynamics. However, a thermodynamic cycle analysis is helpful to derive the performance trends important to scramjet engines. Assuming one-dimensional flow at the entrance and exit of each component, we assess the behavior of the scramjet. It assumes that air, the working fluid, is in its equilibrium state at all times, and combustion is represented as constant pressure heat addition process, equivalent to the energy released by chemical reaction.

2.4.1 Engine Reference Station Numbers

To facilitate their analysis, Heiser and Pratt (1994) introduced engine reference stations at critical axial positions along the scramjet flowpath, as shown in Figure 2.2. Beginning with station 0 to define the undisturbed freestream airflow at the far left, these engine reference stations are located at the junction between major identifiable engine components, marking the beginning of each one. Station 10 is added to account for the external expansion, which is not needed in axially symmetric air-breathing engines that have fully enclosed exhaust nozzles, such as the turbojet or conventional axisymmetric ramjets. Table 2.2 provides further details for the air-breathing engine reference station numbers.

The thermodynamic cycle of a scramjet engine is studied as a series of processes. As depicted in Figure 2.7, the main thermodynamic processes are adiabatic compression, isobaric heat addition, and adiabatic expansion. Each flow property is represented by a single value at each reference station. For example, p_3 represents the static pressure at the end of compression process, while h_{10} represents the specific enthalpy of the flow at the exit plane of the engine. Since the gas flows at high Mach numbers (compared with ramjet and turbojet engines), the static temperature will be lower than the stagnation state throughout the scramjet flowpath.

Table 2.2 Air-breathing engine reference station locations.

Reference station	Engine location
0	Undisturbed or freestream condition. External compression begins.
1	Inlet diffuser entry. External compression ends. Internal compression begins.
3	Inlet diffuser exit. Combustor entry. Internal compression ends.
4	Combustor exit. Nozzle entry. Internal expansion begins.
9	Nozzle exit. Internal expansion ends. External expansion begins.
10	External expansion ends.

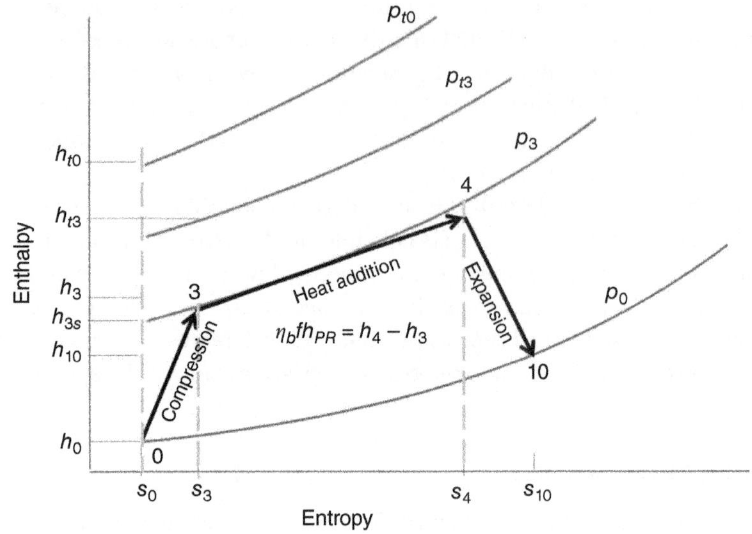

Figure 2.7 Static and stagnation states of the gas in thermodynamic cycle of air-breathing engine.

2.4.2 Flow Thermodynamic Properties

For the compression process, air is represented as a calorically perfect gas having constant properties, with $\gamma_c = 1.36$, while for the expansion process the value of $\gamma_e = 1.24$ is typically used, as the exhaust flow has a larger number of molecular degrees of freedom energized than air under normal conditions. For the heat addition process in the combustor, values of γ_b and R_b are assumed, since changes in chemical composition, molecular weight, and temperature during fuel/air mixing and subsequent combustion all cause γ_b to decrease.

2.4.3 Adiabatic Compression Process 0 → 3

The adiabatic compression process takes place from freestream to combustor entrance condition, $p_0 \rightarrow p_3$ or $T_0 \rightarrow T_3$. The ingested air is first decelerated and compressed from the freestream condition (station 0) to the combustor entry condition (station 3) by a combination of isentropic compression and oblique shock waves in the engine inlet section. In the case of a scramjet, the compression process must be done to maintain a supersonic airflow at the combustor entrance.

The compression process is irreversible because the viscous forces within the shock waves and boundary layers dissipate availability and generate entropy. The change in entropy for adiabatic compression is given by the Gibbs equation (2.25) in the form:

$$s_3 - s_0 = c_{pc} \ln \frac{T_3}{T_0} - R \ln \frac{p_3}{p_0} \tag{2.34}$$

The scramjet inlet must operate to compress the freestream air to a sufficiently large static temperature ratio T_3/T_0 to achieve satisfactory cycle efficiency and also to produce the values of pressure and temperature required to support complete and stable combustion.

This process must occur in a controllable and reliable manner with minimum aerodynamic losses (maximum compression efficiency or minimum entropy increase). The engine static temperature ratio

$$\psi = \frac{T_3}{T_0} \tag{2.35}$$

is a principal determinant of thermodynamic cycle efficiency, and it is typically found to attain values $6.0 < \psi < 8.0$. The upper limit is imposed to the desirable compression temperature T_3 or the cycle static temperature ratio ψ to reduce compression losses; this limit is $\psi < 8.0$.

Since the total temperature is conserved in adiabatic flows, it can be shown that

$$\psi = \frac{T_3}{T_0} = \left[\frac{1 + ((\gamma_c - 1)/2)M_0^2}{1 + ((\gamma_c - 1)/2)M_3^2}\right] \tag{2.36}$$

Hence, the total pressure ratio for the compression process can be written as:

$$\pi_c = \frac{p_3}{p_0}\left(\frac{1}{\psi}\right)^{\gamma_c/(\gamma_c - 1)} \tag{2.37}$$

Moreover, assuming an adiabatic process, air compression is characterized by the adiabatic compression process efficiency which compares the actual or real change in static enthalpy to the ideal or isentropic change in static enthalpy that would accompany the same change in static pressure. The efficiency of compression may also be expressed as:

$$\eta_c = \frac{(p_3/p_0)^{(\gamma - 1)/\gamma} - 1}{(T_3/T_0) - 1} = \frac{(p_3/p_0)^{(\gamma - 1)/\gamma} - 1}{\psi - 1} \tag{2.38}$$

2.4.4 Isobaric Heat Addition Process 3 → 4

The heat addition process in the combustor is assumed to be at constant pressure $p_3 = p_4$ and adiabatic. This results in an increase in total temperature.

In thermodynamic analysis, the fuel contributes only heat to the cycle (no mass). Hence, the heat addition process in the combustor is represented by the equivalent fuel chemical energy release, given by the product $\dot{m}_f\, h_{PR}$, where $\dot{m}_f$ is the mass flow rate of fuel and h_{PR} denotes the heat of reaction or heating value of the fuel. Hydrogen has the highest heat of reaction of any fuel, with $h_{PR} = 119\,954$ kJ/kg, while hydrocarbons have less than half of this value. For example, JP-7 fuel has $h_{PR} = 43\,500$ kJ/kg. The heat input to the engine is approximated as:

$$\dot{Q}_{in} = h_4 - h_3 = \frac{\dot{m}_f \eta_b\, h_{PR}}{\dot{m}_0} = \eta_b f h_{PR} \tag{2.39}$$

where the fuel heating value h_{PR} is multiplied by the combustion efficiency η_b since not all the chemical energy of the fuel can be released in the combustor due to inadequate mixing or reaction time for complete combustion.

Moreover, the change in entropy for isobaric heat addition is given by Eq. (2.25) as:

$$s_4 - s_3 = c_{pb} \ln \frac{T_4}{T_3} \tag{2.40}$$

2.4.5 Adiabatic Expansion Process 4 → 10

The expansion process from the combustor exit to the nozzle exit plane condition is considered adiabatic. The heated gas is further accelerated and expanded in the engine's nozzle from the combustor exit condition (station 4) to the freestream static pressure at station 10, $p_3 \rightarrow p_0$ or $T_4 \rightarrow T_{10}$.

The Mach number of the flow at station 10 will not be as large as the freestream Mach number, i.e. $M_{10} < M_0$. However, the exhaust flow velocity V_{10} must be large enough so that the kinetic energy and velocity at station 10 exceed that of the freestream in order for the scramjet to produce net thrust.

Again, assuming an adiabatic process, air expansion is characterized by the adiabatic expansion process efficiency which compares the actual or real change in static enthalpy to the ideal or isentropic change in static enthalpy that would accompany the same change in static pressure.

The thermodynamic analysis of the scramjet cycle yields the net cycle work output:

$$\dot{W}_{out} = (h_4 - h_{10}) - (h_3 - h_0)$$

By definition, the **thermodynamic efficiency of the cycle** is given by the ratio of net cycle work output to heat added to the engine, that is,

$$\eta_{tc} = \frac{\dot{W}_{out}}{\dot{Q}_{in}} = \frac{(h_4 - h_{10}) - (h_3 - h_0)}{h_4 - h_3} = 1 - \frac{h_{10} - h_0}{h_4 - h_3}$$

where $h_{10} - h_0$ denotes the amount of heat rejected $\dot{Q}_{out}$.

To obtain the highest thermodynamic cycle efficiency, more heat has to be added to the combustor, so we must have $\eta_b f h_{PR} \gg (h_{10} - h_0)$. Consider a hydrogen-fueled scramjet. Knowing that $T_0 = 290$ K, $T_{10} = 1780$ K, assuming stoichiometric fuel-to-air ratio and 90% combustion efficiency, the efficiency of the engine cycle would be approximately 46%, as you can verify:

$$\dot{Q}_{in} = \eta_b f h_{PR} = 0.90(0.0291)(119,954) = 3142\ \text{kJ/kg}$$

$$h_{10} - h_0 = 1978.16 - 290.1 = 1688.06\ \text{kJ/kg}$$

$$\eta_{tc} = 1 - \frac{h_{10} - h_0}{h_4 - h_3} = 1 - \frac{1688.06}{3142} = 0.463$$

2.4.6 Combustor Energy Balance and Combustor Efficiency

Applying the steady flow energy equation, Eq. (2.23) to the CV representing the combustion chamber, obtain

$$\dot{m}_0 h_3 + \dot{m}_f \eta_b h_{PR} = \left(\dot{m}_0 + \dot{m}_f\right) h_4 \tag{2.41a}$$

For calorically and thermally perfect gas, use Eq. (2.18b) to write

$$\dot{m}_0 c_{p3} T_3 + \dot{m}_f \eta_b h_{PR} = (\dot{m}_0 + \dot{m}_f) c_{p4} T_4 \tag{2.41b}$$

Simplify Eq. (2.41b) to

$$c_{p3} T_3 + f \eta_b h_{PR} = (1 + f) c_{p4} T_4 \tag{2.42}$$

divide by $c_{p3} T_3$ and assume $1 + f \approx 1$:

$$\frac{c_{p4}}{c_{p3}} \frac{T_4}{T_3} = 1 + \frac{f \eta_b h_{PR}}{c_{p3} T_3} \tag{2.43}$$

In the analysis of scramjet engines, the gas temperature at the entrance to the combustor, T_3, is determined from the analysis of the inlet (see Chapter 4). In addition, we normally specify h_{PR}, a desired η_b, and the maximum temperature in the combustor to control thermal choking. Thus, Eq. (2.43) is solved for the effective fuel-to-air ratio in the combustor:

$$\eta_b f = \frac{c_{p3}}{c_{p4}} \frac{T_3}{h_{PR}} \left(\frac{T_4}{T_3} - 1 \right) = \frac{c_{p3}}{c_{p4}} \frac{T_3}{h_{PR}} (\tau_b - 1) \tag{2.44}$$

where τ_b denotes the temperature ratio across the combustor. Since the value of T_4 is function of f, the solution of the above expression is iterative.

It is also important to note that in the analysis of turbine jet engines (e.g. turbojet), a similar analysis for the subsonic combustor is done using total quantities (h_t, T_t), rather than static values. Thus, the energy equation is written as $c_{p3} T_{t3} + f \eta_b h_{PR} = (1 + f) c_{p4} T_{t4}$, where T_{t4} is used to establish a design limit for the maximum temperature of the hot gas leaving the combustor and entering the turbine. In other words, T_{t4} is a thermal design input that fixes the maximum allowable turbine inlet stagnation enthalpy as the turbine blades are exposed to severe rotating tensile stresses that impose a limit on the gas temperature to avoid blade failure.

2.5 Air-Breathing Propulsion Performance Measures

The characteristics of the propulsion system that are most important to a flight mission are specific thrust, specific fuel consumption, and specific impulse. These parameters are ratios of total engine properties, independent of the size of the engine, which we use to compare engines of the same type but different sizes. Also, the fuel-to-air ratio appears in performance equations, and its value is an indicator of the conditions in the combustor.

2.5.1 Thermal Efficiency

The thermal efficiency η_{th} of a jet engine is defined as the net rate of organized energy (kinetic energy) out of the engine $\dot{W}_{out}$ divided by the rate of thermal energy available from the fuel in the engine $\dot{Q}_{in}$. For jet engines, $\dot{W}_{out}$ is equal to the net rate of change of the kinetic energy of the fluid flow through the engine. For a jet engine with a single inlet and single exhaust,

$$\dot{W}_{out} = (\dot{m}_0 + \dot{m}_f)\frac{V_{10}^2}{2} - \dot{m}_0\frac{V_0^2}{2} = \dot{m}_0\left[(1+f)\frac{V_{10}^2}{2} - \frac{V_0^2}{2}\right] \tag{2.45}$$

In the ideal propulsion model, the air is heated in a combustion process that releases the chemical energy of the fuel. The fuel's available thermal energy is denoted by $\dot{Q}_{in} = \dot{m}_f h_{PR}$. Thus, the thermal efficiency of an air-breathing propulsion system is expressed as:

$$\eta_{th} = \frac{\dot{W}_{out}}{\dot{Q}_{in}} = \frac{\dot{m}_0\left[(1+f)(V_{10}^2/2) - (V_0^2/2)\right]}{\dot{m}_f h_{PR}} = \frac{\frac{1}{2}\left[(\dot{m}_0 + \dot{m}_f)V_{10}^2 - \dot{m}_0 V_0^2\right]}{f h_{PR}} \tag{2.46}$$

Considering that $\dot{m}_f \ll \dot{m}_0$, the thermal efficiency expression can be simplified to

$$\eta_{th} = \frac{(V_{10}^2/V_0^2) - 1}{f h_{PR}/(V_0^2/2)} \tag{2.47}$$

Clearly, for a given flight speed and fixed heat addition (fh_{PR}), the thermal efficiency of the engine will be greatest when the velocity ratio across the engine is highest.

2.5.2 Propulsive Efficiency

The propulsive efficiency η_p is a measure of how effectively the scramjet engine power $\dot{W}_{out}$ is used to propel the aircraft. Propulsive efficiency is defined as the ratio of the aircraft power (thrust times velocity) to the power out of the engine $\dot{W}_{out}$. The thrust power of an engine is

$$FV_0 = \dot{m}_0[(1+f)V_{10} - V_0]V_0 = \dot{m}_0\left[(1+f)\frac{V_{10}}{V_0} - 1\right] \tag{2.48}$$

Therefore,

$$\eta_p = \frac{FV_0}{\dot{W}_{out}} = \frac{FV_0}{\frac{1}{2}\left[(\dot{m}_0 + \dot{m}_f)V_{10}^2 - \dot{m}_0 V_0^2\right]} = \frac{2[(1+f)(V_{10}/V_0) - 1]}{[(1+f)(V_{10}/V_0)^2 - 1]} \tag{2.49}$$

Considering that $\dot{m}_f \ll \dot{m}_0$, the above expression simplifies to

$$\eta_p \approx \frac{2}{V_{10}/V_0 + 1} = \frac{2M_0}{V_{10}/a_0 + M_0} \tag{2.50}$$

High propulsive efficiency requires that the exhaust speed be approximately equal to the vehicle's flight speed, $V_{10} \approx V_0$. However, propulsion requires that $V_{10} > V_0$ in order to yield positive thrust, Eq. (2.1). Thus, the propulsive efficiency will never exceed 100%. Turbojet engines have high values of the velocity ratio and thus low propulsive efficiency. In fact, when $V_{10} \gg V_0$, then the thrust is maximum but $\eta_p \to 0$. This condition is representative of aircraft takeoff when $V_0 \approx 0$. It is clear that, since the propulsive efficiency of an air-breathing engine drops inversely proportional to V_{10}/V_0, it is important to know how the ramjet and scramjet engines operate at their design condition and how to optimize their propulsive efficiency.

2.5.3 Overall Efficiency

In evaluating a propulsion system, we wish to know how well it uses the energy stored in the fuel ($\dot{Q}_{in}$) in order to provide the thrust power (FV_0) required for a mission. This knowledge comes from the definition of the propulsion system overall efficiency η_0, given by the ratio of the vehicle power (FV_0) and the rate of thermal energy released by the fuel in the engine ($\dot{m}_f h_{PR}$). We combine the thermal and propulsive efficiencies to obtain the overall efficiency η_0 of an air-breathing propulsion system. Multiplying thermal efficiency η_{th} by propulsive efficiency η_p, we obtain the ratio of the vehicle power to the rate of thermal energy released by the fuel in the engine, which is the overall efficiency of the air-breathing engine:

$$\eta_0 = \frac{FV_0}{\dot{m}_f h_{PR}} = \frac{(V_{10}^2/V_0^2) - 1}{fh_{PR}/(V_0^2/2)} \cdot \frac{2}{(V_{10}/V_0) + 1} = \frac{2((V_{10}/V_0) - 1)}{fh_{PR}/(V_0^2/2)} \tag{2.51}$$

A high overall efficiency is needed to achieve the highest mission range (see Chapter 6).

2.5.4 Other Forms of Propulsion Performance Measures

For an air-breathing engine with properly expanded conditions ($p_{10} = p_0$), from Eq. (2.1) the (ideal) specific thrust is expressed as:

$$\frac{F}{\dot{m}_0} = (1 + f)V_{10} - V_0 = (1 + f)\frac{V_{10}}{V_0} - 1 \tag{2.52}$$

The specific thrust and specific impulse given in terms of overall efficiency and fuel heat of reaction can also be expressed as, respectively:

$$\frac{F}{\dot{m}_0} = \frac{fh_{PR}}{V_0} \cdot \eta_0 \tag{2.53}$$

$$I_{sp} = \eta_0 \cdot \frac{h_{PR}}{g_0 V_0} \tag{2.54}$$

In this form, it is clear that a scramjet fueled by hydrogen will have a higher specific impulse as compared with a scramjet fueled by a hydrocarbon. Moreover, the inverse dependence of I_{sp} upon flight speed makes this parameter a very convenient performance measure when comparing different air-breathing propulsion systems. For example, for a flight velocity $V_0 = 3000$ m/s, the specific impulse of a hydrogen fueled scramjet with $\eta_0 = 0.7$ is $I_{sp} = 2853$ seconds. If the fuel is changed to a hydrocarbon with $h_{PR} = 43\,500$ kJ/kg, the specific impulse drops to $I_{sp} = 1035$ seconds, assuming the same overall propulsion efficiency.

The specific fuel consumption is defined as the fuel mass flow rate per unit (uninstalled) thrust. This is the rate of fuel used by the propulsion system per unit of thrust it produces:

$$S = \frac{\dot{m}_f}{F} \tag{2.55}$$

For a commercial cruise engine, a low specific fuel consumption indicates a more economical operation. However, for a military engine, the gross or net thrust of the engine is much more important than the fuel consumption, because aircraft performance parameters, such as stealth, agility, maneuverability, and survivability require highest thrust. In general, low engine fuel consumption can be directly translated into longer range, increased payload, and/or reduced aircraft size.

For preliminary performance analysis of hypersonic air-breathing engine, the stream thrust analysis method is also very helpful because it takes into consideration mass addition, momentum, and kinetic energy fluxes contributed by the fuel. Such method is typically used to perform more extensive parametric analysis. Consult Heiser and Pratt (1994) for an excellent treatment of Stream Thrust Analysis.

2.5.5 Specific Impulse Estimate in Terms of Kinetic Energy Efficiency

A different method to estimate the specific thrust of a scramjet is based on the idea that each major component of the engine performs according to its ability to process the working fluid. For example, the engine inlet is characterized by the kinetic energy efficiency of its compression process, while the nozzle is characterized by the kinetic efficiency of its expansion process. The working fluid is assumed to behave as a calorically perfect gas with the same constants throughout the engine. Thus, the kinetic energy efficiency of the engine inlet, combustor, and nozzle are denoted as $\eta_{KE,\ c}$, $\eta_{KE,\ b}$, and $\eta_{KE,\ e}$, respectively. Therefore, the total kinetic energy efficiency of the scramjet is given by:

$$\eta_{KEO} = \eta_{KE,c} \cdot \eta_{KE,b} \cdot \eta_{KE,e} \tag{2.56}$$

In order to obtain an expression for the specific thrust in terms of η_{KEO}, this method assumes that the nozzle expansion velocity is given by:

$$V_{10} = V_0\sqrt{\tau_b \eta_{KEO}} \tag{2.57}$$

where τ_b denotes the combustor total temperature ratio, which is obtained from the first law analysis of adiabatic combustion process:

$$\tau_b = \frac{T_{t4}}{T_{t3}} = \frac{1}{1+f}\left(1 + \frac{\eta_b f h_{PR}}{c_p T_{t0}}\right) \tag{2.58}$$

Therefore, the specific thrust becomes

$$\frac{F}{\dot{m}_0} = V_0\left[(1+f)\sqrt{\tau_b \eta_{KE,e}} - 1\right] V_0\left[\sqrt{\eta_{KEO}(1+f)\left(1 + \frac{\eta_b f h_{PR}}{c_p T_{t0}}\right)} - 1\right] \tag{2.59}$$

or

$$\frac{F}{\dot{m}_0} = a_0 M_0 \left\{\sqrt{\eta_{KEO}(1+f)\left[1 + \frac{\eta_b f h_{PR}}{c_p T_0(1 + ((\gamma-1)/2)M_0^2)}\right]} - 1\right\} \tag{2.60}$$

Equation (2.60) provides a simple formula to estimate the specific thrust, assuming that the air-breathing engine overall kinetic energy efficiency will be in the range $0.65 < \eta_{KEO} < 0.75$. A well-designed engine inlet may have $\eta_{KE,\ c} \approx 0.90$, and an optimum nozzle is one with $\eta_{KE,\ e} \approx 0.98$. If the combustor has a kinetic energy efficiency $\eta_{KE,\ b} \approx 0.80$, then $\eta_{KEO} \approx 0.7056$.

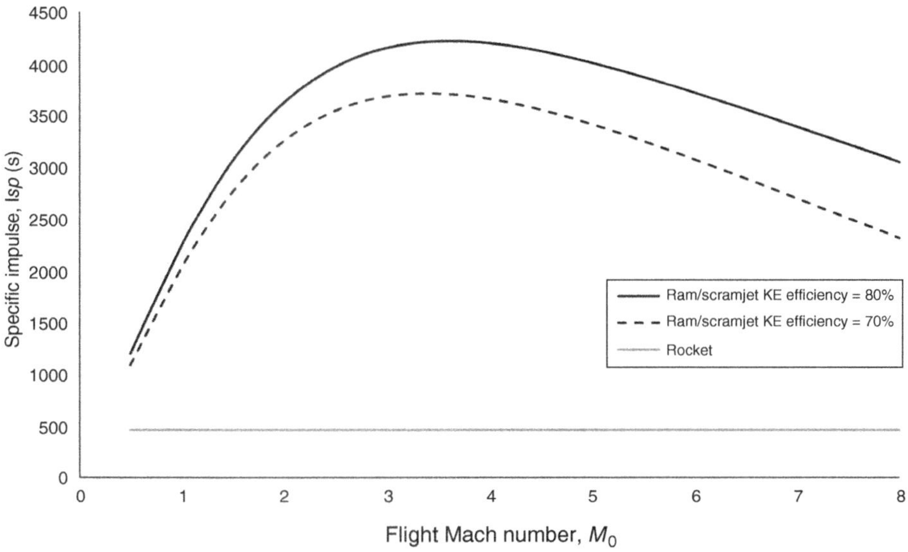

Figure 2.8 Specific impulse as a function of freestream Mach number and kinetic energy efficiency for hydrogen-fueled air-breathing engine.

Figure 2.8 depicts the specific impulse as a function of freestream Mach number for a hydrogen-fueled air-breathing engine, assuming two values of kinetic energy efficiency. Clearly, a lower η_{KEO} results in reductions of I_{sp}. Hence, improvements in overall scramjet engine performance depends on increasing the kinetic energy efficiency of all its components.

2.6 Shock Waves in Supersonic Flow

Oblique shocks are compression waves in a supersonic flow that abruptly turn the flow and compress the gas in the process. Two-dimensional external compression inlets employ external compression ramps to create plane oblique shocks, whereas axisymmetric inlets employ multiple cones to create conical shocks. The following summary reviews oblique shocks only.

In hypersonic flow, compression Mach waves may coalesce to form an oblique shock wave. The fluid temperature, pressure, and density increase dramatically across shock waves. As the Mach number M increases to higher speeds, these increases become more severe and the oblique shock wave moves closer to the surface. For values of $M > 5.0$, the shock wave is very close to the surface, and the flow field between the shock and the body (the shock layer) becomes hot enough to dissociate or even ionize the gas. Aspects of such high-temperature chemically reacting flows are treated by Anderson (2006).

Consider an oblique shock (the wave front is at an angle of other than 90° to the approaching flow) with a representative streamline that abruptly changes direction across the shock, as depicted in Figure 2.9. The flow upstream and downstream conditions are denoted by

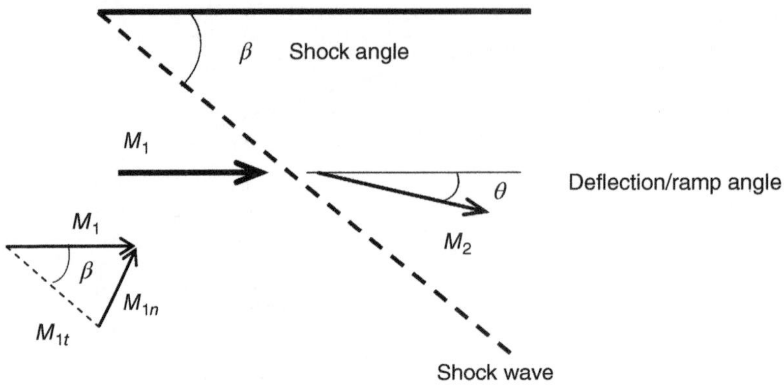

Figure 2.9 Oblique shock wave geometry.

subscripts 1 and 2, respectively. The shock wave angle with respect to upstream flow is denoted as β and the flow turning angle (again with respect to the upstream flow) is θ. A normal shock is a special case of an oblique shock with a wave angle of 90°.

Assuming the geometry of the compression ramp for a given hypersonic propulsion inlet is known (the so-called design problem), each ramp angle represents the flow turning or deflection angle θ, which combined with the information on the local Mach number M, leads to a determination of the wave angle β by using oblique shock charts (or conical shock charts), depending on the inlet geometry of interest. Here, we consider the basic oblique shock geometry.

The shock wave angle β helps establish the strength of the oblique shock, which depends on the normal component of the flow to the shock, M_n. All the flow conditions across an oblique shock are established uniquely by the normal component of the flow to the oblique shock, M_{1n}. With the wave angle β, it yields $M_{1n} = M_1 \sin\beta$. Hence, the flow changes across an oblique shock wave for a calorically perfect gas are given as the ratios of static pressure, static temperature, and total pressure, respectively (for derivation see Farokhi 2014, or Anderson 2020):

$$\frac{p_2}{p_1} = 1 + \left(\frac{2\gamma}{\gamma+1}\right)(M_1^2 \sin^2\beta - 1) \tag{2.61}$$

$$\frac{T_2}{T_1} = \left[\frac{2}{(\gamma+1)M_1^2 \sin^2\beta} + \left(\frac{\gamma-1}{\gamma+1}\right)\right]\left[\left(\frac{2\gamma}{\gamma+1}\right)M_1^2 \sin^2\beta - \left(\frac{\gamma-1}{\gamma+1}\right)\right] \tag{2.62}$$

$$\frac{p_{t2}}{p_{t1}} = \left[\frac{(\gamma+1)M_1^2 \sin^2\beta}{2+(\gamma-1)M_1^2 \sin^2\beta}\right]^{\gamma/(\gamma-1)}\left[\left(\frac{2\gamma}{\gamma+1}\right)M_1^2 \sin^2\beta - \left(\frac{\gamma-1}{\gamma+1}\right)\right]^{-1/(\gamma-1)} \tag{2.63}$$

where is $\gamma = c_p/c_v$ the specific heat ratio of the gas.

The normal component of the Mach number downstream of an oblique shock is

$$M_{2n}^2 = \frac{2+(\gamma+1)M_{1n}^2}{2\gamma M_{1n}^2 - (\gamma-1)} = \frac{2+(\gamma-1)M_1^2 \sin^2\beta}{2\gamma M_1^2 \sin^2\beta - (\gamma-1)} \tag{2.64}$$

From the velocity triangle downstream of the oblique shock, the downstream Mach number M_2 can be obtained as:

$$M_2 = \frac{M_{2n}}{\sin(\beta - \theta)} \tag{2.65}$$

The shock angle β, deflection angle θ, and entering Mach number M_1 are related as:

$$\frac{\tan\beta}{\tan(\beta - \theta)} = \frac{(\gamma + 1)M_1^2 \sin^2\beta}{(\gamma - 1)M_1^2 \sin^2\beta + 2} \tag{2.66}$$

and an explicit solution for $\theta = f(M, \beta, \gamma)$ is given by:

$$\tan\theta = 2\cot\beta\left[\frac{M_1^2 \sin^2\beta - 1}{M_1^2(\gamma + \cos 2\beta) + 2}\right] \tag{2.67}$$

Now let us examine this expression for the extreme values of the shock angle β that might accompany any given flow Mach number. The equations for the oblique shock yield as special cases, the strongest shock possible (normal shock) and the weakest shock possible (no shock), as well as all other intermediate-strength shocks. For any given deflection angle θ, there are two possible shock angles for a specified Mach number. For example, if $M = 9.5$ and $\theta = 4.5°$, the two values for the shock angle are $\beta_1 = 9.359°$ and $\beta_2 = 89.03°$ for the weak and strong shocks, respectively. This holds true for any Mach number and deflection angle combination.

Figure 2.10 provides a graphical representation of the relation between β and the flow Mach number M_1 for various deflection angles (θ). The dashed vertical line at any arbitrary Mach number starts at the top of the plot with the normal shock ($\beta = 90°$, $\theta = 0°$), which is the strongest possible shock. Moving downward, the shock angle decreases to $\theta_{min} = \mu$ (Mach angle), indicating that the shock strength is decreasing continually.

For any given M_1, there is a maximum deflection angle θ_{max}. For example, for $M_1 = 5.0$, $\theta_{max} \approx 41°$, and for $M_1 = 15.0$, $\theta_{max} \approx 45°$. If the physical geometry were such that $\theta > \theta_{max}$, then no solution exists for a straight oblique shock wave. Instead, the shock will be curved and detached.

For any given Mach number and deflection angle $\theta < \theta_{max}$, there are two values of β predicted by the θ, β, M relation, Eq. (2.67). Because changes across the shock are more severe as β increases (see Eqs. (2.61) and (2.62)), the large value of β is called the strong shock solution; a stronger shock results in a higher pressure ratio. The small value of β is called the weak shock solution, and it results in a lower pressure rise across the shock. For the design of the hypersonic air-breathing inlet, the weak shock solution is sought. In the strong shock solution, M_2 is subsonic. In the weak shock solution, M_2 is supersonic except for a small region near θ_{max}, (see Figure 2.10).

The following conclusions are important in the design of a forebody/inlet for a hypersonic vehicle. For the weak shock solutions at a fixed Mach number, as θ is increased, β, p_2, and T_2, increase while M_2 decreases. However, if θ increases beyond θ_{max}, the shock wave becomes detached. Alternatively, for a fixed θ, as M_1 increases from unity, the shock wave is first detached, then becomes attached when M_1 equals that value for which $\theta = \theta_{max}$. As the Mach number is increased further, the shock remains attached, β decreases, and p_2 and

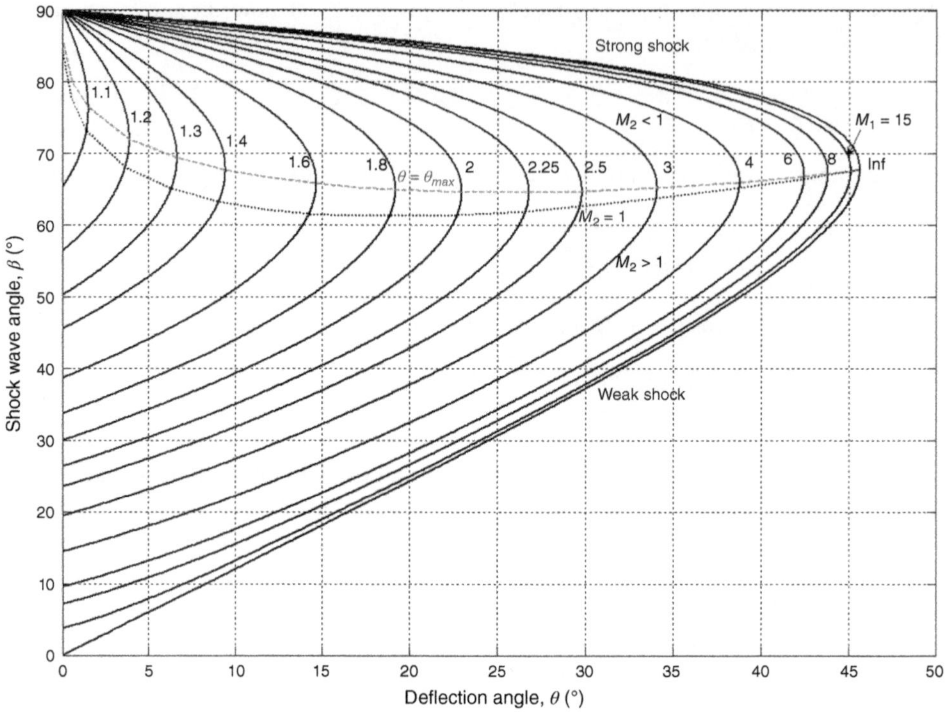

Figure 2.10 Oblique shock relations. The upper dotted line separates the strong and weak solutions. The lower dotted line represents the point when the downstream Mach number becomes sonic.

T_2 increase. The solution with a lower shock angle yields a weaker shock with a lower pressure rise across the shock.

In the hypersonic limit, as $M_1 \to \infty$, $M_1^2 \sin^2\beta \gg 1$. When the deflection angle θ is small, the shock angle is also small. Thus, $\sin\beta \approx \beta$, $\cos 2\beta \approx 1$, and $\tan\theta \approx \sin\theta \approx \theta$, and Eq. (2.67) reduces to

$$\theta = \frac{2}{\beta}\left[\frac{M_1^2\beta^2 - 1}{M_1^2(\gamma + 1) + 2}\right] \tag{2.68}$$

In the hypersonic limit, $M_1 \to \infty$,

$$\frac{\beta}{\theta} = \frac{\gamma + 1}{2} \tag{2.69}$$

which, for $\gamma = 1.4$, is simply $\beta = 1.2\theta$. In the hypersonic limit for a slender body, the wave angle is only 20% larger than the wedge angle. Therefore, in the hypersonic limit, we also have

$$\frac{p_2}{p_1} = \frac{2\gamma}{\gamma + 1}\left(M_1^2 \sin^2\beta\right) \tag{2.70}$$

$$\frac{\rho_2}{\rho_1} = \frac{(\gamma + 1)}{(\gamma - 1)} \tag{2.71}$$

$$\frac{T_2}{T_1} = \frac{2\gamma(\gamma - 1)}{(\gamma + 1)^2} M_1^2 \sin^2\beta \tag{2.72}$$

Most compressible flow textbooks derive and summarize compressible flow relations based on the calorically perfect gas model gas with $\gamma = 1.4$. These include charts and tables for the shock wave angle as a function of the deflection angle and upstream Mach number for a perfect gas. The accuracy of these equations is only as good as the assumption of a constant γ.

2.7 One-Dimensional Flow with Heat Addition

Rayleigh flow is an ideal model describing a frictionless flow with heat transfer through a duct of constant cross-sectional area. Although not accurate in real situations, the Rayleigh model offers a simple and reasonably equivalent process for a constant area combustion chamber. Thus, we use it to probe the behavior of the flow, assuming that the heat added in the combustor is significant and friction can be ignored.

The governing differential equations of Rayleigh flow are usually developed and solved to describe Rayleigh flow property ratios with respect to the values at the choking location. The Rayleigh flow functions are given for an entrance condition Mach number and an exit Mach number of 1. A flow that reaches a sonic state is said to be choked, and the effect of heat transfer may lead to an exit choking condition, which is referred to as a thermally choked flow. The sonic or choked condition is used as a reference state in compressible flow problems. For example, the total temperature and pressure ratios are given by the following Rayleigh flow functions:

$$\frac{T_t}{T_t^*} = M^2\left(\frac{\gamma + 1}{1 + \gamma M^2}\right)^2 \left[\frac{1 + ((\gamma - 1)/2)M^2}{(\gamma + 1)/2}\right] \tag{2.73}$$

and

$$\frac{p_t}{p_t^*} = \frac{\gamma + 1}{1 + \gamma M^2}\left[\frac{1 + ((\gamma - 1)/2)M^2}{(\gamma + 1)/2}\right]^{\gamma/(\gamma - 1)} \tag{2.74}$$

These two Rayleigh flow functions are plotted in Figure 2.11. (For derivation of Rayleigh flow functions see Anderson (2020).)

It is even more revealing to examine the thermodynamics of the Rayleigh flow on a temperature–entropy diagram, which is called a Rayleigh line. The addition of heat causes the entropy of the fluid to increase and continue to add heat until the fluid reaches a state of maximum entropy. The Gibbs equation for this process states that

$$ds = c_p\frac{dT}{T} - R\frac{dp}{p} \tag{2.75}$$

where the incremental pressure change is related to the Mach number variation. Thus,

$$ds = c_p\frac{dT}{T} + R\frac{2\gamma M^2}{1 + \gamma M^2}\left[\frac{1}{2}\frac{dT}{T}\frac{(1 + \gamma M^2)}{(1 - \gamma M^2)}\right] = c_p\frac{dT}{T}\left[1 + (\gamma - 1)\left(\frac{M^2}{1 - \gamma M^2}\right)\right] \tag{2.76}$$

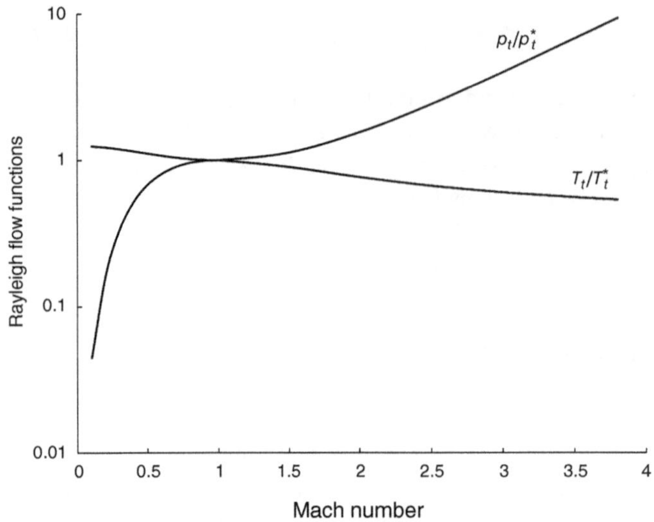

Figure 2.11 Ratios of stagnation or total pressure and temperature as a function of Mach number in flow with heat addition in a constant area duct (Rayleigh flow model).

This equation simplifies to a function $s(T, M)$ for a Rayleigh line that may be plotted on the T–s diagram, where the parameter along the T–s curve is the local Mach number M,

$$ds = c_p \frac{1-M^2}{1-\gamma M^2} \frac{dT}{T} \tag{2.77}$$

As shown in Figure 2.12, the critical points of this function are at $ds/dT = 0$ and $dT/ds = 0$, representing the maximum entropy and the maximum temperature on the Rayleigh line, respectively. The point of maximum entropy occurs at $ds/dT = 0$, which Eq. (2.77) indicates the sonic condition at this point, that is,

$$\frac{ds}{dT} = 0 \quad \text{at}\, M = 1 \tag{2.78}$$

and the maximum temperature point is at $dT/ds = 0$, which again Eq. (2.77) requires

$$\frac{dT}{ds} = 0 \quad \text{at}\, M = \frac{1}{\sqrt{\gamma}} \tag{2.79}$$

There are two branches of the Rayleigh line, namely a subsonic and a supersonic branch. These two branches are separated at the sonic point, labeled (*) in Figure 2.12. Also, the path of heating tends to choke both the subsonic and supersonic flows, and the cooling path drives the flow Mach number down in subsonic flows and up in supersonic flows. Interestingly, the point of maximum temperature occurs at a subsonic Mach number given by Eq. (2.79) and further heating actually reduces the static temperature toward the exit sonic condition. The stagnation temperature continues to increase, however, with heating. The heating that leads to a choked exit condition is called the critical heating Q^*_{in}. Since the point of maximum entropy, s_{max}, occurs at the sonic point, the sonic condition has to occur at the exit of the duct.

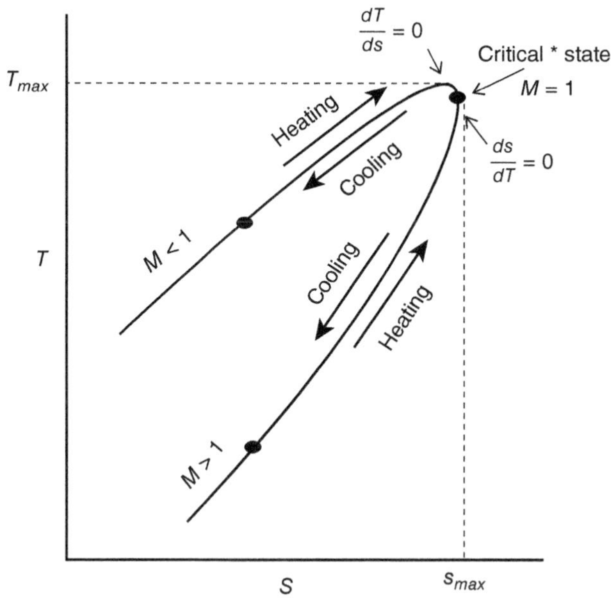

Figure 2.12 Rayleigh line. States of air in a frictionless, one-dimensional flow in constant area duct with heat addition.

Now consider the effects of heat addition on the flow through a scramjet combustor where $M > 1$, assuming the flow area is constant. The differential governing equations are

$$\text{Continuity:} \quad \rho V = \text{const} = \rho_4 V_4 \tag{2.80}$$

$$\text{Momentum:} \quad \rho V \frac{dh}{dx} = -\frac{dp}{dx} \tag{2.81}$$

$$\text{Energy:} \quad \rho V \frac{dh}{dx} = V \frac{dp}{dx} + \frac{dQ_{in}}{dx} \tag{2.82}$$

where dQ_{in}/dx denotes the rate at which energy is added in the combustion process.

Assume that $h = c_p T$, and that the specific heat ratio is constant. From the equation of state $p = \rho RT$, we have, using Eq. (2.80) twice,

$$\frac{dp}{dx} = R\left(\rho \frac{dT}{dx} + T \frac{d\rho}{dx}\right) = R\left(\frac{\rho_4 V_4}{V} \frac{dT}{dx} - \frac{\rho_4 V_4}{V^2} T \frac{dV}{dx}\right)$$

And we eliminate dp/dx by using Eq. (2.81) to obtain

$$(\rho_4 V_4) V \frac{du}{dx} = -R\left(\rho_4 V_4 \frac{dT}{dx} - \frac{\rho_4 V_4}{V} T \frac{dV}{dx}\right)$$

and from Eq. (2.82),

$$(\rho_4 V_4) c_p \frac{dT}{dx} = R\left(\rho_4 V_4 \frac{dT}{dx} - \frac{\rho_4 V_4}{V} T \frac{dV}{dx}\right) + \frac{dQ_{in}}{dx}$$

Solve this expression for dT/dx and write in terms of Mach number, since $M = V/\sqrt{\gamma RT}$:

$$\frac{dT}{dx} = \frac{1-\gamma M^2}{1-M^2}\frac{1}{\rho_4 V_4 c_p}\frac{dQ_{in}}{dx} \tag{2.83}$$

and finally obtain

$$\frac{1}{M}\frac{dM}{dx} = \frac{1-\gamma M^2}{2(1-M^2)}\frac{1}{\rho_4 V_4 c_p T}\frac{dQ_{in}}{dx} \tag{2.84}$$

Equation (2.84) implies that in supersonic flow ($M > 1$), adding heat to the flow lowers the Mach number. Hence, heat addition always drives the Mach number toward the choked condition, $M \rightarrow 1$, as shown in the Rayleigh line, Figure 2.12.

In the Rayleigh flow model, we found that the total pressure increases. Multiply Eq. (2.81) by V, and then add the result to Eq. (2.82) to obtain

$$\rho_4 V_4 \frac{d}{dx}\left(c_p T + \frac{V^2}{2}\right) = \rho_4 V_4 c_p \frac{dT_t}{dx} = \frac{dQ_{in}}{dx} \tag{2.85}$$

Now use the total or stagnation pressure, Eq. (2.33) to write its differential form:

$$\frac{1}{p_t}\frac{dp_t}{dx} = \frac{1}{p}\frac{dp}{dx} + \frac{\gamma}{\gamma-1}\frac{(\gamma-1)M}{1+((\gamma-1)/2)M^2}\frac{dM}{dx}$$

and with the help of Eqs. (2.81) and (2.84) obtain

$$\frac{1}{p_t}\frac{dp_t}{dx} = -\frac{\gamma M^2/2}{\rho_4 V_4 c_p T_t}\frac{dQ_{in}}{dx} \tag{2.86}$$

Equation (2.86) implies that adding heat to a flow always lowers the total pressure. The decrease of p_t is much larger for supersonic flow. Unfortunately, this is a characteristic of supersonic combustion. Another representation of the above equation is

$$\frac{dp_t}{dx} = -\frac{\gamma M(1-M^2)}{(1+\gamma M^2)(1+((\gamma-1)/2)M^2)}dM \tag{2.87a}$$

which can be integrated to yield the total pressure ratio across the duct

$$\frac{p_{t5}}{p_{t4}} = -\frac{1+\gamma M_4^2}{1+\gamma M_5^2}\left[\frac{1+((\gamma-1)/2)M_5^2}{1+((\gamma-1)/2)M_4^2}\right]^{\gamma/(\gamma-1)} \tag{2.87b}$$

And the same can be done to obtain expression for static pressure ratio, and for static and total temperature ratios across the constant are duct:

$$\frac{T_{t5}}{T_{t4}} = 1 + \frac{Q_{in}}{\rho_4 V_4 c_p T_{t4}} = \left[\frac{1+((\gamma-1)/2)M_4^2}{1+((\gamma-1)/2)M_5^2}\right]^2\left(\frac{M_5}{M_4}\right)^2 \tag{2.88}$$

Now we have a set of equations that are helpful to estimate the conditions in a supersonic flow combustor either when thermal choke conditions are expected, or to limit the heat addition (or fuel-to-air ratio) to prevent choking.

For example, consider a hydrogen-fueled constant area combustor and assume that compressed air enters at Mach 2.75. If the total pressure and total temperature at the entrance are 45 kPa and 1800 K, respectively, determine the total temperature at the thermally choked exit of the combustor, the critical or maximum heat release per unit mass, Q_{in}^*, and the fuel-to-air ratio. For simplicity, assume the gas mixture in the combustor is a perfect gas with constant specific heat ratio $\gamma = 1.3$, and $R = 0.287$ kJ/kg · K.

From the Rayleigh flow functions for $M_3 = 2.75$, obtain

$$\frac{T_{t3}}{T_t^*} = 0.6329, \frac{p_{t3}}{p_t^*} = 3.096$$

Therefore, the total temperature at the exit of the combustor is

$$T_{t4} = \frac{1800\ \text{K}}{0.6329} = 2844.05\ \text{K}$$

From the energy balance across the combustor, we obtain the critical heating rate with

$$c_p = \frac{\gamma R}{\gamma - 1} = \frac{1.3\,(0.287\ \text{kJ/kg}\cdot\text{K})}{0.3} = 1.2437\ \text{kJ/kg}\cdot\text{K}$$

$$Q_{in}^* = c_p\left(T_t^* - T_{t3}\right) = c_p T_{t3}\left(\frac{T_t^*}{T_{t3}} - 1\right) = 1.2437(1800)\left(\frac{1}{0.6329} - 1\right) = 1298.48\ \text{kJ/kg}$$

For this critical heat input, the fuel-to-air ratio f is found from Eq. (2.34) assuming perfect combustion efficiency, $fh_{PR} = Q_{in}^* = 1298.48$ kJ/kg That is, to thermally choke the combustor exit the fuel-to-air ratio is

$$f = \frac{1298.48\ \text{kJ/kg}}{120\,000\ \text{kJ/kg}} = 0.0108$$

The total pressure loss across the combustor is given by:

$$\Delta p_t = 1 - \frac{p_t^*}{p_{t3}} = 1 - \frac{1}{3.096} = 0.67$$

Will the obtained values in this exercise be acceptable for a real scramjet combustor?

2.8 Closing Remarks

The concepts and physical models we have reviewed were selected to provide some insight into the technical aspects of hypersonic air-breathing propulsion. These simplified models are not sufficient to understand all the complexities of the processes occurring in a scramjet engine. For example, the discussion of scramjet performance should include an assessment of the stagnation pressure loss and its impact on thrust production. The loss in stagnation pressure is due to complex interaction of shock waves, combustion, and skin friction. This loss rapidly increases with the flight Mach number and, at any hypersonic Mach number, results in very low thrust.

Moreover, the combustion process represented by an external heat source Q_{in}^* in the simplified analysis of Rayleigh flow is not representative of the actual chemical kinetics and aerothermodynamics that occurs in a scramjet engine. However, we used this model to convey the importance of thermal choking. In Chapters 5, 6, and 7, we shall review some aspects of scramjet combustors and the importance of fuel selection for different hypersonic air-breathing propulsion applications.

This overview is by no means complete. The reader is referred to many various textbooks that provide introductions to compressible flow and air-breathing propulsion. For more advanced studies, please consult Heiser and Pratt for their outstanding treatment of *Hypersonic Air-breathing Propulsion* (1994), together with Anderson's *Hypersonic and High Temperature Gas Dynamics* (2006), and Bertin's *Hypersonic Aerothermodynamics* (1994) who emphasize the practical design of hypersonic aerospace vehicles and provides an abundance of data relating to their external flow fields. These and the other books listed in the References include a comprehensive treatment of classical derivations, numerical techniques, and fundamental data that are necessary to fully appreciate the topics we address in subsequent chapters.

Questions

1. How much airflow must be ingested by an air-breathing engine to propel a hypersonic vehicle ($M_0 > 5.0$) flying at 100 kft altitude?

2. What is the stagnation temperature a vehicle will experience flying at Mach 10?

3. What is the optimum altitude for a reusable Mach 7 cruise vehicle?

4. What was the dynamic pressure on the X-51A Waverider when it flew at 70 kft (21 km) altitude?

5. What design parameters must be optimized to increase the thrust of a scramjet engine?

References

Anderson, J.D. Jr. (2006). *Hypersonic and High Temperature Gas Dynamics*, 2e. American Institute of Aeronautics and Astronautics (AIAA).

Anderson, J.D. (2020). *Modern Compressible Flow: With Historical Perspective*, 4e. McGraw-Hill Series in Aeronautical and Aerospace Engineering.

Anonymous (1976). *U.S. Standard Atmosphere, 1976*. Washington, DC: U.S. Government Printing Office.

Bertin, J.J. (1994). *Hypersonic Aerothermodynamics*, AIAA Education Series. Washington, DC: American Institute of Aeronautics and Astronautics (AIAA).

Engelund, W.C., Holland, S.D., Cockrell, Jr. C.C., and Bittner, R.D. (2000). Aerodynamic database development for the hyper-X airframe integrated scramjet propulsion experiments. AIAA 2000–4006. AIAA 18th Applied Aerodynamics Conference, Denver, Colorado (14–17 August 2000).

Farokhi, S. (2014). *Aircraft Propulsion*, 2e. Wiley.

Heiser, W.H. and Pratt, D.T. (1994). *Hypersonic Airbreathing Propulsion*, AIAA Education Series. Washington, DC: American Institute of Aeronautics and Astronautics (AIAA).
Kerrebrock, J.L. (1992). *Aircraft Engines and Gas Turbines*. The MIT Press.
Oates, G.C. (1997). *The Aerothermodynamics of Gas Turbine and Rocket Propulsion*, 3e. Washington, DC: AIAA Education Series.

3

Aerothermodynamics of Vehicle-Integrated Scramjet

In 1950, American aerodynamicist Edward Van Driest stated that the two major problems encountered in aeronautics are the determination of skin friction and skin temperatures of high-speed aircraft. He was referring to the practical impact viscous flow has on hypersonic vehicles, as he developed correlations for aerodynamic heating (skin temperature) and shear stress (skin-friction drag) that were needed for advancing hypersonic vehicle design. These two problems remain at the forefront of hypersonics R&D.

A hypersonic vehicle encounters an aerothermodynamic environment characterized by strong shocks and aerodynamic heating, with temperatures high enough to overheat and even destroy the vehicle. The severity of the environment strongly depends on the mission profile and the vehicle configuration. For air-breathing propulsion vehicles (Figure 3.1), the aerodynamic and aerothermodynamic loads on the airframe prescribe the performance requirements not only for the propulsion flowpath (inside the engine) but also for various vehicle subsystems, such as the thermal protection system (TPS), materials technology, and the control system.

For an air-breathing hypersonic vehicle, the propulsion system and airframe, including the TPS and hot and warm structures, are highly integrated, and thus we deal with highly complex and multiphysical aerothermodynamic phenomena. This close integration of air-breathing engine and vehicle airframe is very different from the configurations for vehicles powered by rocket propulsion, and these differences have significant impacts on their design, especially thermal management.

While in atmospheric flight, rocket vehicles accelerate, but do not cruise. Conventional rocket space launch systems lift off vertically, accelerating fast to leave the atmosphere quickly, and they typically fly at low dynamic pressures ($q_{0_{max}} \approx 14$ kPa). In contrast, hypersonic airbreathers are intended to accelerate and cruise within the atmosphere for longer periods, and fly at high dynamic pressures ($q_0 \approx 50$ kPa) (low altitude within the atmosphere) to capture the necessary amount of air required by the engine. Moreover, the air-breathing hypersonic cruiser is a slender (relatively low-drag) vehicle. Thus, while the entire vehicle body is subjected to atmospheric heating due to drag, the thin, sharp leading edges — including the nose region, tail edges, and scramjet forebody/inlet — are exposed to intense, highly localized stagnation-point heating and a severe aerothermal environment.

This chapter is intended to give an overview of flow phenomena related specifically to scramjet propulsion as it integrates with its airframe. We wish to address some

Scramjet Propulsion: A Practical Introduction, First Edition. Dora Musielak.

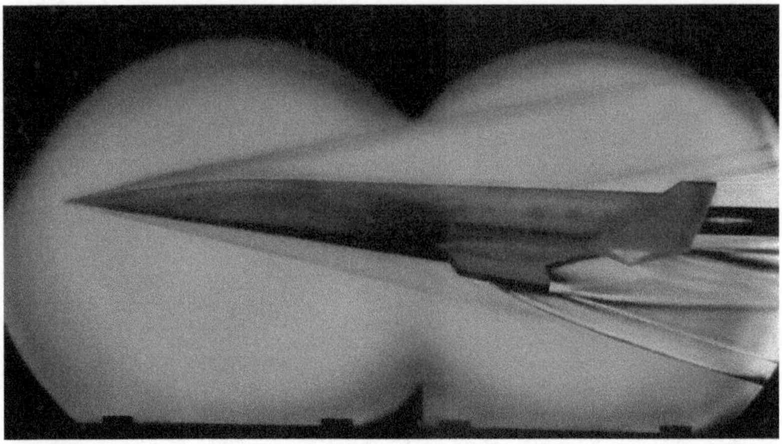

Figure 3.1 Wind tunnel test of a hypersonic vehicle at NASA Langley. Source: Photograph published in Winds of Change, 75th Anniversary NASA publication by James Schultz.

characteristics of the hypersonic flowfield, paying attention to phenomena such as boundary-layer transition (BLT), inlet starting, inlet spillage, thermal choking, and combustor–inlet interaction. There are a number of questions we wish to answer, some as basic as *What is shock-on-lip heating*? Other questions are more challenging, and to answer them we must rely on R&D work carried out to support the development of the hypersonic experimental flight vehicles built to date. BLT is one of the most complex physical processes in hypersonic flow aerothermodynamics. *Can we determine where boundary layer will transition in the forebody of the vehicle-integrated scramjet propulsion system*? Figure 3.1 shows a NASA model of a hypersonic aircraft undergoing tests in the 20 Inch Mach 6 Tunnel. It illustrates some of the hypersonic flow physics addressed in this chapter.

The design of hypersonic vehicles involves numerous engineering disciplines to carry out the necessary aerothermodynamic analysis. Aerothermodynamic analysis requires accurate modeling of shock wave–boundary-layer interaction, and shock–shock vehicle interactions, as we need to assess how those interactions affect the vehicle boundary layers, since they can lead to regions of enhanced aerothermodynamic loading.

These and many other topics are covered in excellent books devoted to hypersonic aerothermodynamics such as those by Anderson (2006), Bertin (1994), Hirschel (2005), and in many fine research papers published in the last two decades. From this body of knowledge, we will draw some important concepts to review in this chapter, with the goal of highlighting the characteristics of hypersonic flowfields, especially those of interest to hypersonic air-breathing vehicles.

3.1 Aerothermodynamic Environment

Hypersonic air-breathing vehicles powered by scramjet engines, such as the X-43A and X-51A experimental aircraft, must fly within an aerothermodynamic environment in which the high-speed aerodynamic flow properties are coupled with the thermodynamic

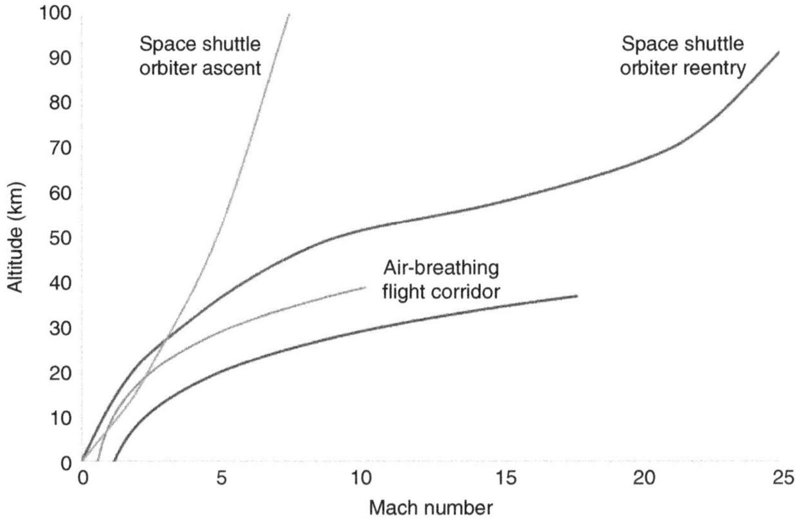

Figure 3.2 Representative flight trajectories for various hypersonic vehicles.

properties of the constituent gases through which the vehicle is moving, or the flow that passes through the engine. The aerothermodynamic environment include phenomena such as shock–flow interactions, turbulence, BLT, viscous–inviscid flow interactions, separated flows, imperfect gas, nonequilibrium chemistry, and high-speed reacting flows (Fletcher 2004).

Hypersonic flight within Earth's atmosphere can be defined over a wide range of altitude and speed, depending on applications and the type of propulsion that powers the aircraft. Figure 3.2 shows several flight trajectories that are appropriate for different types of vehicles, contrasting the trajectories of air-breathing vehicles with the rocket-launched Space Shuttle (SS) Orbiter with its almost vertical ascent and unpowered gliding entry.

The characteristics of the hypersonic environment depend on the mission requirements and vehicle size constraints. The aerothermodynamic phenomena important for hypersonic cruise aircraft or missiles may not be as important to rocket-powered ascent and re-entry vehicles. In fact. the hypersonic environment is extremely demanding for reusable cruise aircraft and spaceplanes that must operate reliably over many flights during their lifetime. However, an expendable hypersonic missile intended for only one short flight would not be subject to the same creep and low cycle fatigue challenges.

Aerodynamic heating refers to the heat transfer from the hot boundary layer to the cooler vehicle surface. Also called convective heating, this process is aggravated when the shock-layer temperature is high enough that the thermal radiation emitted by the gas itself gives rise to a radiative flux to the surface—called radiative heating. For the slender hypersonic vehicles, aerodynamic heating is highest at leading edges, in the regions with shock interactions, in stagnation regions, and at locations where the flow transitions to turbulence.

Thermo-structural designs for a hypersonic missile could use well characterized, reliable, existing high-temperature materials, coatings, and processes, in combination with passive or active cooling. But for a future reusable cruiser the design requirements are very stringent. Hence, the aerothermodynamic phenomena critical to the design differ for the different types of vehicles. The differences are due to differences in the trajectories, in the different airframe-propulsion design configurations, and the reusability required for each application.

3.1.1 Air-Breathing Hypersonic Cruise

A hypersonic cruise vehicle powered by air-breathing propulsion system must fly at altitudes which are low enough that there is sufficient oxygen, i.e., relatively high density, to capture sufficient air for a combustion rate of fuel that will produce thrust levels significantly greater than the drag on the vehicle. At the same time, long range hypersonic cruisers will be exposed to intense aerodynamic heating and drag. Hence, these aircraft must operate within a narrow range of dynamic pressures, and cruise in the range $5.0 < M_0 < 12.0$ will occur at relatively low altitudes, $20\ \text{km} < H < 40\ \text{km}$, for extended periods of time. The optimal flight altitude of a hypersonic cruise aircraft depends on the chosen Mach number and the required dynamic pressure, q_0. For example, for $M_0 = 10.0$, flying at 30 km altitude the vehicle would experience a dynamic pressure of ~82 kPa, but moving just 6 km higher would reduce q_0 to 34 kPa.

As we discussed in Chapter 2, the dynamic pressure imposes a limit that corresponds to the maximum stagnation pressure which the vehicle structure can bear in the presence of high acceleration loads. Hence, sustained (steady) flight at hypersonic speeds must be confined to occur within a restricted corridor of altitudes at each speed (M_0), and the airframe must be designed to operate within the range of dynamic pressures $23\ \text{kPa} < q_0 < 95\ \text{kPa}$, (500–2000 lbf/ft^2). Above this corridor the vehicle cannot provide enough lift; while flying below this corridor, the surface heating and structural loads on the vehicle would be excessive. In contrast, for rocket powered vehicles, the maximum allowed dynamic pressure during launch is approximately (e.g., SS Orbiter ascent) $q_{0_{max}} \approx 14\,000\ \text{Nm}^{-2}$.

The combinations of Mach number and dynamic pressure mean flight environments with high temperatures and heating rates, and large aerodynamic forces. The thermal environment (the temperature of the incoming airflow on a radiatively cooled surface) ranges from temperatures of about 595 °C (1100 °F) at Mach 4 to about 2316 °C (4200 °F) at Mach 8.

A hypersonic vehicle propelled by air-breathing engines must be slender, designed with a relatively high lift/drag (L/D) ratio since skin-friction drag at hypersonic speeds is dominant. This is in contrast to blunt body vehicles for which the drag is mostly wave drag due to the high pressures behind the strong bow shock wave. The waverider is another type of hypersonic cruise vehicle designed for relatively high L/D to achieve efficient, long-range hypersonic flight. In general, for slender hypersonic cruise vehicles, skin-friction drag has a major impact on their design.

The aerothermodynamic environment on these hypersonic cruise vehicles is dominated by turbulence, and by shock-shock and shock–boundary layer interactions. The shock interactions occur inside the scramjet flowpath (Section 3.8), and also on forebody, aftbody, engine cowl, and other external aerodynamic control surfaces. These interactions lead to localized heating that are much higher than the surrounding areas. In addition, the flow may be transitional over portions of the vehicle, and this may complicate its TPS. Aerodynamic heating will require aggressive management to ensure reusability and sustained and reliable flight. There may also be vibrational excitation of air molecules behind a shock wave, which may impact the aerothermal environment for these vehicles.

Because the static temperature of Earth's atmosphere is relatively low, the sound speed is lower than at sea level. This means that higher Mach numbers can be achieved flying at lower speed at high altitude. Hence, we will focus on the hypersonic speed itself, since it can also give an indication of the kinetic energy involved in the trajectory. Consider a freestream velocity of 3.74 km/s (~Mach 12), the kinetic energy per unit mass is about 7×10^6 J/kg. As the air particles pass through the shock wave ahead of the vehicle, they slow down, converting the flow kinetic energy into internal energy. Thus, since enthalpy is a measure of kinetic energy, we can estimate the level of flight enthalpy using the vehicle's velocity. This is important since the amount of aerothermodynamic heating that the hypersonic vehicle will experience is linearly dependent on its kinetic energy.

The thermal environment (freestream stagnation enthalpy) must be characterized in terms of thermal gradients, heat fluxes, and stresses to which a hypersonic vehicle will be exposed. The freestream total temperature T_{t0} allows a first estimation of maximum wall temperatures expected on vehicle surfaces. As speed increases, hypersonic vehicles experience regions of high stagnation temperatures especially near the leading edges, and these are aggravated by the interactions with shock waves. For example, at Mach 10, the stagnation enthalpy is about 21 times the static temperature of the atmospheric air around the vehicle. Real gas effects become significant as vibrational excitation and some dissociation of oxygen will occur due to strong shocks at leading edges and due to viscous dissipation in the boundary layer. Thermal analysis for the NASA X-43A vehicle designed for Mach 10 flight test predicted temperatures that would approach 2204 °C (4000 °F) at the nose tip. This temperature required careful selection of materials and coatings that would survive even for the short duration (seconds), single flight.

3.1.2 Air-Breathing Access to Space Vehicles

For access-to-space vehicles, real gas effects become significant as vibrational excitation and gas dissociation will occur due to strong shocks at leading edges, and also due to viscous dissipation in the boundary layer. The hypersonic aerodynamic environment is extremely harsh and hypersonic vehicles function near the edge of system and technology capability. At the highest hypersonic velocities, the aerodynamic heat transfer to the vehicle is affected by phenomena that are not normally present at low velocity, resulting in huge and destructive heat transfer rates especially at leading edges. Air dissociation and ionization due to the higher static temperature in the shock layer are possible, especially behind shock waves.

A trans-atmospheric vehicle (TAV) is a single-stage-to-orbit (SSTO) spaceplane designed to take off horizontally from an existing runway and then literally blast its way into Earth orbit mainly on the strength of its air-breathing propulsion. A TAV using air-breathing propulsion must fly in the denser part of the atmosphere, at altitudes low enough that there is sufficient oxygen for the propulsion system to operate and provide adequate vehicle acceleration to reach orbital speed. Such TAV will be exposed to a wide range of aerodynamic phenomena that includes the ascent trajectory from the dense regions of the atmosphere to the low-density region in low Earth orbit. The elements of severe aerothermodynamic environment are therefore coupled with the requirement of low aerodynamic drag, and thus the vehicle must be slender and must have a relatively sharp nose and thin wing leading edges, designed to minimize wave drag at very high speeds.

The combination of high heating rates experienced by surfaces with small curvature leading edges and the long ascent times will result in huge total heat loads. The heat load is the integration of the heat rate with time over a given trajectory. Moreover, air-breathing hypersonic vehicles must fly at altitudes that are low enough that there is sufficient oxygen, i.e., relatively high density, for the propulsion system to operate effectively. However, the convective heat transfer and the drag increase as the density increases, i.e., as the altitude decreases. Thus, the optimal trajectory of an air-breathing hypersonic vehicle must represent a compromise between the propulsion requirements and the heat-transfer/drag requirements.

For two-stage to orbit (TSTO) configurations in which the first or booster stage is powered by air-breathing engines, the maximum flight Mach number and maximum altitude depend on the trajectory staging condition. The aerothermodynamic environment for the ascent and reentry vehicles (the orbiters) will be dominated by partly viscous effects, transition laminar- turbulent boundary layers, low-density effects, surface radiations, and strong real gas effects.

The uncrewed, fully reusable NASA TSTO launch concept is envisioned to consist of an air-breathing-propelled booster first stage, which can be considered equivalent to a cruise vehicle, and a rocket-powered second stage (the orbiter), which is the ascent to orbit and re-entry vehicle. For this TSTO launch system, staging is planned to occur at Mach 10 (to release the orbiter at its most advantageous state). The aerothermodynamic phenomena critical to the design differ for the two vehicle stages. These differences are due to differences in the trajectories and in the different vehicle-propulsion configurations. The booster stage powered by TBCC air-breathing engines will separate from the upper stage at an altitude of about 51.8 km (170 kft). The booster vehicle's trajectory will include ascent, pull-up maneuvers, and flyback with a powered cruise segment.

The ascent trajectory for the SSTO Skylon Spaceplane is designed for the SABRE engine to operate in air-breathing mode up to Mach 5.2; transition to rocket mode will occur at about 28.5 km (93.5 kft) altitude. As it ascends to low Earth orbit, the Skylon will be subject to real gas/high enthalpy effects with air dissociation. Other SSTO vehicle concepts similar to the National Aerospace Plane (NASP) studied in 1990s will require air-breathing propulsion flight at the highest Mach number possible, flying in an environment where complete oxygen dissociation and even some nitrogen dissociation would likely occur.

Unpowered reentry winged vehicles, such as the former SS Orbiter, descend through an aerothermal environment dominated by high enthalpy effects. This environment is characterized by ionization and result in significant radiative heating and high heat shield ablation rates. Moreover, the air particles in the shock layer of a re-entry vehicle undergo vibrational excitation, dissociation, and even ionization. These chemical processes absorb energy, limiting the temperature increase and converting the kinetic energy of the high velocity air particles crossing the bow shock wave into thermal energy. For a treatment of aerothermodynamic phenomena encountered in atmospheric entry flight, please consult specialized texts such as Bertin (1994) and Anderson (2006).

3.1.3 Reynolds Number and Air-Breathing Hypersonic Cruise

The viscous effects experienced by a vehicle in flight require accurate prediction in order to determine the wall skin friction and boundary-layer thickness, shape, BLT, and separation. These viscous effects are assessed through the freestream Reynolds number $Re_{L,0}$, which represents the ratio of inertia to viscous forces in the airflow:

$$Re_{L,0} = \frac{\rho_0 V_0 L}{\mu_0} = \frac{\rho_0 a_0 M_0 L}{\mu_0} \tag{3.1}$$

where μ_0 denotes the absolute viscosity of freestream air, and L is the length scale of the vehicle.

The relation between the dynamic viscosity and the absolute temperature of an ideal gas is given by Sutherland's law. This law is based on kinetic theory of ideal gases and an idealized intermolecular-force potential. The most useful form of Sutherland's law is expressed by the following formulas:

$$\mu_0 = 2.27 \times 10^{-8}\left(\frac{T^{3/2}}{T + 199}\right) \tag{3.2a}$$

where μ_0 is in $\text{lbf} \cdot \text{s/ft}^2$ and T in °R; or

$$\mu_0 = 1.46 \times 10^{-6}\left(\frac{T^{3/2}}{T + 111}\right) \tag{3.2b}$$

where μ_0 is in $\text{N} \cdot \text{s/m}^2$ and T in K.

- **Aerothermodynamic environment**: the flight environment in which a high-speed vehicle's aerodynamic properties are coupled with the thermodynamic properties (including heat transfer) of the constituent gases through which the vehicle is moving.
- For vehicles powered by air-breathing propulsion, the **aerodynamic environment extends to the propulsion flowpath** such as forebody/inlet, isolator, combustor, and afterbody/exhaust nozzle.

3.2 Hypersonic Viscous Flow Phenomena

Hypersonic flow is that regime where certain physical flow phenomena become increasingly more problematic as the Mach number increases. The flow phenomena around a hypersonic vehicle include shock layers, entropy or vorticity interaction layers, viscous interaction, high-temperature flow, and low-density flow. Some of these physical phenomena may not be important to the hypersonic cruise vehicles powered by scramjet engines flying at relatively low altitudes and Mach numbers below 12. However, for access to space and entry vehicles, all hypersonic flow phenomena are important (Cockrell 2003).

For $M_0 \geq 1$, bow shocks form ahead of the vehicle and radiate obliquely from the sides of a vehicle's leading edges, as illustrated in Figure 3.3. Pressure disturbances, which allow the downstream conditions to communicate with the upstream conditions), are no longer able

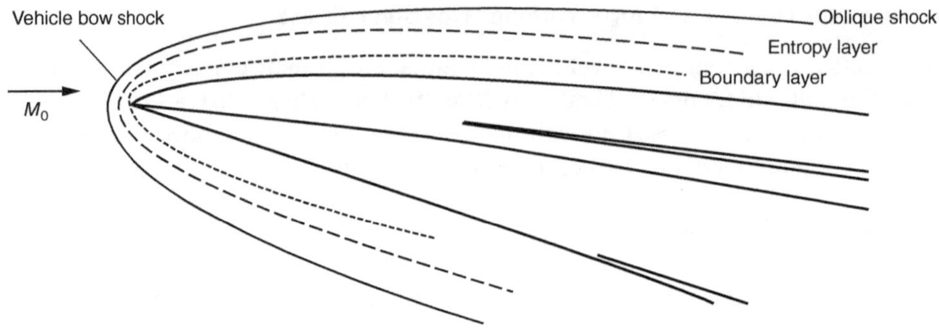

Figure 3.3 Characteristics of hypersonic flow over vehicle forebody.

to propagate against the flow, and this leads to discontinuities (jumps) in some of the flow properties immediately upstream and downstream of the shock line.

Hypersonic aircraft encounters such severe heating that a significant part of the design and development effort is concerned with providing sufficient thermal protection, especially for long-duration flight reusable vehicles. In hypersonic flow, the molecules behind a shock wave become vibrationally excited, partially or completely dissociated depending on their bond energy, and, at very high speeds, partially ionized. These characteristics of hypersonic flow are typically called real gas effects.

The flow around a hypersonic vehicle is predominantly three dimensional and is dominated by aerodynamic heating, thin shock layers, and viscous effects. Aerodynamic heating is the result of the hot boundary layer adjacent to the vehicle's surfaces. This thermal energy arises because the high kinetic energy hypersonic flow in the boundary layer is slowed down by viscous effects near the surface, which dissipate the kinetic energy and transforms some of the energy into internal energy of the gas in the boundary layer. Figure 3.3 depicts some of the characteristics of hypersonic flow over a vehicle forebody. While the entire vehicle airframe is subjected to aerodynamic heating due to air drag, the leading edges — including the nose region, tail edges, forebody, and scramjet inlet — are exposed to even more intense, highly localized stagnation-point heating and a more severe thermal environment.

3.2.1 Shock Layer

The shock layer is composed of the flowfield between the oblique shock wave and the vehicle. The thickness of this layer can be very thin at hypersonic speeds. For instance, the shock angle for Mach 5 flow of a calorically perfect gas over a wedge of 10° half angle is 19°. As the Mach number increases to 12 over the same wedge, the shock angle reduces to 13.7°.

This thinning of the shock layer occurs because, from oblique shock theory, for a given flow deflection angle θ, the density increases across the shock, becoming larger as the flow Mach number increases (see Table 3.1). For hypersonic flow over a vehicle, at higher density, the mass behind the shock can squeeze through smaller areas, and thus the distance between the body and the shock is small.

It is a basic characteristic of hypersonic flows that shock waves lie close to the body and that the shock layer is thin. If we add the effect of a high temperature chemically reacting

Table 3.1 Oblique shock properties for flow of a calorically perfect gas with $\gamma = 1.4$ over a wedge of 10-deg half angle.

M_0	β	M_1	ρ_1/ρ_0	p_1/p_0	p_{t1}/p_{t0}	T_1/T_0
5	19.37	3.99	2.13	3.04	0.87	1.43
12	13.77	7.67	3.72	9.35	0.37	2.51

flow, the shock angle will be smaller. One of the consequences of having the shock wave so close to the vehicle is the merging of the shock itself with a thick, viscous boundary layer growing from the body surface.

3.2.2 Viscous Interaction Layer

Hypersonic boundary layers grow more rapidly than at slower speeds. The thicknesses of hypersonic boundary layers are orders of magnitude greater than those developed at low speed, and frictional shearing stresses within this thick boundary layer are considerable due to large velocity gradient across the flow.

Consider a laminar boundary layer over a flat plate. The self-similar solution yields the following result for the boundary-layer thickness δ:

$$\delta \propto \frac{x}{\sqrt{Re}} = \frac{x}{\sqrt{\frac{\rho_w V_e x}{\mu_w}}} \tag{3.3a}$$

where Re is evaluated using the conditions at the wall, and ρ_w, μ_w, and V_e denote the velocity at the edge of the boundary layer. Thus, we can write

$$\delta \propto \frac{x}{\sqrt{\frac{\rho_e V_e x}{\mu_e}}} \sqrt{\frac{\rho_e}{\rho_w}} \sqrt{\frac{\mu_w}{\mu_e}} \tag{3.3b}$$

Using the equation of state, we get

$$\frac{\rho_e}{\rho_w} = \frac{p_e}{p_w}\frac{T_w}{T_e} = \frac{T_w}{T_e} \tag{3.4}$$

where we assumed constant pressure through the boundary layer, $p_e = p_w = p = const$. Also assume that the viscosity depends on temperature only so that

$$\frac{\mu_w}{\mu_e} = \frac{T_w}{T_e} \tag{3.5}$$

Now combine Eqs. (3.4) and (3.5) to rewrite (3.3b) as

$$\delta \propto \frac{x}{\sqrt{Re}}\left(\frac{T_w}{T_e}\right) \tag{3.3c}$$

where the Reynolds number is evaluated with the properties at the outer edge of the boundary layer. This is an important result that implies the thickness of the hypersonic boundary layer is proportional to the temperature gradient in it.

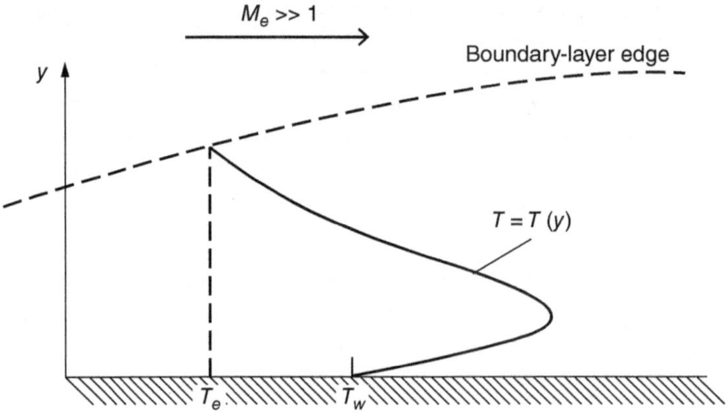

Figure 3.4 Temperature profile through a boundary layer.

Figure 3.4 depicts a typical temperature profile within the boundary layer, a profile that develops because of viscous dissipation. In this sketch, T_w denotes the wall temperature. A hypersonic flow contains a large amount of kinetic energy, so when the flow is slowed by viscous effects within the boundary layer, that kinetic energy is converted into internal energy of the air, which causes its temperature to increase.

Moreover, the viscosity of the gas increases with temperature, and this by itself will cause the boundary layer to grow thicker. And since we assume that the gas pressure is constant in the normal direction through the boundary layer, according to the equation of state an increase of temperature must result in a decrease in density. This means that to pass the required mass flow through the layer at reduced density, the boundary-layer thickness must be larger. These phenomena make hypersonic boundary layers grow more quickly than at lower speeds.

Now, assume that the wall is adiabatic with a recovery factor $r = 1$ (see Section 3.4.1) so that

$$\frac{T_w}{T_e} = \frac{T_{aw}}{T_e} = 1 + \frac{\gamma - 1}{2} M_e^2 \tag{3.6}$$

which for large M_e this expression reduces to

$$\frac{T_w}{T_e} = \frac{T_0}{T_e} = \frac{\gamma - 1}{2} M_e^2 \tag{3.7}$$

If we substitute Eq. (3.7) into (3.3c), we discover that

$$\frac{\delta}{x} \propto \frac{M_e^2}{\sqrt{Re}} \tag{3.8}$$

This result implies that since δ varies as the square of the Mach number, the thickness of the boundary layer becomes huge at hypersonic speeds (at the same Reynolds number). The thick hypersonic boundary layer modifies the free stream flow around the vehicle. How thick is "thick"? One example is given by the NASA X-43A flight program. Berry et al. 2008 estimated the vehicle forebody boundary-layer thickness for expected variations of the Mach 7 trajectory, angle-of attack, and wall temperature in flight to have a maximum and minimum thickness of $\delta_{max} = 0.205$ in (5.207 mm) and $\delta_{min} = 0.173$ in (4.3942 mm).

Equation (3.3c) suggests one way to thin the boundary layer is to cool the wall. For the X-43A flight at Mach 7, the boundary-layer thickness at the trip location was calculated to be 0.122-in (3.10 mm) for a wall temperature of 800 °R (444.44 K), see Section 3.4.3.

Considering both the velocity and temperature gradients, the hypersonic boundary-layer edge is often defined as the location where the total enthalpy gradient $\Delta H(y)$ goes to zero since the inviscid part of the shock layer is usually assumed to be adiabatic (perfectly insulated). Hence, the edge of the boundary layer is the location where $H(y) = 0.99H_0$.

The interaction between the inviscid freestream and the viscous boundary layer affects the wall-pressure distribution, the vehicle's skin friction, and the overall heat transfer. These physical characteristics near the wall affect the vehicle's lift, drag, stability, and heating loads. Since the boundary layer can be laminar, turbulent, or transitional, each type must be considered in the design of the hypersonic vehicle.

Viscous Interaction: The state of the boundary layer has first-order effects on heating, viscous drag, and control surface effectiveness. At hypersonic speed, the thick boundary layer deflects the external inviscid flow, creating strong, curved shock waves trailing downstream from leading edges. Inviscid flow is one in which friction (shear resistance) is negligible, i.e., frictionless flow. The surface pressure is much higher than the freestream p_0 (far downstream $p \rightarrow p_0$). The high pressure increases the aerodynamic heating at the vehicle's leading edges.

3.2.3 Entropy or Vorticity Interaction Layer

Another characteristic of hypersonic flow is the entropy increase near the surface of a vehicle, especially at the nose where the curved shock wave will yield a larger entropy growth. The stronger the shock wave, the larger the entropy of the flow increases across the shock. The entropy layer that forms around the vehicle (see Figure 3.2) flows downstream, and the boundary layer may grow inside this entropy layer, causing a strong vorticity interaction that is difficult to model with standard boundary-layer theory. Boundary-layer thickness may or may not be of the same magnitude as the shock layer thickness.

For a perfect-gas (gas for which the composition is frozen and one for which γ is constant), the entropy increase across the shock wave is given by:

$$(\gamma-1)\frac{s_2-s_1}{R} = \ln\left[\frac{2\gamma M_1^2 \sin^2\theta_{sw} - (\gamma-1)}{\gamma+1}\right] - \gamma \ln\left[\frac{(\gamma+1)M_1^2 \sin^2\theta_{sw}}{(\gamma-1)M_1^2 \sin^2\theta_{sw} + 2}\right] \tag{3.9}$$

where the local wave inclination is θ_{sw}, and M_1 is the Mach number of the flow upstream.

The entropy layer can interact with the boundary layer, creating some problems on the surface of a hypersonic vehicle. This can occur when, as the air flows over the body, all of the high entropy gas initially processed by the bow shock wave (the entropy layer), is eventually entrained, or swallowed, by the boundary layer. However, additional gas subsequently entrained by the boundary layer is cooler (has lower-entropy) since it has traversed only the weaker portions of the shock wave. This phenomenon is referred to as *entropy layer swallowing*, and if it occurs it will change the condition at the edge of the boundary layer, and it can cause a significant increase in turbulent heat transfer on the body's surface.

Hence, including entropy-layer analysis will yield more accurate predictions of quantities such as the edge Mach number, the skin friction, and heat transfer processes.

3.3 Laminar to Turbulent Transition in Hypersonic Flows

The first and most important question in the design process of a hypersonic vehicle is whether the boundary layers are likely to be laminar, transitional, or turbulent, whether we are referring to the leeward or windward parts of the vehicle. The onset of boundary-layer turbulence results from instability waves originating in the laminar boundary layer. However, not all disturbances produce boundary-layer transition. In some cases, the disturbances attenuate and the boundary layer remains laminar. In other cases, the disturbances grow as they travel downstream, until they finally produce a turbulent boundary layer.

Laminar boundary layers develop when the flow is highly stable, and disturbances are not amplified. We can characterize laminar boundary layers with more certainty, and they cause considerably less skin friction and wall heat transfer. However, laminar boundary layers are more susceptible to separate in the presence of adverse pressure gradients or shock waves. A separated boundary layer produces increased local values of pressure drag and wall heat transfer, and it can have a much larger displacement thickness.

Turbulent boundary layers are characterized as having a large-scale turbulent motion; thus, energy is transmitted more readily in turbulent boundary layers. Turbulent boundary layers are less susceptible to separation caused by flow phenomena such as shock/boundary layer and shock–shock interactions. On the other hand, the skin friction and heat transfer are larger for turbulent in comparison to laminar flows.

Consider flow over a flat surface, as sketched in Figure 3.5. When the freestream velocity is uniform and laminar, as it approaches the plate, the boundary layer is first laminar and stable. But farther downstream from the leading edge, the flow becomes unstable. A gust of air, surface roughness or other disturbances from the freestream now become amplified and the flow begins to show characteristics of turbulence. The transitional flow region occurs when the velocity profile begins to change and the boundary layer becomes enlarged. If the disturbances continue, the flow over the plate will become fully turbulent.

For hypersonic air-breathing vehicles, knowing where the flow transitions on the forebody/inlet is crucial, in order to ensure that turbulent flow enters the engine. A turbulent boundary layer is less susceptible to separation resulting from shock–boundary layer interactions. Turbulent boundary layer improves mass capture and fuel/air mixing.

For analysis purposes, we assume that transition takes place at a point within the flow transition region on the surface, which we denote x_T. For the accurate prediction of skin friction and aerodynamic heating on the hypersonic air-breathing vehicle, knowledge of the transition Reynolds number is critical. Normally, we determine whether transition from laminar to turbulent boundary-layer flow will take place by comparing the expected freestream Reynolds numbers with a critical (or transition) Reynolds number, which we denote as Re_T. We define a transition Reynolds number as

$$Re_T = \frac{\rho_e V_e x_T}{\mu_e} \tag{3.10}$$

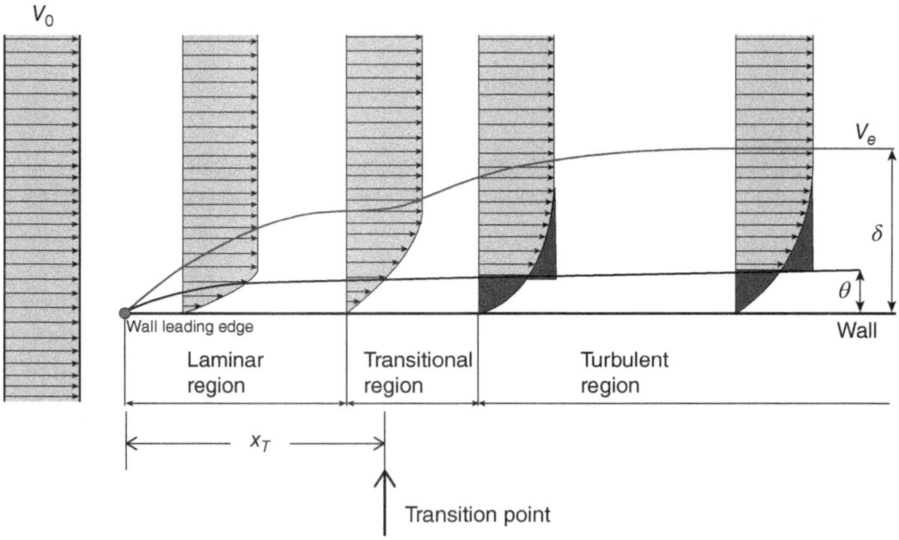

Figure 3.5 Transitional boundary layer showing growing thickness (δ) and momentum thickness (θ).

The value of the transition Reynolds number for a given situation is based on experimental data and empirical relationships. However, selecting the most appropriate value is difficult because the transition process is one of the most complex found in nature, and especially so for hypersonic flows. Viscous–inviscid interactions near the vehicle surfaces, and interactions between a shock wave and the boundary layer depend on the character of the boundary layer, i.e., whether it is laminar, transitional, or turbulent. When and if boundary-layer transition occurs depends on many, interrelated flow parameters.

Effects on boundary-layer transition (BLT) include (i) Leading-edge bluntness, (ii) surface wall temperature, (iii) surface roughness, (iv) adverse pressure gradient in the vehicle's compression ramp, (v) freestream Mach number, (vi) Reynolds number, and (vii) angle of attack. Three-dimensional configuration effects on transition is studied experimentally at quiet ground test facilities. Engineering codes are used to determine BLT during hypersonic flight.

3.3.1 Transition Correlations Based on Boundary-Layer Momentum Thickness

The momentum thickness, denoted by θ, is another measure of boundary-layer thickness, that is used to estimate hypersonic flow transition. The boundary-layer momentum thickness is defined as the loss of momentum flux that is due to the presence of the growing boundary layer. Illustrated in Figure 3.5, the momentum thickness (for compressible flow) is defined as:

$$\theta = \int_0^{\delta} \frac{\rho V}{\rho_e V_e}\left(1 - \frac{V}{V_e}\right) dy \tag{3.11}$$

Hence, we reference the transition Reynolds number to the value of the momentum thickness at transition, which we denote θ_T:

$$Re_{\theta_T} = \frac{\rho_e V_e \theta_T}{\mu_e} \tag{3.12}$$

Prediction of BLT is difficult. A number of empirical correlations for hypersonic transition were developed in the past decades in terms of momentum thickness, studying the effects of key geometry and flow parameters, including wall cooling and nose bluntness, to determine the location of transition. In the 1970s, researchers sought reliable predictions for flight conditions by establishing the effect of wind tunnel disturbances. One such effort studied the variation of transition onset with local Mach number and with local unit Reynolds number to determine the minimum operational Reynolds numbers for a quiet tunnel, (Beckwith 1975). This study considered free-flight data for transition *Re* using sharp cones at small angle of attack, given the variation of transition onset with local Mach number and local unit Reynolds number, denoted $Re_{x,t}$. For example, for $M_e = 6$, the transition Reynolds number using wind tunnel data correlation was determined to be $Re_{x,t} = 3 \times 10^6$. The predicted flight transition $Re_{x,t}$ values were generally higher than the correlation line for the wind-tunnel data.

In 1990, Elias and Eisworth published flight transition correlation for cone shapes, where the transition Reynolds number Re_θ was given as a function of the local Mach number. Using linear stability predictions (*see* Reshotko 1976) and quiet tunnel transition tests for Mach numbers $10 \leq M_0 \leq 20$, Elias and Eisworth concluded that transition would occur when

$$\frac{Re_{\theta_T}}{M_e} > 150 \tag{3.13}$$

where M_e is the local Mach number at the edge of the boundary layer. This is similar to a correlation used for the preliminary design of the SS Orbiter, $Re_{\theta_T}/M_e = 100$.

Another rule of thumb for flow in which transition is caused by the generation of Goertler vortices with a momentum thickness θ is:

$$G_e = Re_{\theta_T}\sqrt{\frac{\theta}{R_c}} = 8-9 \tag{3.14}$$

where G_e is the Goertler number, and R_c is the radius of curvature of the compression surface.

An empirical correlation (based on cone data) was proposed for a study of hypersonic waverider vehicles (Bowcutt and Anderson 1987):

$$\log_{10}(Re_T) = 6.421 \exp\left(1.209 \times 10^{-4} M_e^{2.641}\right) \tag{3.15}$$

Neither of the above correlations may be appropriate for new hypersonic flow/configuration conditions outside the data set on which they are based.

3.3.2 Surface Roughness Effect on Boundary-Layer Transition

The ability to accurately predict when BLT will occur on a reusable hypersonic flight vehicle is important when designing its TPS. At the same time, the roughness of the TPS material will affect the transition process. Laminar-to-turbulent transition on any hypersonic vehicle determines the heat loads and limits the allowable flight trajectories. Transition during the peak heating phase of the ascent or entry may dictate higher temperature materials on the vehicle afterbody. Vehicle weight, payload capacity, and mission are directly affected by these constraints. Unfortunately, predicting transition onset is very difficult.

One such case was the former NASA SS Orbiter. Its windward (lower) surface was protected with a ceramic TPS to ensure it survived during its atmospheric entry and descent flight. The tiles were made up of a rigidized fibrous ceramic with a reaction-cured glass (RCG) coating to provide a high-temperature reusable insulation. During each flight, some of this insulation became damaged or detached either at liftoff or upon reentry.

BLT on the SS Orbiter was dominated by surface roughness in the form of launch-induced damage and/or protruding gap fillers (Bouslog et al. 1997). The random nature of this roughness resulted on a wide range of free-stream conditions for the onset of transition, e.g., Mach numbers between 6 and 18 and length Reynolds numbers between 2.5 and 13 million (Berry et al. 1999). Such scatter was not acceptable as prediction of hypersonic BLT.

While developing the TPS for the NASA X-33, a reusable rocket-powered launch vehicle (RLV), researchers examined the smooth body transition patterns, the effect of discrete roughness on and off windward centerline, and the effect of distributed bowed panels on the SS Orbiter during its atmospheric entry and descent flight. Both vehicles had similar requirements. For descent, the X-33 would also fly at angles of attack near 40-deg. Descending at this angle results in a moderately blunt flowfield that produces similar boundary-layer edge conditions on the windward body surface (Berry et al. 1999).

For the X-33 program, NASA conducted experiments in the Langley 20-Inch Mach 6 Air Tunnel to examine boundary layer and aeroheating characteristics of several X-33 configurations. The experiments assessed the effects of discrete and distributed roughness elements on BLT, which included trip height, size, location, and distribution, both on and off the windward centerline. The discrete roughness results on centerline were used to provide a transition correlation for the X-33 flight vehicle that was applicable to the range of reentry angles of attack. The roughness elements were fabricated from Kapton tape to simulate a raised TPS tile. The height of the roughness element (k) was varied ($k = 0.0025$, 0.0050, and 0.0075-inch) and compared with boundary-layer thickness to obtain a roughness parameter in the range of $0.2 < k/\delta < 1.2$.

To estimate the onset of transition in flight, researchers selected the value of $Re_\theta/M_e = 250$ at $x/L = 0.8$ on the windward centerline, based on a conservative view of the smooth body and discrete centerline roughness results, as well as experience from the SS Orbiter. This corresponded to a $k/\delta = 0.2$ from the discrete correlation curve, which was used to estimate the allowable roughness for the X-33 flight vehicle. Using results for all the discrete trip data along the X-33 model centerline for angles of attack of 20, 30, and 40-deg, Berry et al. (1999) obtained the following transition correlation:

$$\frac{Re_\theta}{M_e} = C\left(\frac{k}{\delta}\right)^{-1.0} \tag{3.16}$$

where the value of C determines incipient or effective transition.

The maximum Re_θ at which laminar flow was maintained behind the trip identifies the *incipient* value. The minimum Re_θ where the transition front is fixed at the roughness element identifies the *effective* value. The Re_θ where significant nonlaminar flow first appears

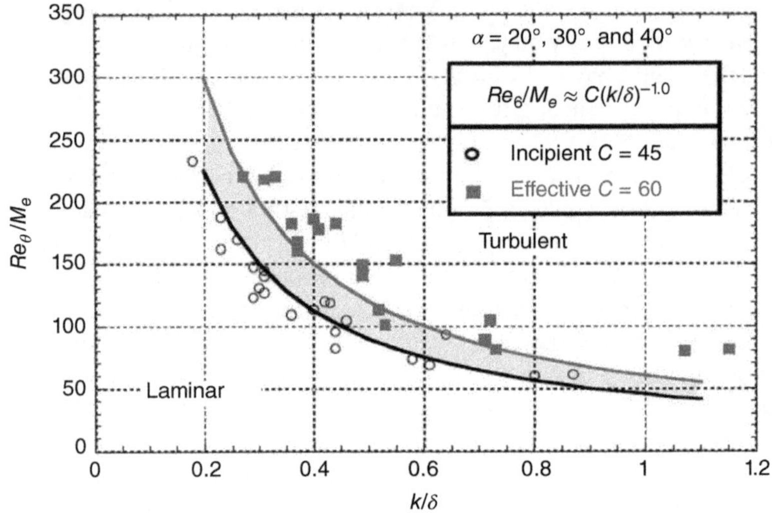

Figure 3.6 Transition correlation for the X-33 windward centerline. Source: From Berry et al. (1999)/ American Institute of Aeronautics and Astronautics.

downstream of the roughness element identifies the critical value. The correlations shown in Figure 3.6 approximate the well-behaved patterns of incipient and effective transition data.

Berry et al. (1999) also developed a numerical tool that predicts when transition will occur on the windward centerline of a vehicle for a given altitude, velocity, and angle of attack condition. By coupling this tool with trajectory simulations enabled the team to make modifications of the flight profile and ensure TPS design constraints were not exceeded.

3.4 Hypersonic Flowfield for Propulsion-Integrated Vehicles

At hypersonic speeds, the engine inlet compresses the ingested atmospheric airflow by the deceleration of the ram flow, which converts the freestream kinetic energy into pressure energy, using a supersonic diffuser. In a scramjet inlet, the flow is not diffused to sonic condition as the ramjet inlet does. This is because in a ramjet, diffusing the air to such total or stagnation condition would be impractical at hypersonic speeds, as it would develop unsustainable levels of aerothermodynamic loads and tremendously high heat flux in the flowpath, and the propulsion efficiency would be too low. Thus, in a scramjet engine, the flow at the combustor must be partially diffused and the result is that the fuel–air reaction occurs at supersonic conditions.

For the scramjet to be truly integrated into the vehicle, the forebody is designed to provide substantial flow compression, while the aftbody is designed to provide the bulk of the exhaust expansion to produce thrust. Sizing and shaping of forebody and aftbody must be optimized to ensure the vehicle's hypersonic performance.

3.4.1 Sources of Viscous Interactions and Shock–Shock Interactions

The flowfield shock–shock interactions and viscous–inviscid interactions produce locally severe heating on slender vehicles cruising at hypersonic speeds. As we noted earlier, the three-dimensional hypersonic flowfields contain numerous viscous interactions, and shock–shock interaction that affects the flight trajectories and airframe design. As shown in Figure 3.7, the hypersonic air-breathing propulsion vehicle has thin control and tail surfaces with very sharp leading edges – all designed to minimize drag.

Other critical aerodynamic phenomena are caused by viscous and shock–shock interactions on various parts of hypersonic air-breathing vehicles. These interactions include

- **Boundary-Layer Transition.** The effects of hypersonic flow BLT are particularly important to characterize aeroheating on different parts of the exposed surfaces, as these will determine the measures for thermal management that must be included in the design of reusable aircraft. BLT affects hypersonic flight performance and ascent/entry heating. Hence, BLT plays a primary role in TPS selection, trajectory shaping, and upon vehicle design. For hypersonic air-breathing vehicles, it is more imperative to determine where flow transitions on the forebody/inlet, as turbulent flow must enter the engine to increase its airflow mass capture and overall propulsive efficiency.
- **Shock Wave–Boundary-Layer Viscous Interactions** (SWBLI). These interactions result when the shocks interact with the boundary layer. When a shock impinges and reflects from the boundary layer, the boundary layer thickens and, for higher shock strengths (equivalent to high wedge angles), it will separate. Examples of shock-wave–boundary layer interactions are (i) shock impingement, (ii) ramp flow, and (iii) transonic flow over a bump. This phenomenon will develop on the vehicle's forebody, aftbody, and its control surfaces.
- **Shock–Shock Interactions.** Shock waves can interact with other shock waves in several parts of the vehicle. These interactions can be more pronounced on the inlet engine cowl as a result of the "shock-on-lip" condition when the vehicle's bow shock intersects the shock wave generated by the cowl leading edge. Shock–shock interactions can also develop when the vehicle's bow shock intersects the shocks produced by aerodynamic

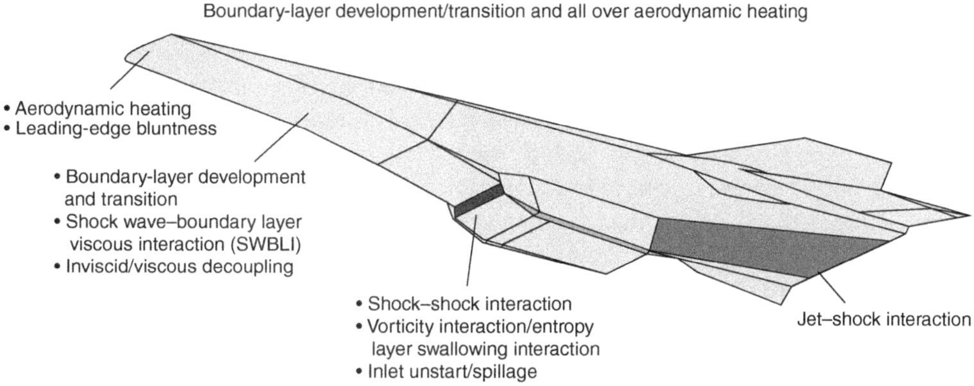

Figure 3.7 Aerothermodynamic phenomena for scramjet integrated vehicle.

control surfaces. The shock–shock interactions produce a variety of flowfield perturbations around the hypersonic aircraft.

- **Vorticity Interaction/Entropy Layer Swallowing**. Conical or blunted bodies in hypersonic flow generate high entropy fluid (forming the entropy layer), which is subsequently entrained in the boundary layer, as the boundary layer grows on the surface of the body and alter the flowfield in a detrimental manner. Entropy layer swallowing results in a variable entropy layer that produces increased heating rates.

The viscous–shock wave interactions produce complex flowfields that can cause other interactions, especially with the jets of hot gases exhausted by the propulsion system, and can produce regions of high pressure and locally high heat transfer. Some interactions can result in flow relaminarization, that is, reversing a turbulent flow to laminar. Under some flight conditions, the cooling of the boundary layer by the inlet/forebody walls, for example, can cause the turbulent boundary layer to return to a laminar state, which greatly increases its displacement thickness. Relaminarization can also happen in the exhaust nozzle/aftbody, although the cause there is primarily rapid flow acceleration rather than cooling of the wall. All these effects can cause flight performance degradation and even catastrophic failures.

3.4.2 Forebody/Inlet Flowfield

The flowfield features that affect the scramjet inlet performance and operability include

- Boundary-layer development/transition,
- Forebody and cowl leading-edge bluntness,
- Shock–boundary layer interaction,
- Shock–shock interaction,
- Inviscid–viscous coupling,
- Flow profile at the throat.

All of these phenomena must be addressed by extensive research and analysis. In the following section, we review BLT by focusing on the work carried out by the NASA Hyper-X program (Cockrell et al. 2002). The X-43-A vehicle was designed to have turbulent flow in the scramjet engine flowpath. This means that the forebody had to be designed to ensure that the boundary layer approaching the scramjet inlet should be turbulent. Ingesting a turbulent boundary layer increases inlet operability and thus enhances overall propulsion performance. However, a subscale vehicle might require to force BLT, using passive or active trips, to properly scale the scramjet flight tests to future full-size vehicles. Passive methods for boundary-layer control include discrete roughness elements such as surface cavities, surface weavings, square or rectangular protrusions, and swept ramp vortex generators. Active methods for tripping the boundary layer include mass addition or gas blowing. Surface blowing can be done through holes or slots and porous material.

The X-43A vehicle structure had to be designed to survive the increased heating and skin friction due to this turbulent flow. The subscale vehicle design requirements raised concerns regarding the accuracy of the turbulent flow prediction techniques available at the time, applicable to slender, planar configurations flying at Mach 7 and Mach 10 conditions. Although the main objective of the Hyper-X program was to demonstrate a scramjet powered in flight, BLT was a major component of the R&D program. Smooth wall (on the upper surface) and roughness (on the inlet flowpath) BLT data were to be obtained, using thermocouples (type-S) embedded just under the surface of the tiles covering the vehicle.

3.4.3 NASA Hyper-X: A Case Study for Forebody Boundary-Layer Transition

The vehicle forebody must provide a naturally turbulent boundary layer for the scramjet inlet. Turbulent boundary layer improves mass capture and fuel/air mixing. A turbulent boundary layer is less susceptible to separation resulting from a shock–boundary layer interaction. If a separated boundary layer develops in the forebody, less airflow rate can be captured by the engine inlet. Separation of boundary layers in the combustor can have detrimental effects on the fuel/air mixing and chemical reactions.

An analysis of the NASA Hyper-X forebody using the hypersonic BLT criteria developed during the NASP program indicated that the subscale X-43A vehicle forebody would remain laminar during the flight. Using a boundary-layer code, the Hyper-X team computed laminar values of the momentum thickness Reynolds number ($Re_\theta = (\rho u_e \theta)/\mu$) over the boundary-layer edge Mach number (M_e) for a sharp-nose wedge with 4.5-deg of turning. To estimate the length to natural transition onset, the team used the NASP "smooth wall" sharp planar transition criterion of $x_{cr} = Re_\theta/M_e = 305$. This criterion was deemed acceptable, concluding that nose bluntness would have a stabilizing influence that would further delay transition onset.

For the X-43A Mach 7 vehicle, the critical length was $x_{cr} > 9$ ft (2.74 m), and for the Mach 10 vehicle, $x_{cr} > 25$ ft (7.62 m). Since the forebody was 6 ft (1.8288 m) long, the predicted critical length for transition indicated that boundary-layer flow would be laminar. Hence, over 200% more running length would be required for transition to occur on the first ramp prior to the X-43A vehicle's compression corner, which was beyond the engine inlet's entrance (Berry et al. 2008).

Thus, to provide the required turbulent boundary layer for the scramjet inlet, the forebody required a means to promote transition. Without forcefully promoting transition with boundary-layer trips, the potential for a laminar layer separation existed at the first ramp break that could generate lateral flow spillage, potentially reducing the mass capture, and affecting overall scramjet performance.

Since very little ground-based experimental or flight data was available at the time on forced transition, a wind tunnel test program was initiated at the NASA Langley Research Center (LaRC) to develop passive (inert) boundary-layer trips for the Hyper-X vehicle (Berry et al. 2000). A lateral vortex generators array on the first ramp of the Hyper-X lower surface forebody was selected as the approach for tripping the boundary layer for the Mach 7 and 10 flight vehicles. Over the course of the experimental program, five trip configurations were tested and compared. The trips included a "diamond" shape and swept ramp configurations that were based on prior experience at LaRC during transition studies of the SS Orbiter, X-38, and X-33 vehicles, and trip designs considered for the NASP, Hyflite, and HySTP programs.

The experimental investigation of the boundary-layer trip effectiveness and the effect of the trips on the aeroheating characteristics for a 33% scale Hyper-X forebody model was conducted in the 20-Inch Mach 6 Air Tunnel, the 31-Inch Mach 10 Air Tunnel, and the

Hypulse RST Tunnel at GASL. Five trip configurations were screened, and the results indicated that all provided adequate transition enhancement.

Data from the exhaustive experimental program were used to select a final trip configuration and height for the X-43A flight vehicles. A lateral array of swept ramp vortex generators configuration was chosen based on a minimization of entrained vorticity within the turbulent region and consideration of the thermal survivability of the trip. The boundary-layer thickness was the primary dimension with which to size and scale each unit-vortex generator. For the Mach 7 flight, they sized the final trip design with a height of 0.125 in. (3.175 mm) by considering variations of vehicle's flight trajectory, its angle of attack, and the estimated forebody temperature. For the Mach 10 vehicle, the trip geometry was scaled up from the Mach 7 design to be 0.26-in (6.604 mm) high. Therefore, to provide transition onset prior to the end of the first ramp, the team determined the effective trip height-to-boundary-layer thickness ratio as $k/\delta = 0.6$ for Mach 7, and as $k/\delta = 1.0$ for the Mach 10 flight (Berry et al. 2000, 2008). This simple approach was deemed acceptable due to the ground-based data being at a similar length Reynolds number as flight.

The predicted boundary-layer thickness δ and edge Mach number M_e for the NASA LaRC facilities and the HYPULSE RST are given in Table 3.2, along with the corresponding flight values of these parameters. Note the thick boundary layer for the Mach 10 flight (0.283 in. or 7.188 mm), more than twice as thick as the boundary layer measured in the 31-inch tunnel.

Figure 3.8 shows a sketch of the boundary-layer trip sized for the Mach 7 flight. The left side of this figure illustrates the strip installed on a full-scale, prototype engine vehicle, which was ground tested at the Mach number and enthalpy matching the Mach 7 flight. The spanwise array was added on the first ramp of the X-43A forebody, positioned at 18.5 in. (47 cm) from the nose leading edge, to trip the boundary layer ahead of the scramjet entrance. This trip array did promote transition, ensuring turbulent flow at the engine inlet face, while at the same time providing flow spillage relief at the first ramp break (Huebner et al. 2001).

The Hyper-X team carried out extensive CFD solutions of the scaled model, assuming a constant wall temperature of 540 °R (300 K) for the wind tunnel tests as a result of the relatively short run times. However, the flight vehicles had an axially varying wall temperature caused by the carbon–carbon leading edge and tungsten block forebody. This temperature variation was modeled with CFD for the flight cases (Berry et al. 2000).

Table 3.2 Calculated boundary-layer parameters at trip location of NASA X-43A vehicles.

Tunnel or flight	M_∞	Re_∞ (× 10^6/ft)	δ, in.	M_e
20-In. Mach 6	6.0	2.2	0.081	3.1
31-In. Mach 10	9.9	2.2	0.125	4.4
HYPULSE	7.3	1.4	0.075	4.2
Mach 7 flight	7.0	0.9	0.180	3.4
Mach 10 flight	10.0	0.6	0.283	4.5

Source: Berry et al. (2000)/NASA/Public Domain.

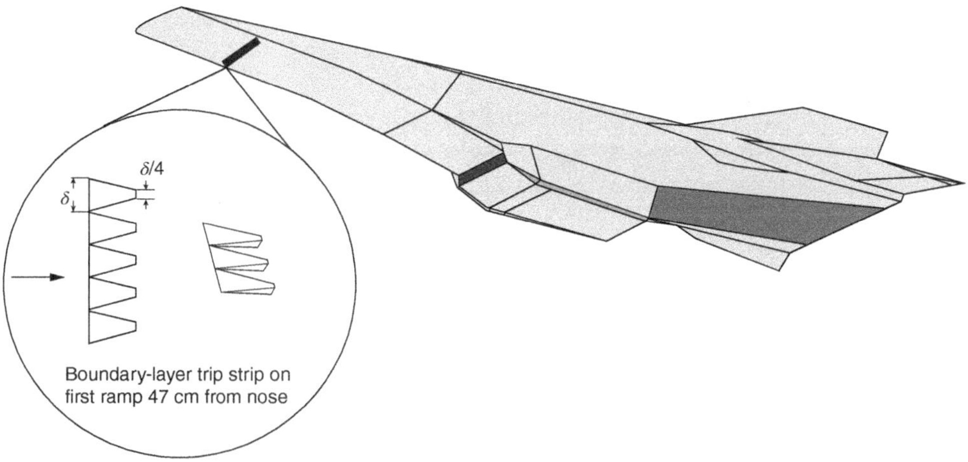

Figure 3.8 NASA X-43A vehicle showing the boundary-layer trip strip on the lower surface forebody. The detail sketch represents the swept ramp boundary-layer trips tested for the Mach 7 flight.

Figure 3.9 is a sketch of the lower surface of the X-43 forebody, showing the location of pressure and temperature instrumentation for both vehicles. The eight thermocouples along the centerline on the second and third ramps (T201, T003, T204, T207, T005, T208, T006, and T209) were all imbedded just slightly below the surface of the TPS tiles. The thermocouple on the nose (T002) was monitored for local surface temperature along

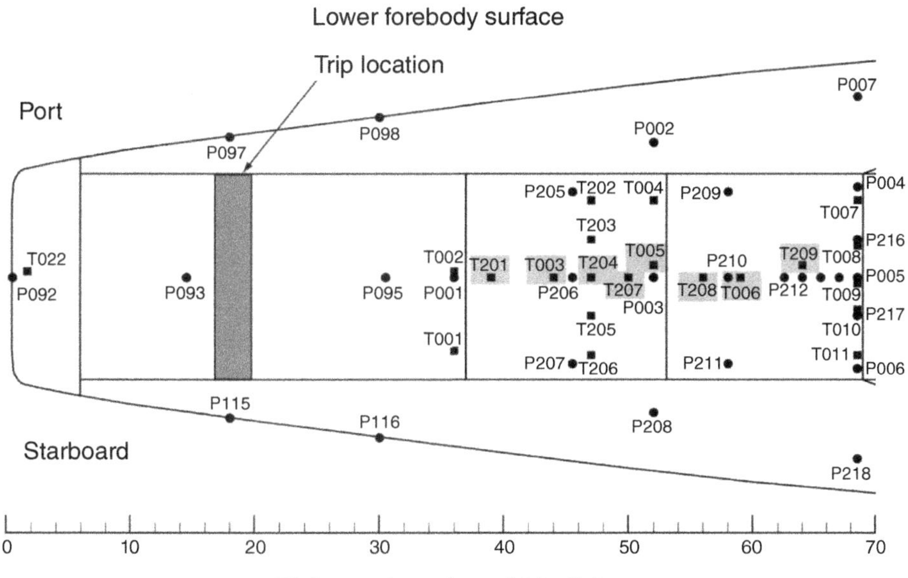

Figure 3.9 NASA X-43A lower surface forebody showing the boundary layer trip strip. Source: From Berry et al. (2008)/American Institute of Aeronautics and Astronautics.

the tungsten nose section (see Chapter 9). For each flight, the trajectory points (TP) of interest for the determination of momentum thickness Reynolds number over edge Mach number ratio (Re_θ/M_e) are indicated in the plots that follow.

From the measured temperature profiles in flight, it is possible to assess the state of the boundary layer over the vehicle. Figure 3.10 plots Re_θ/M_e as a function of axial distance for the Mach 7 ascent trajectory points denoted TP2, TP3, TP4, and TP5. Points TP2 and TP4 correspond to when gauges T201 and T003 indicate the beginning and end to brief reduction in heating (interpreted as a reduction in tripping effectiveness), while TP3 represented a trajectory point in between where the effectiveness was the least. Trajectory Point 5 (uppermost curve) is an unclassified representative of the nominal test point (similar angle of attack and flow conditions but before the cowl opened), $M_0 = 6.975$. As shown, Re_θ/M_e is equal to 127, 121, 125, and 136 for the indicated TP.

Compared to the preflight natural transition criterion ($Re_\theta/M_e = 305$ from NASP smooth wall correlation), these local values are a factor of three less than that required for natural transition; thus, the first ramp of the forebody would likely have a laminar boundary layer, with possible separation at the ramp corner during the Mach 7 flight ascent trajectory. The data are for a wall temperature of 800 °R (444.44 K). Although not included on the plot, researchers reported that when the range of measured surface temperatures were investigated, they noted only a 3% difference in Re_θ/M_e (Berry et al. 2008).

For the Mach 10 ascent trajectory, the calculated Re_θ/M_e are plotted in Figure 3.11 as a function of the lower surface axial distance, just for the TP associated with thermocouple T201, the furthest forward measurement location (see Figure 3.9). TP2 and TP10 correspond to when the trips were fully effective (turbulent at T201, highest values of Re_θ/M_e), while TP6 and TP8 correspond to when the trips were not effective (not turbulent at T201, lowest

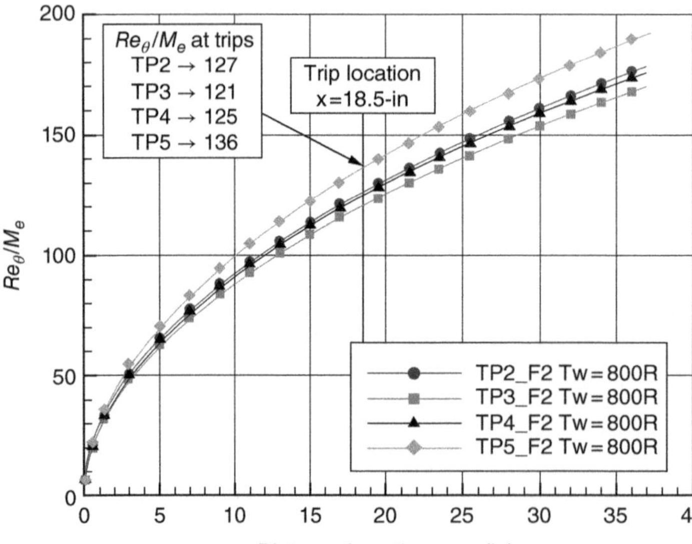

Figure 3.10 NASA X-43A lower surface Re_θ/M_e at boundary-layer trips for the Mach 7 ascent trajectory points. Source: From Berry et al. (2008)/American Institute of Aeronautics and Astronautics.

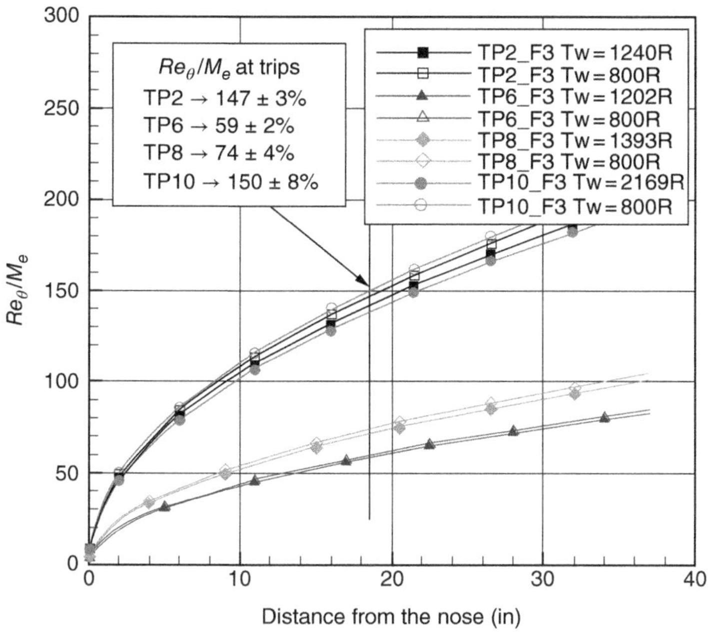

Figure 3.11 NASA X-43A lower surface Re_θ/M_e at boundary-layer trips for the Mach 10 ascent trajectory points showing effect of wall temperature. Source: From Berry et al. (2008)/American Institute of Aeronautics and Astronautics.

values of Re_θ/M_e). As shown in Figure 3.11, the calculations reveal a distinct shift in Re_θ/M_e between those two groupings. The actual values for the trip location, based on a wall temperature of 800 °R, were 147, 59, 74, and 150 for TP2, TP6, TP8, and TP10, respectively. Researchers reported an 8% difference between calculations and measured wall temperature variations (Berry et al. 2008).

We can also deduce the state of the flowfield over the X-43A vehicle in flight by examining the measured temperature profiles (Figure 3.12). The lower and upper surfaces used the centerline gauges to measure temperature during the first 800 seconds of the Mach 10 flight trajectory. Times are based on the release of the Hyper-X Launch Vehicle from the B-52 aircraft. Separation from the booster occurred at about 88 seconds after drop, while cowl open and engine ignition was at about 92 seconds. All the sensors, both upper and lower, indicate significant temperature fluctuations. Beginning at around 80 seconds, the lower (windward) surface temperatures rise sharply while the upper (leeward) surface temperatures decrease slightly (Berry et al. 2008).

The upper (leeward) surface for the Mach 10 (three bottom curves) shows a distinct departure from the quickly rising temperatures (initially of the same magnitude as the lower surface) that stabilizes roughly around 1000 °R (555.56 K) before increasing at 400 seconds into the trajectory, reaching a maximum of ~1200 °R (666.67 K). This suggests that the upper surface was fully laminar when the scramjet was on in flight.

The lower forebody surface for the Mach 10 flight (uppermost curves) shows different temperature profiles. T006, a gauge near the beginning of the third ramp, measured the highest temperature, 2880 °R (1600 K), followed by T208 and T209 with similar trends.

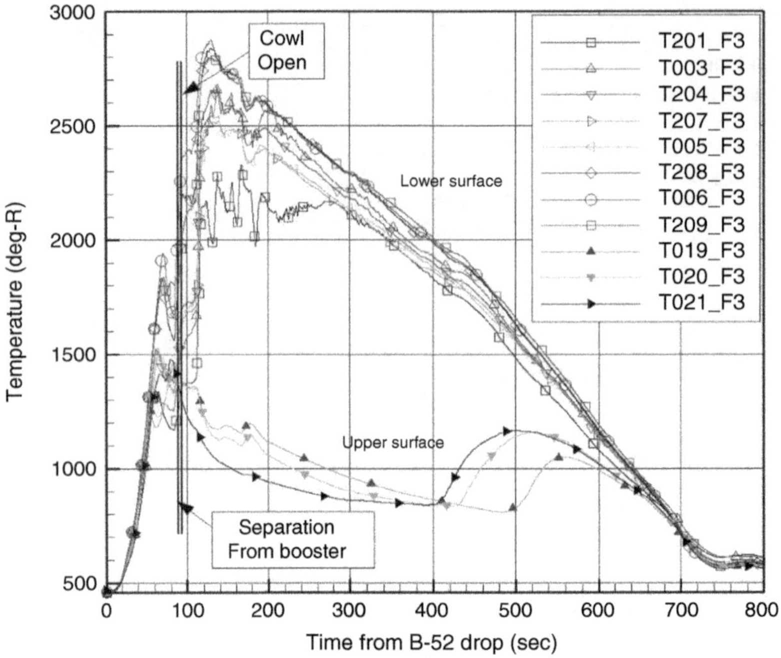

Figure 3.12 Surface temperature measurements for X-43A Mach 10 flight. Source: From Berry et al. (2008)/American Institute of Aeronautics and Astronautics.

However, gauge T201 (the furthest forward sensor, Figure 3.9) displays a different profile, with a peak temperature of about 2300 °R. One can argue that, being closest to the boundary-layer trips, this sensor captured the incipient transition to turbulent flow. However, since the scramjet was on for only 10.5 seconds, it is difficult to judge whether it ingested fully turbulent airflow.

For the Mach 7 flight, the upper (leeward) surface was predicted to be laminar, based on a preflight trajectory using the classical correlation, $Re_\theta/M_e = 305$. The upper surface temperature time histories during the entire Mach 6.83 flight trajectory from the point of release from the B-52 are shown in Figure 3.13. The three surface thermocouples on the upper surface were evenly spaced along the centerline, starting about midpoint (T/C#19) and ending near the trailing edge (T/C#21). As shown, the maximum temperature is about 1500 °R (833 K). By the time the cowl opened and the scramjet ignited, the entire upper surface appears to be laminar, as shown by the temperature decrease beginning at ~70 seconds for the farthest forward T/C#19 and 85 seconds for the farthest aft T/C#21. At about 240 seconds, the transition from laminar to turbulent flow happened as the X-43A vehicle decelerated (Voland et al. 2005).

The preflight predictions were accurate in estimating transition onset on the upper surface ($Re_\theta/M_e = 300 \pm 12$). Thus, in the absence of discrete roughness (trips), transition onset was consistent with the established NASP criteria. However, earlier in the flight, the time of transition (~70 seconds) from turbulent to laminar corresponded to $Re_\theta/M_e = 400$ (based on laminar predictions). When using a consistent computational approach,

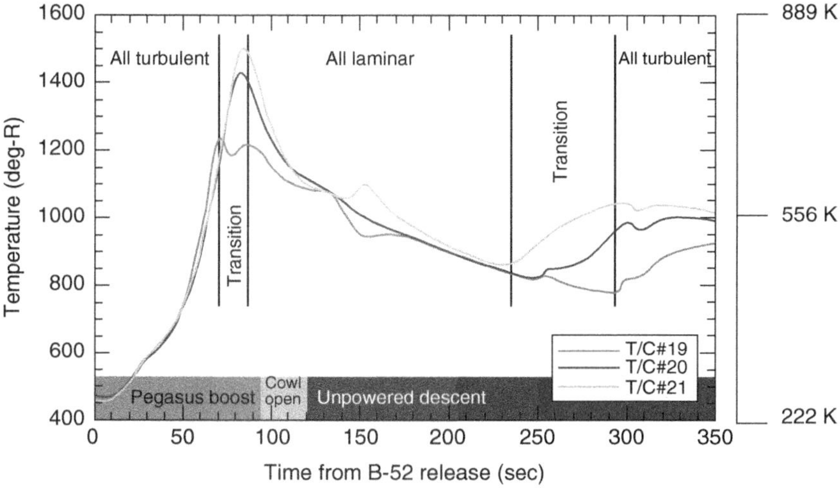

Figure 3.13 X-43A upper surface temperature history during the Mach 6.83 flight. Source: Voland et al. (2005)/With permission of Elsevier.

the laminar-to-turbulent and turbulent-to-laminar smooth wall transition criteria are not the same, and the X-43A flight data can be used to quantify this hysteresis effect.

The boundary-layer trips performed as designed, maintaining turbulent flow for the engine over the Mach 7 trajectory. The sensors closest to the trips measured a very brief and slight drop in temperature during maneuvering right before the scramjet lighted, suggesting a threshold for fully effective tripping at Mach 7 based on $k/\delta = 0.7$ and $Re_\theta/M_e = 110$. Comparing these values with the X-33 discrete roughness effective data in Figure 3.6, the X-43A flight-derived correlation would fall within the 20% scatter shown with a curve coefficient of $C = 77$ in Eq. (3.16) (Berry et al. 2008). A report on a post-flight analysis of the BLT data from the X-43A flights is provided in Berry et al. 2010. Their main conclusion is that the flow transition on both the upper surface (smooth) and lower surface (tripped) behaved as expected based on the ground-based data and scaling for flight.

A similar study was undertaken for the U.S. Air Force/Boeing Scramjet Engine Demonstrator-Waverider flight program. In order to support analysis of BLT on the X-51A vehicle, experiments on a 20% scale model were carried out. The model includes the forebody up to the engine inlet cowl. To ensure a turbulent boundary layer entering the X-51A inlet, an array of discrete roughness elements was used to trip the boundary-layer upstream of the compression corner on the vehicle's windward surface, Figure 3.14. Sizing the boundary-layer trips is a balance between competing effects: trips must be just large enough to cause transition. Too large trips could cause excessive momentum loss, possibly invalidating propulsion measurements, and they add unnecessary drag to the vehicle, adversely impacting flight performance (Borg 2007).

For the X-51A model, the maximum ramp height on the other strip is 0.038 cm (0.015 in.). The diamond-roughness strip has diamond-shaped roughness elements that are 0.305 cm (0.120 in.) wide across the diagonal and 0.152 cm (0.060 in.) high. The

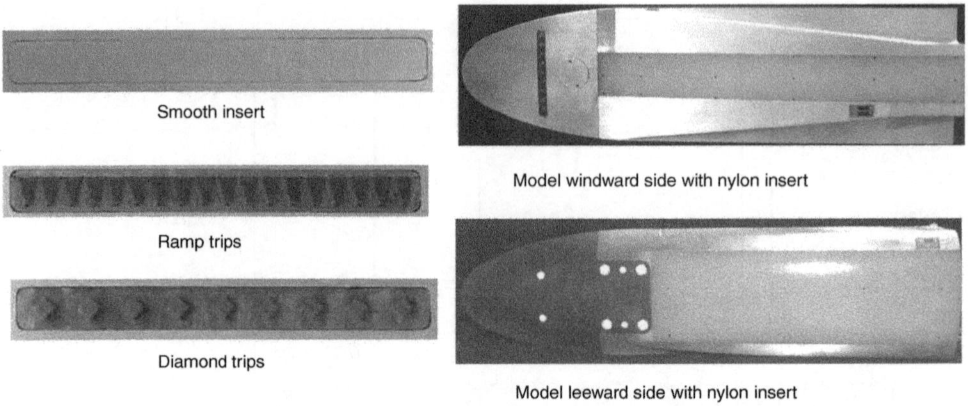

Figure 3.14 Photographs of the 20% scale model of the X-51A and trips. Source: Borg (2007), Purdue University.

diamond strip was flush with the model surface. The model was tested in the Boeing/AFOSR Mach-6 Quiet tunnel using an inviscid Mach number of 6.0 at 4° angle of attack with an initial stagnation temperature of 433 K (780 °R). Tunnel freestream noise had an impact on transition on the windward and leeward surfaces. The effect of noise on transition was observed with a smooth wall and for roughness-induced transition on both vehicle surfaces (Borg 2007).

It is very difficult to trip a hypersonic boundary layer in planar flow, as demonstrated by the relatively large trip size for X-43A flight tests. *Have we made substantial progress towards a new transition prediction method that we can apply with confidence to future hypersonic air-breathing propulsion vehicles*? At hypersonic speeds, where effects of the external stream gradients on stability are as yet uncertain, the knowledge of the freestream Reynolds number or Re_θ along the body surface may not be enough to predict the onset of transition.

Accurate prediction of transition at hypersonic speeds is one of the most important problems of hypersonics. Its ultimate solution may come when we obtain better understanding of turbulence itself. Although an in-depth discussion of turbulence in hypersonic flow is beyond the scope of this chapter, our intent here is to indicate trends, and to discuss some pertinent results from scramjet research and development efforts to date to serve as a reference for the future work.

3.4.4 Cowl Leading-Edge Shock Interactions and Shock-on-Lip Heating

The scramjet engine cowl leading edge is designed for maximum air capture to maximize thrust. This requires that the vehicle is configured and operated such that the compression shocks produced by the forebody ramps intersect the cowl leading edge. This is the ***shock-on-lip condition*** we illustrate in Figure 3.15. The forebody-induced shock wave impinges on the leading edge of the cowl, and this causes one or more interactions that produce extremely high convective heating to the cowl. Hence, the design of the cowl leading edge requires high temperature materials and thermal management.

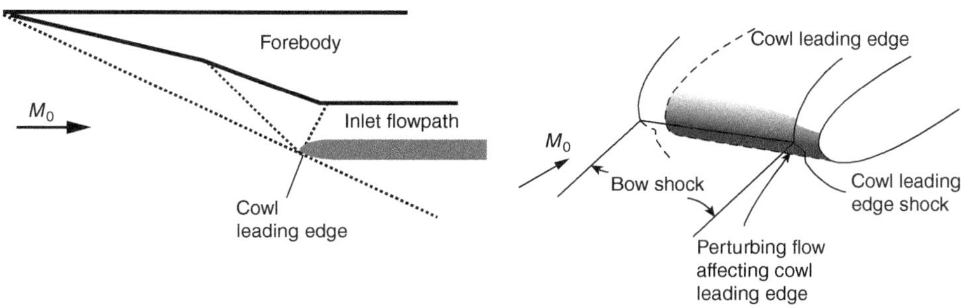

Figure 3.15 Shock-on-lip: compression shocks produced by forebody intersect cowl leading edge.

Shockwave interference heating was seen in a flight test of NASA X-15 hypersonic research aircraft. The study of viscous shock-on-shock interactions intensified after the devastating damage to the dummy ramjet flown on the NASA X-15, which was caused by oblique shocks interacting with the bow shock of the engine support during the Mach 6.7 flight. Hence, when the NASP vehicle was conceived, designers expected the most intense heating rates would occur on the blunted lip of the engine-inlet cowl.

Barry Edney was the first to give the most complete description of the shock impingement phenomena in 1968. Edney investigated the impingement problem for a variety of model geometries in a blowdown tunnel at Sweden's *Flygtekniska Försöksanstalthen* (FFA) for Mach 4.6–7.0 air flows (Edney 1968). Combining heat transfer and surface measurement techniques, Edney classified the spectrum of interactions that developed in those experiments. The six shock interaction patterns described by Edney are sketched in Figure 3.16.

The leading edge shows the approximate angular regions, and the type of interaction pattern that occurs when a weak oblique shock wave interacts with the bow shock at the leading edge. As he noted, interactions I, II, and V result in shock-wave–boundary-layer interactions. Type III interaction results in an attaching shear layer that produces high, localized pressure and heating that depend on the state of the impinging shear layer. Type VI interaction results in an expansion fan–boundary layer interaction.

Type IV interaction is the most severe shock interaction; it occurs when an oblique shock wave intersects near the normal portion of a bow shock, producing a supersonic jet that impinges on the leading edge. The maximum pressure and heating rate occur when the jet impinges perpendicular to the leading-edge surface. Hence, efforts to mitigate interaction heating now focus on controlling the shock interaction. These shocks can produce extremely high pressure and heating on the cowl leading edge, depending on where they intersect with the cowl bow shock. The peak pressure and convective heating rate depend on the flight Mach number, the Reynolds number, the gas composition, and on the impinging shock strength.

Shock impingement on a surface at hypersonic speed leads to extreme heating. Therefore, the cowl leading-edge design is driven by prediction of shock-on-lip heating.

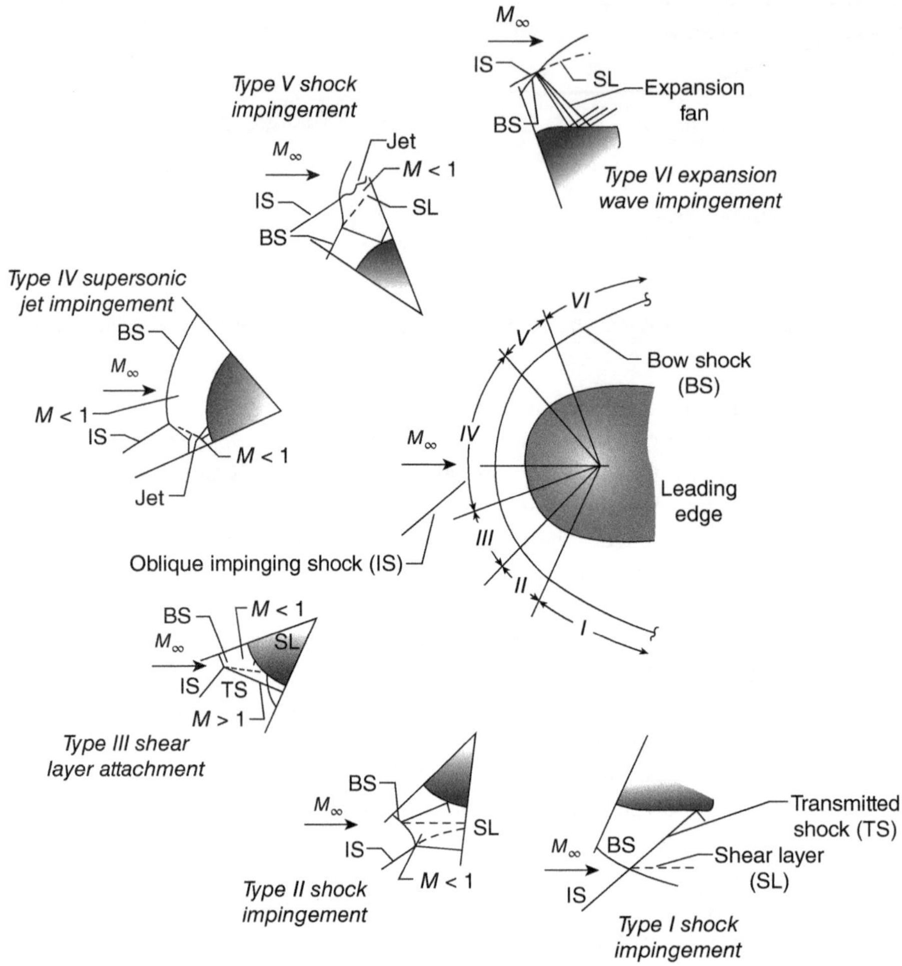

Figure 3.16 Six interaction patterns described by Edney. Source: Adapted from Edney (1968).

3.5 Convective Heat Transfer or Aerodynamic Heating

The structure of air-breathing propelled hypersonic vehicles will be exposed to high temperature, high enthalpy, and in some cases disassociated gas. Adiabatic wall temperatures are predicted to reach 2778–8333 K (5000–15 000 °R), depending on the flight conditions, shocks interactions, and vehicle aerodynamic design. These temperatures are higher than the allowable working temperatures of most materials (see Chapter 9). Hence, vehicle surfaces experiencing those high temperatures must be cooled. Wall cooling yield temperature gradients, and these in turn will generate thermal stresses that could easily exceed the strength of the materials.

Consider the simple case of a flat plate in a high enthalpy flow. Heat transfer to the plate will be due to radiation to or from the surroundings, and by conduction to or from the boundary-layer gas. This is a physical process we call convective heat transfer because it is coupled to the overall motion of the fluid over the surface. In the case of a hypersonic vehicle, the thermal flux depends on the flight velocity and atmospheric conditions, which vary with altitude. The relationship between Mach number and altitude is established by the dynamic pressure at which the engine is designed to operate.

3.5.1 Heat Flux Over a Flat Surface

The heat flux on the flat portion of the vehicle wall can be evaluated by analyzing the simple case of a flat plate in a high enthalpy flow. Preliminary estimates of convective heat transfer for thermal boundary layers with variable properties are based on an equation of the form

$$\dot{q}_w = St \cdot \rho_e V_e (h_{aw} - h_w) \tag{3.17}$$

where the coefficient St is the Stanton number, a dimensionless heat transfer quantity (ratio of heat transferred into a fluid to the fluid's heat capacity) that depends on the Reynolds number Re and the Prandtl number Pr, while ρ_e and V_e denote the density and velocity of the fluid in the adjacent inviscid flow at the outer edge of the boundary layer, respectively. The term in brackets represents the enthalpy difference at the wall that drives the convective heat transfer, or heat flow through the boundary-layer gas, where h_w denotes the actual wall enthalpy, and h_{aw} represents the adiabatic wall enthalpy.

As a property of the gas, the Pr is proportional to the ratio of energy dissipated by friction to the energy transported by thermal conduction,

$$Pr = \frac{\text{momentum transport}}{\text{heat transport}} = \frac{\mu c_p}{k} \tag{3.18}$$

where k is the thermal conductivity of the gas.

For air at standard conditions, $Pr = 0.71$. For a nonreacting gas, Pr is a function of temperature only. If the gas is chemically reacting, then Pr also depends on the local chemical composition, which in turn depends on the local temperature and pressure for an equilibrium flow and on the history of the upstream conditions for a nonequilibrium flow.

The static enthalpy at the surface h_w is given by

$$h_w = c_{p_w} T_w \tag{3.19}$$

where c_{p_w} is the temperature-dependent specific heat of the wall material operating at wall temperature T_w. This term represents the selected design value of the wall enthalpy and must correspond to a temperature lower than the allowable working temperature of the material.

The adiabatic wall enthalpy is given by

$$h_{aw} = c_{p_0} T_e + r \frac{V_e^2}{2} \tag{3.20}$$

where r is a recovery factor to account for dissipation effects, and c_{p_0} is the temperature-dependent specific heat of the airflow at the temperature on the boundary layer T_e.

Also known as the recovery enthalpy, h_{aw} represents the enthalpy that the bounding- or zero-velocity streamline would attain if the wall were perfectly insulated. Equation (3.20) states that, if the fluid had no thermal conductivity, according to the first law, h_{aw} must be the same as the stagnation enthalpy of the adjacent flow, meaning that the net effect of viscosity would be to convert all the original freestream kinetic energy to internal or thermal energy. However, the real fluid has finite thermal conductivity, and thus some of the thermal energy is conducted away from the higher temperature region to the cooler fluid, thus lowering the adiabatic wall enthalpy in accordance with the recovery factor r, which must be less than 1. The value of r depends on the fluid properties and whether the boundary layer is laminar or turbulent, and thus we use the Prandtl number Pr to determine r.

The Prandtl number is a dimensionless quantity that correlates the viscosity of a fluid (μ) and specific heat (c_p) to its thermal conductivity (κ). It therefore assesses the relation between momentum transport and thermal transport capacity of a fluid.

The Prandtl number of a fluid gives the relative importance of the momentum boundary layer to the thermal boundary layer in the transfer of heat. The Prandtl number is defined as the ratio of momentum diffusivity (kinematic viscosity) to thermal diffusivity, and to determine the recovery factor we use

$$\text{Laminar flow } r = \sqrt{Pr} = \sqrt{\frac{\mu c_p}{\kappa}} \tag{3.21a}$$

$$\text{Turbulent flow } r = \sqrt[3]{Pr} = \sqrt[3]{\frac{\mu c_p}{\kappa}} \tag{3.21b}$$

The Prandtl number for air is given as a function of temperature by Hansen (1957). For equilibrium, nondisassociating air, $0.6 < Pr < 1.0$, and we should expect $0.7 < r < 0.99$.

Since heat transfer for hypersonic systems depends on the thermal conductivity κ of the gas, the following formula is often used to determine the thermal conductivity of air:

$$\kappa = 2.39 \times 10^{-7}\left(\frac{T^{3/2}}{T + 202}\right) \tag{3.22a}$$

where κ is in BTU/(s · ft · °R) and T in °R, or

$$\kappa = 1.99 \times 10^{-3}\left(\frac{T^{3/2}}{T + 112}\right) \tag{3.22b}$$

where κ is in J/(s · m · K) and T in K.

The formulas for thermal conductivity are valid in the range of static pressures encountered in atmospheric flight, for temperatures up to at least 3000 K (3600 °R).

3.5.2 Stagnation-Point Heat Flux

Consider the hypersonic flowfield near a blunted body with its detached bow shock, Figure 3.17. The stagnation point at the nose is located within the boundary layer behind the bow shock where x is the distance measured along the body surface, and the local surface radius of curvature at the stagnation point is the body's R_N. The boundary-layer thickness is finite at the stagnation point, and V_e is the velocity in the x direction at the outer edge of the boundary layer. In the stagnation region, V_e is very small and $h_e = h_0$.

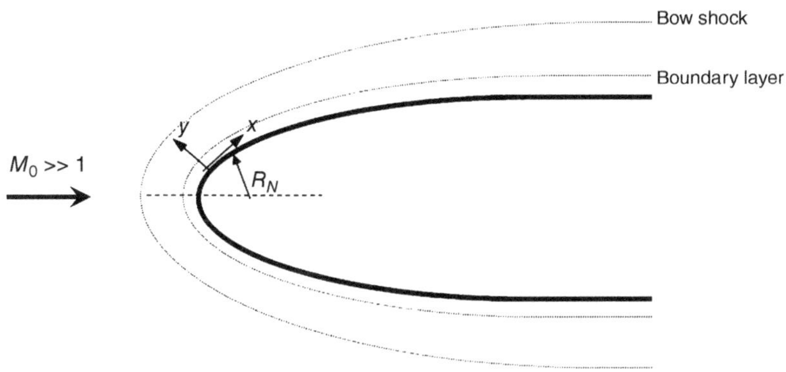

Figure 3.17 Stagnation region for a blunted body.

In the early stages of the space program, while developing an understanding of the heating rates to be experienced by space vehicles on reentry, it was initially assumed that the nose should be sharply pointed. However, it was later determined that a blunt shape would be most appropriate, as such bluntness forces a detached shock wave, and most of the heat moves away from the surface and into the flowfield, not the vehicle. For determining the stagnation-point heating, pioneering engineering theories were developed in the 1950s. In the suborbital velocity regime, the correlation formula of Fay and Riddell (1958) is considered the most reliable and is widely used. Several other engineering correlations for flight applications were also developed.

Fay and Riddell studied the convective heat transfer at the stagnation point for an equilibrium boundary layer, for the case of high-speed flight where the external flow is dissociated. They considered numerical solutions for spherical-shaped bodies and used Sutherland's equation for viscosity and equilibrium air. Fay and Riddell found that their individual heat transfer calculations could be correlated by this expression:

$$\dot{q}_w = \frac{0.763}{(Pr)^{0.6}}(\rho_w\mu_w)^{0.1}(\rho_e\mu_e)^{0.4}(h_{te}-h_w)\left[1+\left(Le^{0.52}-1\right)\frac{h_d}{h_{te}}\right]\left[\left(\frac{dV_e}{dx}\right)_t\right]^{0.5} \tag{3.23}$$

where ρ_e and μ_e are evaluated for the flow external to the boundary layer at stagnation conditions, h_{te} is the stagnation point total enthalpy, h_w is the enthalpy at the wall, and h_d is the average atomic dissociation energy times the atom mass fraction in the external flow. The Lewis number Le and Prandtl number Pr were treated as independent parameters.

Calculations for a frozen air assumption were performed, and it was concluded that "for a Lewis number not too far from unity, there is little difference in heat transfer for the frozen as opposed to the equilibrium boundary layer." It was also found that with $Pr = 0.71$, $Le = 1.4$, and Sutherland's viscosity equation, Eq. (3.23) gave good agreement with the experimental shock tube data for velocities between 1768 m/s (5800 ft/s) and 6949 m/s (22 800 ft/s) and at altitudes between 7620 m (25 000 ft) and 36 576 m (120 000 ft). Evaluation of the stagnation-point velocity gradient $(dV_e/dx)_t$ used to obtain the good agreement was not discussed, i.e. whether the gradient was computed assuming a real gas or a perfect gas.

Bertin (1994) derived an expression for the inviscid flow velocity gradient at the stagnation point from modified Newtonian theory:

$$\left(\frac{dV_e}{dx}\right)_t = \frac{1}{R_N}\sqrt{\frac{2(p_e - p_0)}{\rho_e}} \tag{3.24}$$

Fay and Riddell equation was a significant advance in determining the heat transfer near the nose of a vehicle, but it still requires many quantities that are not readily available to the designer.

For simple engineering calculations, the empirical correlation of Detra et al. (1957), which is based on experimental shock tube data, is more amenable to calculation:

$$\dot{q}_w = \frac{11\,030}{(R_N)^{0.5}}\left(\frac{\rho_0}{\rho_{SL}}\right)^{0.5}\left(\frac{V_0}{V_{co}}\right)^{3.15} \tag{3.25a}$$

where R_N is the nose radius in m, ρ_{SL} is the density at sea level in kg.m^3, and V_{co} is the circular orbit velocity in m/s (7950 m/s). The heat transfer is in W/cm^2. Bertin modified Eq. (3.25a) to account for finite wall temperature as

$$\dot{q}_w = \frac{11\,030}{(R_N)^{0.5}}\left(\frac{\rho_0}{\rho_{SL}}\right)^{0.5}\left(\frac{V_0}{V_{co}}\right)^{3.15}\left(1 - \frac{h_w}{h_{te}}\right) \tag{3.25b}$$

where h_w is the wall enthalpy, and h_{te} is the total enthalpy at the stagnation point.

All the theoretical and semi-empirical formulations and experimental data agreed that the stagnation-point heat transfer for blunt bodies in continuum flow is inversely proportional to square root of nose radius R_N at the leading edge, i.e.

$$\dot{q}_w \propto \frac{1}{\sqrt{R_N}}$$

Thus, the blunter the body, the lower the convective heat transfer rate of the stagnation point. That is why, Earth entry vehicles, such as the Mercury and Apollo capsules, were designed with very blunt geometries to overcome the aerodynamic heating of reentry. The winged SS Orbiter entered the atmosphere at a very high angle of attack so that it became in effect a blunt body.

For the slender hypersonic vehicles propelled by air breathing, the most severe heating will occur during ascent to orbit and during sustained cruise. The entire vehicle is subjected to aerodynamic heating; however, the leading edges (nose region, tail edges, inlet/forebody, and scramjet cowl leading edge) will experience intense, highly localized stagnation-point heating. For example, the NASA Hyper-X forebody had a 0.762 mm leading-edge radius (Mach 7 vehicle), creating exceptional thermal challenges that required innovative solutions, even for the short duration (seconds) demonstration test flight.

In the 1990s, shock wave interference heating was vigorously investigated to support the design of the TPS and load carrying structure of hypersonic air-breathing vehicles such as the NASP, which are designed to have sharp leading edges. In addition to the extreme heat transfer rate gradient that occurs over the narrow impingement regions, thermal stresses are of concern, as these limit the useful life of the structural component. In 1987, Wieting and Holden initiated a set of experiments for a cylindrical leading edge with its axis parallel to the plane of the impinging shock. These experiments provided data for the design of cowl leading edges for rectangular hypersonic engine inlets, providing the first experimental pressure and heat transfer rates on cylindrical leading edge typical of a hypersonic inlet cowl.

In 1992, Wieting studied the effect of two impinging oblique shocks on the leading edge, using a 2-D forebody, 7.5° wedge followed by 5° turn, immersed in Mach 8 flowfield. The test conditions were primarily for a Type IV pattern (see Figure 3.16), since this type of shock interaction represents the most severe pressure and heat transfer rate condition. For the condition where the two impinging oblique shock waves coalesced just prior to intersecting the cylindrical bow shock wave produced the highest local heating rate (Wieting 1992). Wieting measured circumferential heat transfer amplification (local heat transfer referenced to undisturbed stagnation-point heat transfer) and found that the peak transfer rate was 38 times the undisturbed flow stagnation-point level ($\dot{q}/\dot{q}_{st} \sim 38$), and it occurred when the two oblique shock waves coalesced before intersecting the body's bow shock. These high heat transfer rates represent a huge design problem as the leading edge may not survive such high heat load.

Van Wie et al. (1990) estimated the heating levels encountered in a National Aero-Space Plane (NASP) inlet for a typical ascent trajectory. The heating rates on the vehicle nose (50 mm radius) and cowl lip (2.5 mm radius) were calculated with the Fay and Riddell engineering analysis. Their results showed the highest heat transfer rate at the cowl lip when it was subjected to shock–shock interactions. Even for the Mach 10 condition, the heating rate at the cowl was estimated to be as high as 8.0×10^7 W/m^2 and it rose to 3.0×10^8 W/m^2 at Mach 15. As a point of reference, it would take a fraction of a second to melt a slab of copper if subjected to those heating rates. Hence, for a vehicle to survive such hypersonic flow environment, it will require unique thermal management designs for the leading edges of future hypersonic vehicles on sustained flight.

Another value for the NASP during ascent was reported in 1991 (Bertin 1994), estimating an even greater shock-on-lip heat transfer rate of 6.24×10^8 W/m^2 (55 000 BTU/ft$^2 \cdot$ s). This heating rate is orders of magnitude greater than the heating rates experienced by the SS Orbiter on entry. The severity of the shock-on-lip heat transfer is due to a combination of factors, including the relatively high density of the air, the small radius of the cowl lip, and the existence of a Type IV shock–shock interaction. Even for the scramjet engine combustor, the heat transfer rate of 1.70×10^7 W/m^2 (1500 BTU/ft$^2 \cdot$ s) exceeds the Shuttle Orbiter heat transfer rates by an order of magnitude (Bertin 1994).

The following is a first-order approximation of the heat flux $\dot{q}_{st}$ for a cold surface at the stagnation point that we can use to estimate heat transfer at a leading edge (expressed in W/cm^2):

$$\dot{q}_{st} = 1.83 \times 10^{-8} \left(\frac{\rho_0}{R_{LE}}\right)^{1/2} V_0^3 \tag{3.26}$$

where ρ_0 (kg/m^3) and V_0 (m/s) are the density and velocity of the freestream, respectively, and R_{LE} (m) is the radius of the leading edge. This expression indicates that the heat flux is a very strong function of the vehicle speed. It also increases with reduction in altitude as the air density increases.

Consider a vehicle with a nose radius equal to 50 mm, flying a trajectory where the dynamic pressure is constant and equal to 95.76 kN/m^2. We can use Eq. (3.26) to estimate the order of magnitude heat transfer rates at the nose, and also help us compare the heat transfer for two cowl lip radii: 2.5 and 3.0 mm. At an altitude of 30 km, the air static conditions are $p_0 = 1196.648$ Pa; $\rho_0 = 0.018412\ \text{kg/m}^3$; $a_0 = 301.71$ m/s. So, using Eq. 2.10 for the Mach number, the flight velocity is

$$V_0 = a_0 \cdot M_0 = a_0 \left(\frac{2q_0}{\gamma p_0}\right)^{1/2} = (301.71)\left[369.96\left(\frac{1}{1196.648}\right)^{1/2}\right] = 3226.72\ \text{m/s}$$

The heat transfer rate at the cowl leading edge with radius $R_N = 25$ mm is

$$\dot{q}_{st} = 1.83 \times 10^{-8}\left(\frac{\rho_0}{R_{LE}}\right)^{1/2} V_0^3 = 1.83 \times 10^{-8}\left(\frac{0.018\,412\ \text{kg/m}^3}{0.0025\ \text{m}}\right)^{1/2} (3228.3\ \text{m/s})^3$$
$$= 1671\ \text{W/cm}^2$$

Figure 3.18 shows the heat transfer rate as a function of flight Mach number. As shown, the highest heat rate is experienced by the cowl with a lip radius of 25 mm. The obtained results are of the same order of magnitude as those obtained by more sophisticated analysis.

The external flow properties are much more important than the wall values in determining the heat transfer rate. The uncertainty in the heat transfer is about 40% of the uncertainty in the external viscosity. The physical reason for the importance of the external viscosity is that the growth of the boundary layer and, hence, the heat transfer to the wall depend mostly upon the external properties.

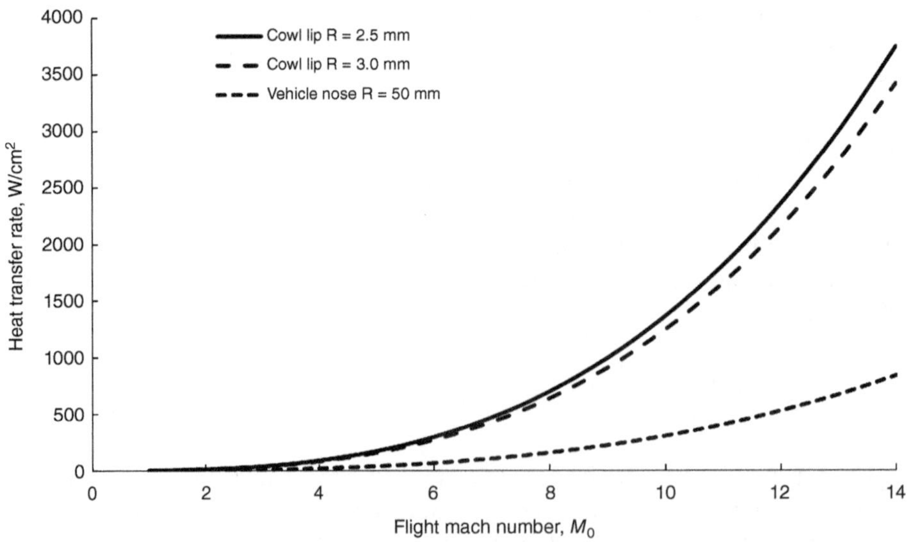

Figure 3.18 Heat transfer rate as a function of flight Mach number according to Eq. (3.26).

3.5.3 Effect of Dynamic Pressure on Aerodynamic Heating

The dynamic pressure $\left(q_0 = \rho_0 V_0^2/2\right)$ and freestream velocity V_0 affect the convective heat transfer on the hypersonic vehicle. Since heat flux to a surface is proportional to the free-stream energy flux $(\rho_0 V_0^3/2)$, this can be very crudely estimated in the following form (Heiser and Pratt 1994):

$$\text{Laminar flow } \dot{q}_w \propto \frac{\rho_0 V_0^2 V_0}{(\rho_0 V_0)^{1/2}} \propto q_0^{1/2} V_0^{3/2} \tag{3.27a}$$

$$\text{Turbulent flow } \dot{q}_w \propto \frac{\rho_0 V_0^2 V_0}{(\rho_0 V_0)^{1/5}} \propto q_0^{4/5} V_0^{6/5} \tag{3.27b}$$

These expressions clearly show that convective heat transfer increases with both dynamic pressure and flight velocity. Hence, a trajectory of lower q_0 and higher altitude are recommended in order to reduce convective heat transfer on the hypersonic vehicle.

3.6 NASA X-43A Leading-Edge Flight Hardware

Aerodynamic heating and skin-friction drag have a major impact on the design of slender hypersonic vehicles such as the NASA Hyper-X experimental aircraft. Figure 3.7 gives an indication of the sharp nose. The leading-edge hardware for the X-43A Mach 10 flight vehicle consisted of eleven pieces; a flat nose, two forward chines, two aft chines, two horizontal tailpieces, two upper vertical tailpieces, and two lower vertical tail pieces. The vehicle windward forebody had a thin leading edge with a radius $R_N = 0.030$ in (0.762 mm) for the Mach 7 vehicle and $R_N = 0.050$ in (1.27 mm) for Mach 10, and 3 flat ramps are designed to provide a series of discrete, nonisentropic flow compressions for the scramjet. Figure 3.19 shows a photograph of the X-43A nose materials and construction.

Thermal analysis for the Mach 10 vehicle predicted temperatures that would approach 2204 °C (4000 °F) at the nose tip. This temperature greatly exceeds the temperatures of

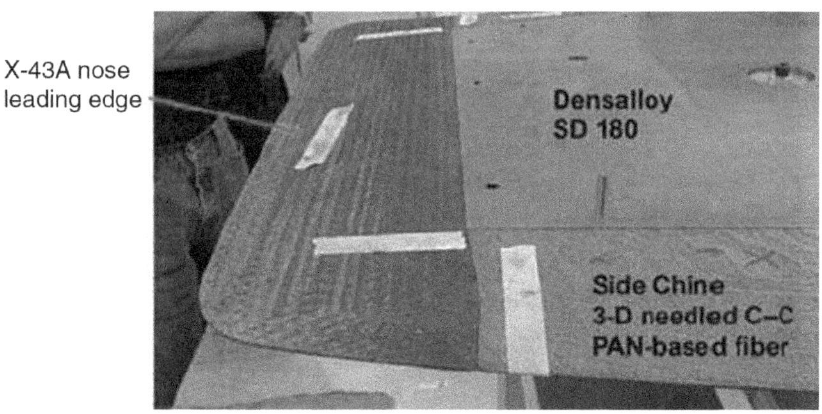

Figure 3.19 NASA X-43A Mach 7 C/C nose. Source: From Ohlhorst et al. (2005), NASA, Public domain.

silicon carbide (SiC)-based oxidation coatings even for a short duration, single flight. Major concerns for the nose were tip temperature and high thermal gradients leading to high compressive stresses (Rivers and Glass 2006). Aerothermal heating on sharp leading edges such as this produces high temperatures and high thermal gradients. The high thermal gradient at the tip of the nose leading edge leads to huge thermal stresses in the spanwise direction, parallel to the tip of the leading edge.

The design of the X-43A flight demonstrator vehicles incorporated hot structures. Hot structures refer to any part of the airframe that can operate efficiently up to 3300 K without active cooling or insulation. In an effort to reduce the nose tip temperature and minimize thermal gradients, the nose leading edge was constructed with high thermal conductivity carbon fibers woven in an unbalanced weave to give more fibers perpendicular to the leading edge. A K321 fiber woven was baselined for the nose leading edge of the Mach 7 vehicle, and with this material thermal analysis predicted that the maximum temperature would only reach 3000 °F (1649 °C), so a SiC oxidation coating was acceptable. Only seven of the Mach 7 flight vehicles leading-edge pieces were fabricated of carbon/carbon (C/C). The four vertical tailpieces were made with Haynes alloy since analysis predicted these would not be subjected to high enough temperatures to require C/C (Rivers and Glass 2006).

Nose temperature profiles for the Mach 7 flight compared to analysis data were published (Amundsen et al. 2004). The nose maximum temperature was about 1600 °F (1144 K), measured 0.5 inch aft of the leading edge (sensor embedded within the carbon fibers). For the Mach 10 flight, Figure 3.20 indicates the maximum nose temperature rose to about 2000 °F (1093 °C or 1366 K) within two minutes (Marshall et al. 2005) after launch.

Albertson and Venkat (2005) carried out an experimental study to qualitatively determine the effectiveness of stagnation-region gas injection in protecting a scramjet cowl leading edge from the intense heating produced by Type III and Type IV shock interactions. Their

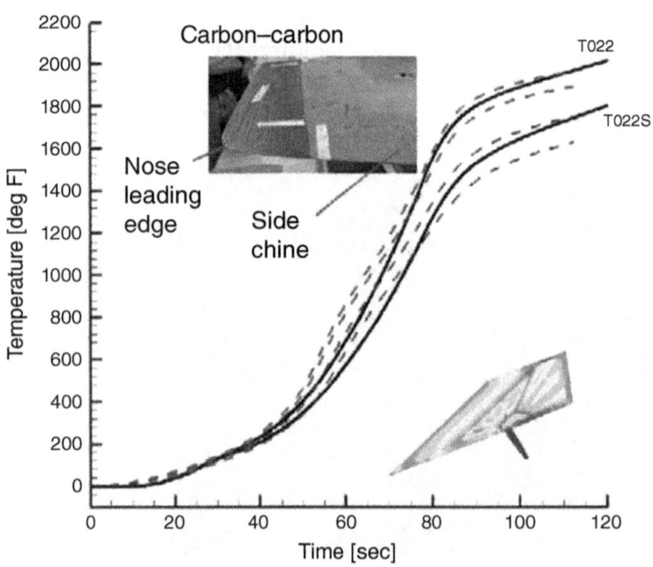

Figure 3.20 X-43A nose leading edge heating for Mach 9.68 flight. Source: Voland et al. (2005)/ With permission of Elsevier.

model consisted of a two-dimensional leading edge, representative of that of a scramjet cowl, testing it at a nominal freestream Mach number of 6. They injected gaseous nitrogen through the leading-edge nozzles at various mass flux ratios and with the model pitched at angles of 0° and −20° relative to the freestream flow. Schlieren images were obtained of the shock interaction patterns. Their results indicate that large shock displacements can be achieved, and both the Type III and IV interactions can be altered such that the interaction does not impinge on the surface of the leading edge.

For hypersonic vehicles, some degree of nose bluntness is needed to reduce the convective heat transfer and to alleviate asymmetric vortex effects associated with the subsonic portion of the flight. As a result, the bow shock wave near the nose is curved. This means that the entropy increase (or, equivalently, the total pressure decrease) is proportional to the local inclination of the shock wave and to the freestream Mach number.

For air-breathing hypersonic vehicles, the inlet forebody and cowl lip must also be blunted to decrease surface temperature and obtain acceptable heating levels at hypersonic speeds. However, bluntness can degrade the performance of the scramjet engine inlet. Even small amount of bluntness in leading edges cause curved bow shocks that generate large entropy layers. The entropy layers modify the development of the boundary layer, including transition, and can create significant changes in the inviscid flowfield, affecting shock positioning, and engine inlet air capture. In addition, a blunt leading edge can alter the shock structure and pressure distribution in the scramjet inlet and decreases the pressure recovery.

3.7 Inlet Blunt Leading-Edge Effects and Entropy Layer Swallowing

After crossing the curved shock wave at some angle, the airflow particles maintain the entropy and stagnation pressure associated with the shock inclination as they move through the inviscid shock layer. At the same time, downstream from the stagnation point, the boundary layer grows into the rotational, inviscid flow. The flow that enters the boundary layer is initially the hot, high entropy gas that was stagnated after passing through the almost normal portion of the shock wave upstream from the nose. Therefore, as the airflow moves over the blunted body, all of the high-entropy gas is entrained or swallowed by the thick boundary layer at some point downstream. The entropy is constant along a streamline for a steady, inviscid adiabatic flow, and thus entropy will vary continuously through the shock layer. Having crossed the curved, bow shock wave at a particular location, a fluid particle retains the entropy and the stagnation pressure associated with the shock inclination at the point where it crossed the shock layer (Bertin 1994). Figure 3.21 illustrates the effect of entropy layer (streamline) being entrained or swallowed by the boundary layer over a blunted body.

And since additional gas subsequently entrained by the boundary layer is cooler, possessing lower entropy (because this flow passed through weaker portions of the shock wave), entropy gradients develop in the inviscid shock layer flow. These entropy gradients can remain for a considerable length downstream. These physical processes affect the condition of the boundary layer and thus heat transfer.

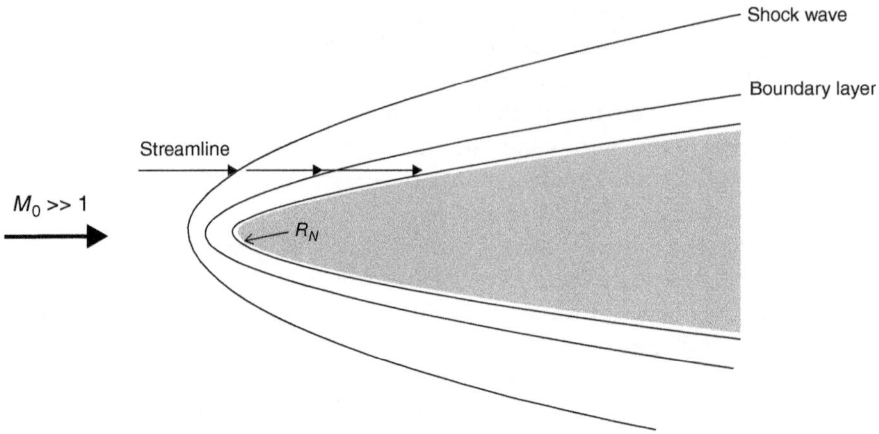

Figure 3.21 Blunted body shock geometry showing entropy layer swallowing.

Therefore, the effect of body bluntness on BLT can be examined in terms of the entropy swallowing distance. These effects on the local boundary-layer edge conditions have been studied on flows over spherically blunted and sharp cones. For blunt cones, Bertin (1994) presented a correlation developed by F. G. Blottner, showing the effect of nose radii on the swallowing distance:

$$\left(\frac{x}{R_N}\right) \propto (Re_{0,R_N})^{1/3}$$

They found that thin cones have a longer swallowing distance, i.e. the swallowing length decreases rapidly as the conical half-angle is increased.

The effect of bluntness and entropy layer on transition for the inlet/forebody of air-breathing propulsion-integrated systems require further investigation, since any entropy increase or entropy gradients will lead to pressure loss, heat gradients, and other inefficiencies that will be detrimental to overall propulsion performance.

3.8 Inlet Shock-On-Lip Condition or Inlet Speeding

The beginning of the scramjet inlet is the nose of the hypersonic vehicle. As illustrated in Figure 3.22, the hypersonic vehicle forebody is a ramp-like structure that generates oblique shocks which decelerate, turn, and vertically compress the approaching freestream air flow. The lower surface of the forebody is the compression system for the scramjet inlet, and as such its design must be optimized to support the performance of the overall propulsion system. The flow moving parallel to the forebody lower surface is ingested into the internal portion of the scramjet at the cowl leading edge plane. This surface is designed to provide precompressed airflow to the engine relatively free of flow gradients. The precompression is achieved from the shock waves generated by the forebody surface. We require that this precompression process is done with minimum drag, heat transfer, and flow stagnation pressure losses.

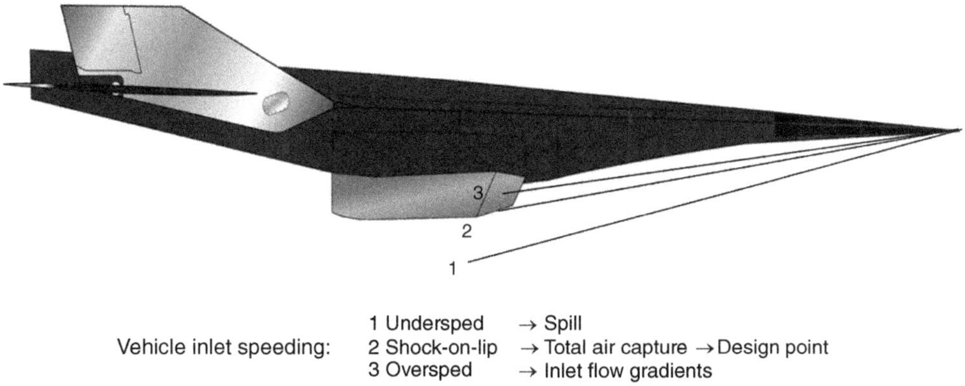

Figure 3.22 Inlet speeding.

The main determination of the inlet design point is the speed at which the bow shock sweeps back to the cowl lip, a condition referred to as *shock-on-lip*. In principle, the maximum operating Mach number is the shock-on-lip Mach number. Czysz and Vandekerckhove (2000) proposed the following expression for the cowl Mach number to define shock-on-lip cowl flow:

$$M_{cowl} = 2.37 \ln (M_0) - 0.43 \tag{3.28}$$

At $M_0 > 2.7$, the Mach number at the inlet cowl lip is two or greater, and there is a shock attached at the cowl lip entering the inlet. Hence, at $M_{cowl} \geq 2$, the cowl shock is expected to be attached to the cowl lip.

To minimize stagnation pressure losses during shock wave precompression in the inlet, the forebody is constructed with multiple wedges. It is more efficient to turn flow to same net angle with several wedges, thus minimizing strength of any individual shock. Hence, the amount of precompression the forebody provides to the airflow entering the scramjet is a function of the wedge angle(s) relative to the flight direction. Therefore, the vehicle angle of attack plays an important role in the inlet efficiency and overall propulsion system performance.

The fixed geometry scramjet inlet is most efficient when it captures all compressed airflow, which occurs at the shock-on-lip condition (the design point for the scramjet engine), as illustrated in Figure 3.22. However, the shock angle for a constant forebody wedge angle increases with decreasing flight Mach number.

Consider for an example an 18.8°-wedge forebody designed to operate with the shock attached to the cowl lip at Mach 10. For operation at a lower Mach number, given the same geometry and vehicle angle of attack, the shock wave would pass outside the cowl lip; this condition is referred to as an *undersped* inlet (represented by shock 1 in Figure 3.22). In this configuration, the inlet is less efficient because the precompressed airflow would be turned past the inlet cowl, spilling out; thus, the spilled airflow would be wasted. On the other hand, if the Mach number is greater than the design value, the shock would end up inside the inlet duct, resulting in an operational condition known as *oversped* inlet (represented by shock 3). In this condition, the inlet is also inefficient since the airflow would enter the engine with incomplete compression, yielding uneven pressure and temperature gradients that would reduce the performance of the combustor.

The airflow must enter the scramjet inlet with optimum compression and minimum flow gradients, to reduce unacceptable loading and heating patterns. Uneven air mass flow gradients can lead to uneven fuel–air distributions in the combustor.

3.9 Shock–Boundary Layer Interactions in the Propulsion Flowpath

Another flow phenomenon that affects the performance of the hypersonic air-breathing propulsion system is the shock–boundary layer interactions that develop within the inlet isolator, downstream from the engine cowl. The isolator is a constant area diffuser duct that contains a precombustion shock train, a part of the propulsion system added to ensure that the optimum operation of the engine in a wide flight Mach number range, say, $3 < M_0 < 12$. If this isolator diffuser is not long enough, the shock train can migrate upstream far enough to unstart the inlet. Inlet unstart is an unwelcome and undesirable instability phenomenon that can happen in hypersonic mixed compression inlets, one of the most difficult operational issues to control. Inlet unstart causes a large drop of thrust and specific impulse, and under extreme conditions, it may cause catastrophic damage or the inlet may fail to restart.

3.9.1 Scramjet Operation at High Hypersonic Speed

At flight speeds above Mach 7, the scramjet engine flowpath is characterized by predominantly supersonic turbulent flow, as the engine operates with predominantly supersonic combustion. The critical ratio of heating rate caused by fuel–air combustion in the burner to the entering total enthalpy rate is very small. Moreover, since the airflow through the flowpath has a high kinetic energy, the pressure rise generated by heat addition in the combustor is not high enough to cause adverse effects on the boundary layer, which remains attached and supersonic throughout. It is true that the combustor may have small recirculation zones caused by fuel injectors, but these are not strong enough and the engine operates as a pure scramjet.

In such high-speed flow combustion conditions, there is little interaction with the inlet. The inlet exit conditions are then essentially decoupled from the flow field within the combustor. At the high hypersonic flight speeds, the degree of diffusion throughout the inlet system is not sufficient to cause the flow Mach number to approach unity. Thus, flow choking and inlet unstart are not concerning issues for the scramjet operating at high hypersonic speeds.

3.9.2 Scramjet Operation at Low Hypersonic Speed

As the flight Mach number is lowered below Mach 7, significant effects drastically change the flow character and scramjet performance and operability. Specifically, a scramjet that must operate in the flight regime that includes both low and high hypersonic regime, say Mach 4–12, is characterized by dual-mode combustion. Figure 3.23 shows a schematic of a scramjet-powered vehicle operating in this manner.

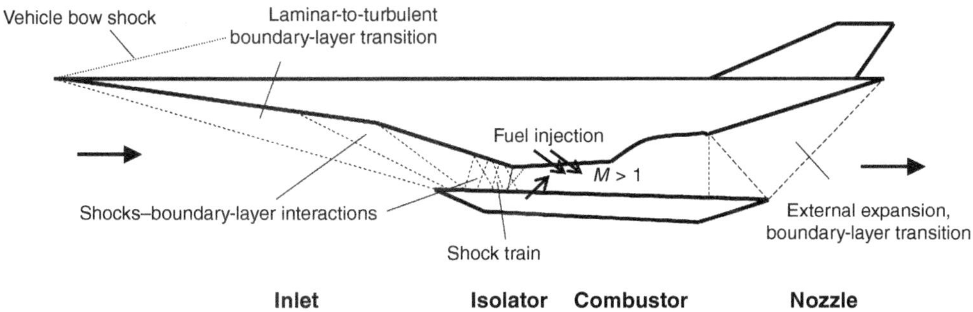

Figure 3.23 Dual-mode scramjet engine.

Dual-mode combustion can produce large pressure levels in the combustor. This occurs when the ratio of heating rate (due to fuel–air combustion) to the entering total enthalpy rate becomes large. The pressure rise due to heat release is coupled with the Mach number reduction in the combustor, and if it reaches Mach 1, it will result in thermal choking. If this happens, the flow can back pressure enough that significant interaction can occur with the inlet. This interaction develops as an oblique shock train upstream of the combustor, where large recirculation regions adjacent to the walls will develop.

To contain the shock train and stop it from disrupting the operation of the inlet, a short length duct is added to the scramjet flowpath between the inlet and the combustor. This diffuser duct is called isolator. The shock system that develops within the isolator will serve to stabilize the flowfield, allowing started engine operation. At low speeds, therefore, the combination of diffusion in the isolator and heat release in the combustor decelerates the core flow to subsonic conditions. Thus, the core flow must then re-accelerate through Mach 1 via a thermal throat since the dual-mode scramjet flowpath does not have the physical throat that a ramjet engine does.

3.9.3 Inlet Isolator Shock Train

The isolator is added between the inlet proper and the combustor as a buffer, to isolate the inlet from the disturbances caused by the high-pressure combustion process. The isolator ensures that the inlet of a dual-mode scramjet propulsion system operates without adverse interactions generated by the combustion process.

The shock train depicted in Figure 3.24 begins with an oblique shock that separates the boundary layer and the shock is then reflected on the opposite wall as an expansion wave. In

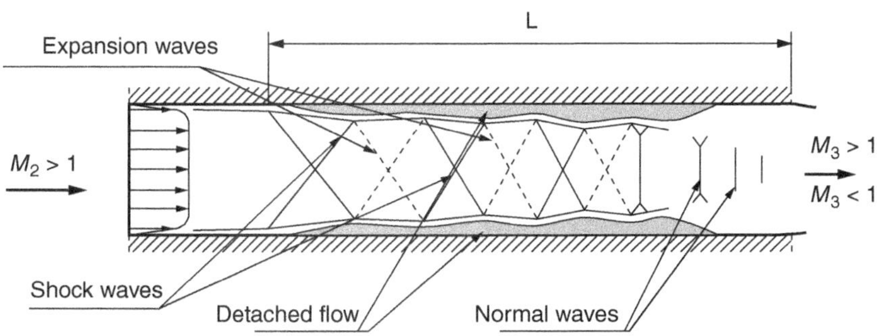

Figure 3.24 Inlet isolator with oblique shock train.

this manner, the oblique shocks continue propagating downstream, alternating between compression and expansion waves, causing a gradual ziz–zag increase in static pressure. The boundary layer slims and thickens in response to the expansion and compression shock waves, and it remains separated.

The shock train that develops inside the isolator (normal shocks or oblique shocks) provides a mechanism for supersonic flow to adjust to a specific backpressure. The type of shock train that develops in the isolator determines whether the flow is decelerated to subsonic velocity or remains supersonic as it enters the combustor.

The maximum air compression is achievable by a normal shock train (neglecting viscous effects), provided that the isolator duct is sufficiently long. However, if the back pressure exceeds this limit, the shock train is disgorged and the scramjet unstarts. An un-started inlet captures less airflow with lower efficiency and higher aerodynamic and thermal loads compared to started inlets. The self-starting ability is a very important parameter of the hypersonic inlet, so when removing factors resulting in unstart, the inlet can be restarted.

3.10 Inlet Unstart

Inlet unstart can be caused by a flow-retarding disturbance such as a large internal area contraction ratio (design issue), or by shock–boundary layer interactions. Unstart may be the result of the operating conditions, such as flying at lower Mach number than designed for, and high combustor backpressure. Unstart conditions are dependent on the mass flow rate balance between inlet entrance and combustor exit.

If a flow-retarding disturbance develops, the internal shock system can move abruptly upstream and reposition itself outside the inlet duct, causing an abrupt and severe drop in thrust due to lower mass flow and lower recovery. The disturbance could be small, such as caused by a gust of wind, or by a rapidly changing the angle of attack. However, if the disturbance is relatively large, the transient response could be very severe, especially at high Mach numbers.

To prevent such undesirable behavior, inlets require some form of stability control system. In low-speed engines (turbojets, pure ramjets), inlet stability can be aided by self-actuating bleed valves located in the inlet nacelle. The valves open in response transient excursion of the terminal shock from its required position. When the inlet of a ramjet or turbojet unstart, a normal shock is expelled with spilled air flow. And since the thickness of the boundary layer at the low-speed inlet cowl is small, bleeding helps stabilize the flowfield.

However, this approach is not practical for hypersonic inlets. Remember, the boundary layer of a hypersonic inlet is rather thick. And if the inlet unstarts, the expelled shock system can be strong enough to separate the boundary layer and can expel a substantial part of the flow. Moreover, at high hypersonic speeds, the high gas enthalpy causes additional difficulties. The bleed air taken from the scramjet flow path is at both high temperature and low pressure. The bleed ductwork becomes hot and large, and bleed air increases vehicle drag. In recent years, a NASA GRC team investigated a novel concept using liquefying "bleed" flow. Using cooling from a cryogenic fuel and a bleed air heat exchanger, the volume of the cooled or liquefied bleed air is reduced dramatically thereby reducing bleed drag, improving propulsion integration, and enabling high-speed bleed to improve propulsion efficiency (Saunders et al. 2014).

3.11 Closing Remarks

Hypersonic flows are highly complicated, strongly coupled, and contain complex turbulence mechanisms, shock and boundary layer interactions, and complicated physics. These flows present us with tremendous challenges for analysis, numerical simulation, or experimental research. Computational aerothermodynamics has become a powerful tool for improving our understanding of physical phenomena in hypersonic air-breathing propulsion. For example, to solve the shock–shock interaction flowfield a most accurate prediction of the heat transfer within the interaction region is performed with a sophisticated codes that include turbulence and shock wave models to capture the complex flowfield developed, especially at the thin leading edge of the scramjet engine cowl.

The boundary layer on a hypersonic vehicle can become so thick that it essentially merges with the shock waves. When this happens, the shock layer must be treated as fully viscous, and the conventional boundary-layer analysis must be completely abandoned. The extreme viscous dissipation that occurs within hypersonic boundary layers can create very high temperatures – high enough to excite vibrational energy internally within molecules and to cause dissociation and even ionization within the gas.

There is a strong interaction between aerothermodynamics and other disciplines. Prediction of boundary-layer transition is an enabling technology in air-breathing hypersonic propulsion development. Hypersonic BLT is a critically important discipline with potential impact on all future designs for high-speed applications. The ability to accurately predict when BLT will occur on a flight vehicle is important when considering the sizing of a TPS.

BLT plays a primary role in the vehicle's TPS selection, trajectory shaping, and upon airframe configuration (Walton 1992). To date, however, we do not have a theory for the accurate prediction of the Reynolds number for transition. Thus, we must rely on experimental data and use the power of CFD models to help us determine the value of Re_T for a given flight situation (Berry et al. 2011).

Although transition remains a challenging topic for hypersonic flight, it is now reasonable to generate transition correlations for arbitrarily complex vehicle geometries using boundary-layer edge conditions extracted from high-fidelity viscous CFD solutions.

We require extensive engineering estimates of these affects to design the scramjet. It should be clear that modeling the complex three-dimensional flowfield of this integrated system requires a "tip-to-tail analysis of the flowfield over the vehicle in hypersonic flight, which presents tremendous challenges as it must include viscous–inviscid interactions, and the effects of the physical processes taking place within the internal propulsion flowpath.

It should be clear that aerothermodynamics of hypersonic air-breathing propulsion must be analyzed with validated computational tools that account for nonequilibrium chemistry, streamline spreading, entropy swallowing, turbulence, and chemical kinetics in order to arrive to valid models that can serve design purposes.

Most of the research in air-breathing hypersonics has been concerned with the ascent and cruise phase. However, reusable SSTO and TSTO vehicle concepts that integrate air-breathing propulsion must also characterize the aerothermodynamics of atmospheric entry and descent trajectories and develop the appropriate TPS to survive those crucial phases of the mission.

We envision that SSTO vehicles will enter the atmosphere at high angles of attack. This creates strong shock waves to dissipate heat and increase drag. Although there are sources of entry flight data from conventional entry reusable vehicles (RVs), these are not totally useful as all RV data are from conical ablating vehicles that introduce contamination into a flowfield lacking in crossflow and adverse pressure gradients. And for those, much uncertainty exists in hypersonic boundary-layer behavior. However, the slender design of air-breathing reusable vehicle may require a completely new approach to thermal management on its return flight. Hence, the design of a launch vehicle powered by air-breathing propulsion presents significant challenges to the designers.

Questions

1. BLT is one of the most complex physical processes in hypersonic flow aerothermodynamics. How does BLT affect the performance of a vehicle/integrated scramjet propulsion system?

2. How can one predict where the boundary layer will transition in the forebody of a full-size Mach 10 cruise vehicle?

3. Knowing that shock interactions can enormously increase local convective heat transfer rates, which part(s) of a reusable hypersonic air-breathing engine must incorporate an efficient TPS? Why?

4. Describe shock-on-lip heating and give viable engineering methods to alleviate it.

5. The forebody of a vehicle/integrated scramjet must provide a naturally turbulent boundary layer for the engine. How would you promote transition from laminar to turbulent boundary layer in a small-scale experimental scramjet?

6. How can you determine the thickness of the boundary layer developing over a forebody of a hypersonic vehicle wind tunnel model?

7. Review the literature and assess whether substantial progress has been made towards a new transition prediction method that one can apply with confidence to future reusable hypersonic vehicles.

References

Albertson, C.W. and Venkat, V.S. (2005). Shock interaction control for scramjet cowl leading edges. *AIAA Paper 2005-3289. AIAA/CIRA 13th International Space Planes and Hypersonics Systems and Technologies Conference*, Capua, Italy (16–20 May 2005).

Amundsen, R.M., Leonard, C.P., and Bruce, W.E. (2004). Hyper-X Hot Structures Comparison of Thermal Analysis and Flight Data. Fifteenth Annual Thermal and Fluids Analysis Workshop (TFAWS), Pasadena, CA (30 August–3 September 2004).

Anderson, J.D. Jr. (2006). *Hypersonic and High Temperature Gas Dynamics*. AIAA Education Series.

Beckwith, I.F. (1975). Development of a high reynolds number quiet tunnel for transition research. *AIAA Journal* 13(3): 300–306.

Berry, S.A., Horvath, T.J., Hollis, B.R. et al. (1999). X-33 Hypersonic boundary layer transition. *Paper AIAA 99-3560. 33rd AIAA Thermophysics Conference*, Norfolk, VA (28 June to 1 July 1999).

Berry, S.A., Auslender, A.H., Dilley, A.D., and Calleja, J.F. (2000). Hypersonic boundary layer trip development for hyper-X. *Paper AIAA 2000-4012. 18th AIAA Applied Aerodynamics Conference*, Denver, CO (14–17 August 2000).

Berry, S., Daryabeigi, K., and Wurster, K. (2008). Boundary layer transition on the X-43A. *Paper AIAA-2008-3736. 38th Fluid Dynamics Conference and Exhibit*, Seattle, WA (23–26 June 2008).

Berry, S., Daryabeigi, K., Wurster, K., and Bittner, R. (2010). Boundary layer transition on X-43A. *Journal of Spacecraft and Rockets* 47 (6): 922–934.

Berry, S.A., Kimmel, R., and Reshotko, E. (2011). Recommendations for hypersonic boundary layer transition flight testing. *AIAA Paper 2011-3415. 41st AIAA Fluid Dynamics Conference*, Honolulu, HI (27–30 June 2011).

Bertin, J.J. (1994). *Hypersonic Aerothermodynamics*, AIAA Education Series. American Institute of Aernautics and Astronautics.

Borg, M. (2007). Instability and transition on the X-51A. Doctoral Dissertation. Purdue University, School of Aeronautics and Astronautics. Approved for public release by the AFOSR X-51A program.

Bouslog, S.A., Bertin, J.J., Berry, S.A., and Caram, J.M. (1997). Isolated Roughness Induced Boundary-Layer Transition: Shuttle Orbiter Ground Tests and Flight Experience. Paper AIAA 97-0271. *35th Aerospace Sciences Meeting and Exhibit*, Reno, NV (6–10 January 1997).

Bowcutt, K.G. and Anderson, J.D. (1987). Viscous Optimized Hypersonic Waveriders. AIAA Paper 87-0272. AIAA 24th Aerospace Sciences Meeting, Reno, NV (12–15 January 1987).

Cockrell, C.E., Jr. (2003) Aerosciences, aero-propulsion and flight mechanics technology development for NASA's next generation launch technology program. *AIAA Paper 2003-6948. AIAA 12th International Space Planes and Hypersonic Systems and Technologies Conference*, Norfolk, VA (19 December 2003).

Cockrell, C.E., Jr., Auslender, A.H., White, J.A., and Dilley, A.D. (2002) Aeroheating predictions for the X-43 hyper-X cowl-closed configuration at Mach 7 and 10. *AIAA Paper 2002-0218. 40th AIAA Aerospace Sciences Conference & Exhibit*, Reno, Nevada (14–17 January 2002).

Czysz, P. and Vandernkerckhove, J. (2000). Transatmospheric Launcher Sizing. Chapter 16 in *Scramjet Propulsion* (eds. E.T. Curran and S.N.B. Murthy). AIAA Progress in Astronautics and Aeronautics, Vol. 189.

Detra, R.W., Kemp, N.H., and Riddell, F.R. (1957). Addendum to heat transfer to satellite vehicles reentering the atmosphere. *Jet Propulsion* 27 (12): 1256–1257.

Edney, B. (1968). Anomalous Heat Transfer and Pressure Distributions on Blunt Bodies at Hypersonic Speeds in the Presence of an Impinging Shock. *FFA Report 115*, The Aeronautical Research Institute of Sweden, February, 1968.

Elias, T.I. and Eisworth, E.A. (1990), Stability studies of planar transition in supersonic flows. *AIAA Paper 90-5233. 2nd International Aerospace Planes Conference*, Orlando, FL (29–31 October 1990).

Fay, J.A. and Riddell, F.R. (1958). Theory of stagnation point heat transfer in dissociated air. *JAS* 25 (2): 73–85.

Fletcher, D.G. (2004), Fundamentals of hypersonic flow – aerothermodynamics. *Paper presented at the RTO AVT Lecture Series on "Critical Technologies for Hypersonic Vehicle Development"* (10–14 May 2004), held at the von Kármán Institute, Rhode-St-Genèse, Belgium, and published in RTO-EN-AVT-116.

Hansen, C.F. (1957). Approximations for the Thermodynamic and Transport Properties of High-Temperature Air, NACA TN R-50, November 1957. Washington, DC.

Heiser, W. H. and Pratt, D.T., *Hypersonic Airbreathing Propulsion*, AIAA Education Series, 1994.

Hirschel, E.H. (2005). *Basics of Aerothermodynamics*. Berlin: Springer Verlag.

Huebner, L.D., Rock, K.E., Ruf, E.G. et al. (2001). Hyper-X flight engine ground testing for flight risk reduction. *Journal of Spacecraft and Rockets* 38 (6): 844–852.

Marshall, L., Bahm, C., Corpening, G., and Sherrill, R. (2005) Overview with results and lessons learned of the X-43A Mach 10 flight. *Paper AIAA 2005-3336. AIAA/CIRA 13th International Space Planes and Hypersonics Systems and Technologies Conference*, Capua, Italy (16–20 May 2005).

Ohlhorst, C.W., Glass, D.E., Bruce, W.E., et al. (2005). Development of X-43A Mach 10 Leading Edges. Paper IAC 05-D22.5.06. 56th International Astronautical Congress, Fukuoka, Japan (17–21 October 2005).

Reshotko, E. (1976). Boundary layer stability and transition. *Annual Review of Fluid Mechanics* 8: 311–349.

Rivers, H.K. and Glass, D.E. (2006) Advances in hot-structure development. *5th European Workshop on Thermal Protection Systems*, Noordwijk, The Netherlands (1 January 2006).

Saunders, J.D., Davis, D., Barsi, S.J. et al. (2014). Liquefied Bleed for Stability and Efficiency of High Speed Inlets. *NASA Report E-18916*. Moffett Field, CA.

Van Wie, D.M., White, M.E., and Corpening, G.P. (1990). NASP Inlet Design and Testing Issues. *John s Hopkins APL Technical Digest* 11 (3 and 4): 353–362.

Voland, R.T., Huebner, L.D., and McClinton, C.R. (2005). X-43 Hypersonic vehicle technology development. *56th Intl. Astronautical Congress of the International Astronautical Federation, the International Academy of Astronautics, and the International Institute of Space Law*, Fukuoka, Japan (17–21 October 2005). (IAC-05-D2.6.01).

Walton, J.T. (1992). Aerothermodynamic flow phenomena of the airframe-integrated supersonic combustion ramjet. NASA-TM-4376, E-6751, NAS 1.15:4376.

Wieting, A.R. (1992). Multiple shock-shock interference on a cylindrical leading edge. *AIAA Journal* 30 (8): 2073–2079.

4

Scramjet Inlet/Forebody and Isolator

4.1 Introduction

At hypersonic speeds, a very large amount of airflow is required for the engine to produce sufficient thrust. The airflow requirements of a scramjet engine are best met by utilizing the precompression obtained from the vehicle forebody and locating the engine on the underside of the vehicle toward the aft end. The inlet area of the scramjet propulsion system is therefore restricted to the space between the vehicle undersurface and the bow shock, and it is several times wider than it is high.

The airframe-integrated scramjet powered vehicle depicted in Figure 4.1 uses its forebody to partially compress the atmospheric air before the airflow enters the engine properly. The airflow is further compressed inside the inlet by a series of shock waves in the engine flowpath, before the air is mixed with a fuel and ignited in the combustor region. The highly integrated vehicle must be designed to provide both maximum inlet capture area while maintaining minimum losses in order to generate optimum thrust.

Because hypersonic air-breathing propulsion performance is highly dependent on its air compression system, the objective of this chapter is to highlight important aspects of the hypersonic inlet design and operability issues. Questions that must be answered include:

- How much air compression a scramjet inlet must do?
- Is there an optimum number of oblique shocks to achieve adequate compression performance for a given flight condition?
- How can the inlet maintain shock-on lip condition at off-design conditions?
- What causes engine unstart and how to prevent it?

4.2 Engine Inlet Function and Design Requirements

The inlet of an air-breathing engine must collect and process large amounts of air from the atmosphere in order to generate thrust, as this fundamental force is directly proportional to air mass flow rate. The freestream mass flow per unit area primarily varies inversely with

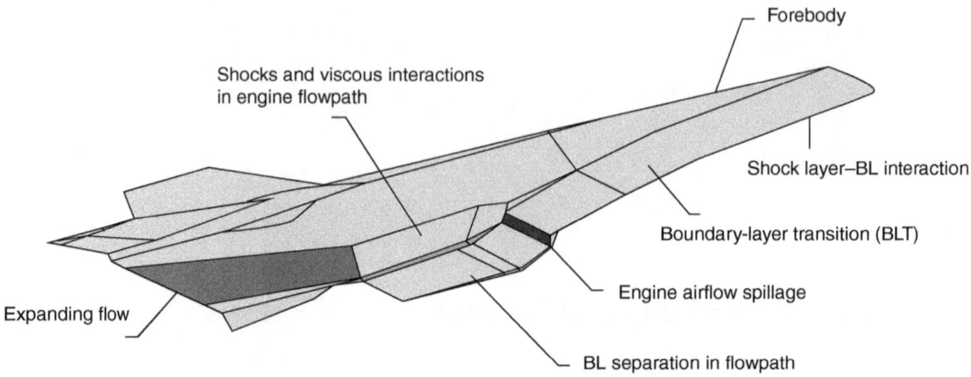

Figure 4.1 Inlet flow aerodynamic characteristics of airframe-integrated scramjet hypersonic vehicle.

Mach number along any trajectory of constant dynamic pressure, i.e. $\rho_0 V_0 = 2q_0/(a_0 M_0)$. Hence, long forebody compression surfaces are needed to make the freestream capture area much larger than the physical opening of the engine inlet. In addition, the inlet must compress the airflow entering the engine, increasing the static pressure to that required by the combustion system. Specifically, the inlet must provide the desired cycle static temperature ratio $\psi = T_3/T_0$ over the entire range of vehicle operation in a controllable and reliable manner, with minimum aerodynamic losses, i.e. maximum compression efficiency with minimum entropy increase. Thus, to attain optimum overall flight performance, the hypersonic air-breathing aircraft must carefully integrate the propulsion system with the vehicle airframe.

A mixed external/internal compression inlet is shown in Figure 4.2. As before, the subscript 0 denotes the undisturbed freestream flow conditions far ahead of the vehicle as seen from the reference frame of the vehicle, and the subscript 3 represents the condition of the flow at the end of the compression process. We will use the term *inlet contraction* to denote the geometric degree to which the airstream captured by the engine is compressed.

The lower surface forebody turns the airflow toward the engine flowpath using a combination of a blunted wedge and isentropic turning. The forebody is characterized by the nose bluntness R_N, initial wedge angle θ_i, and final ramp angle θ_f; the number of ramps depends on the optimum amount of flow turning required. At the design point, the forebody shock waves focus at the cowl lip at a prescribed freestream Mach number M_0. In Figure 4.2, the

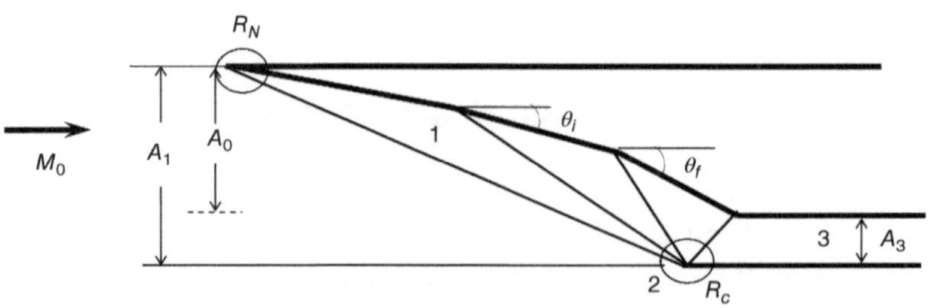

Figure 4.2 Scramjet inlet forebody cowl geometry with shock-on-lip.

flat, horizontal cowl generates a single cowl shock, which turns the flow toward the inlet centerline. The engine cowl is designed with lip bluntness R_c. The leading edge of the engine cowl must be designed to maximize engine thrust by maximizing air capture.

Moreover, the inlet flowfield and shock position are determined by the distance measured by nose diameter on the forebody. Bluntness also has a strong effect on transition, boundary-layer heating, and displacement of the forebody flow entering the cowl (Schneider 2004). Thus, even the vehicle's nose affects the propulsion performance.

The hypersonic air-breathing propelled vehicle must be designed and operated such that compression shocks produced by the forebody ramps intersect the cowl leading edge to achieve the shock-on-lip condition. These compression shocks may produce extremely high pressure and heating on the cowl leading edge. Peak pressure and heating also depend on the flow Mach number, Reynolds number, gas composition, and the impinging shock strength. Optimal cowl inlet configuration is the "shock-on-lip" condition or shock-on-cowl.

4.2.1 Captured Airflow and Capture Area

The freestream mass flow per unit area for any flight altitude and Mach number can be written in terms of dynamic pressure, $q_0 = \rho_0 V_0^2/2$, as:

$$\frac{\dot{m}_0}{A_0} = \rho_0 a_0 M_0 = \frac{2\,q_0}{V_0} = \frac{2\,q_0}{a_0 M_0} \tag{4.1}$$

This expression shows that the freestream mass flow per unit area is proportional to the dynamic pressure at any given Mach number. It also implies that flying at a higher Mach number at constant dynamic pressure severely reduces the available freestream mass flow per unit area, simply because this means flying at a high altitude where the density of the atmosphere is lower.

In Figure 4.2 we identify the air capture area ratio A_0/A_1 related to the amount of air that passes through or into the engine. The aerodynamic contraction ratio A_0/A_3 is the amount that the inlet squeezes the airflow, which is not the same as the geometric contraction factor A_1/A_3 related to inlet design. Since the inlet capture area is the area A_0 associated with the streamtube containing the flow entering the inlet, we represent the spillage by the area difference $A_1 - A_0$.

For subsonic air-breathing engines, the capture ratio A_0/A_1 is a function of the desired external deceleration from the cruise Mach number M_0 to the inlet lip Mach number M_1 and is called the mass flow ratio parameter. Since the external flow (devoid of shocks and/or a center body protruding outside the inlet) is reversible and adiabatic, the ratio of parameters between the flight condition and the inlet face follows the isentropic rule, that is,

$$\frac{A_0}{A_1} = \frac{M_1}{M_0}\left[\frac{1+\dfrac{\gamma-1}{2}M_0^2}{1+\dfrac{\gamma-1}{2}M_1^2}\right]^{(\gamma+1)/2(\gamma-1)} \tag{4.2}$$

For such subsonic flight application, the inlet design imposes limitations on M_1 in order to control the "overshoot" in local speed at the throat to a value below the sonic speed and prevent a sonic bubble formation at the throat, which may terminate in a shock and possibly cause boundary layer separation. Typically, a throat Mach number of 0.75 is given as an upper bound used for the throat sizing of a subsonic inlet.

Using a stream thrust analysis, Heiser and Pratt used as indicator of scramjet performance the capture area ratio $A_3/A_0 = \psi \cdot p_0/p_3 \cdot V_0/V_3$, where ψ denotes the cycle static temperature ratio (which determines thermodynamic cycle efficiency) and is used to impose the limit of maximum allowable compression temperature (see Chapter 2). They found that the area ratio A_3/A_0 changes rapidly for freestream velocities below 3048 m/s (M_0~10). For example, at V_0~1000 m/s, the capture area is A_3/A_0~0.07, decreasing to A_3/A_0~0.03 at V_0~4000 m/s. Such estimate suggests that variable geometry would be required for such range of flight velocity. However, the area ratio changes very little for $M_0 > 10$.

In any case, the air-breathing engine intended for hypersonic flight must ingest the necessary mass flow to maintain adequate thrust as flight speed increases, and thus a long forebody compression surface is needed in order to make the freestream capture area A_0 much larger than the physical opening of the engine inlet. The capture area can be larger or smaller than the physical opening of the engine inlet, and this is determined by the entire flowfield upstream of the engine face.

4.2.2 Air Compression Requirement

For scramjet performance analysis, we first need to know the static pressure that must be at the junction or interface between the engine inlet (air compression component) and the combustor. As we found in Chapter 2, the compression ratio p_3/p_0 is closely related to the thermodynamic efficiency of the air-breathing propulsion system.

For the design of a scramjet inlet, we must know how much air compression it must provide. If the compression ratio is too high, this can lead to large losses and external drag. The compression pressure represents the combustor entry pressure, which is the highest static pressure within the propulsion flowpath. Since mechanical and thermal loads both increase with p_3, the maximum airflow pressure at the combustor entrance must be limited to a value that will lead to acceptable weight, complexity, and cost. On the other hand, if p_3 is too low, it can make adequate ignition and combustion of fuel very difficult and lead to low propulsion cycle efficiency. The consensus is that air compression for effective combustion will be in the range 0.5 atm $< p_3 <$ 10.0 atm.

In practice, the optimal level of air compression required for a particular application depends on numerous conflicting factors, including: (i) overall scramjet cycle efficiency, (ii) nonequilibrium effects in the nozzle, (iii) need for robust combustion, and (iv) operability requirements (e.g. inlet starting and boundary layer separation).

4.2.3 Inlet/Forebody Design Requirements

The hypersonic inlet must be designed to provide the proper quantity and uniformity of air to the engine over a wider range of flight and propulsion conditions. At high velocities, the air-breathing propulsion system is extremely sensitive to losses within the inlet component. Hence, the design of a hypersonic inlet requires to define an inlet configuration that operates efficiently at high speeds but also provides adequate performance at lower speeds.

Because a vehicle in hypersonic flight will experience an extraordinarily wide range of conditions, the design of the inlet is subjected to many requirements to satisfy several interrelated (and frequently conflicting) design requirements simultaneously. For example, the inlet requires to have

- High air mass capture efficiency;
- High pressure recovery (minimum total pressure loss);
- Good operability over a wide range of flight Mach numbers and flow angles;
- Low flow distortion into combustor;
- Low external drag and minimum weight (length).

The inlet design point is determined by the flow speed at which the bow shock sweeps back to the cowl lip, to achieve full or nearly full capture at its design Mach number, which is known as the shock-on-lip condition. At off-design Mach numbers, the positions of the shock waves change, thus affecting the external drag and the efficiency of compression. For example, at lower Mach numbers, the airflow will spill, and this will create additional drag and reduce the thrust produced by the engine. For a given capture area, thrust can be increased by selecting a lower inlet design point.

The inlet decreases the flow velocity relative to the engine from the flight velocity V_0 to some smaller value. This flow deceleration must be accompanied by efficient flow compression within the entire flight envelope. The difference in kinetic energy of the air is converted to an increase in thermal energy, and thus the static temperature of the flow at the exit of the inlet is greater than the static temperature of the freestream air.

The flow of hypersonic inlets is dominated by shocks intersecting, interacting, reflecting, and shock–boundary layer interaction. The flow near the walls is dominated by its boundary layer behavior, and the shocks of high-speed flow impinge on the boundary layer on the compression ramps and the airframe, causing a complex aerothermodynamic flow interaction.

The design of scramjet inlets presents a number of challenges due to the complex characteristics of the hypersonic flow. As the airflow is slowed by the compression process and by viscous effects, the temperature of the air rises. The complicated flow field in a hypersonic inlet can only be analyzed properly with computational fluid dynamics (CFD) tools. However, here we wish to provide some insight into the physical phenomena of inlet flows and provide performance metrics for the high-speed compression system. Hence, using a few fundamental equations describing oblique shocks, we obtain a perspective on the design and operational requirements for hypersonic inlets.

The shape of the forebody shape is derived from one planar shock wave that intersects with the cowl lip for effective mass flow capture, the so-called "shock-on-lip" condition. Since the local Mach number is only a priori known in station 0, that is, the freestream condition M_0, analysis begins with the first oblique shock and continues marching downstream through the multiple ramps and shocks.

Consider the forebody depicted in Figure 4.3, which must be designed to provide the scramjet with a uniform flow at a prescribed back pressure. Its aerodynamic configuration is derived under a set of cruise conditions: M_0, prescribed planar shock wave shape β (to intersect cowl lip) and flight altitude. These data generate the flow field from which

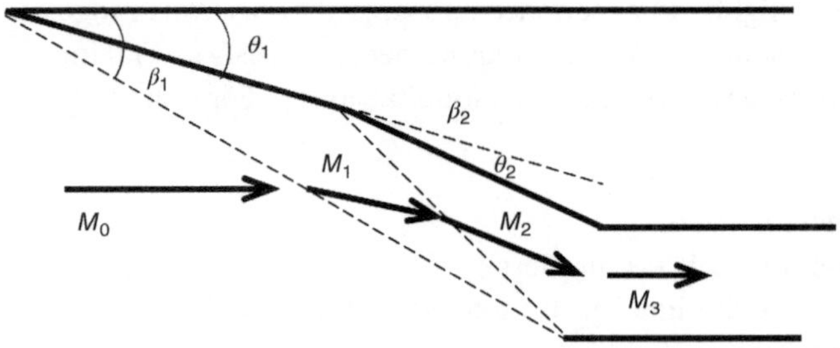

Figure 4.3 Two-ramp scramjet inlet forebody.

the forebody/inlet design can be derived. The forebody compression surfaces are designed by tracing stream surfaces emanating from the prescribed shock waves.

For the case $M_0 = 10.0$, if the ramp angles are 8° and 10°, we can easily calculate the flow conditions and total pressure recovery after each shock.

First oblique shock with $\theta_1 = 8°$ and $M_0 = 10.0$ yields a wave angle $\beta = 12.36°$. Thus,

$$M_{n1} = M_0 \sin\beta = 10.0 \sin 12.36° = 2.1418, \qquad p_2/p_1 = 5.185, \qquad T_2/T_1 = 1.806$$

$$\frac{p_{t2}}{p_{t1}} = 0.6548$$

With M_{n1}, determine the normal Mach number downstream of the first shock (from the normal shock table) $M_{n2} = 0.5551$, and thus,

$$M_2 = \frac{M_{n2}}{\sin(\beta - \theta)} = \frac{0.555}{\sin 4.36^\circ} = 7.289$$

Second oblique shock with $M_2 = 7.289$ and $\theta_2 = 2°$ (flow was already turned by 8° through the first shock, so the net turning angle through the second shock is $10° - 8° = 2°$), obtain $\beta = 9.20°$. Hence,

$$M_{n2} = M_2 \sin\beta = 7.289 \sin 9.2° = 1.1656, \qquad p_3/p_2 = 1.4184, \qquad T_3/T_2 = 1.1063$$

$$\frac{p_{t3}}{p_{t2}} = 0.99566; \quad M_{n3} = 0.8643$$

and the Mach number downstream of the second oblique shock is

$$M_3 = \frac{M_{n3}}{\sin(\beta - \theta)} = \frac{0.8643}{\sin 7.2^\circ} = 6.8948$$

Since the flow is normal to the terminal **normal shock**, $\beta = 90°$, with this M_3 obtain from the normal shock table $M_4 = 0.397\,942\,85$, $p_4/p_3 = 55.2946$, $T_4/T_3 = 10.185$, and

$$\frac{p_{t4}}{p_{t3}} = 0.016\,398\,65$$

The overall total pressure ratio for this inlet model is, therefore, the product of all ratios. Knowing the freestream conditions p_0, T_0, it is possible then to assess whether this geometry provides the required flow conditions for the engine (e.g. air pressure and total pressure recovery). One must keep in mind that for $M_0 > 5.0$, the temperature on the sharp leading edge may exceed practical limits for most structural materials. Hence, the inlet must have blunt leading edges. That is why the design approach for the hypersonic forebody/inlet requires a great degree of engineering refinement.

4.3 Inlet Types

Inlets can be classified into three basic types, characterized by the location of the wave system: internal compression, external compression, and mixed compression. Inlets can be two- or three-dimensional (2D and 3D), planar or rectangular, or two-dimensional (2D) axisymmetric (circular or a portion of a circle).

Since the beginning of high-speed air-breathing propulsion development, a wide variety of engine inlet designs have been sought to obtain the most efficient compression with a minimum of design complexity to reduce inlet size and thus minimize weight and entropy losses. For ramjet inlets, for example, it was found that, if the velocity is reduced from a supersonic speed to a subsonic speed with one normal shock wave, the compression process is relatively inefficient. However, employing several oblique shock waves to decelerate the flow, the compression process is more efficient. The position of the normal shock, also called a terminal shock, is dictated by the inlet back pressure. The optimal location for the normal shock is at the throat of the flowpath where the flow Mach number is the least.

Flow deceleration over multiple shocks is more efficient than the deceleration through a single normal shock. In 1944, Oswatitsch considered the problem of decelerating supersonic flow to subsonic flow by n shocks ($n - 1$ oblique plus one normal shock) in a missile propulsion inlet. Knowing that the results would be different if we take a number of weak oblique shocks and one strong normal shock, or if we take several strong oblique shocks and then a single very weak normal one, the question to answer is how the individual shocks must be arranged in order to obtain a maximum value of total pressure recovery (see Chapter 2 for an example). Oswatitsch (1944) set out to determine what we mean by optimum pressure recovery. He obtained results of the solutions for $n = 1$, 2, 3 and four compression shocks. He found that the total pressure recovery raises significantly by increasing the number of shocks from $n = 2$ to $n = 4$.

4.3.1 Internal Compression Inlet

In this configuration (Figure 4.4), internal compression is achieved through a series of internal oblique shock waves and a terminal normal shock positioned downstream of the diffuser's throat (stable position). It requires variable throat area to allow the inlet to swallow the normal shock (during starting). The internal compression inlet has poor performance with changes of angle of attack, and also the boundary layer is prone to separation due to modest adverse pressure gradients.

4.3.2 External Compression Inlet

In this type of inlet (Figure 4.5), external compression is achieved through one or more oblique shocks generated by external compression surfaces and a terminal normal shock located at the cowl lip (stable position). Deflected flow causes more drag, reducing performance. This inlet has adequate performance and is stable under all flight conditions up to Mach 2.

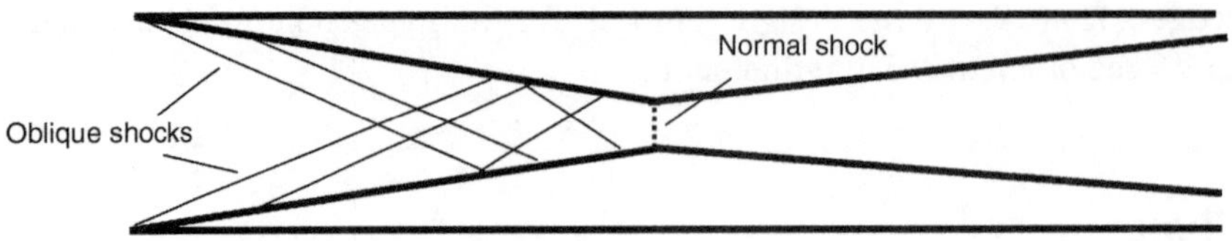

Figure 4.4 Internal compression inlet.

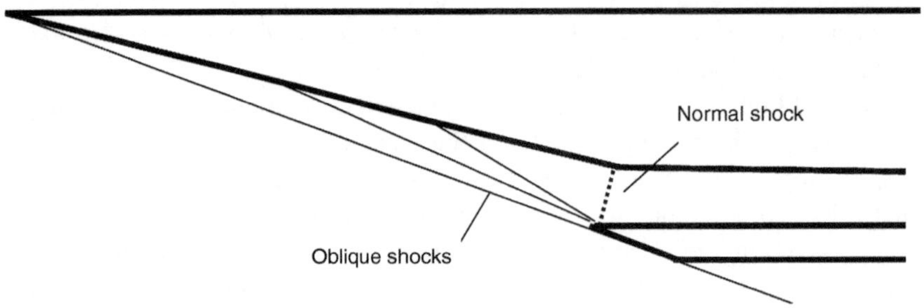

Figure 4.5 External compression inlet.

4.3.3 Mixed External–Internal Compression Inlet

In the mixed compression inlet (Figure 4.6), air compression is achieved through a combination of external oblique shocks, internal reflected oblique shocks, and a terminal normal shock located downstream of inlet throat (stable position). This inlet can have adverse shock–boundary layer interactions causing separation and inlet shock system instability. It requires a mechanism to maintain normal shock in a stable location, and variable throat area to allow the inlet to start by swallowing the normal shock.

An axisymmetric inlet with mixed air compression was designed for the Pratt & Whitney J58 (JT11D-20) afterburning turbojet engines that powered the SR-71 Blackbird Mach 3+ aircraft. This type of inlet (Figure 4.7) was chosen because it offered lower weight and drag and could attain high total pressure recovery, higher than is possible with an external compression inlet at Mach 3.0+. Moreover, the inlet required to match the airflow captured to the airflow required by the engine for all conditions, reduce the velocity of the captured air to about Mach 0.4 at the engine face, and maximize the pressure recovery while reducing

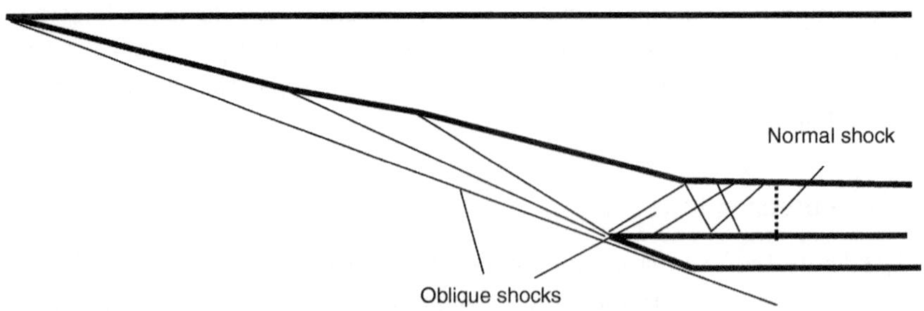

Figure 4.6 Mixed external–internal compression inlet.

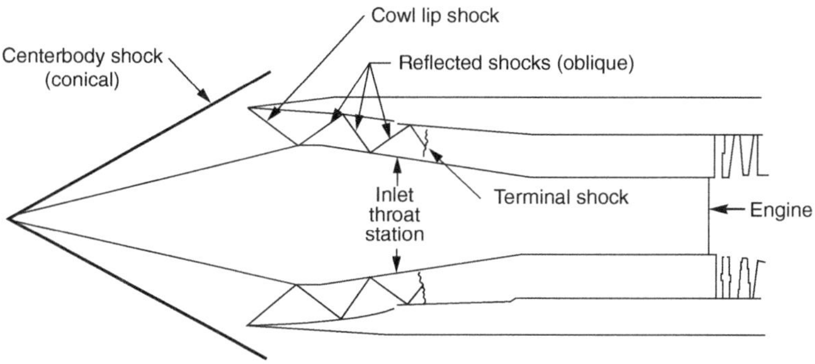

Figure 4.7 Axisymmetric mixed compression inlet concept.

velocity and minimizing the transient flow effects of external disturbances. Since no fixed inlet configuration can simultaneously satisfy all performance requirements over the entire flight envelope and engine operating ranges, a variable inlet geometry was required (Matthews and Jones 2006).

Variable compression inlets are axisymmetric or 2D with collapsing or translating center bodies that provide throat area variation with Mach number. To achieve compression, the center body must turn the airflow to high angles relative to the inlet axis in order to achieve a low throat Mach number. A high cowl lip angle is necessary to capture more airflow, but it will incur a large cowl drag.

The variation in cone angle of the SR-71 spike created additional steeper shocks such that the airflow was gradually decelerated. At the intake lip, the spike had a perforated wall for boundary layer control; it contained a translation mechanism which shifted the spike back and forth to adjust the inlet cross section and to adjust the position of the shock.

Comparing their weight and total pressure ratio, axisymmetric inlets are slightly better than 2-D rectangular inlets. However, 2-D inlets are easier to design and provide a larger variation in inlet airflow. Axisymmetric inlets have the added design problem of getting sufficient boundary layer bleed air out from the center body through the support struts.

Two-dimensional axisymmetric inlets are further divided into outward-turning and inward-turning designs. The Oswatitisch inlet (Figure 4.8a) is an outward-turning axisymmetric inlet, while the Busemann inlet (Figure 4.8b) is an inward-turning axisymmetric design concept. The outward-turning inlets are usually developed using the same approach as the 2-D planar designs, although slightly more turning is required to achieve the same compression ratio. The inward-turning inlets are very different from planar designs. As shown in the schematic, the Busemann inlet consists of an inward-turning isentropic compression that leads to a freestanding conical shock wave that is canceled at a shoulder. This type of inlet provides uniform parallel inviscid flow to a combustor.

Hypersonic missiles conceived in the early 1960s had round scramjet geometries and inward-turning inlets. Round structures integrate well into missile systems and provide structural, cooling, and internal drag advantages. Inward-turning inlets are most suitable for flow transition to the round cross section of the combustor.

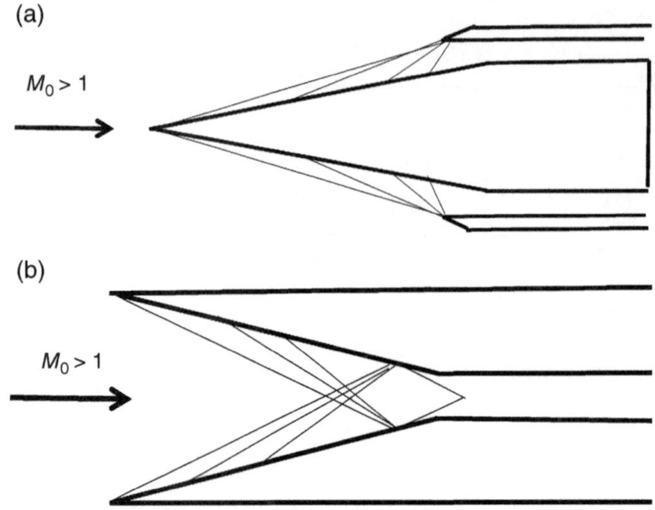

Figure 4.8 Axisymmetric outward-turning and inward-turning inlets. (a) Oswatitisch inlet and (b) Busemann inlet.

4.4 Inlet Compression System Performance

For ramjet engines, the inlet efficiency is primarily dependent on the freestream Mach number. This is because the most significant losses are associated with the configuration of the shock waves. Since the ramjet requires subsonic flow at the entrance of the combustor, the desired subsonic velocity is obtained by the appropriate choice of the area ratio of the subsonic diffuser. This subsonic velocity is fairly low in order to maintain combustor loading within acceptable pressure and temperature limits, while at the same time reducing the losses of internal total pressure. In the ramjet inlet, the losses due to diffusion are relatively small in comparison with those losses that arise from the shock waves, whether these are outside or inside of the inlet. The efficiency of the ramjet inlet is typically given in terms of the total pressure recovery (η_r).

4.4.1 Diffusion Process in Ramjet Inlet

We represent the ramjet inlet compression process with an enthalpy–entropy thermodynamic map, as shown in Figure 4.9. In compressing the airflow from the freestream static pressure p_0 to the combustor entrance static pressure p_3, to achieve the required low-subsonic exit velocities, the inlet diffuses the flow almost to the stagnation condition (represented by $p_{t0} \approx p_{t3}$) so that $V_3 \ll V_0$.

Because of shocks in the ramjet inlet flow, only a portion of the ram total pressure can be recovered, since the inlet flow is always irreversible ($s_3 > s_0$). Hence, the total pressure recovery parameter η_r describes the extent of losses in the inlet. These are due to viscous dissipation in the boundary layer and the shocks. We expect that the total pressure ratio should decrease very rapidly as entropy increases, since entropy is the best indicator of the effects of dissipation or flow losses.

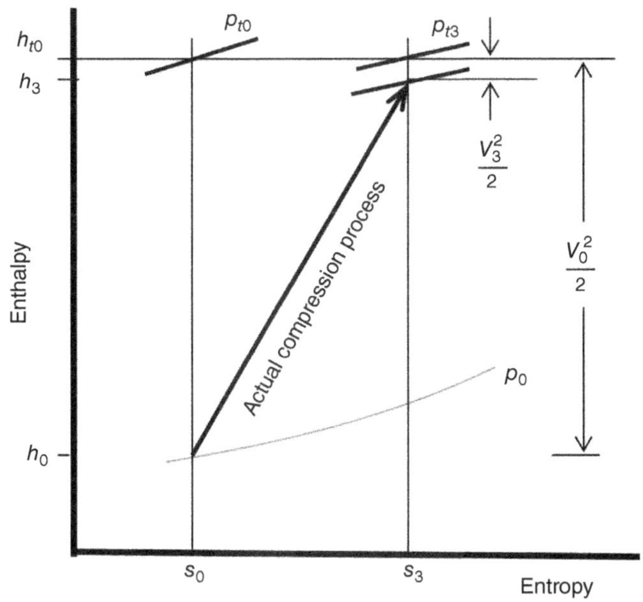

Figure 4.9 Inlet compression process for ramjet engine.

For ramjet inlets, the supersonic flow deceleration is accompanied by shock waves that can produce a total pressure loss much greater than, and in addition to, the wall friction loss. An inlet parameter for turbine engine performance is total pressure recovery, defined as the ratio of engine face total pressure, p_{t1}, to freestream total pressure, p_{t0}. The inlet's overall pressure ratio $\pi_d = p_{t1}/p_{t0}$ is the product of the ram pressure ratio and the diffuser pressure ratio. Because of shocks, only a portion of the ram total pressure can be recovered. Thus, if we denote $\pi_{d\max}$ the portion of π_d that is due to wall friction and define η_r as that portion due to ram recovery, the inlet overall total pressure ratio is written as $\pi_d = \pi_{dmax}\eta_r$.

The performance of the ramjet inlet decreases markedly with increasing flight speeds. Inlets are designed to meet or exceed MIL-E-5008B, the Military standard for ram recovery for the flight Mach range of interest. The pressure recovery for $M_0 > 1$ is given by:

$$\eta_r = \begin{Bmatrix} 1 - 0.075(M_0 - 1)^{1.35}; & 1 < M_0 < 5 \\ \dfrac{800}{M_0^4 + 935}; & 5 < M_0 \end{Bmatrix} \tag{4.3}$$

These expressions clearly show that the inlet total pressure recovery quickly worsens with flight Mach number. For example, for $M_0 = 4$, $\eta_r = 0.6695$, and as the flight Mach number increases to $M_0 = 6$, the ram recovery drops to $\eta_r = 0.3586$. We illustrate this trend in Figure 4.10, where we plot the MIL Standard and the total pressure ratio of the inlet, assuming $\pi_{dmax} = 0.95$. A complicated, multiple shock pattern must be developed by an inlet in order to achieve the high-pressure recoveries indicated by the Military standard. Mixed external–internal compression inlets can in fact attain high-pressure recovery above Mach 2.2.

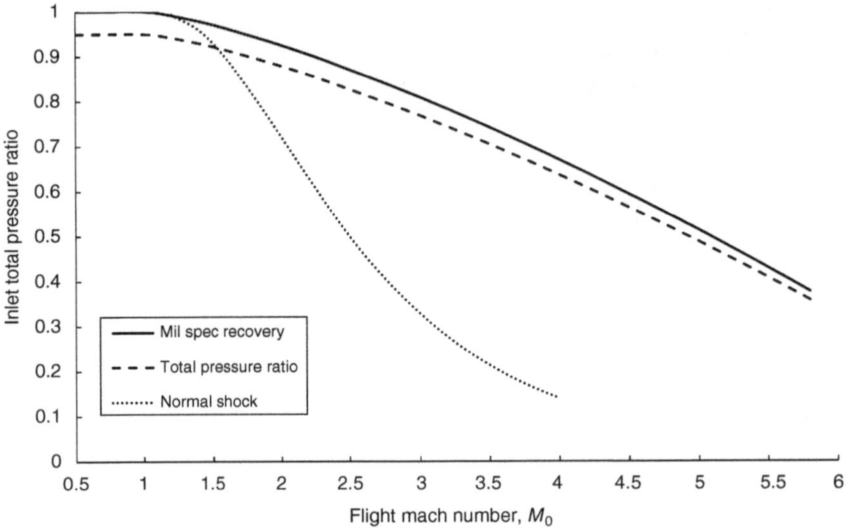

Figure 4.10 Inlet total pressure recovery for supersonic flow.

The terminal normal shock has the most negative impact in total pressure recovery of ramjet inlets; this type of shock is also responsible for the rising gas temperatures in the combustor. For example, for the case of $M_1 = 3.0$, the flow decelerates after a normal shock to $M_2 = 0.4752$, while the static pressure ratio across the shock is $p_2/p_1 = 10.33$; this pressure ratio across a normal shock dramatically increases to $p_2/p_1 = 29$ when $M_1 = 5.0$. However, if the amount of diffusion done by the inlet is reduced, then the diffusion losses can be decreased.

This explains why the concept of supersonic combustion ramjet arose as a solution for hypersonic flight, keeping supersonic flow velocity at the exit of the inlet. The performance of a scramjet depends to a high degree on the efficiency of the inlet as the amount of slow diffusion is reduced. In fact, in the early stages of scramjet development, when there was no experimental data to base performance assessments, a number of empirical relations were developed relating the inlet performance to its exhaust velocity. One such empirical inlet efficiency was the following:

$$\eta_D = 0.94 + 0.06\frac{M_3}{M_0}$$

This expression shows that, as M_3 approaches low-subsonic values, the inlet efficiency η_D tends to an acceptable value (0.94), and, with zero diffusion ($M_3 = M_0$), η_D yields 100%. However, as we stated earlier, for subsonic flow (without the losses from the shocks), friction and flow nonuniformities in the inlet will reduce its efficiency. Military Specification 5008B is set as an ideal goal for ram recovery.

4.4.2 Performance Parameters for Scramjet Inlets

Assuming that the air passing through the scramjet inlet is in its equilibrium state at all times, we consider the compression process taking place by plotting the variation of the flow field in a thermodynamic enthalpy–entropy diagram, as shown in Figure 4.11, where

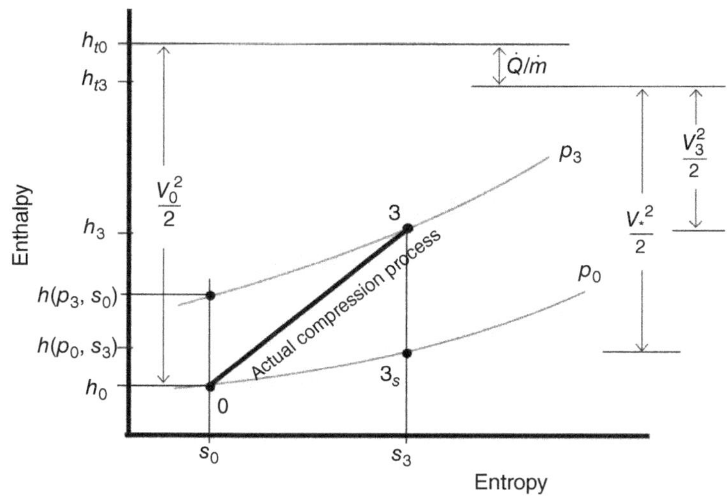

Figure 4.11 Inlet compression process for scramjet engine.

the state variable numbers correspond to the inlet station numbers as depicted in Figure 4.2.

The compression process occurs from the freestream static pressure p_0 to the static compression pressure p_3, also known as combustor entry static pressure, where the airflow velocity remains supersonic, and thus the airflow does not diffuse to the stagnation condition. During the compression process, the kinetic energy is partially converted to the static pressure (rise) and partially it is dissipated into heat; irreversibilities are caused by skin friction and shock waves in the inlet flow.

The dissipation of energy makes the inlet process irreversible and causes the entropy to rise. Entropy increases from the freestream value s_0 to the compression value s_3. Ideally, in the absence of those losses, the compression process would be isentropic. Note that the enthalpy states h_{t0} and h_0 are separated isentropically by an amount $V_0^2/2$, as required by the definition of stagnation state and its relation to the static state ($h_{t0} = h_0 + V_0^2/2$).

The total enthalpy gradient $h_{t0} - h_{t3}$ represents the heat loss of the inlet, which can be significant at high flight Mach numbers. Hence, we define the enthalpy ratio h_{t3}/h_{t0} as the parameter that relates the amount of energy lost from the captured streamtube through heat loss to the inlet/forebody.

By inspection of Figure 4.11, it is clear that the thermodynamic analysis of an engine compression process must consider an evaluation of the pressure ratio (p_3/p_0), the compression efficiency (η_c), the kinetic energy efficiency (η_{KE}), and the magnitude of the entropy increase ($s_3 - s_0$). These parameters, together with a geometric description of the inlet, can be combined to calculate the mass, momentum, and energy of the flow entering the combustor.

For the hypersonic inlet, the definition of performance parameters assumes that the gas is thermally perfect and that the flow is adiabatic, uniform, and one-dimensional (1D). It should be understood that for realistic analysis, we must incorporate heat losses to the inlet. We will also assume that the inlet considered will operate with a specified amount of diffusion, an amount corresponding to a specified V_3/V_0. Thus, the process will yield an increase in enthalpy from h_0 to h_3 while compressing the airflow to remain in the supersonic flow condition required by the scramjet combustor. During the compression process in the inlet, the kinetic energy is partially converted to the static pressure (rise) and partially it is dissipated into heat. The dissipation of energy makes the inlet process irreversible and cause the entropy to rise.

4.4.2.1 Allowable Compression Static Pressure and Temperature

The level of air compression achieved by the inlet is one of the most important parameters required for the design of the combustor. This is the highest static pressure the scramjet experiences, and thus p_3 represents a measure of the mechanical and thermal loads experienced by the combustor, requiring an optimized design that incorporates advanced materials and cooling techniques to sustain those loads. That is why, the maximum pressure of compression p_3, also known as *the burner entry pressure*, is limited to a value that will lead to an acceptable combustor design and still be sufficient to meet the requirements of the combustion process. The *allowable range of static compression pressure* is typically bounded in the range 0.5 atm $< p_3 <$ 10 atm.

By defining the air compression process by the ratio of enthalpy at the two end states, $\psi = h_3/h_0$, we can determine the level of compression the inlet must do since the enthalpy (or static temperature) at the combustor inlet is limited. The compression static pressure p_3 is limited by the combustor entry temperature. If the air temperature is too high, it would cease to behave as a calorically perfect gas and begin to dissociate (see Chapter 2). Hence, by entering the combustor at a lower static temperature (for a given static pressure), a correspondingly lower combustion temperature results, and the air does not experience chemical dissociation.

But combustion reaction rates increase rapidly with static pressure and temperature. So, if the compression pressure and temperature are too low, the length of the combustor required to complete the reaction and consume the available fuel would be unacceptable. Thus, if we limit the air temperature to a range between 1440 and 1670 K, then we can effectively constrain the air compression pressure at the entry of the combustor. Heiser and Pratt (1994) recommend the *allowable compression static temperature* T_3 to be within the region where air behaves as calorically perfect gas, which yields a cycle temperature ratio $6 < \psi = T_3/T_0 < 8$, giving a value that limits the entry combustor Mach number to roughly one-third of the freestream Mach number, i.e. $M_3/M_0 = \sqrt{T_0/T_3}$.

4.4.2.2 Compression Efficiencies

The compression efficiency must give a measure of the effectiveness of the inlet to compress the airflow to the conditions required by the combustion process. The compression efficiency must relate the actual change in static enthalpy to the ideal or isentropic change in static enthalpy that would accompany the same change in static pressure. However, there is no consistency in the published definitions of compression efficiency for hypersonic inlets. While some authors refer to compression effectiveness, inlet compression efficiency, process efficiency, polytropic efficiency, adiabatic compression efficiency, and kinetic energy efficiency, sometimes some of the parameters thus defined are the same. Thus, we will refer to parameters that can be expressed in terms of measurable quantities.

Using the nomenclature from Figure 4.11, the *compression efficiency* can be expressed as Van Wie (2000) did,

$$\eta_B = \frac{h(p_3, s_0) - h_0}{h_3 - h_0} \tag{4.4}$$

Van Wie also defined *process efficiency*, denoted η_{KD}, which is based on the change of kinetic energy in the diffusion process. This is the same definition called *adiabatic compression process efficiency* by Heiser and Pratt (1994) but denoted η_c:

$$\eta_c = \frac{h_3 - h(p_0, s_3)}{h_3 - h_0} \tag{4.5}$$

Since $h = c_{pc}T$, we can rewrite η_c in terms of cycle static temperature ratio $\psi = T_3/T_0$,

$$\eta_c = \frac{c_{pc}(T_3 - T_{p_0,s_3})}{c_{pc}(T_3 - T_0)} = \frac{\psi - \dfrac{T_{p_0,s_3}}{T_0}}{\psi - 1} \tag{4.6}$$

It is clear that for a specified temperature ratio, a higher compression efficiency is obtained when $T_{p_0,s_3}/T_0$ is lowest, close to 1. This will be indicative of a smaller entropy increase for the compression process.

We solve for the isentropic compression temperature from Eq. (4.6),

$$\frac{T_{p_0,s_3}}{T_0} = \psi(1 - \eta_c) + \eta_c \tag{4.7}$$

From the Gibbs equation, the pressure ratio is given by:

$$\frac{p_3}{p_0} = \left(\frac{T_3}{T_{p_0,s_3}}\right)^{c_{pc}/R_c} = \left(\psi\frac{T_0}{T_{p_0,s_3}}\right)^{\gamma_c/(\gamma_c - 1)} \tag{4.8}$$

Substituting (4.7), we obtain the adiabatic compression pressure ratio in terms of adiabatic efficiency:

$$\frac{p_3}{p_0} = \left[\frac{\psi}{\psi(1 - \eta_c) + \eta_c}\right]^{\gamma_c/(\gamma_c - 1)} \tag{4.9}$$

Now, to get a sense of the maximum static pressure of compression, which is needed to establish if the combustor entry pressure is adequate to sustain combustion reaction, we write an expression for p_3 by replacing p_0 in Eq. (4.9) with the freestream dynamic pressure, Eq. (2.7), and obtain

$$p_3 = \frac{2q_0}{\gamma_0 M_0^2\left[1 - \eta_c\left(1 - \dfrac{1}{\psi}\right)\right]^{\gamma_c/(\gamma_c - 1)}} \tag{4.10}$$

Hypersonic vehicles are designed to operate within a narrow range of dynamic pressure, 23 000–95 000 N/m^2 (500–2000 lbf/ft^2). In general, a vehicle will fly along a trajectory of its highest allowable q_0. Thus, we examine the behavior of p_3 by plotting Eq. (4.10) assuming a constant value of $q_0 = 47\,880\ \text{N/m}^2$ (1000 lbf/ft^2), and a cycle temperature ratio $\psi = 7.0$. As shown in Figure 4.12, the compression static pressure p_3 is greatest when the compression

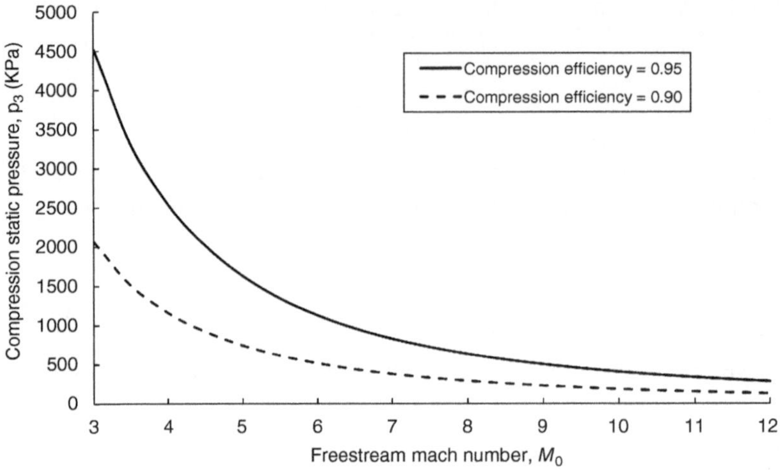

Figure 4.12 Compression static pressure as a function of flight Mach number and compression efficiency.

efficiency is higher, and for all values of η_c the pressure becomes smaller as the freestream Mach number increases.

As deduced from Eq. (4.10), for a given value of q_0 and ψ, the pressure at the entrance of the scramjet combustor varies inversely with the square of the flight Mach number. Thus, in order to ensure the pressure is adequate for maintaining complete combustion, a high compression efficiency is crucial.

4.4.2.3 Kinetic Energy Efficiency

One performance measure we need for the high-speed inlet (HSI) is the kinetic energy efficiency, which is related to the preservation of the kinetic energy (or velocity or momentum) that must be available to produce thrust. The kinetic energy efficiency η_{KE} is defined as the ratio of the available or useful kinetic energy remaining in the airstream after the diffusion compression process to the available kinetic energy of the flow before the diffusion compression. Using the nomenclature of Figure 4.11, an expression for the inlet kinetic energy efficiency is

$$\eta_{KE} = \frac{h_{t3} - h(p_0, s_3)}{h_{t0} - h_0} = \frac{\frac{1}{2}V_*^2}{\frac{1}{2}V_0^2} \tag{4.11}$$

Hence, we can think of η_{KE} as the ratio of the kinetic energy, the compressed flow would achieve if it were processed isentropically to the freestream pressure p_0, relative to the kinetic energy that the freestream possesses. Moreover, when the flow entering the engine is compressed from p_0 to p_3, there is heat loss to the forebody/inlet structure, which is accounted for by the enthalpy ratio h_{t3}/h_{t0}. However, if we assume an adiabatic process, we can use instead the adiabatic kinetic energy efficiency:

$$\eta_{KEad} = \frac{h_{t0} - h(p_0, s_3)}{h_{t0} - h_0} = \frac{\frac{1}{2}V_*^2}{\frac{1}{2}V_0^2} \tag{4.12}$$

For scramjet performance calculations, researchers typically determine the properties at the inlet throat using empirical correlations for η_{KE} or η_{KEad} and relay on CFD numerical simulations of the flow field developing in the scramjet forebody/inlet. A frequently quoted correlation for kinetic energy efficiency is that of Billig, et al. (1968). In their inviscid analysis (neglected skin friction), the researchers used an optimized set of four oblique shock waves that produce the most efficient compression, regardless of geometry. The resulting correlation relates the kinetic energy efficiency to the inlet capability parameter M_3/M_0, where M_3 is the inlet throat Mach number, which is valid for the range of flight Mach numbers $4 < M_0 < 14$:

$$\eta_{KE} = 1 - 0.2\left(1 - \frac{M_3}{M_0}\right)^5 \tag{4.13}$$

This correlation affirms that, for hypersonic velocities, the kinetic energy efficiency decreases as the amount of deceleration increases and the amount of compression increases.

Waltrup et al. (1982) developed another viscous correlation for η_{KEad} for $4 < M_0 < 14$:

$$\eta_{KEad} = 1 - 0.4\left(1 - \frac{M_3}{M_0}\right)^4 \tag{4.14}$$

Equation (4.14) provides a first-order accurate model for inlet performance. Using this correlation, Van Wie found that most 2-D inlets performed better at high Mach numbers, while most 3-D designs at lower Mach numbers perform slightly worse. No enough data was available to determine whether this effect is caused by differences in the freestream Mach number or differences in the inlet designs.

Smart (2012) compared Eq. (4.14) against published data for a range of inlet geometries and determined that it is conservative for flight Mach numbers of 8 and above. Hence, he proposed the following alternative correlation, which better matches experimental data and computational solutions for flight Mach numbers of 6, 8, 10, and 12:

$$\eta_{KEad} = 1 - \left(\frac{9}{M_0}\right)^{0.7}\left[0.018\left(1 - \frac{M_3}{M_0}\right) + 0.12\left(1 - \frac{M_3}{M_0}\right)^4\right] \tag{4.15}$$

The different empirical correlations are plotted in Figure 4.13.

Assuming adiabatic compression process, Heiser and Pratt derived a form of a kinetic energy efficiency for hypersonic inlets amenable for computation:

$$\eta_{KE} = 1 - \frac{2}{(\gamma - 1)M_0^2}(\psi - 1)(1 - \eta_c)$$

In this form, we note immediately that the inlet kinetic energy efficiency varies linearly with its adiabatic compression efficiency and is very sensitive to freestream Mach number. The above expression suggests that $\eta_{KE} \to 1$ as the flight Mach number increases. So, for example, for a flight Mach number $M_0 = 5$, $\psi = 7.0$, for an inlet with compression efficiency $\eta_c = 0.95$, the kinetic efficiency is $\eta_{KE} = 0.933$. For the same adiabatic efficiency and cycle temperature ratio, when $M_0 = 10$ the kinetic energy efficiency increases to $\eta_{KE} = 0.991\ 67$.

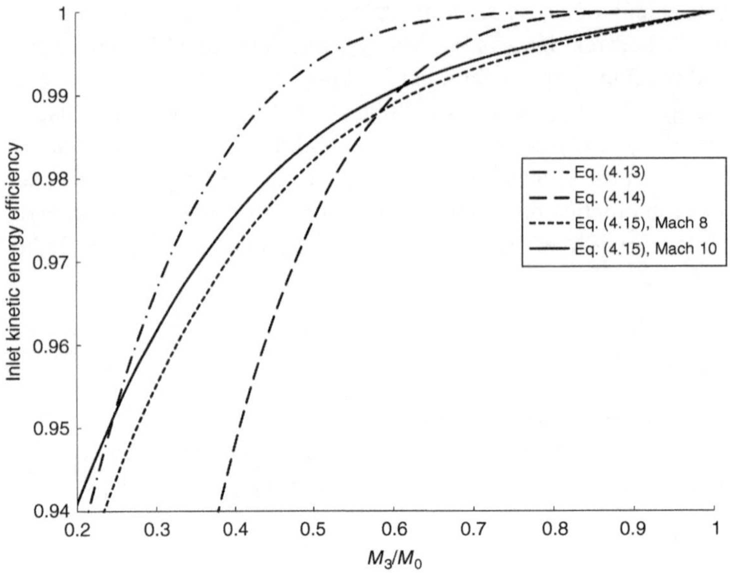

Figure 4.13 Inlet kinetic efficiencies with various empirical correlations.

4.4.2.4 Total Pressure Recovery

Now consider the total pressure ratio across the inlet p_{t3}/p_{t0}, a parameter also known as "total pressure recovery efficiency," which is a measure of the total pressure loss. This total pressure ratio, denoted by π_c, depends exponentially on the entropy increase across the inlet. Assuming 1-D flow without energy interaction, we can write the total pressure ratio in terms of static pressure and Mach numbers as follows:

$$\pi_c \equiv \frac{p_{t3}}{pt_0} = \frac{p_3}{p_0}\left[\frac{1+\dfrac{\gamma_c-1}{2}M_3^2}{1+\dfrac{\gamma_c-1}{2}M_0^2}\right]^{\gamma_c/(\gamma_c-1)} \tag{4.16}$$

where the term in the square brackets is the inverse of ψ the cycle temperature ratio:

$$\psi = \frac{T_3}{T_0} = \left[\frac{1+\dfrac{\gamma_c-1}{2}M_0^2}{1+\dfrac{\gamma_c-1}{2}M_3^2}\right]$$

Hence, we rewrite Eq. (4.16) as:

$$\pi_c = \frac{p_3}{p_0}\left[\frac{1}{\psi}\right]^{\gamma_c/(\gamma_c-1)} \tag{4.17}$$

Substitute (4.9) into (4.17) to obtain the total pressure ratio across the inlet:

$$\pi_c = \left(\frac{1}{\psi(1-\eta_c)+\eta_c}\right)^{\gamma_c/(\gamma_c-1)} \tag{4.18}$$

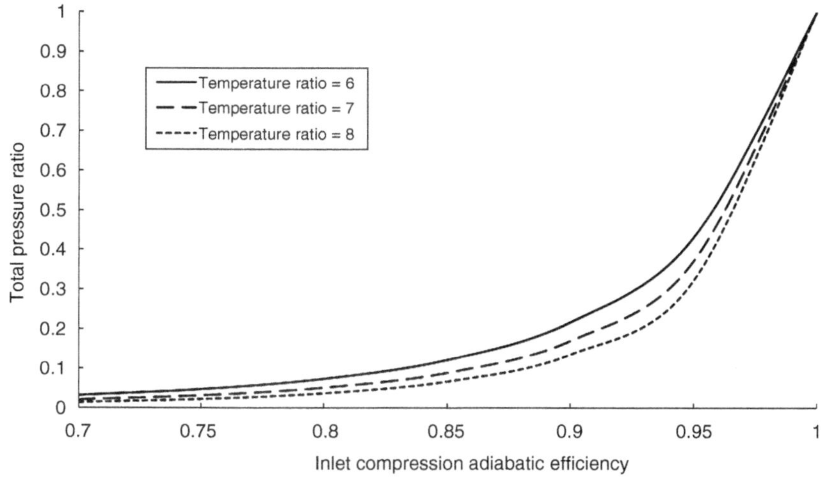

Figure 4.14 Inlet total pressure ratio as a function of adiabatic compression efficiency.

From this expression, we determine the adiabatic compression efficiency:

$$\eta_c = \frac{\psi - \left(\dfrac{1}{\pi_c}\right)^{(\gamma_c - 1)/\gamma_c}}{\psi - 1} \tag{4.19}$$

The total pressure ratio π_c is plotted as a function of η_c in Figure 4.14, for three typical cycle temperature ratios. As shown for all cases, a HSI must achieve a high compression efficiency in order to maintain the maximum total pressure recovery. It is known that, for a perfect gas, the maximum total pressure recovery will occur when the total pressure ratio across each shock is identical (i.e. the normal component of Mach number is the same for all shocks).

4.4.2.5 Dimensionless Entropy Gradient

We have emphasized that the design of the inlet must include at least two or three oblique shock waves in order to achieve adequate compression performance. However, the compression process in a hypersonic engine inlet is irreversible due to the viscous forces and the shock waves and boundary layers generate entropy. Hence, to assess the performance of the inlet, we can use the dimensionless entropy gradient, $\Delta s/c_{pc}$, where c_{pc} is the specific heat of air at the conditions of the compression process.

Entropy is related to the flow properties through the Gibbs equation, $T\,ds = dh - dp/\rho$. For a calorically perfect gas, we can use Eq. (2.19), which we integrate according to the compression process depicted in Figure 4.11, and obtain

$$\frac{s_3 - s_0}{c_{pc}} = ln\,\frac{T_3}{T_0} - \frac{R}{c_{pc}}\,ln\,\frac{p_3}{p_0} \tag{4.20}$$

This expression can be written in terms of the compression efficiency η_c and the cycle static temperature ratio ψ:

$$\frac{s_3 - s_0}{c_{pc}} = ln\left[\psi(1-\eta_c) + \eta_c\right] \tag{4.21}$$

Observe, the magnitude of Δs is directly proportional to the compression temperature T_3 and to the degree of adiabatic compression efficiency η_c. Although the dimensionless entropy gradient appears to be independent of the freestream Mach number M_0, the design of the inlet will become more critical as M_0 increases.

Using more than two shocks to compress the freestream airflow entering the engine would result in a more efficient inlet. For example, consider the two-ramp external compression inlet shown in Figure 4.3 at its Mach 9.5 design condition (shock-on lip), where the first ramp angle is $\theta_1 = 4.5°$. We can estimate the flow conditions after the first shock by using elementary shock wave relations, and this result will help in selecting the next ramp angle θ_2.

The equations for the oblique shock yield as special cases the strongest shock possible (normal shock) and the weakest shock possible (no shock) as well as all other intermediate-strength shocks. For the given deflection or turning angle θ, there are two possible shock angles for any given Mach number. For an efficient hypersonic inlet, we desire weak shocks. From a chart with $M_0 = 9.5$ and $\theta = 4.5°$, we find two values for the shock angle: 9.359° and 89.03°. Since we seek the weak solution, we select $\beta = 9.359°$,

$$M_{0n} = M_0 \sin\beta = 9.5 \sin 9.359^{\circ} = 1.5449$$

From the shock table with $M_{0n} = 1.5449$, we find

$$M_{1n} = 0.6857, \qquad \frac{T_1}{T_0} = 1.35, \qquad \frac{p_1}{p_0} = 2.61, \qquad \frac{p_{t1}}{p_{t0}} = 0.9149$$

Therefore, the flow conditions after the first shock are

$$T_1 = (1.35)(233.7) = 315.5\ \text{K}; \qquad p_1 = (2.61)\left(6.63 \times 10^2\right) = 1.73 \times 10^3\ \text{N/m}^2$$

$$M_1 = \frac{M_{1n}}{\sin(\theta - \delta)} = \frac{0.6857}{\sin(9.359^{\circ} - 4.5^{\circ})} = 8.09$$

We continue this procedure with the remaining forebody ramps and estimate the overall compression properties of such inlet. As illustrated by Anderson et al. (2000) in Figure 4.15, achieving a prescribed static pressure level of air compression, an inlet with high efficiency is required where compression is comprised mainly of isentropic turning and weak shock waves, as this results in a much lower static temperature T_3 than that generated by a less efficient inlet employing stronger shock waves (Anderson et al. 2000).

The inlet must be designed to keep the air temperature low to avoid air dissociation, especially at higher Mach numbers. For example, at $M_0 = 14$, a four-shock inlet is more effective than a two-shock inlet, yielding inlet outlet temperatures in the ranges 1167 K $< T_3 <$ 1333 K (four-shock inlet) and 1222 K $< T_3 <$ 2000 K (two-shock inlet), as shown in Figure 4.15. Note the less efficient two-shock inlets produce air static temperatures up to 3600 °R, a value much higher than the recommended maximum allowable combustor entrance temperature of $T_3 = 1560$ K (2800 °R).

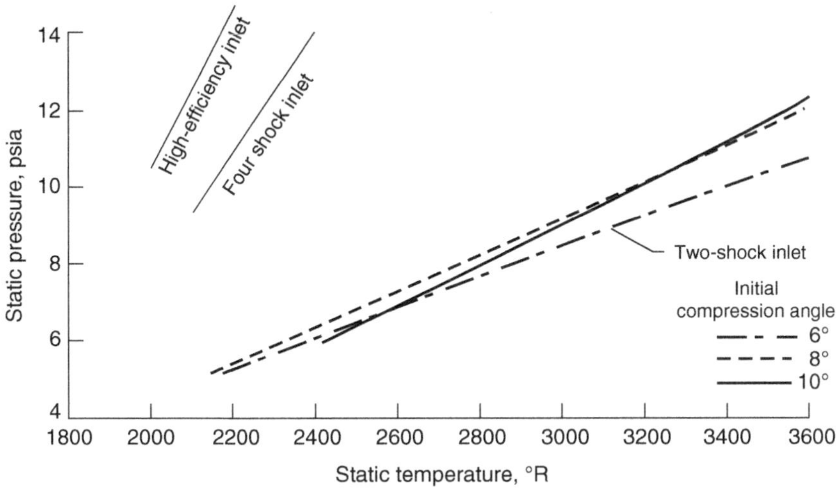

Figure 4.15 Inlet static compression and static temperature. Source: Anderson et al. (2000).

Moreover, if the air entering the combustor has a lower static temperature, it will ensure a lower combustion temperature and less chemical dissociation. Air dissociation restricts compression temperature to the range 2600–3000 °R (1440–1670 K). Hence, the scramjet cycle static temperature ratio $\psi = T_3/T_0$ will be in the range $6.0 < \psi < 7.5$. Overall, for $M_0 = 14$, an optimum high efficiency inlet would yield compression temperatures between 1111 K (2000 °R) and 1222 K (2200 °R). These values are lower than the temperatures leading to air dissociation.

The compression efficiency of the inlet has a dramatic effect on the scramjet performance. The associated higher static temperature rise resulting from an inefficient air compression process can limit engine performance, since higher gas temperatures can result in excessive chemical dissociation, and this in turn significantly reduces the combustion heat release.

4.5 Hypersonic Inlet Designs

In general, a high compression efficiency inlet is one where compression is obtained by a series of isentropic turning and weak shock waves. An optimal inlet design for hypersonic flight is one that includes external and internal compression, e.g. flow deceleration through multiple shocks and their reflections inside the engine flowpath.

Classifying HSI designs by application, we consider inlets for conventional ramjets, inlets for dual-mode scramjet (DMRJ), inlets for pure scramjet engines, and inlets for combined cycle air-breathing propulsion systems. The scramjet inlet can have a wide range of geometries and wave compression arrangements. The main requirement for a scramjet inlet is that the Mach number of the processed air be supersonic at the combustor entrance. This Mach number $M_3 > 1$ should be low enough to allow correct injection of propellant and complete ignition before the flow has reached the end of the combustion chamber. As we determined before, $M_3 \sim 0.3\ M_0$.

The design of the inlet affects the performance of the scramjet in a number of ways: the level of compression achieved through both the external and internal sections of the inlet, degree of flow uniformity, combustor entrance flow field distortion caused by shocks and expansions, and total pressure losses resulting from shock waves, wall friction, and shock/viscous interactions.

4.5.1 Axisymmetric

The first inlet designs were axisymmetric to fit the pod-type scramjet engines. The NASA Hypersonic Research Engine (HRE), for example, had axisymmetric configuration with a controlled translating spike inlet connecting to an annular combustor fueled by hydrogen. The component development program culminated with the ground tests of a full-scale HRE concept, designated the Aerothermodynamic Integration Model (AIM) (Figure 4.16).

The full-scale HRE/AIM was 87 in long (2.2098 m) and 18 in (0.4572 m) in diameter at the cowl. It had a mixed compression inlet with a translating spike that enabled the close-off of the engine. The spike could be moved fore and aft from inlet close-off to full open and to intermediate positions. At Mach 6, shock-on-lip occurred, and from Mach 6 to 8 the spike was designed to translate to maintain shock-on-lip over this Mach number range. An upsloping throat was incorporated to enable the inlet to maintain shock-on-lip with spike translation and also to have increased inlet contraction ratio with increased Mach number.

In 1994, NASA contracted with the Central Institute of Aviation Motors (CIAM) to perform a flight test of an axisymmetric H2-fueled scramjet, using the Russian Hypersonic Flying Laboratory (HFTB). The NASA contract required ground and flight tests at a Mach number of 6.5 for a modified CIAM axisymmetric scramjet, which included internal–external mixed compression, Mach number 6 design shock-on-lip inlet, and a short ($L/H = 6$) isolator section. The isolator is a duct designed to contain a shock train that

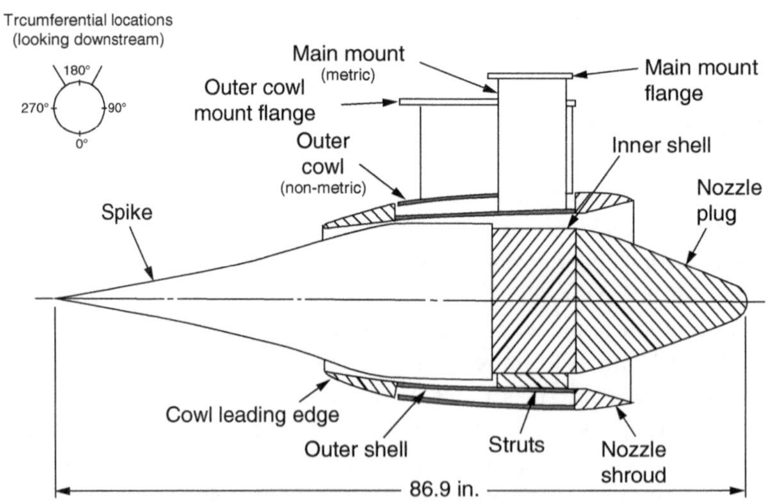

Figure 4.16 NASA HRE aerothermodynamic integration model. Source: Andrews and Mackley (1976)/NASA/Public Domain.

develops upstream of the combustor, used to isolate it and prevent disruptions to the inlet operation (see Section 4.9).

The fixed geometry spike inlet consisted of three conical section compression ramps set at 10°, 15°, and 20° ramp one-half angles. This geometry generates a single shock-on-lip condition at Mach 6 on the inlet leading edge or cowl lip, allowing for a slightly overspeed flow condition at Mach 6.5. In high-speed propulsion, *overspeed* means to operate above the maximum flight Mach number defined for the design condition.

For the US Air Force HIFiRE experiment, the 2-D inlet limited the design to be planar in the spanwise direction and to turn the airflow in either an outward or inward direction (see Figure 4.8). To limit the forebody (and shroud) length, designers selected a single compression angle from the nose (Ferlemann 2008). The outward-turning concept could use a shock trap (flow exhausted laterally) for the top flowpath, or a cowl shock could be used to cancel at the body shoulder for an on-design Mach number optimum for the bottom flowpath. Ultimately, the inward-turning inlet concept was chosen to meet the flowpath scale and angle of attack requirements of the HIFiRE experiment. It resulted in a longer shroud, a sizable cantilevered forebody structure, and an increased nozzle length.

The HIFiRE forebody is comprised of symmetrical opposing 7-deg ramps, 0.76 mm (0.030 in) radius leading edges spanning 127.0 mm (5 in) straight across the flow (Figure 4.17). The forebody capture height at the leading edge is 228.6 mm (9 in). The inlet entrance width is 121.9 mm (4.8 in). Each inlet sidewall produces 3-deg compression and requires 193.8-mm (7.63-in) streamwise length to produce the final flowpath width of 101.6 mm (4 in).

Drayna et al. (2006) studied the influence of the design choices on the efficiency of inward-turning inlets for Mach 10. They found that the inlet truncation angle should be kept between 2° and 4° to balance the increased skin friction and heat transfer losses with the shock losses. Presumably, a vehicle system optimization would favor the higher end of the truncation angle range to reduce the cooling requirements. Moreover, they recommended that the leading edge radius be kept as small as possible, i.e. small radii (~1 mm) give better performance than the ideal sharp leading edge inlets due to improved coalescence of the compression waves. These guidelines were used in the design of the HyCAUSE and ASET inward-turning inlets under the DARPA Falcon program (Candler and Drayna 2008).

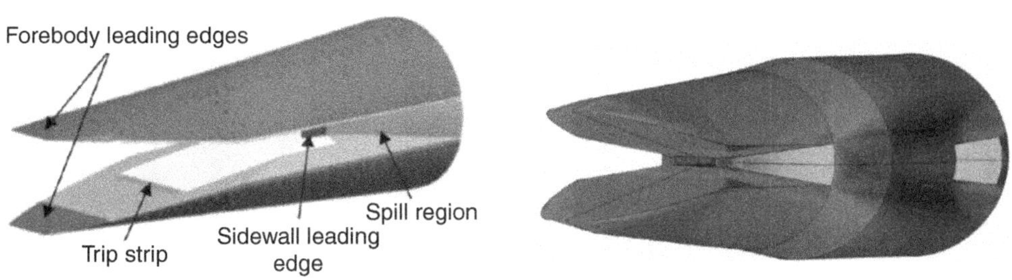

Figure 4.17 HIFiRE forebody/inlet design. Source: From Jackson et al. 2011; Ferlemann 2008.

4.5.2 Two-Dimensional Fixed Geometry Inlet

Planar inlet designs are easy to design and test. They can easily be adapted to have variable geometry features such as movable ramps and cowls, allowing the engine to function over a wide range of flight and propulsion operating conditions. Two-dimensional (2-D) inlet designs are those whose primary flow is characterized by 2-D analysis. Three-dimensional (3-D) features of the flow field may exist, depending on the size of the cross section.

Van Wie and Ault (1996) designed a 2-D inlet to investigate its internal flow field characteristics operating at Mach 10. The mixed external/internal compression inlet includes a forebody, sized to turn the flow outward using a combination of a blunted wedge and isentropic turning. The forebody was characterized by a blunt nose. They calculated the shock wave angles to focus at the cowl lip at a prescribed Mach number. The inlet included a flat, horizontal cowl to generate a single cowl shock to turn the flow back toward the inlet centerline. They determined that control of the positioning of the strong cowl shock within the inlet was a dominant issue in the design of the 2-D inlet. The positioning of this shock relative to an inner body shoulder significantly influenced the inlet efficiency and uniformity of the flow being supplied to a combustor.

Sidewalls can be used as additional compression surface. Hohn and Gülhan (2011) investigated sidewall compression effects on the flow field of the inlet GK-01 of the GRK scramjet configuration, which is a 2-D mixed compression inlet designed for a flight Mach number of 7.5, and compared sidewall compression by swept wedge inserts in the external part of the inlet. Their experimental results show that external sidewall compression induces strong separation and vortex structures in the external part which impair the starting behavior and inhibit any gains of the compression ratio. With internal sidewall compression, strong increases of the pressure ratio could be achieved and the inlet still started at internal contraction ratios well above the Kantrowitz limit (Kantrowitz and Donald 1945). This contraction area ratio limit is a criterium for inlet starting, relating the flow Mach number to the inlet contraction ratio.

4.5.3 Three-Dimensional Sidewall Compression Inlet

In a full-size hypersonic vehicle, the airframe-integrated propulsion system may be made up of a group of rectangular modules attached to the bottom surface. The inlet is a 3-D design that utilizes lateral compression from the sidewalls to complete the vertical compression initiated by the vehicle forebody (Holland and Perkins 1991; Holland 1995). The cowl leading edge (bottom surface) is typically located aft near the inlet throat to allow air to spill at low speeds; this is done to ensure the fixed geometry inlet design can operate over a wide speed range (Trexler and Souders 1975). In such configuration, the vehicle bow shock performs the initial compression, and the capture shape for the inlet of each scramjet module is required to form three sides of a rectangle so that the modules may be mounted side by side.

The sweep angle of the sidewall compression surfaces affects the airflow characteristics and therefore the performance of this type of inlet. As shown in Figure 4.18, aft sweep and forward sweep inlets have distinct characteristics since, when flow is compressed by a swept sidewall, a flow component is generated in the direction of the sweep, which turns the flow down or upward toward the cowl. Sidewalls are desirable from a structural and heat management perspective. However, the sweep angle must be chosen carefully to ensure required performance.

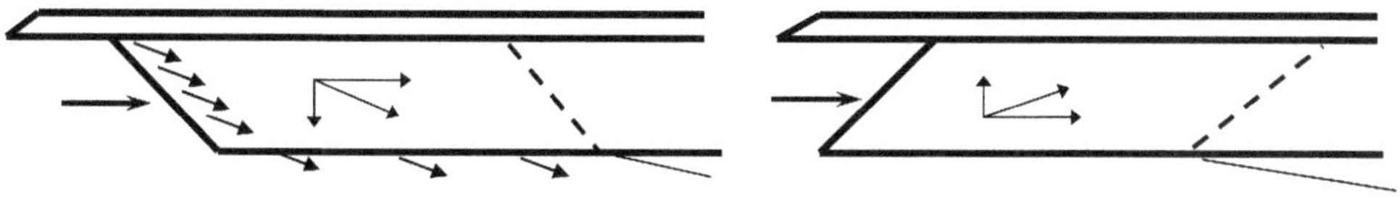

Figure 4.18 Aft swept and forward swept inlet designs. Source: Based on Anderson et al. (2000).

In general, aft sweep compression surfaces allow increased flow spillage, which make inlets easier to start, since by more air spill the normal shock passes and establishes supersonic flow inside the inlet. On the other hand, using forward sweep causes flow to be turned upward ahead of the cowl and reduces spillage and flow distortion, which is desirable to have at the entrance of the combustor. And although it has increased mass capture, the inlet with forward sweep is harder to start, and consequently variable geometry may be necessary.

The NASA X-43A vehicle shown in Figure 1.11 was a subscale design based on a Mach 10 cruise, global-reach mission configuration, and the extensive National Aerospace Plane (NASP) database available in the 1990s. As we noted in Chapter 1, the X-43 could not be photographically scaled from the 200-ft (60.96-m)-long Mach 10 cruise vehicle. Hence, engineers determined the minimum subscale engine size for flight testing was 12 ft (3.6576 m).

The inlet of the X-43A vehicle had a planar configuration with compression sidewalls. The lower forebody surface flowpath was intended to provide a nominally 2-D inflow into the scramjet inlet. The vehicle design incorporated an inlet cowl with a hinged moveable door for engine close-off and inlet starting. During the X-43A flight boost to takeoff speed, the door was closed; the inlet door opened for the powered flight.

The 6-ft (1.8288-m) forebody had a thin leading edge with a 0.03-in (0.762-mm) nose radius for the Mach 7 vehicle, and 0.05 in (1.27 mm) for Mach 10, and three flat ramps designed to provide a series of discrete, non-isentropic flow compressions for the scramjet. While flying at the nominal angle of attack of 2 deg (original design point for the Mach 7 vehicle, later changed to 2.5 deg), the first forebody ramp provided an initial 4.5 deg of compression, followed by the second ramp with an additional 5.5 deg, and the third ramp with the final 3 deg of additional compression. It had chines outboard of the flat ramps, which were designed to minimize 3-D effects and flow spillage.

4.5.4 Three-Dimensional Rectangular-to-Elliptical Shape Transition Inlet

In some scramjet applications, it is a desirable to have an inlet with a cross-sectional shape that transitions from the rectangular-like capture to an elliptical throat. Such inlet could be used in combination with an elliptical combustor. In 1999, M. Smart introduced the design of streamtraced hypersonic inlets with rectangular-to-elliptical shape transition (REST). The streamtracing design process provides a powerful and relatively simple technique for determining the inviscid shape of an inlet with predetermined capture shape and pressure ratio (Smart 1999).

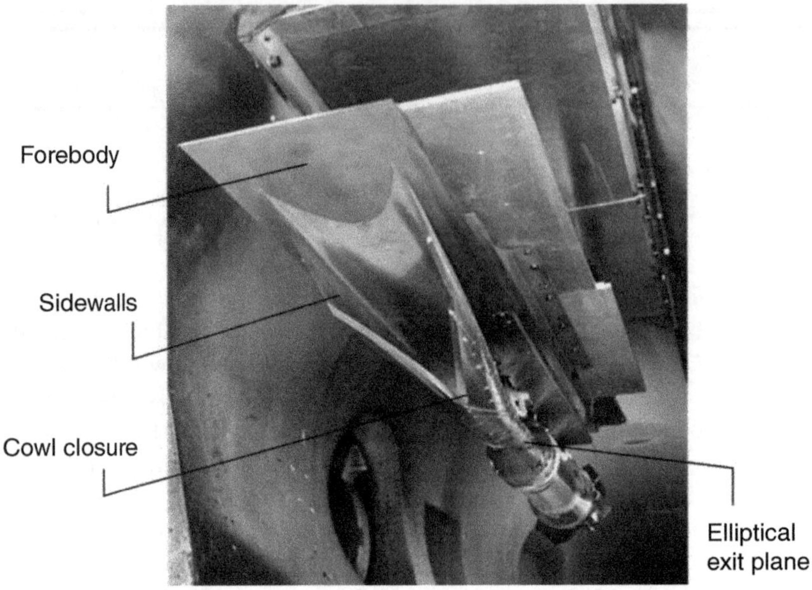

Figure 4.19 REST inlet model installed in NASA LaRC AHSTF. Source: Adapted from NASA photo, for test described by Smart and Ruf (2006).

The three-dimensional inlets with REST (3-D REST) inlets (Figure 4.19) are of interest for use on scramjet engines, because they integrate well with the forebody of a planar vehicle. A tight integration is desirable where the angle of the airflow approaching the inlet is not affected by the vehicle angle of attack. Using quasi-streamline-traced design, Smart demonstrated that REST inlets are efficient and can be connected to various combustor geometries and can be arranged in a modular form to have more than one engine flowpath.

Figure 4.19 shows a REST inlet model installed in the NASA LaRC Arc Heated Scramjet Test Facility (AHSTF). Designed for a vehicle with a short 6-degree forebody, the flight design point for the inlet was Mach 7.1. Testing at Mach 6.2 showed that the REST inlet was highly efficient and self-started with an internal contraction ratio well above the Kantrowitz limit. However, the inlet would not start at Mach 4.7. Further testing was performed on several modified designs at Mach 4.0 to assess changes to the overall design procedure intended to extend the inlet operational regime to lower Mach numbers (Smart and Ruf 2006). Based on testing and CFD analysis, the REST inlets are a viable fixed-geometry configuration for airframe-integrated scramjets in the Mach 4.5 to 8.0 flight regime (Turner and Smart 2013).

According to Gollan and Ferlemann (2011), a limitation of the design method is that it produces inlet geometries for only a restricted portion of the design space for this class of inlet. Hence, they developed tools that allow the inlet geometry to be manipulated in a parametric way that is meaningful from an inlet design perspective. Their new toolset provides a means to move rapidly from altering surface geometry to the generation of a volume grid ready for a Navier–Stokes CFD calculation.

4.5.5 Variable Geometry or Dual-flowpath Inlets

Variable geometry inlets are needed for combined cycle propulsion that have more than one engine capture flow requirement. Many hypersonic cruise vehicles have been conceived that utilize a ram/scramjet propulsion system for high speeds and turbojets for subsonic flight and low-speed acceleration. Many studies conducted in the past decades have considered combined cycle engines configured as a wraparound turboramjet engine, or as separate turbojet and ram/scramjet dual flowpath that requires a highly variable inlet to accommodate the two separate and very distinct flowpaths. The separate engine configuration is also known as over-under turbine-based combined cycle (TBCC) concept because the turbojet and its ducting are embedded in the body of the vehicle with the ram/scramjet engine mounted underneath. Using a DMRJ, the over-under TBCC design is considered for two-stage access to Earth orbit vehicles (see Chapter 10).

The wraparound turboramjet is an integrated propulsion concept that consists of a core turbojet inside an annular duct which forms the ram/scramjet proper. The single flowpath results in a ram/scramjet engine that has a large surface area.

In addition, the diameter at the engine face of the combined engine requires a large subsonic diffuser ahead of the engine to meet the requirements of both the turbojet and ram/scramjet. Thus, the wraparound turboramjet engine leads to large, wetted, internal duct areas and heat loads at cruise, and because of its integrated nature, may require the simultaneous development of both the turbojet and ram/scramjet parts of the engine.

The separate engine or over-under concept has separate subsonic diffusers (for turbojet and ram/scramjet), but both engines remain within the confines of a single nacelle. In this configuration, the engine better isolates the turbojet from the ram/scramjet (turbojet does not require special close-off mechanisms) and avoids the circumferential heat load imposed by a wraparound ram/scramjet. However, the separate flowpath requires more variable geometry for the inlet.

When considered for a Mach 5 cruise application that does not require supersonic combustion operation, an additional performance advantage of the turboramjet inlet geometry is a reduced amount of external compression when both engines are operating, at the transition point (~Mach 3), which results in an increase in mass flow ingested. This improvement is caused by nearly eliminating the second external compression surface when the turbojet inlet is opened.

The inlet design requirements for such combined cycle propulsion systems depend on the configuration it takes (wraparound or separate dual flowpaths) as well as the mission-related factors (cruise or accelerator). The relative location of the turbojet and ram/scramjet on the vehicle and their location to each other are important factors. Many early studies considered the possibility of using a common inlet to operate a separate turbojet and ramjet or scramjet, and they suggest certain advantages for separating the two engines.

While studying inlets for a turboramjet to power a Mach 5 cruise vehicle, Weidner (1979) found some differences between the separate and wraparound turbojet/ramjet concepts. He concluded that separate inlets with diffusers from an external compression surface of a basic 2-D inlet becomes impractical for separate turbojet and ramjet engines when the turbojet inlet size approaches the size of the ramjet cruise inlet. In general, the two-inlet approach provides the shortest inlet/diffuser system. However, a detailed vehicle design and mission

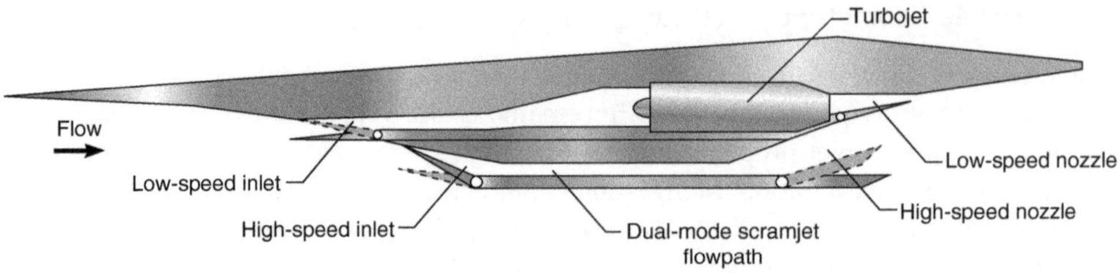

Figure 4.20 NASA TBCC engine integration with a dual flowpath inlet in over-under arrangement. Source: Albertson et al. (2006)/NASA/Public Domain.

analysis would be required to properly assess differences in propulsion system performance on overall vehicle performance.

A major concern regarding separate flowpath TBCC propulsion is the potential for flow interference during mode transition, when both low-speed (turbojet) inlet and the high-speed (ram/scramjet) inlet must operate simultaneously. An over-under TBCC propulsion system developed by NASA places a turbojet over a DMRJ, as depicted in Figure 4.20. This TBCC engine configuration requires a variable geometry inlet consisting of a low-speed inlet (LSI) (to feed the turbojet) and a HIS (for the ram/scramjet flowpath). Adjustable doors match the airflow requirements. The upper door is designed to close off the turbojet flowpath above the Mach 3–4 range when the scramjet ignites. The bottom door isolates the scramjet when it is not operating. The inlet performance must be coupled with the requirement to fully close the inlet for terminal ascent operation and during re-entry.

To evaluate the performance of this TBCC dual-flow inlet concept, an experimental study was carried out with an 8% scale model (Albertson et al. 2006). The main objectives were to identify interactions between the LSI and the HSI during the mode transition phase in which both inlets must operate simultaneously, and also to determine the effect of the LSI operation on the HSI performance. Tests were conducted at a nominal freestream Mach number of 4.

Figure 4.21 shows the shock wave configurations for the low-speed cowl (LSC) in the closed and fully opened positions, and with the high-speed cowl (HSC) set at zero angle relative to the freestream flow. As shown in Figure 4.21a, with the LSC closed, the LSC leading edge produces a shock well upstream of the HSC. When the LSC is opened (Figure 4.21b, it produces a compression corner at the LSC hinge, the angle of which increases with increasing LSC open angle. For this case, the shock produced at the LSC hinge impacted the HSC near the HSC kink. Test results showed the competing effects of the LSC hinge shock and the LSI boundary layer diversion on the HSI operability (Albertson et al. 2006).

Testing showed that the LSI was in fact beneficial to the operability of the HIS for HSI throat heights of 60 and 80% of the original design height, due mainly to the removal of the boundary layer, which outweighed the effect of the secondary shock created at the joint of the LSC flap (Albertson et al. 2006). Interestingly, the Mach number ahead of the HSI also increased as the LSC was opened, leading to an increase of the contraction ratio limit.

The position of the doors in the TBCC dual-flow inlet is also a major design concern. For the TSTO vehicle concept developed for the NASA/AF Joint System Study (JSS), the booster stage is powered by a variant of an over-under TBCC propulsion concept (turbine

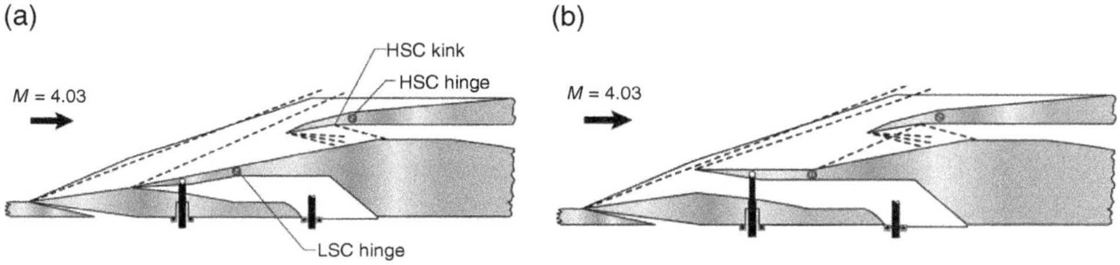

Figure 4.21 Inviscid, shock wave patterns for NASA dual-flow inlet. (a) LSC closed and (b) LSC fully opened. Source: Albertson et al. (2006)/NASA/Public Domain.

→ ram/scramjet, with rocket assist for takeoff and transonic push-through). At Mach 3, the air-breathing engine undergoes mode transition from gas turbine to ramjet/scramjet, which subsequently accelerates the vehicle to Mach 8 with a pull maneuver at Mach 10. The low-speed, variable, mixed compression inlet is highly integrated with the forebody, and it uses boundary layer bleed to ensure efficient turbine airflow capture. The system is designed with sidewalls that extend into and partition the low-speed propulsion area. Integration requires constraining the size of the turbines to fit between the high-speed module sidewalls (Snyder and Espinosa 2013).

Boundary layer bleed is considered as a means to control or modify the boundary layer. It involves the removal of the low momentum flow adjacent to the wall via a bleed system. This has the effect of moving high momentum flow closer to the surface, resulting in a boundary layer that is more resistant to separation.

The higher and farther forward the LSI door is on the forebody, the greater the length, volume, and weight required by the low-speed propulsion system. But this also reduces the amount of flow turning and boundary layer growth for the flow captured by the LSI, which reduces the amount of inlet bleed and the amount of flow turning for turbine engine airflow. Conversely, positioning the LSI door lower and farther aft on the forebody increases its interaction with the HSI ramp, potentially disrupting the high-speed propulsion airflow, especially during the critical propulsion mode transition from turbine to ramjet/scramjet. In addition, there are heat loads and sealing requirements on the inlet doors. These requirements become more stringent as the LSI door location is moved further aft on the booster forebody.

In 2011, Aerojet (now Aerojet Rocketdyne) introduced a rocket-based combined cycle propulsion system design concept for reusable hypersonic vehicles identified as "TriJet" engine. The name implies that the propulsion system is comprised of three engines – turbojet, ejector ramjet (ERJ), and dual-mode ramjet – to achieve a seamless transition from a standing start to Mach 7 plus. The TriJet engine is comprised of a jet turbine engine, a DMRJ engine that has a converging inlet, a compression zone, a combustion zone, an expansion zone, and an outlet. A secondary turbine has an inlet and an outlet. The inlet of the secondary turbine is fluidly coupled to the compression zone, and the outlet of the secondary turbine is fluidly coupled to the turbine engine.

Powered by TriJet engines, a hypersonic aircraft would take off on turbine power, then ignite the rocket-augmented ERJ to push through the transonic drag rise and accelerate to the take-over speed for the DMRJ to achieve the final push to hypersonic flight. The Trijet operating envelope of three engine cycles overlaps to provide seamless propulsion. Interestingly, one inlet feeds all three engines. While DMRJ has an unobstructed flowpath, the turbojet and ERJ are concealed behind doors that open and close, depending on the phase of flight. The door settings for operating or closing the inlet are determined. According to published reports, the innovative inlet minimizes air spillage and enables a degree of combined thrust to be used throughout mode transitions when one engine shuts down while another takes over.

4.6 Inlet Operation: Start and Unstart

A scramjet must operate with its inlet started. Starting refers to the initiation of supersonic flow through the inlet duct. The Hypersonic inlet must operate in a *started* mode to capture and compress the required amount of air for engine processing. Started means that the inlet operates under conditions where flow phenomena in the internal inlet duct do not change the air capture characteristics of the inlet proper.

Starting of the inlet is one of the major constraints in the design phase, especially when the inlet is designed to operate in a wide flight Mach number range. This is because at the lower Mach number limit, the starting condition leads to a low contraction ratio, whereas, at the upper limit, high contraction ratio is mandatory to satisfy combustor entry requirements (slowing down of the flow). Inlet starting depends on the local flight Mach number, internal contraction ratio, diffuser flow field pressure recovery, and the time-dependent details of the starting process (Van Wie 2000).

When the normal operation of inlet deviates from its started condition we say that the inlet becomes unstarted (Wagner et al. 2007). The term *unstarted* means that the inlet operation is disturbed by either overcontraction, which causes the flow to choke at the throat of the inlet, or by an increase in the back pressure, which could occur if the chemical energy release in the combustor increases beyond the level appropriate to the inlet.

For a given inlet geometry and a set of freestream conditions, an inlet will start at a certain internal contraction. After the inlet starts, variable geometry features can be adjusted to increase the contraction ratio up to a maximum contraction ratio, at which point further increases would cause the inlet to unstart (Van Wie 2000; Wagner et al. 2007).

As we will see in another section, the function of the engine isolator is to prevent inlet unstart. It must provide sufficient adiabatic compression above its entry pressure p_2 to match or support whatever back pressure p_3 the burner may impress upon it.

4.6.1 Contraction Ratio Limit for Inlet Starting

A preliminary estimate of internal contraction ratio ($CR = A_2/A_3$) for the inlet to start is obtained with the Kantrowitz–Donalson limit. This is a value obtained by assuming a normal shock at the beginning of the internal contraction, and calculating the 1-D, isentropic, internal area ratio that will produce sonic flow at the throat.

For an arbitrary inlet diffuser, there are two equations that relate the Mach number to the contraction ratio: first, isentropic compression to sonic flow at the inlet diffuser throat limits the intake's contraction ratio to a maximum for a given Mach number. Second, a normal shock covering the inflow plane of a quasi-one-dimensional inlet flow is swallowed when the following holds. In a converging duct, the subsonic portion downstream of the shock is accelerated. If the contraction is equal to or less than a contraction that would choke the flow, that is to cause $M = 1$ at the throat, then the shock is swallowed. If the contraction is larger, the shock detaches and unstarts the diffuser flow. This criterion for inlet starting is due to Kantrowitz and Donaldson (1945), widely known simply as the "Kantrowitz limit."

For a perfect gas, and denoting the entrance to the internal part of the inlet by A_2, and A_3 the inlet throat, the Kantrowitz limit is given by the contraction ratio:

$$CR_K = \left(\frac{A_2}{A_3}\right)_K = \frac{1}{M_2}\left[\frac{(\gamma+1)M_2^2}{(\gamma-1)M_2^2+2}\right]^{\gamma/(\gamma-1)}\left[\frac{1+\frac{\gamma-1}{2}M_2^2}{\frac{\gamma+1}{2}}\right]^{(\gamma+1)/2(\gamma-1)}\left[\frac{(\gamma+1)}{2\gamma M_2^2-(\gamma-1)}\right]^{1/(\gamma-1)} \tag{4.22}$$

Please note that I assume that the inlet is internally contracting from station 2 to the inlet throat at station 3. Clearly, if the internal contraction begins at a different point in the diffuser, we must use the Mach number at that point instead of M_2. Moreover, if a given inlet design has $A_2/A_3 < CR_K$, this indicates that the inlet will always start.

The isentropic limit for the diffuser area contraction ratio is given in terms of M_0:

$$\left(\frac{A_2}{A_3}\right)_I = M_0\left[\frac{\gamma+1}{2}\right]^{(\gamma+1)/2(\gamma-1)}\left[1+\frac{\gamma-1}{2}M_0^2\right]^{-(\gamma+1)/2(\gamma-1)} \tag{4.23}$$

Van Wie published an empirical limit on achievable CR for the range $2.5 < M_0 < 10$:

$$\left(\frac{A_2}{A_3}\right)_{VW} = 0.05 - \frac{0.52}{M_0} + \frac{365}{M_0^2} \tag{4.24}$$

Smart (2010a) showed that the Kantrowitz limit is relatively accurate for 2-D inlet geometries but is conservative for the 3-D inlets he studied. Some experiments have shown that inlet can restart at internal contraction ratios (CR) beyond the Kantrowitz limit (Van Wie 2000).

In 2010, Sun and Zhang made the important observation that the wide range of contraction ratios published include very different kinds of inlets such as 2-D planar, some with sidewall compression, inward-turning Busemann, and REST inlets. Moreover, the unstart requirements imposed on those inlets were also quite different. Unstart was induced mechanically by: 1) increasing the internal contraction ratio by variable geometry, 2) raising the back pressure, and 3) increasing cowl lip shock strength to produce large separation flow. For example, Pan Jin (2005) and Jacobsen, et al. (2006) obtained the maximum self-starting internal contraction ratio at a given M_0 by translating or rotating the cowl, respectively. Jin used an inlet with sidewall compression; Jacobsen, et al. used an inward-turning inlet with a highly swept leading edge (beneficial to supersonic spillage), and thus it had a larger maximum self-starting internal contraction ratio. On the other hand, Smart (2001) used a REST inlet with a close self-starting internal contraction ratio.

Thus, using the data from Jin (2005), Smart (2001), and Jacobsen et al. (2006) (both experimental and numerical), Sun and Zhang (2010) found a polynomial law between the maximum self-starting internal contraction ratio and the average Mach number of the cowl lip

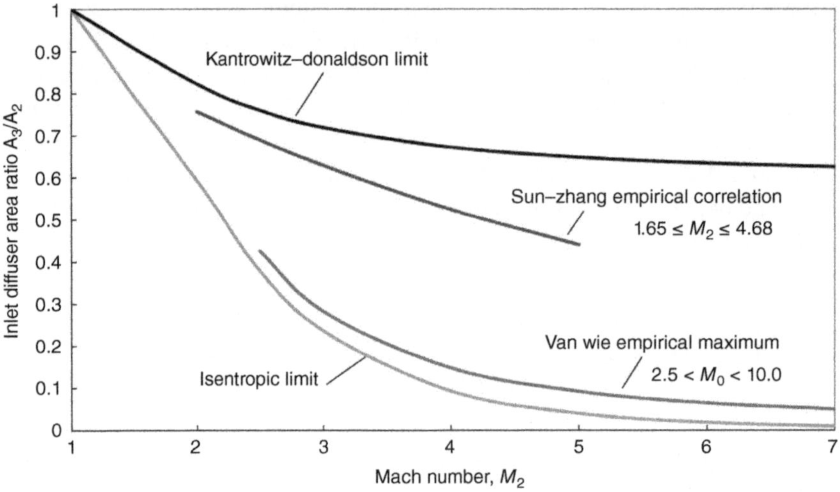

Figure 4.22 Criteria for inlet starting and maximum contraction limits.

section, an empirical correlation valid for $1.65 \leq M_c \leq 4.68$, where M_c is the Mach number at the cowl lip. For consistency with the notation in the previous correlations, their correlation is expressed as:

$$\left(\frac{A_2}{A_3}\right)_{S-Z} = 0.933 - \frac{M_2}{6.87} + \frac{M_2^2}{40.9} \tag{4.25}$$

Figure 4.22 plots the inlet contraction ratios given by these expressions, where the Kantrowitz limit gives the upper bound for any entrance flow Mach number. The maximum contraction ratio corresponds to the minimum attainable throat Mach number with inlet started (i.e. shock train downstream of throat).

The self-starting limits of particular inlet classes are determined through analysis and experimental testing. In addition to considering the range of operating Mach numbers, the design analysis must consider effects of cowl bluntness, cowl length, cowl angle of attack, boundary layer thickness, freestream Reynolds number, and wall temperature. To date, self-starting remains a challenging constraint for scramjet inlet design.

4.7 Inlet Aerodynamics

The flow field of the hypersonic inlet is complex due to the variety of developing aerothermodynamic phenomena, including shock waves, thick, transitional boundary layers, viscous interactions of shocks with boundary layers, bow shock/cowl shock interaction, and aerodynamic heating with enormous thermal loads at the leading edges, and even separated flow. Figure 4.1 indicates some of the inlet flow phenomena that make the design of the inlet more challenging. This is especially important for inlets designed to operate over large ranges of speed and altitude.

4.7.1 Inlet Boundary Layer

The boundary layer over the lower surface of the forebody should be turbulent. This is because ingesting a turbulent boundary layer reduces the potential for flow separations, increasing inlet operation stability and thus ensuring efficiency and overall propulsion performance. Hence, the forebody of a highly integrated hypersonic scramjet-powered vehicle should be designed to provide a turbulent boundary layer for the inlet (Berry et al. 2001).

The size of a full-scale vehicle is likely to have sufficient forebody length to provide a naturally turbulent boundary layer. However, a subscale prototype will probably require a mechanism to force the boundary layer to transition from laminar to turbulent. This is in fact what happened while planning the NASA Hyper-X (X-43A) and the US Air Force Scramjet Engine Demonstrator-WaveRider (X-51A) flight demonstration programs.

As we discussed in Chapter 3, an analysis of the Hyper-X forebody using the hypersonic boundary layer transition criteria developed during the NASP program indicated that the subscale X-43A vehicle forebody would remain laminar during the flight. Using a boundary layer code, the Hyper-X team computed laminar values of the momentum thickness Reynolds number ($Re_\theta = (\rho u_e \theta)/\mu$) over the boundary layer edge Mach number (M_e) for a sharp-nose wedge with 4.5 deg of turning. Then they used the NASP sharp planar transition criterion of $Re_\theta/M_e = 305$ to estimate the onset of transition.

Based on their initial estimate, the team concluded that transition would not occur on the forebody's first ramp prior to the X-43A vehicle's compression corner. They determined that over 200% more running length would be required for transition to occur on the first ramp, which was beyond the inlet. The vehicle's forebody length is about 6 ft (1.8288 m). The estimated length to natural transition onset is over 9 ft (2.7432 m).

Hence, a passive boundary layer trip array was added to the lower surface of the X-43A vehicle forebody to ensure a turbulent boundary layer entering the inlet for both Mach 7 and 10 flights (Berry et al. 2001). Without forcefully promoting transition with those boundary layer trips, the potential for a laminar layer separation existed at the first ramp break that could generate lateral flow spillage, potentially reducing the mass capture and affecting overall scramjet performance. The X-51A subscale waverider and the HIFiRE scramjet also used boundary layer trips on the lower surface forebody (see Chapter 3).

The successful X-43A, X-51A, and HIFiRE hypersonic flights have provided the opportunity to assess the rationale and methodologies utilized during the design of the flight vehicles. The published reports from those programs conclude that the boundary layer trips performed as expected and that the approach taken to scale the trips to flight is acceptable. For the X-43A, the upper surface analysis revealed that the pre-flight hypersonic transition criteria of $Re_\theta/M_e \sim 300$ is a reasonable estimate of transition onset (see Chapter 3).

4.7.2 Boundary Layer Growth in Lower Forebody Surface

At hypersonic speeds, the oblique shock wave generated by the vehicle's forebody nose is very close to the surface. The shock layer between the shock wave and the vehicle's surface is very thin. The thin shock layer can cause adverse interaction between the inviscid flow

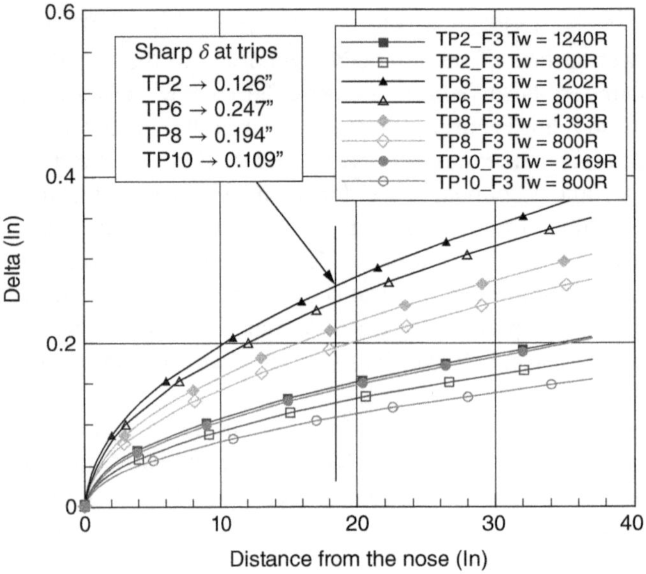

Figure 4.23 X-43A Mach 10 boundary layer thickness growth. Source: Berry et al. (2008)/American Institute of Aeronautics and Astronautics.

behind the shock and the viscous boundary layer on the surface. Hypersonic vehicles fly at high altitudes where density is low. Hence, the boundary layer is thick. As we noted in Chapter 3, at hypersonic speeds, the boundary layer thickness (δ) on slender bodies is proportional to the Mach number on the edge of the boundary layer (M_e). Boundary layer thickness may be of the same magnitude as the shock layer thickness.

Figure 4.23 shows the growth of the boundary layer over the forebody surface of the NASA X-43A vehicle, for its Mach 10 flight. In this figure, T201 indicates a thermocouple located at the furthest forward measurement location (see Figure 3.9), downstream from the boundary layer trip. TP2, TP6, TP8, and TP 10 represent trajectory points during the ascent flight; TP2 and TP10 are points when the trips were fully effective (turbulent flow measured by T201), while TP6 and TP8 correspond to when the trips were not effective (no turbulent flow at T201) (see Berry et al. 2008 for further discussion).

This example clearly shows that at the trip location the boundary layer thickness is about 0.28 in. (7.112 mm) and it grows to 0.32 in. (8.128 mm) at 32 inches (0.8128 m) from the vehicle nose. Other values of δ are indicated in Figure 4.23 for the four trajectory points. The effect of wall temperature on boundary layer thickness is also evident in this plot. Observe at the farthest axial location, when the temperature is T_w = 1202 °R (666.67 K), $\delta \approx 0.35$ in (8.89 mm), but when T_w = 800 °R (444.44 K), $\delta \approx 0.33$ in (8.38 mm).

At hypersonic speed, the thick boundary layer deflects the external inviscid flow, creating a strong, curved shock wave trailing downstream from the leading edges. There the surface pressure is much higher than the freestream pressure (far downstream $p \rightarrow p_0$). High pressure increases the aerodynamic heating at the vehicle's leading edges. The interaction between the inviscid freestream and the viscous boundary layer affects the wall pressure distribution, the skin friction, and the heat transfer.

4.8 Isolator

In the flight regime $4 < M_0 < 12$, the scramjet operates with dual-mode combustion. That is, the propulsion system must operate in ramjet mode and as a scramjet proper. In the ramjet mode, there is a large pressure rise associated with both subsonic diffusion and combustion processes, and disturbances from the heat release in the combustor can propagate and affect the performance of the inlet. Billig et al. (1972) recognized the need for inlet isolation while testing a hydrogen combustor. They used boundary layer bleed to stabilize the shock system at the entrance. However, the boundary layer bleed enhanced the pressure drop immediately downstream of the combustor entrance. Thus, the solution was to add a duct section to isolate and contain the high-pressure flow.

Two critical questions that must be answered in preliminary design: *how the flow within the inlet compression system can be arranged to change from subsonic to supersonic as the vehicle accelerates* and *how efficient air compression can be obtained over the entire range of freestream Mach numbers*.

We could consider an inlet with variable geometry so that the captured airflow changes with Mach number. But we know that the inlet has to get started at the slowest flight velocity. But inlet starting is very sensitive to the shape of the internal flow field, that is, the contraction ratio and back pressure. In fact, lateral walls and cowl leading edges may be designed to support the starting mode of the inlet. Also, the inlet of the dual-mode engine could use a short diffuser (internal compression section) and use thermal choke for controlling the transition from subsonic to supersonic condition.

To stabilize the operation of the ramjet, back pressure is generated by choking in the exhaust nozzle throat. To minimize total pressure loss across the normal shock, the flow Mach number at the beginning of the shock train is restricted to low values, ~ Mach 1.4–1.8. However, this requires a relatively large inlet contraction ratio, and also boundary layer bleed to stabilize the shock system and prevent inlet unstart. This means that the inlet total pressure loss is greater than that of a scramjet, since the latter operates at a lower inlet contraction ratio.

A fixed geometry scramjet inlet decelerates the airflow through oblique shocks, and this process is optimized for a given flight Mach number. Hence, operation at a lower or higher Mach number than designed for, the inlet will not operate efficiently as it will be either underspeed or overspeed, as explained in Chapter 3. The shock in an underspeed inlet misses the cowl and the airflow spills out. In an overspeed inlet, the shock impinges in the inner part of the inlet, yielding an airflow to the combustor with adverse gradients. The optimal operation is with the shock-on-lip (see Figure 4.2). Therefore, a fixed geometry inlet must incorporate some safety margin in its design to minimize those operational issues.

At low speeds or conditions with high heat release in the combustor, this forces the pre-combustion shock system into the constant area inlet throat, downstream of the engine cowl. Due to the relative thick boundary layer in the scramjet inlet, it requires an isolator duct to contain the pressure rise accompanying the shock system. The isolator in fact separates the inlet proper from the combustor. The isolator section starts at the minimum geometric cross-sectional area of the inlet, "inlet throat," and it extends to the combustor section in the form of a constant area (or nearly constant area) duct.

4.8.1 Isolator Shock Train

Heiser and Pratt (1994) give schematics of the types of shock trains that may develop in the isolator. They show that the characteristics of the shock train depend on the conditions at the entrance of the isolator, in particular the Mach number and the boundary layer thickness. A normal shock train tends to occur at lower inlet Mach numbers and have thicker boundary layers. An oblique shock train tends to occur at higher inlet Mach numbers and have thinner boundary layers. Since the constant area duct produces a static pressure rise, the isolator is also called constant area diffuser. Moreover, the pressure ratio profile for each type of shock train reflects the development of the distinct flow field within the duct.

Consider the normal shock train. At the entrance, the initial normal shock wave decomposes into oblique shock waves near the wall as it interacts with the boundary layer separating at the wall. The highest air compression occurs immediately after due to the first normal shock wave. A jump in pressure ratio is indicative of a dramatic air compression. Downstream of the jump the boundary layer thickness grows and evens the pressure rise over the subsequent normal shock waves, developing into a continuous compression process.

Now let us turn our attention to the oblique shock train. It begins with an oblique shock that separates the boundary layer, and the shock is then reflected on the opposite wall as an expansion wave. In this manner, the oblique shocks continue propagating downstream, alternating between compression and expansion waves, causing a gradual zigzag increase in static pressure. The boundary layer slims and thickens in response to the expansion and compression shock waves, and it remains separated.

In both cases, the shock train provides the mechanism for a supersonic flow to adjust to a static back pressure higher than its inlet static pressure. The shock train is considered to be fully contained within the isolator if further addition of isolator length does not continue to increase the pressure level.

Isolator: Scramjet engine duct section added between the inlet and the combustor to maintain stable operation at low hypersonic speeds. The isolator diffuser contains the inlet shock train.

Shock train: System of oblique or normal shocks, where pressure rises and Mach number decreases. The increase of static pressure occurs within the initial duct region, after which the boundary layer grows as the flow moves downstream.

A very lucid explanation regarding the aerothermodynamics of the dual-mode combustion system (inlet isolator and combustor) is given by Heiser and Pratt (1994). They used a thermal energy versus kinetic energy plot to analyze the complex interactions between the isolator and combustor, concluding that the nature of interaction between the two components is different for ramjet and scramjet operations. In the ramjet mode, the constant area isolator must contain a shock system consistent with subsonic combustion pressure rise. In the scramjet mode, heat addition in a constant area combustor occurs in a separated core flow at nearly constant pressure equal to the maximum pressure rise at the combustor exit.

4.8.2 Isolator Length

The key feature of the isolator is the length required to contain the shock train and isolate the inlet from the flow influences propagating upstream from the combustor. In early studies attempting to predict the length scale of this flow structure, Waltrup and Billig (1973) observed that for a given imposed pressure rise ($\Delta p/p$), the length over which the shock train spread varied with $(D\theta)^{1/2}$ and inversely with $(M_2^2 - 1)(Re_\theta)^{1/4}$, where D is the duct diameter, M_2 is the Mach number of the inflow, and Re_θ is the Reynolds number based on the momentum thickness θ of the boundary layer.

Based on experiments in round ducts where $1.5 < M_2 < 2.7$, Waltrup and Billig found that thicker boundary layers led to longer shock trains, and they developed an empirical correlation for the distance over which the shock structure (or pressure rise) is spread. This correlation was published by Heiser and Pratt (1994) in a form amenable to calculation, as the ratio of shock train axial length L to duct height H, written in terms of back pressure ratio:

$$\frac{L}{H} = \frac{\left(\frac{\theta}{H}\right)^{1/2}}{(Re_\theta)^{1/4}} \cdot \frac{\left[50\left(\frac{p_3}{p_2} - 1\right) + 170\left(\frac{p_3}{p_2} - 1\right)^2\right]}{(M_2^2 - 1)} \tag{4.26}$$

where p_3/p_2 is the maximum static pressure ratio obtainable for a given inlet Mach number. This ratio is obtained by substituting the static pressure ratio for a normal shock wave at the isolator inlet Mach number M_2:

$$\frac{p_3}{p_2} = \frac{2\gamma}{\gamma + 1} M_2^2 - \frac{\gamma - 1}{\gamma + 1} \tag{4.27}$$

Using representative values of the inlet conditions and back pressure ratios, Heiser and Pratt determined that L/H can be large compared to 1 when M_2 is small and the back pressure ratio is large and determined that $L/H \geq 10$ for typical flight conditions. Moreover, Heiser and Pratt concluded that, in order to prevent unstart, one may require a relatively high L/H for the isolator operating at relatively low supersonic freestream Mach numbers. Then, "when the height of the duct is reduced in order to match the freestream capture area at high freestream Mach numbers, the isolator surface area divided by throughflow area may become so large that the internal wall friction losses are unacceptable." (Heiser and Pratt 1994)

Determining the length of the isolator requires modeling of separated, diffusing flows in internal ducts. Ortwerth (2000) carried out an extensive study using experimental data at different Mach numbers, Reynolds numbers, and in different duct geometries. He determined that the rate of pressure rise (diffusion) in a duct is directly proportional to the dynamic pressure of the incoming flow and the skin friction coefficient at the initial point of separation in the duct and inversely proportional to the duct hydraulic diameter.

Ortwerth's diffuser model provides a correlation to determine a length scale over which pressure rise must be spread in a constant area duct:

$$\frac{dp}{dx} \approx \frac{89}{D_H} c_f \left(\frac{\rho V^2}{2}\right) \tag{4.28}$$

where D_H is the hydraulic diameter of the duct and c_f is the friction coefficient at the initial separation point.

The skin friction coefficient can be determined using a semiempirical equation such as that of Mahoney (1990), which is valid for compressible turbulent flow up to Mach 4:

$$c_f = \frac{0.472}{(\log_{10} Re_e)^{2.58}\left(1 + \frac{\gamma - 1}{2} M_e^{0.467}\right)^{0.467}} \tag{4.29}$$

where the Reynolds number Re_e is based on the condition of the boundary layer edge.

Ortwerth's correlation is widely used as it provides a formula to determine a length scale over which pressure rise must be spread in a duct. Smart (2010b) implemented this model as part of a quasi-one-dimensional cycle code for the calculation of isolator flows. Considering the distribution of heat release in the combustor and the isolator/combustor geometry, Ortwerth predicted the length of the upstream influence and hence the required isolator length.

The isolator duct has two functions: first, the isolator must provide a buffer zone between the inlet and combustor in order to impede or, at least, to minimize interferences between components. This requires that the isolator allows the inlet to operate continuously and effectively over the specified speed range, while at the same time withstanding the high peak pressure rises that originate in the combustor. Second, the isolator must diffuse the supersonic flow to a subsonic condition, while maximizing recovery of the total pressure, a requirement that is vital to efficient operation of both the inlet and combustor in the ramjet mode and preventing inlet unstart.

Hence, if the length of the isolator is shorter than the shock train length, the maximum value of back pressure before inlet unstart decreases rapidly with decreasing isolator length. On the other hand, if the isolator length is longer than the shock train length, the additional viscous losses decrease the pressure recovery gradually with increasing isolator length, and it adds more structural and heat loads. Therefore, the optimal length for the isolator must be determined early on the design process, a length that yields a large percentage of the normal shock pressure rise at the inlet throat with short length scales. This determination will result from trade-off studies of integrated components over the flight trajectory of a propulsion system. Surface roughness can significantly increase boundary layer thickness and growth, and it must be considered in defining isolator performance.

Much research has been carried out since Curran and Stull (1964) first described DMRJ operation, including development of correlations based on experimental work and analysis by Billig and other researchers. Their studies identified the importance and degree of downstream pressure rise, total temperature and enthalpy ratios, and inflow boundary layer characterization in order to predict the isolator length. These correlations were developed from simple geometries such as axisymmetric and two-dimensional configurations, and thus they may not be applicable to more complex configurations. This has been demonstrated by more recent studies involving experimental tests and numerical simulations.

The challenge remains to design DMRJ capable of operation over a wide speed range, which requires an optimized isolator capable of performing its function at low speeds without penalizing propulsion performance at high hypersonic speeds. Most recently, numerous CFD studies of dual-mode combustion scramjets are providing additional insights in the flow physics within the isolator (Fiévet et al. 2015; Baurle and Axdahl 2017). Chapter 7 will address other aspects of dual-mode combustion.

Questions

1. What type of inlet compression system would you select for a scramjet engine that will power a reusable Mach 10 cruise vehicle?

2. What is the function of the Isolator when the scramjet operates in pure supersonic combustion mode?

3. What causes scramjet engine unstart? Explain how to prevent it.

4. Is there an optimum number of oblique shocks to achieve adequate compression performance for a given hypersonic flight condition?

5. At high flight velocities, the leading edges of the scramjet inlet are blunted to decrease surface temperature. Leading edge bluntness can result in significant decrease of pressure recovery and mass flow rate. Explain how leading edges bluntness affects the flow characteristics at the scramjet engine entrance.

References

Albertson, C.W., Emami, S., and Trexler, C.A. (2006). Mach 4 test results of a dual-flowpath, turbine based combined cycle inlet. *Paper AIAA 2006-8138. 14th AIAA/AHI International Space Planes and Hypersonics Systems and Technologies Conference*, Canberra, Australia (1 January 2006).

Anderson, G.Y., McClinton, C.R., and Weidner, J.P. (2000). Chapter 6 in scramjet propulsion. In: *Scramjet Performance* (ed. E.T. Curran and S.N.B. Murthy). AIAA Progress in Astronautics and Aeronautics.

Andrews, E.H. Jr. and Mackley, E.A. (1976). Hypersonic research engine/aerothermodynamic integration model, experimental results. Vol. 1: Mach 6 component Integration. NASA TM-X-72821.

Baurle, R.A. and Axdahl, E.L. (2017). Uncertainty quantification of CFD data generated for a model scramjet isolator flowfield. *JANNAF Joint Subcommittee Meeting* (4–8 December 2017). Newport News, VA. Unclassified; Publicly available paper.

Berry, S.A., Auslender, A.H., Dilley, A.D., and Calleja, J.F. (2001). Hypersonic boundary layer trip development for Hyper-X. *Journal of Spacecraft and Rockets* **38** (6): 853–864.

Berry, S., Daryabeigi, K., Wurster, K., and Bittner, R. (2008). Boundary layer transition on X-43A. *AIAA-2008-3736. 38th Fluid Dynamics Conference and Exhibit*, Seattle, Washington (23–26 June 2008).

Billig, F.S., Orth, R.C., and Lasky, M. (1968). Effects of thermal compression on the performance estimates of hypersonic ramjets. *Journal of Spacecraft and Rockets* **5** (9): 1076–1081.

Billig, F.S., Dugger, G.L., and Waltrup, P.J. (1972). Inlet-combustor interface problems in scramjet engines. *Proceedings of the 1st International Symposium on Air-Breathing Engines*, Marseilles, France (June 1972).

Candler, G.V. and Drayna, T.W. (2008). Design and optimization of the ASET inward-turning scramjet inlet. *JANNAF 30th Airbreathing Propulsion Subcommittee Meeting*, Boston (2008).

Drayna, T.W., Nompelis, I., and Candler, G.V. (2006), Hypersonic Inward Turning Inlets: Design and Optimization. *AIAA Paper 2006-0297. 44th AIAA Aerospace Sciences Meeting and Exhibit*, Reno, Nevada (9–12 January 2006).

Ferlemann, P.G. (2008). Forebody and inlet design for the HIFiRE 2 flight test. *JANNAF Airbreathing Propulsion Subcommittee Meeting*, Boston, Massachusetts (12–16 May 2008).

Fiévet, R., Koo, H., and Raman, V. (2015). Numerical simulation of a scramjet isolator with thermodynamic nonequilibrium. *AIAA 2015-3418. 22nd AIAA Computational Fluid Dynamics Conference*, Dallas, TX (22–26 June 2015).

Gollan, R.J. and Ferlemann, P.G. (2011). Investigation of REST-class hypersonic inlet designs. *AIAA-2011-2254, 17th AIAA International Space Planes and Hypersonic Systems and Technologies Conference*, San Francisco, CA (11–14 April 2011).

Heiser, W.H. and Pratt, D.T. (1994). *Hypersonic Airbreathing Propulsion*. AIAA.

Hohn, O.M. and Gülhan, A. (2011). Experimental investigation on the influence of sidewall compression on the flowfield of a scramjet inlet at Mach 7. *AIAA 2011-2350. 17th AIAA International Space Planes and Hypersonic Systems and Technologies Conference*, San Francisco, California (11–14 April 2011).

Holland, S.D. (1995). Mach 10 computational study of a three-dimensional scramjet inlet flow field. Technical Memorandum TM–4602, NASA, March 1995.

Holland, S.D. and Perkins, J.N. (1991). Contraction ratio effects in a generic three-dimensional sidewall compression scramjet inlet: a computation and experimental investigation. *AIAA 1991-1708. 22nd AIAA Fluid Dynamics, Plasma Dynamics, and Lasers Conference* (June 1991).

Jackson, K.R., Gruber, M.R. and Buccellato, S. (2011). HIFIRE Flight 2 overview and status update 2011. *58th JANNAF Propulsion Meeting*, Arlington, VA (18–22 April 2011).

Jacobsen, L.S., Tam, C.J., Behdadnia, R., and Billig, F.S. (2006). Starting and operation of a streamline-traced busemann inlet at Mach 4. *AIAA Paper 2006-4508. 42nd AIAA/ASME/SAE/ASEE Joint Propulsion Conference & Exhibit*, Sacramento, CA (9–12 July 2006).

Kantrowitz, A. and Donald, C. (1945). *Preliminary Investigation of Supersonic Diffusers*, NACA WRL-713, 1–25. National Advisory Committee for Aeronautics. Langley Aeronautical Lab, Langley Field, VA.

Mahoney, J.J. (1990). *Inlets for Supersonic Missiles*, AIAA Education Series. American Institute of Aeronautics and Astronautics.

Matthews, A. and Jones, T. (2006). Design and test of a modular waverider hypersonic intake. *Journal of Propulsion and Power* **22**: 913–920.

Ortwerth, P. (2000). Scramjet flowpath integration. Chapter 17 in. In: *Scramjet Propulsion*, Progress in Astronautics and Aeronautics, vol. **189** (ed. S.N.B. Murthy and E.T. Curran). AIAA.

Oswatitsch, K. (1947). Pressure recovery in missiles with reaction propulsion at high supersonic speeds (the efficiency of diffusers). NACA TM 1140.

Pan, J. (2005). Tests and numerical simulation on characteristics of self-starting for sidewall compression inlets. M.S. Dissertation. College of Energy and Power Engineering, Nanjing Univ. of Aeronautics and Astronautics, Nanjing, PRC(in Chinese).

Schneider, S. (2004). Hypersonic laminar-turbulent transition on circular cones and scramjet forebodies. *Progress in Aerospace Sciences* **40**: 1–50.

Smart, M.K. (1999). Design of three-dimensional hypersonic inlets with rectangular-to-elliptical shape transition. *Journal of Propulsion and Power* **15** (3): 408–416.

Smart, M.K. (2001). Experimental testing of a hypersonic inlet with rectangular-to-elliptical shape transition. *Journal of Propulsion and Power* **17** (2): 408–416.

Smart, M.K. (2010a). Scramjet inlets. *RTO-EN-AVT-185-09*. Presented at the AVT-185 RTO AVT/VKI Lecture Series held at the von Karman Institute, Rhode St. Genèse, Belgium (13–16 September 2010).

Smart, M.K. (2010b). Scramjet isolators. *RTO-EN-AVT-185-10*. Presented at the AVT-185 RTO AVT/VKI Lecture Series held at the von Karman Institute, Rhode St. Genèse, Belgium (13–16 September 2010).

Smart, M.K. (2012). How much compression should a scramjet inlet do? *AIAA Journal* **50** (3): 610–619.

Smart, M.K. and Ruf, E.G. (2006). Free-jet testing of a REST scramjet at off-design conditions. *AIAA Paper 2006-2955. 25th AIAA Aerodynamic Measurement Technology and Ground Testing Conference*, San Francisco, CA (5–8 June 2006).

Snyder, C.A. and Espinosa, A.M. (Jose) (2013). Lessons learned during TBCC design for the NASA-AFRL joint system study. NASA/TM-2013-218100.

Sun, B. and Zhang, K. (2010). Empirical equation for self-starting limit of supersonic inlets. *J. of Propulsion and Power Technical Notes* **26** (4): 874–875.

Trexler, C.A. and Souders, S.W. (1975). Design and performance at a local Mach number of 6 of an inlet for an integrated scramjet concept. NASA TN-D-7944.

Turner, J.C. and Smart, M.K. (2013). Mode change characteristics of a three-dimensional scramjet at Mach 8. *Journal of Propulsion and Power* **29** (4): 982–990.

Van Wie, D.M. (2000). Scramjet inlets. In: *Scramjet Propulsion* (ed. E.T. Curran and S.N.B. Murthy). AIAA Progress in Astronautics and Aeronautics.

Van Wie, D.M. and Ault, D.A. (1996). Internal flowfield characteristics of a scramjet inlet at Mach 10. *Journal of Propulsion and Power* **12** (1): 158–164.

Wagner, J., Valdivia, A., Yuceil, K. et al.(2007). An experimental investigation of supersonic inlet unstart. *AIAA Paper 2007-4352. 37th AIAA Fluid Dynamics Conference and Exhibit*, Miami, FL (25–28 June 2007).

Waltrup, P. J., and Billig, F. S., Prediction of precombustion wall pressure distributions in scramjet engines, *Journal of Spacecraft and Rockets*, Vol. **10**, No. 9, Sep. 1973. pp. 620–622.

Waltrup, P.J., Billig, F.S., and Stockbridge, R.D. (1982). Engine sizing and integration requirements for hypersonic airbreathing missile applications. AGARD-CP-207, No. 8.

Weidner, J.P. (1979). *Conceptual Study of a Turbojet/Ramjet Inlet*, NASA Technical Memorandum 80141. Hampton, Virginia: NASA Langley Research Center.

5

Scramjet Combustor

Supersonic combustion is a topic of intense research, aimed at advancing scramjet propulsion. Combustor technology development main focus is on understanding the requirements for fuel injection, fuel/air mixing, and chemical reaction (ignition and flameholding), and their performance in the scramjet combustor environment. Such effort takes a multidisciplinary approach, using experimental technology and computational tools to address the complexity of high-speed combustion phenomena. In the scramjet design process, the extensive application of computational fluid dynamics (CFD) tools is essential. The flowpaths may be highly three-dimensional, requiring the application of CFD codes that are specifically developed for high-speed flows, including complex turbulent and chemical kinetic models to accurately describe relevant physical and chemical effects.

In the 1960s, NASA Langley began research efforts to establish a technology base for scramjet engine components, focusing on the Mach 4 to 7 speed range to be within the enthalpy limits of combustion ground facilities. At the same time, efforts to develop the computational tools to model the combustion process began with cold flow studies of single and multiple fuel injectors to characterize fuel/air mixing, followed by reacting coaxial flow configurations to characterize the supersonic combustion phenomena. In those early studies, various finite-rate chemistry models were developed and applied for hydrogen-air and silane-hydrogen-air mixtures. Results from those studies provided valuable insights into the highly complex combustion flow fields.

Building on the early findings, further improvements on both experimental and computational methods led to greater advances in the understanding of supersonic combustion phenomena, extending the Mach speed range and the type of fuels injected in the combustor. However, many challenges remain to optimize the scramjet combustor design, especially when requiring to operate in a dual mode. For example, the combustor section of a combined cycle propulsion system must be designed to provide high combustion efficiency and maximum energy release over a wide range of flight conditions, from subsonic to supersonic velocities. While combustion technology is relatively mature for rocket, ramjet, and turbojet engines, this is not the case for dual-mode scramjet combustion, and more so for multimode, combined cycle propulsion operation.

A great deal of research has been carried out to understand and characterize supersonic combustion, making it difficult to address all topics in this chapter. Hence, an effort is made

Scramjet Propulsion: A Practical Introduction, First Edition. Dora Musielak.

to provide the reader sufficient material as it is necessary for understanding basic operating principles, expected performance parameters, and the design limitations of the scramjet combustor. To develop a more thorough background, the reader will be referred to the relevant literature.

Key concepts related to scramjet combustors are highlighted, striving to answer basic questions such as:

- *What are the optimum air conditions* (static pressure and temperature) *at the combustor entrance?*
- *What are the ignition times for hydrocarbon fuels injected into supersonic airflow?*
- *How fuel can be injected to maximize mixing and ignition in a short residence time?*
- *Why silane gas was injected into the X-43A scramjet combustor?*

The scramjet combustor may have planar two-dimensional (2-D), axisymmetric, or highly three-dimensional (3-D) geometry, and its design becomes more complicated if the combustion process transitions from subsonic to supersonic conditions. Combustor geometries also vary from high aspect ratio (width/height) rectangular and divergent to circular (see Section 5.5). In the next sections, we provide an overview of fundamental concepts and discuss the technical challenges we face to design combustors for scramjet propulsion.

5.1 Combustor Process Desired Properties

Combustion is an exothermic chemical process which requires fuel, oxidizer, initiation energy, and time for molecular mixing and chemical reactions to take place. At high hypersonic speeds, the internal static pressure and temperature of the air compressed in the inlet become extremely high. That is why, at flight speeds beyond Mach 5, air entering the combustor must be supersonic throughout to keep thermal loads within acceptable limits, and to avoid excessive dissociation of nitrogen and oxygen species.

At high flight Mach numbers, combustor residence times are very small. Residence time refers to the time available for fuel injection, fuel–air mixing, and chemical reaction to take place. For example, the residence time in a 0.5-m combustor is about 2.3×10^{-4} seconds (assuming air velocity of 2165 m/s). This very short residence time combined with flow compressibility effects may suppress the required molecular mixing of fuel and air species.

Considerable research and development work has established the performance requirements for the scramjet combustor. The desired properties of the combustor include:

- Complete combustion, i.e. combustion efficiency $\eta_b \rightarrow 100\%$;
- Low total pressure loss, i.e. combustor total pressure ratio $\pi_b \rightarrow 1.0$;
- Combustion stability, i.e. perturbations do not grow or enlarge enough to alter the required features of the flow in the combustor;
- Adequate temperature distribution at the exit plane, i.e. small thermal gradients in the flow entering the engine exhaust nozzle;
- Short length and small cross section, i.e. reduced internal drag and minimum weight;
- Re-lightability, i.e. capable of re-igniting if flameout occurs;

- Operation over wide range of mass flow rates, pressures, and temperatures to accommodate to variation of flight conditions.

For stability, the overall combustion process should be mixing controlled. However, at high Mach numbers combustion in the upstream portion of the combustor is kinetically controlled (Drummond 2014). This means that the ignition delay times are on the same order as the fluid scale. In general, both mixing and combustion timescales must be considered in studies of mixing and chemical reaction to better characterize the flow processes and achieve an optimum scramjet design. To achieve optimum performance at all flight conditions, the combustor design must consider many interrelated processes, including:

- Fuel injection;
- Fuel/air mixing;
- Ignition;
- Flameholding, propagation, and stabilization;
- Transition from subsonic to supersonic flow combustion;
- Thermal choking;
- Interaction with engine inlet operation (inlet pressures, spillage, and unstart), and exhaust nozzle (expanding flow to produce thrust).

5.2 Combustor Entrance Conditions

The combustor entry conditions, such as Mach number and thermodynamic state of the gas, are defined by taking into consideration propulsion cycle efficiencies, temperature limitations imposed by materials, and by-product gas dissociation. Let us review the entrance conditions, by referring to Figure 5.1, which shows the combustor section (flowpath stations 3 and 4) connected to other parts of the scramjet engine.

5.2.1 Combustor Entrance Pressure

The inlet component of the scramjet must supply air to the combustor at a pressure that ensures completion of the chemical reaction of air and fuel within a reasonable length. This

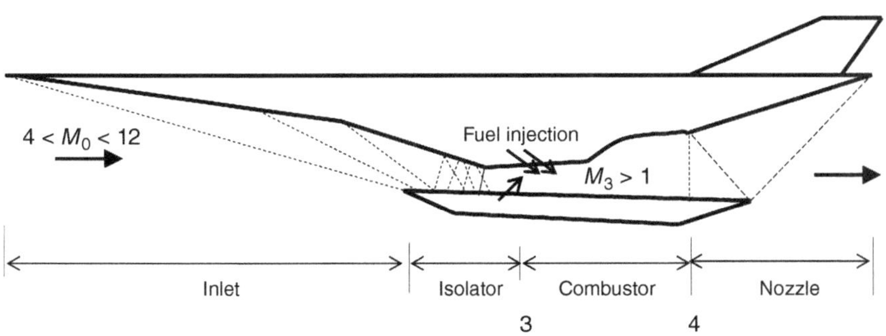

Figure 5.1 Scramjet flowpath.

length should be as short as possible, since heat and structural loads are higher in the combustor. The static pressure required at the combustor entrance, p_3, is one of the most important parameters that determine the efficiency and completeness of combustion, affecting the design of the combustor. Combustion reaction rates increase rapidly with static pressure and temperature. On the other hand, if p_3 is too low then the length of the combustor required to complete the reaction and consume the available fuel will become excessive. Hence, the combustor requires the correct air compression from the engine inlet.

Using a one-dimensional, steady flow model, Heiser and Pratt (1994) developed an expression for the required combustor entry pressure p_3 that incorporates the effect of engine inlet compression efficiency η_c, cycle temperature ratio $\psi = T_3/T_0$ (the principal determinant of thermodynamic engine efficiency), and flight conditions (dynamic pressure q_0 and flight Mach number M_0):

$$p_3 = \frac{2\,q_0}{\gamma_0 M_0^2[1 - \eta_c(1 - (1/\psi))]^{\gamma_c/(\gamma_c - 1)}} \tag{5.1}$$

where the specific heat ratios take average values $\gamma_0 = 1.4$, $\gamma_c = 1.36$

According to this expression, the entry pressure to the combustor varies inversely with the square of the flight Mach number. We plot Eq. (5.1) to determine the trends in combustor entry static pressure. First, examine the effect of the cycle temperature ratio, choosing $\psi = 7.0$, and 8.0, and consider a compression efficiency of 90% for a scramjet flying at an altitude where the dynamic pressure is $q_0 = 47.88$ kPa. As shown in Figure 5.2, this model predicts p_3 attaining values that are very large, especially at the low Mach numbers. For example, for $M_0 = 5.0$, $\psi = 7$,the required combustor entrance pressure is $p_3 = 722$ kPa (7.12 atm), while for $M_0 = 7.0$, $\psi = 7$, the value is $p_3 = 368.4$ kPa (3.63 atm). When the cycle temperature increases to $\psi = 8.0$, the pressures at the same flight Mach numbers increase to $p_3 = 951$ kPa (9.38 atm) and $p_3 = 485$ kPa (4.78 atm), respectively.

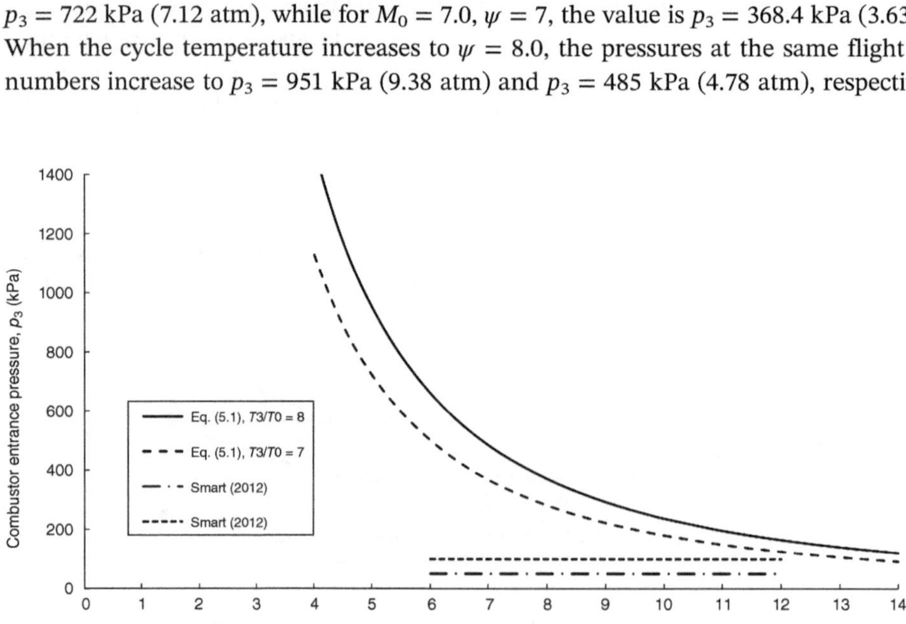

Figure 5.2 Combustor entry static pressure as a function of freestream. Mach number and cycle static temperature ratio.

In 2012, M. Smart published research aimed at determining the level of air compression required by a scramjet engine. He found that the overall cycle efficiency did not improve directly with compression level but reached an optimum value at a compression pressure ratio $p_3/p_0 \sim 50 - 100$ over the entire Mach range he investigated (Mach 6, 8, 10, and 12). Moreover, nonequilibrium effects in the nozzle and inlet operability constraints related to flowpath starting and boundary layer separation suggested a need for keeping compression level as low as possible. Not surprisingly, the minimum limit on compression level was given by the combustor pressure needed to complete the combustion reaction in the available length scale of the engine combustor. Smart concluded that none of these effects indicated a need for compression ratio higher than 100, even at the maximum flight Mach number of 12.

Based on this study, Smart (2012) recommended to set the inlet compression ratio to the minimum that enables the combustion reaction to be completed in the length scale of the combustor. For example, if the combustor scale is such that the allowable reaction length is 0.1 m, then the inlet must supply airflow at approximately 80 kPa. For typical wind-tunnel-scale hydrogen-fueled scramjet combustors (with allowable reaction length of approximately 0.3 m), the combustor flow pressure is approximately 50 kPa (~0.5 atm). Extending the allowable length scale to 1.0 m (for a larger-scale flight vehicle) lowers the required combustor supply pressure to 20 kPa (~0.2 atm). These values are indicated in Figure 5.2.

Although the specific value of entrance pressure chosen for a given combustor may also depend on other parameters, including the type of fuel, the method of fuel injection, the combustor geometry, and the length of the actual vehicle, we expect the combustor entrance static pressure to be in the range 0.2 atm (2.94 psi) $< p_3 <$ 10 atm (147 psi).

5.2.2 Combustor Entrance Temperature

Another question we must answer is, *what is the combustor entry temperature for the optimum scramjet engine cycle*? For hypersonic air-breathing engines, the maximum allowable combustor entry temperature T_3 requires a comprehensive analysis, as its value depends on many interrelated variables, including flight altitude, flight Mach number M_0, inlet losses, fuel type, fuel-to-air ratio, combustor, and nozzle geometry. However, to limit the value of the static temperature in the scramjet, we can examine the behavior of air as it is subjected to different temperatures.

First, consider the need to ignite the fuel. If the air temperature is too low, it is difficult to sustain a hydrogen or hydrocarbon fuel flame. Thus, the static temperature must be high enough to achieve ignition. For example, the minimum temperature to achieve hydrogen gas ignition is about 845 K. For fast ignition of SiH_4 (<50 μs at 1.25 atm), the combustor entrance air temperature should be approximately 1000 K (Koshi et al. 1991).

The cycle static temperature ratio T_3/T_0 is a principal factor in thermodynamic cycle efficiency and can be used to impose a limit of maximum allowable compression temperature, or equivalently the combustor entry temperature T_3 of the flow in the scramjet. At the same time, the combustor entry temperature T_3 must be controlled due to air dissociation during the heat addition (enthalpy increase) process, which will reduce thermodynamic cycle efficiency.

The efficiency of the scramjet inlet affects the combustion process. An efficient inlet can achieve a given static pressure level of air compression with isentropic turning and weak shock waves, yielding a lower static temperature rise. A less efficient inlet will compress the flow with very strong shock waves, and the static temperature rise will be much larger. This can limit scramjet performance because higher static temperature can lead to excessive chemical dissociation of the air, which in turn significantly reduces the combustion heat release.

Therefore, the air entering the combustor should have a lower static temperature (for a given static pressure) so that the combustion temperature is lower than the chemical dissociation temperature. The dissociation of molecular oxygen (O_2) into atomic oxygen (O) begins at approximately 2000 K (3600 °R), and the rate of change of dissociation is greatest at about 2800 K (~5000 °R). This establishes a maximum allowable compression temperature to be in the range $1440\ K < T_{3,\ max} < 1670\ K$ (2600–3000 °R).

Cain and Walton (2006) plotted inlet–outlet temperature (or combustor entry temperature) as a function of flight Mach number for a number of inlet designs (theoretical and actual inlets), assuming freestream air static temperature of 220 K. They obtained the theoretical curves in Figure 5.3 by considering air compression to a pressure that is twice the dynamic pressure q_0 since that is close to the limit for a self-starting inlet. The CFD data point from Smart (2016) is superimposed on this figure for a Mach 8 REST combustor analysis where $T_3 = 932$ K, $p_3 = 47.9$ kPa. For all cases in the flight regime $5 < M_0 < 11$, the air compression temperature is in the range $650\ K < T_3 < 1100\ K$, values that are lower than the upper limit to avoid dissociation before air enters the combustor. As noted by Cain and Walton, the theoretical curves in Figure 5.3 do not include skin friction losses, which are significant for hypersonic inlets, but the values are given to provide trends and gain a sense of the expected performance.

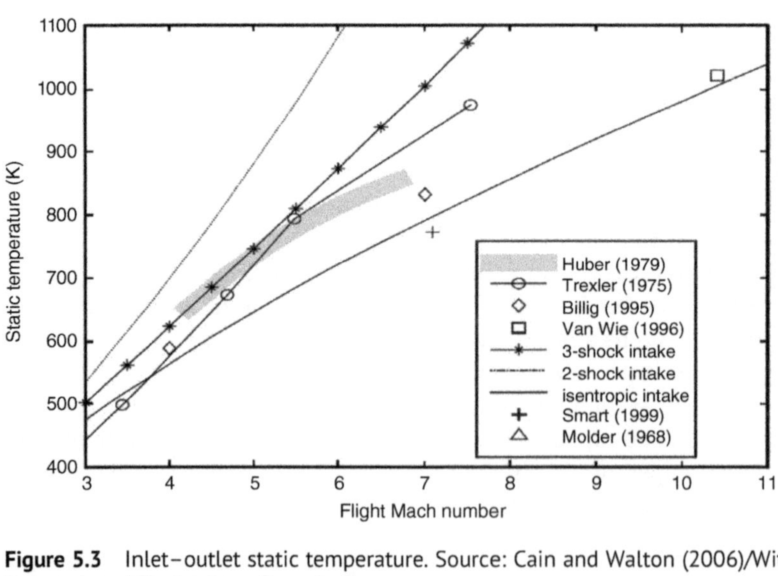

Figure 5.3 Inlet–outlet static temperature. Source: Cain and Walton (2006)/With permission of Research and Technology Organisation.

Anderson et al. (2000) emphasized the requirement to have an efficient inlet where the compression process results mainly from isentropic turning and weak shock waves, since a less efficient inlet that employs stronger shock waves will heat the flow beyond acceptable values. Although some dissociation will occur in the combustor due to chemical reaction, it is expected that an equilibrium expansion process in the nozzle may allow the dissociated species to recombine and recover the dissociation energy before leaving the engine. However, because the integrated hypersonic vehicle/propulsion system requires rapid expansion area in the initial nozzle contour, this would generate a correspondingly rapid decrease in static pressure, which would result in a frozen flow. In such case, the dissociation energy is lost and the net thrust is reduced.

5.2.3 Required Combustor Entry Mach Number

Since the air compression static temperature and pressure are limited for optimum combustor operation, we expect that this state will determine the required combustor entry velocity. Assuming one-dimensional flow in a control volume and no energy interactions with the surroundings, the energy equation yields an expression for the stagnation temperature of the inlet flow:

$$T_t = T_0\left(1 + \frac{\gamma_c - 1}{2}M_0^2\right) = T_3\left(1 + \frac{\gamma_c - 1}{2}M_3^2\right) \tag{5.2}$$

from where one can solve for M_3, which is the **combustor entry Mach number:**

$$M_3 = \sqrt{\frac{2}{\gamma_c - 1}\left[\frac{T_0}{T_3}\left(1 + \frac{\gamma_c - 1}{2}M_0^2\right) - 1\right]} \tag{5.3}$$

First note in this expression that when the flight Mach number M_0 is

$$M_0 < \sqrt{\frac{2}{\gamma_c - 1}\left(\frac{T_3}{T_0} - 1\right)}$$

Equation (5.3) has no solution because the allowable compression temperature T_3 would be higher than the stagnation temperature of the freestream flow.

However, when

$$M_0 > \sqrt{\frac{2}{\gamma_c - 1}\left[\left(\frac{\gamma_c + 1}{2}\right)\frac{T_3}{T_0} - 1\right]} \tag{5.4}$$

the combustor entry Mach number is supersonic so that the allowable temperature remains below the limit, that is, $M_3 > 1$, so that $T_3 < T_{3,\ max}$. This result yields a relationship between the combustor entry temperature and the static temperature ratio of the engine cycle:

$$\frac{M_3}{M_0} = \sqrt{\frac{T_0}{T_3}} = \sqrt{\frac{1}{\psi}} \tag{5.5}$$

Within the framework of the first law analysis, the cycle static temperature ratio for the scramjet will be in the range $0.6 < \psi < 8.0$. For example, if the compression temperature ratio is $\psi = T_3/T_0 = 8$, then $M_3 \sim 0.35\, M_0$. Thus, Eq. (5.5) serves as a guide to help us estimate the required Mach number at the entrance of the scramjet combustor.

From the limits thus presented, the conflicting requirements of high cycle thermal efficiency and dissociation of the working fluid at excessively high static temperatures dictate that the combustion process must be subsonic for flight Mach numbers less than about 5, and supersonic for $M_0 > 5$.

5.3 Combustion Stoichiometry

Combustion occurs when a fuel reacts with the oxygen in the air, and the chemical reaction produces thermal energy (heat). Hydrocarbon fuels are composed primarily of carbon and hydrogen, with a chemical identification C_xH_y. For example, methane is a hydrocarbon fuel with chemical formula CH_4 (one atom of carbon and four atoms of hydrogen). When vaporized hydrocarbons mix with air at the molecular level, if the temperature and pressure in the reaction zone are sufficiently high, the reaction rate will be fast and the fuel vapor will react upon contact with the oxygen molecules. The maximum combustion temperature occurs when hydrocarbon fuel molecules are mixed with the correct (stoichiometric) amount of air. Carbon dioxide (CO_2) and water (H_2O) are the main chemical products of the chemical reaction, which are formed from the reactants carbon (C_x) and hydrogen (H_y) in the fuel and the oxygen (O_2) in the air. Energy is released as a product of combustion in the form of heat. This process is represented by a general chemical equation for complete combustion called the stoichiometric equation:

$$C_xH_y + \left(x + \frac{y}{4}\right)\left(O_2 + \frac{79}{21}N_2\right) \rightarrow x\,CO_2 + \frac{y}{2}H_2O + \frac{79}{21}\left(x + \frac{y}{4}\right)N_2 + \text{energy} \tag{5.6}$$

In this stoichiometric reaction, all oxygen O_2 in the air is consumed in the reaction, and thus O_2 does not appear in the products. Also, the nitrogen N_2 in the air is treated as remaining unreacted (or inert in chemical terms) in the combustion process; that is, nitrogen acts merely as an inert diluent, absorbing some of the sensible thermal energy released by combustion (due to its specific heat capacity). The simplest example of hydrocarbon fuel combustion is the (ideal) reaction of methane (CH_4) with O_2 in the air. For stoichiometric (balanced) combustion, each molecule of methane reacts with two molecules of O_2 producing one molecule of CO_2 and two molecules of H_2O:

$$CH_4 + 2O_2 \rightarrow CO_2 + 2H_2O + \text{energy}$$

In the actual combustion process, other products are also formed. Figure 5.4 indicates a hydrocarbon fuel injected in the combustion chamber and reacting with air entering from the left; the combustion creates the products shown on the right, while subjected to a number of molecular changes and phenomena that contribute to the combustion process.

Assuming ignition and optimum molecular mixing occur for combustion, the oxygen in the air and the carbon in the fuel combine to form carbon dioxide (CO_2) and generate heat in a complex process that requires the right mixing turbulence, sufficient activation temperature, and enough time for the reactants to come into contact and combine. Moreover,

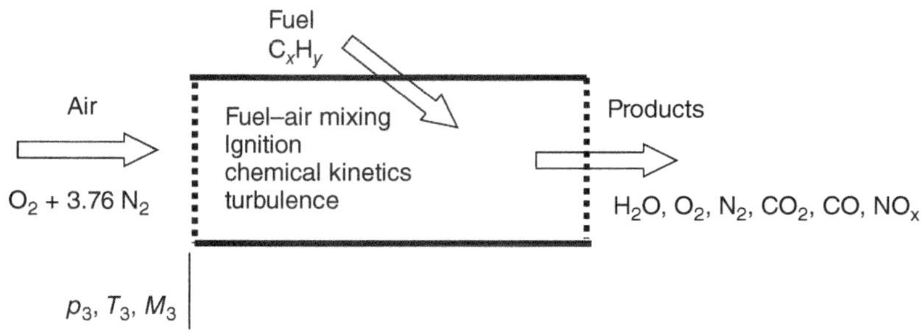

Figure 5.4 Hydrocarbon–air combustion reactants and products.

unless combustion is properly controlled, high concentrations of undesirable products can form. For example, carbon monoxide (CO) and soot result from poor fuel and air mixing, or when there is insufficient air for reaction. Other undesirable products, such as nitrogen oxides (NO, NO_2), form in excessive amounts when the flame temperature is too high. If a fuel contains sulfur, the reaction will yield sulfur dioxide (SO_2).

Nitrogen oxides, principally nitric oxide (NO) and nitrogen dioxide (NO_2), are pollutant gases that contribute to the formation of acid rain, ozone, and smog. Nitrogen oxides result when oxygen combines with nitrogen. At high flame temperatures, NO is generated first, then it oxidizes further to form NO_2 at cooler temperatures after being exhausted. The nitrogen oxide gas concentrations are typically combined and referred to as the NO_x concentration.

5.3.1 Stoichiometric Fuel-to-Air Ratio

The stoichiometric fuel-to-air ratio f_{st} is a unique proportion of fuel to air for combustion that results in neither excess oxygen nor any excess fuel. Any more fuel would result in unburned fuel in the products of combustion, and any more air would result in excess oxygen in the products of combustion. The general expression for the stoichiometric fuel-to-air ratio is obtained from the left-hand side of Eq. (5.6):

$$f_{st} = \frac{\dot{m}_f}{\dot{m}_0} = \frac{36x + 3y}{103(4x + y)} \tag{5.7}$$

This expression incorporates the atomic masses of the four participating elements: H = 1, C = 12, N = 14, and O = 16. Thus, for hydrogen fuel, for which $x = 0$, $y = 2$, we calculate the stoichiometric fuel-to-air ratio as $f_{st} = 0.0291$. For JP-7 fuel, with a chemical formula $C_{12.5}H_{26}$ ($x = 12.5$, $y = 26$), we obtain $f_{st} = 0.06745$.

If we use proportionately less fuel than the stoichiometric ratio, combustion is said to be fuel-lean, and the opposite is called fuel-rich combustion. To describe the combustor fuel-lean or fuel-rich condition, we use the equivalence ratio ϕ. This parameter is defined as the actual fuel-to-air ratio divided by the fuel-to-air ratio required for complete combustion (stoichiometric fuel-to-air ratio):

$$\phi \equiv \frac{f}{f_{st}} \tag{5.8}$$

Hence, fuel-rich, fuel-lean, and stoichiometric mixtures are defined by $\phi > 1$, $\phi < 1$, and $\phi = 1$, respectively. Moreover, the combustion reaction for off-stoichiometric mixtures and possibly incomplete combustion can be represented as:

$$\phi C_xH_y + \left(x + \frac{y}{4}\right)\left(O_2 + \frac{79}{21}N_2\right) \rightarrow \sum_{i=1}^{NS} n_i A_i \tag{5.9}$$

where n_i denotes mole number of i-th gas molecule and A_i represents its chemical formula.

The highest combustion temperature (adiabatic flame temperature) is obtained very near the stoichiometric ratio. For many fuels, the maximum temperature occurs at the equivalence ratio between 1 and 1.1, which occurs because the product-specific heat is reduced with a slightly fuel-rich mixture ratio. In the scramjet engine, ignition occurs only after fuel and air are micromixed to flammable proportions, which requires to operate at an equivalence ratio in the range $0.2 < \phi < 2.0$. Most importantly, turbulent supersonic combustion involves complex reactions of fuel and air involving sequences of many reactions and a number of different reactive species present in a reactive mixture, which are not obvious in Eq. (5.7). In fact, actual supersonic combustion analysis and modeling require solution of many chemical reaction equations in addition to the thermo-fluid dynamic equations that characterize the flow in the combustor. The combustor chemical kinetics is addressed in Section 5.4.7.

The fuel of choice for scramjets has been hydrogen, mainly because of its high specific energy, rapid chemical reaction rate, mixing rate, and excellent cooling capacity (per unit mass). Hydrocarbons are also considered because these fuels have a much higher liquid density, which makes them very practical since hydrocarbon-fueled vehicles have a higher volumetric efficiency, compared with hydrogen. Today, the interest focuses on endothermic hydrocarbon fuels which have a high cooling capacity (per unit mass), making these fuels as suited for practical hypersonic vehicles as hydrogen is. Chapter 6 gives a general overview of endothermic fuels for scramjet engines.

5.4 Combustion Flowfield

As a propulsion device, the scramjet is geometrically simple compared with any other engine. However, the supersonic, turbulent combustion flow field that develops in the scramjet flowpath is quite complex. This is especially evident when we consider the simultaneous injection of fuel from multiple injector configurations, air/fuel molecular mixing developing over thick boundary layers, interacting shocks, ignition and reaction governed by interdependent chemical kinetics, and aerothermodynamic process. Although the core flow of the scramjet combustor will be supersonic, there will be regions of subsonic flow near walls, fuel injectors, and wall cavities. Owing to the very short residence time, the injected fuel must mix and burn efficiently and almost instantaneously.

To describe the combustor flow field, we must consider a number of interrelated processes, including fuel injection, fuel jet penetration, fuel/air mixing, ignition and chemical reaction, turbulent burning, flame structure development, flameholding, and thermal choking. Thermal choking can be caused by chemical reaction and by fuel injection.

The combustor flow field will be characterized by shocks, turbulence, vorticity, and boundary layer separation, and this requires extensive use of computational tools. We address this topic in Chapter 12.

5.4.1 Fuel Injection and Injector Devices

The performance capability of a scramjet engine relies on efficient fueling throughout the combustor. Due to the short residence time in the combustor, the fuel injected into the supersonic airflow must efficiently mix within microseconds and quickly react in order to release its energy. The fuel can be injected into the supersonic airflow through a wide variety of devices. The fuel jet can emerge from orifices or slots flush on the combustor wall. The fuel can be injected normal or transverse to the main flow direction. Injectors can be protruding into the main flow, or be imbedded in wall cavities. The main requirements for the scramjet fuel injection system are summarized in Table 5.1.

The amount of heat release in a supersonic combustor is largely dependent on the efficiency and effectiveness of its fuel injection system. Efficiency is determined by the degree of fuel/air mixing achieved; effectiveness is associated with minimization of the combustor exit stream thrust losses incurred in the mixing process, and the extent of the additional wall cooling or thermal protection risk associated with certain type of fuel injectors (Kutschenreuter 2000).

An effective fuel injector is one that distributes the fuel uniformly and promotes rapid mixing with the air to ensure fast reaction in the shortest length possible. The injector system should provide a region for flameholding while minimizing the total pressure losses and internal drag. Flameholding refers to the process needed to maintain continuous combustion in the flowing mixture. Since the fuel flow provides a significant fraction of the scramjet thrust, the velocity components of the injected flow must be carefully aligned with the main flow vector in the engine.

Let us first consider the simplest, nonintrusive fuel injector. Injecting the fuel through an orifice on a flat wall in normal direction to the supersonic airstream results in a relatively complex flow structure in the immediate vicinity of the jet, as depicted in Figure 5.5. Note

Table 5.1 Fuel injection system requirements.

- **Optimum mixing**. Fuel injection system must be designed to mix fuel and air and ensure the mixture reacts quickly and sustains a stable flame.
- **Fuel uniformity distribution**. Fuel injection system should distribute the fuel throughout the combustor cross section to take advantage of all available oxygen in the airstream.
- **Scalable**. Injector size must accommodate to changes in combustor cross section.
- **Optimum performance**. Injection performance measures include fuel dilution into the airstream, fuel/air mixture flammability, and total pressure loss.
- **Minimum total pressure loss**. Total pressure losses created by the injectors and the injection processes must be minimized since these losses reduce the thrust of the scramjet.
- **Tailored fuel jet angles**. At moderate flight Mach numbers ($M_0 \leq 10$), fuel injection may have a normal component into the airflow mainstream. At high Mach numbers ($M_0 > 10$), the injection must be nearly axial since the fuel momentum provides a significant portion of the engine thrust.

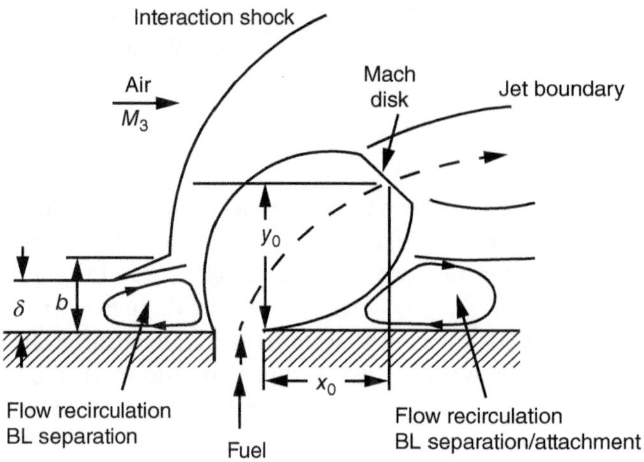

Figure 5.5 Fuel jet physical structure of a sonic fuel jet injected into airflow.

how the supersonic airflow is displaced by the fuel jet, and the interaction results in a detached normal shock wave forming just upstream of the fuel stream, causing the upstream boundary layer to separate. The main features in this flowfield include (i) the jet expansion fan and barrel shock, (ii) the upstream and downstream separated boundary layer regions (including recirculation), and (iii) the bow shocks off the separation and reattachment regions, which are evident in both sides of the fuel jet in Figure 5.5. The barrel shock (Mach disk) height y_0 from the surface wall provides a measure of the mean fuel jet penetration for perpendicular injection.

Adequate penetration by the fuel across the combustor is required for good mixing. The jet penetration is governed by the dynamic pressure ratio J, Eq. (5.10), the jet geometry (see Figure 5.5), and the angle of injection measured relative to the axial direction of the airstream. If α is the angle between the injector fuel jet and the main flow or airstream in the combustor, then $\alpha = 0^\circ$ and $\alpha = 90^\circ$ denote injection parallel and normal to the stream, respectively. Many studies have been done with circular hole injectors at angles ranging from $\alpha = 0^\circ$ to $\alpha = 150^\circ$ (angled upstream against the main flow). Slot injectors are usually oriented so that the fuel and main airflow velocity vector are aligned.

Many strategies for injecting fuel and enhancing the mixing of the fuel and air in high-speed flows have been proposed. Two traditional options for injecting fuel include injection from the combustor walls and in-stream injection from struts. Injectors which are flush with the wall do not protrude into the main flow, and thus the fuel jet emerges from an orifice on the wall and must penetrate the thick boundary layer in order to intersect the airflow stream. This method of injection generates stagnation pressure losses due to the strong bow shock generated by the transverse fuel jet.

To enhance the mixing of the injected fuel with the main combustor flow, a number of techniques are possible. The actual injector orifice can be lifted from the wall into the core of the main supersonic flow using a strut or a pylon. Struts allow a deeper penetration because the fuel is directly injected into the airflow, above the boundary layer. Injecting the fuel downstream from a rearward facing step or ramp is also done to enhance mixing. Other techniques have also been proposed for the same reason, and the devices designed are known as hypermixers.

Initial experimental and computational studies sought to understand the mechanism of jet penetration so that the resulting jet trajectory and fuel–air mixing could be predicted. Experimental data yielded correlations for different injectors, usually relating the injectant to freestream pressure and momentum ratios. For fuel to penetrate into the core flow, the fuel stagnation pressure must exceed the surface static pressure. A number of correlations were proposed to determine the penetration characteristics of fuel injection. For example, Kutschenreuter (2000) used a penetration parameter defined by the ratio of vertical distance from the combustor wall (Y) to the sonic round hole diameter of the injector ($D^* = 4\sum A^*/\pi$):

$$\left(\frac{Y}{A^*}\right)^2 = K\left(\frac{q_f}{q_a}\right) = KJ$$

where K is a constant and J is the momentum flux ratio (subscripts f and a denote fuel and air, respectively):

$$J \equiv \frac{q_f}{q_a} = \frac{(\rho v^2)_f}{(\rho v^2)_a} \tag{5.10}$$

Correlating experimental data based on schlieren-determined penetration height, Kutschenreuter found that in spite of differences in defining and determining penetration height, the penetration characteristics for normal injection with round sonic holes agree well with the simple expression $(Y/A^*)^2 = 16J$. Therefore, for round sonic holes, the penetration of the fuel jet increases with momentum ratio and/or sonic hole diameter D^*.

Based on extensive ground testing at ONERA, Leuchter proposed the following law characterizing fuel jet penetration based on experiments conducted with hydrogen (heated to 1000 K) injected in a Mach 1.5 flow with different values of momentum flux ratio J:

$$\frac{Z_{max}}{d} = 0.78\,J^{0.5}\cdot\left(\frac{x}{d}\right)^{0.35};\ \frac{Z_a}{d} = 1.45\,J^{0.5}\cdot\left(\frac{x}{d}+0.5\right)^{0.35}$$

where Z_{max} is the line of maximum hydrogen concentration and Z_a is the external jet boundary (1% hydrogen concentration) (Falempin 2000).

The simplest method for wall injection uses wall orifices. The orifices can be elliptically-shaped or round. To control reaction and heat release, a scramjet combustor can combine transverse and streamwise injection, which is helpful to vary fuel injection and mixing requirements over the flight Mach number range (Shenoy et al. 2018). Parallel injection allows fuel momentum to contribute to axial thrust. However, since fuel mixing and spreading are caused by interacting shear layers, long combustors are required for complete mixing. On the other hand, parallel injection reduces wall heat load as the fuel can provide film cooling. Perpendicular injection allows for deep fuel penetration and rapid mixing; however, it creates a large flow disturbance, and fuel momentum may not be recovered for axial thrust.

Since vorticity addition to the airstream provides more significant mixing enhancement of fuel and air, wall injection can use geometrical shapes such as wedges on the combustor

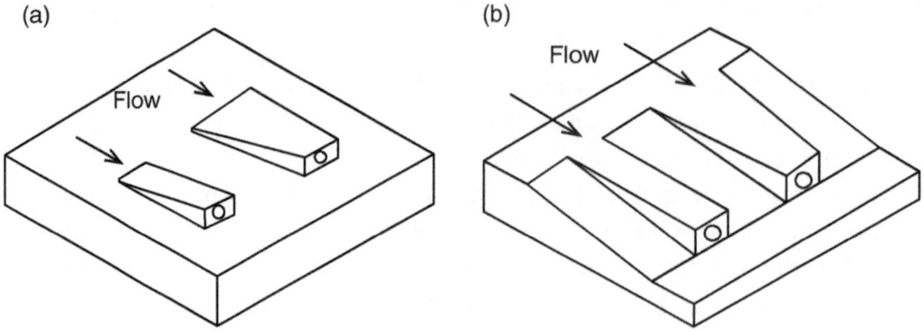

Figure 5.6 Hypermixer ramp fuel injection (a) unswept and swept compression ramp, and (b) unswept and swept expansion ramp.

wall to add axial vorticity into the airflow field. Vorticity can be also induced in the fuel jet using convoluted surfaces or small tabs at the exit of the fuel injector.

Fuel can be injected from the base of a swept or an unswept ramp mounted on the wall of the combustor (Figure 5.6); ramp injectors generate streamwise vortices which increase the fuel penetration and the fuel/air contact surface. The streamwise vortices entrain a large amount of the freestream air into the vortex region due to their large-scale vortex motion. This process acts to rapidly increase the fuel–air interface area by the action of small-scale turbulent eddies evolving from their breakdown to enhance molecular mixing. The ramps must be designed to introduce streamwise vortices with large circulation to increase mixing performance. Unswept ramps provide nearly streamwise fuel injection, adding a thrust component. Flow separation at the base of the ramp forms a region for flameholding and flame stabilization by generating a pool of radical species. The ramp itself produces streamwise vorticity as the airstream sheds off of its edges, improving the downstream mixing. Swept ramps appear to have better axial vorticity generation and thus improved mixing. Pairing expansion and compression ramps can be done to generate pairs of counter-rotating streamwise vortices.

Early researchers found that adequate penetration of the injected fuel into the airstream is a necessary but not a sufficient requirement for achieving adequate fuel/air mixing. It is difficult to achieve effective fuel distribution at high flight speeds in reasonable combustor length while matching the local inlet airflow profile, and this is more so if the fuel is injected only from the walls of the combustor. Hence, intrusive fuel injectors such as struts and pylons may be required (Doster 2008).

Fuel can be injected from the sides and the base of in-stream struts (Figure 5.7). The angle of injection can be varied, from being perpendicular to the airstream flow to being at an angle, as indicated. The flow field of the strut injector includes a bow shock that forms as the freestream flow traverses the mounting plate that supports the strut fuel injection system. In addition, shock waves form at the leading edge of the strut, and these interact with the mixing plume developing downstream of the base of the fuel jet. As a result of baroclinic torque, the interaction of shock waves and fuel–air mixing field generates vorticity. In mixing, nonreacting studies, helium is typically used to represent the fuel, as this approach isolates the mixing effects from the combustion process. Di Stefano et al. (2021) carried out analysis to study turbulence model uncertainty on strut flow fields.

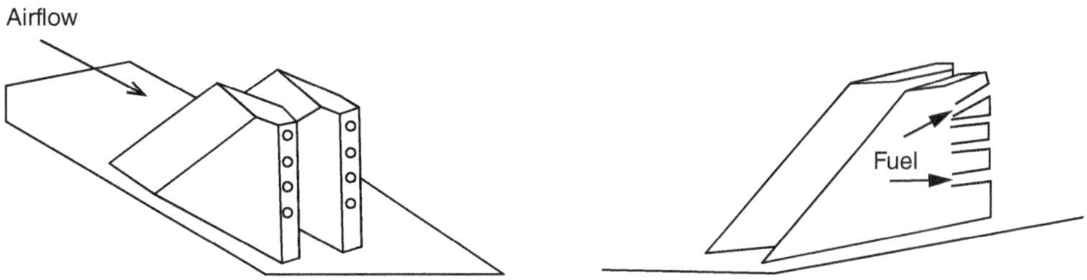

Figure 5.7 Strut injectors.

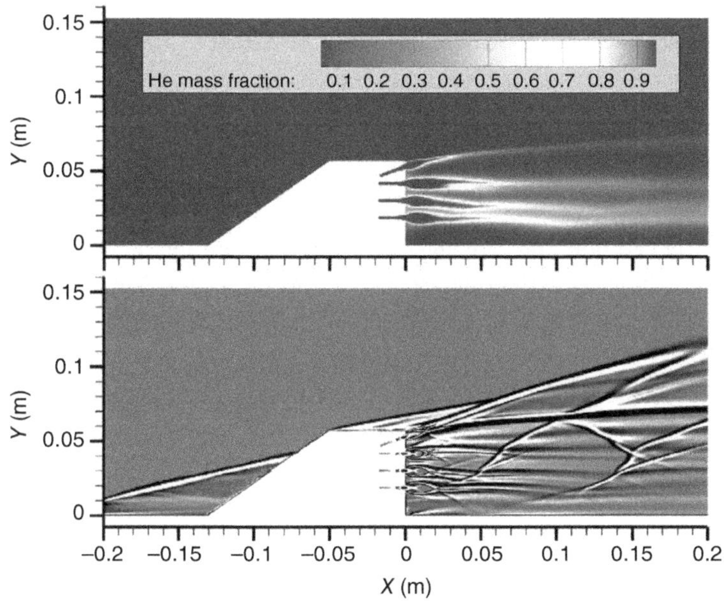

Figure 5.8 Scramjet strut injector flow structure visualized with contours of injected helium mass fraction (top) and numerical schlieren (bottom). Source: Di Stefano et al. (2021)/With permission of Elsevier.

In this study, Figure 5.8 is used to illustrate the mixing plume, shown as a contour plot of the helium mass fraction. Note the complex shock structure of the strut injector visualized with numerical schlieren (approximated as the magnitude of the density gradient).

Intrusive injectors introduce drag and flow blockage. In addition, they must be made of advanced high-temperature materials and may require cooling in order to survive the high heat loads in the combustor. In such scenario, a design trade study is made to strike a balance between intrusive injector losses and reduced combustor length.

Since the combustor requires rapid fuel/air mixing and ignition, a flameholder mechanism is often used, which generates turbulence and shock waves and maintains a recirculation region through geometric effects in the flowpath. For example, fuel injectors can be integrated with wall cavities or steps in the flowpath. As shown in Figure 5.9, an injector can be integrated in the cavity, positioned upstream of the cavity, or downstream of a step.

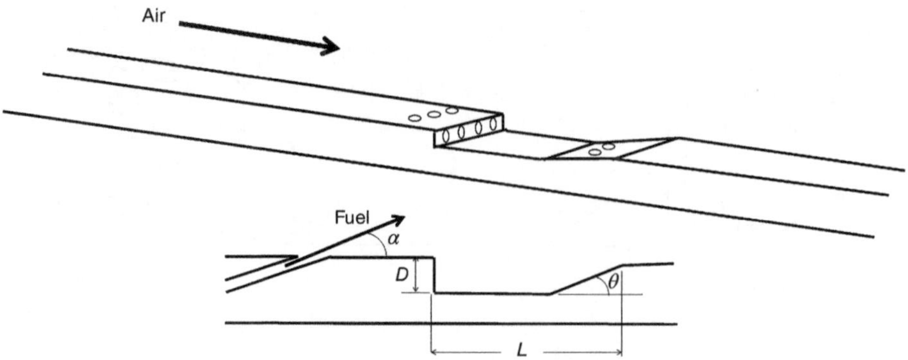

Figure 5.9 Flameholder cavity geometry. Ovals indicate possible locations for fuel injectors.

Many other fuel injection designs have been conceived and studied. Pulsed injection (using either mechanical devices or fluidic oscillation) techniques, for example, are proposed to control injection and improve mixing (Murugappan et al. 2005; Kouchi et al. 2010). However, it is important to note that no single fuel injector concept is optimum. In selecting a fuel injection method or strategy, consider the flow speed, the type of fuel, the geometry of the combustor cross section, and many other design and operational constraints.

5.4.2 Combustion Performance Parameters

A number of parameters are proposed to measure combustor performance, including mixing and combustion efficiency, thermodynamic losses, and thrust performance. Let us review some of these parameters.

Total Mass-Flux-Weighted **Pressure Recovery**. Typically denoted P_t^{rec}, the total mass-flux-weighted pressure recovery is used as an indicator of thermodynamic losses, mechanical stirring, drag, turbulence, mixing, and shock wave losses. This parameter is defined as:

$$P_t^{rec} = \frac{\int p_t \rho u \, dA}{\int p_{t0} \rho u \, dA}$$

where p_t and p_{t0} are the local and reference (e.g. freestream and flowpath entrance) values of the total pressure, respectively, ρ is the static density, u is the streamwise velocity, and dA is the incremental area projected in the streamwise direction.

In simulating nonreacting fuel/air mixing, researchers use P_t^{rec} to quantify the losses due to drag on injector bodies (struts and ramps) and on the combustor walls, the mechanical stirring induced by injector bodies, flow turbulence, and molecular mixing. For reacting flow simulations, the total pressure recovery is further reduced by the entropy increases due to the chemical reactions. Hence, the values of P_t^{rec} obtained from the mixing simulations are considered as the maximum achievable for a given injector design.

Mass-Flux-Weighted **Mach Number**. The mass-flux-weighted Mach number is a parameter used to reveal the global behavior of the flow, and researchers use it to determine the extent of margin with respect to choked flow conditions in the flowpath. The mass-flux-weighted Mach number is obtained using the following expression:

$$\overline{M} = \frac{\int M\rho u \, dA}{\int \rho u \, dA}$$

where M is the Mach number, and the overbar denotes a one-dimensional property.

Thrust Potential. This is a parameter used to more objectively quantify the potential performance of a combustor flowpath. The thrust potential determines the momentum portion of ideal potential net thrust that could be obtained when a flowpath is truncated at a given streamwise location and coupled at that location to an ideal thrust nozzle. The thrust potential is defined by:

$$TP = \dot{m}_e u_e + (p_e - p_i)A_e - \dot{m}_i u_i$$

where $\dot{m}, p, u, A$ are the mass flow rate, static pressure, velocity, and area, respectively, with subscripts e and i to denote conditions at the nozzle exit plane and the flowpath entrance, respectively.

When the flow expands to the inflow stream static pressure, the second term in the above equation is identically zero. If the flow in the scramjet nozzle is assumed to be chemically "frozen" (starting at the point of expansion), the thrust potential will not account for any mixing (losses) and reaction (thrust increases) during the expansion process, which could either further decrease or increase the thrust. Also, in a numerical simulation usually the combustor entrance static pressure is not the actual static pressure, the thrust potential only accounts for a fraction of the expected thrust. However, this performance parameter serves to account for the total pressure, shown as a decrease in the value of the thrust potential.

Dividing TP by the inflow air mass flow rate, one obtains the specific thrust potential, $TP/\dot{m}_i$, which is useful for comparing flowpaths of different sizes, different geometries, or that have fuel injectors designed to fuel different areas of the flowpath per injector.

Combustion Efficiency. This parameter accounts for the possibility that some of the chemical energy of the fuel is not released due to inadequate mixing or reaction time for complete combustion. There exists a number of combustion efficiency definitions that include combinations of the total temperature, the enthalpy or enthalpy of formation as a function of the local equilibrium composition, and the fuel/oxidizer depletion or combustion product formation. For numerical simulations, the simplest definition is based on the fuel mass fraction depletion:

$$\eta_c = 1 - \frac{\dot{m}_f}{\dot{m}_{f,total}}$$

where $\dot{m}_f$ and $\dot{m}_{f,total}$ denote the integrated mass flow rates of fuel at a streamwise location of interest and the total injected fuel flow rate, respectively.

Clearly, for air/fuel mixing simulations, $\eta_c = 0$. However, for simulations of reacting flows, the value of η_c increases monotonically, and $\eta_c \rightarrow 1$ when all of the fuel is depleted. For fuel-rich simulations, the oxidizer mass fraction depletion can also be used.

5.4.3 Fuel/Air Mixing Efficiency

One of the greatest challenges of supersonic combustion is the requirement to achieve rapid fuel/air mixing. At high speeds, the residence time for the atmospheric air ingested through the scramjet inlet and exiting from the propulsion nozzle is on the order of milliseconds. Moving at supersonic speeds through the combustor, the amount of time available for fuel injection, mixing at the molecular level with the airstream and achieving complete chemical reaction, is severely limited. Thus, fuel injected into the high-speed air must efficiently mix and quickly react to release its energy in the combustor, all that within tens of microseconds.

Both mixing and combustion timescales must be considered in a detailed study of mixing and chemical reactions in a scramjet combustor to understand the underlying high-speed reacting flow processes and to ultimately achieve a successful design. In fact, by enhancing the rate at which the fuel and air mix, it is possible to both reduce the combustor length and achieve significant gains in engine performance (Kutschenreuter 2000).

Numerous experiments were and are still conducted to study the mixing and combustion of fuel and air intended to validate combustion CFD codes employed in scramjet design, and to support the development of improved turbulence and combustion models. In those studies, mixing efficiency is used to quantify injector mixing performance which provides a true mass-flux-weighted measure of mixing completeness.

Fuel mixing efficiency, denoted η_m, is defined as the fraction of the least available reactant (either O_2 or fuel) that would react if the fuel/air mixture were brought to chemical equilibrium without further local or global mixing. This means that in fuel-rich regions, all the local O_2 is considered mixed, while in fuel-lean regions all the fuel is mixed. Mao et al. (1990) proposed the following definition of mixing efficiency that takes into consideration the two possibilities: one for flows which are globally fuel-rich and one for flows which are lean. For the fuel-lean flows they studied, using H_2–air mixtures, mixing efficiency is defined as:

$$\eta_m = \frac{\dot{m}_{H_2,mix}}{\dot{m}_{H_2,total}} = \frac{\int_{A_{\alpha=0}} \alpha_R \rho u \, dA}{\int_{A_{\alpha=0}} \alpha \rho u \, dA} \tag{5.11}$$

where α denotes the fuel mass fraction, and

$$\alpha_R = \begin{cases} \alpha, & \alpha \leq \alpha_{st} \\ \left(\dfrac{1-\alpha}{1-\alpha_{st}}\right)\alpha_{st}, & \alpha > \alpha_{st} \end{cases} \tag{5.12}$$

In this formulation, α_{st} is the H_2 stoichiometric mass fraction (0.0285), $A_{\alpha=0}$ is the area enclosed by the zero H_2 contour defining the extent of the mixing region, $\dot{m}_{H_2,mix}$ is mixed H_2 mass flow rate, and $\dot{m}_{H_2,total}$ is total H_2 mass flow rate from the flow field integration.

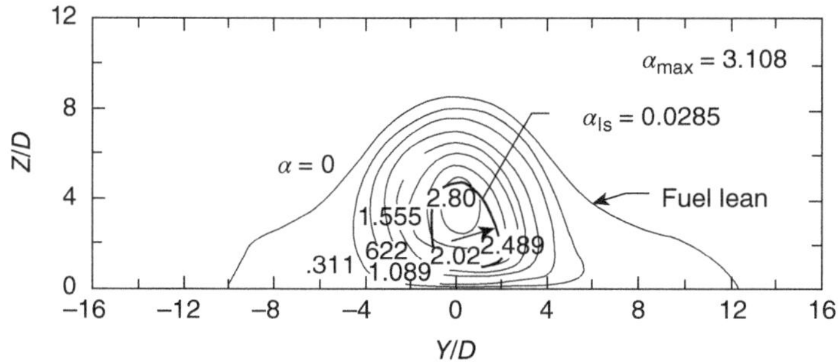

Figure 5.10 Experimental determination of H_2–air mixing efficiency. Source: Mao et al. (1991)/ NASA/Public Domain.

Hence, $\eta_m = 1.0$ when $\alpha_{max} < \alpha_{st}$. To evaluate experimental mixing efficiency, Mao et al. determined by integration the H_2 and airflow contours in Figure 5.10. Note the stoichiometric H_2–air contour ($\alpha_{st} = 0.0285$) overlaid on the H_2 and airflow contours to provide integration limits. The total hydrogen mixed in the fuel-lean region was determined by integration of H_2 flow rate in that region.

The Enhanced Injection and Mixing Project (EIMP) is a fuel/air mixing study currently underway at the NASA Langley Research Center (Cabell et al. 2014; Drozda et al. 2016). The EIMP investigates the fundamental physics relevant to fuel injection and mixing for scramjets considered to power vehicles flying at high flight Mach numbers ($M_0 > 8$). The main goal of EIMP is to develop mixing enhancement strategies that minimize the total pressure loss incurred during the fuel/air mixing process at high Mach numbers.

5.4.4 Combustion and Ignition Time

In a multidimensional flow simulation of supersonic combustion, there are different ways to determine when combustion occurs. One approach to determine the ignition location for a given fuel/air mixture is as follows:

- Determine the time of maximum rate of temperature change,
- Find the maximum rate of OH production,
- Define the threshold of OH concentration (e.g. 0.005 mol/m^3),
- Define a threshold ignition temperature.

The combustion process at high Mach numbers is kinetically controlled (Drummond 2014). This means that the ignition delay times are of the same order as the fluid scale. The ignition delay time refers to the induction period or the time interval immediately following some form of homogeneous bulk ignition, and it occurs between the creation of a combustible mixture and the onset of a flame. A sudden pressure increase occurs due to combustion; the induction period ends as the mixture temperature begins to rapidly increase. Ignition delay time is a function of initial temperature, pressure, and composition of a reactant mixture. For hydrogen/air, the induction time or ignition delay time is on the order of 2×10^{-4} seconds. However, hydrocarbon fuels have much longer ignition delay time (see Chapter 6). Hence, to improve the ignition of hydrocarbon fuels, the scramjet must effectively integrate fuel injection, piloting, and flameholding schemes. For example, a cavity can be used to establish a pilot flame and thus reduce the induction time of hydrocarbon–air mixtures.

5.4.5 Ignitors and Ignition Promoters

At certain conditions, if the injected fuel does not reach its self-ignition temperature, an ignition source or ignition promoter may be required to overcome slow ignition chemistry. Ignition promoters include fuel additives, pilot flames, passive wall cavities, and spark plugs, which are especially helpful for ground-testing of scramjet combustors.

Silane gas is used for piloting supersonic H_2-air combustion to reduce characteristic reaction times (Morris et al. 1987; Koshi et al. 1991). Silane (SiH_4) is a pyrophoric gas that is added to hydrogen fuel to decrease the ignition delay time. Silane is very reactive and burns on contact with air or oxygen even at room temperature and in very dilute mixtures. Silane concentrations between 5% and 20% (by volume) are typically used. This type and quantity of ignition promoter is useful when combustion chambers are either too short or the gas temperature is below that temperature where spontaneous combustion would normally occur.

Jachimowski (1988) studied the reaction mechanism for silane/hydrogen combustion to theoretically investigate the ignition characteristics of silane/hydrogen mixtures. The results revealed that over the entire range of temperature examined (800–1200 K), substantial reduction in ignition delay time is obtained when silane is added to hydrogen. Thus, during the NASA HyperX program it was decided to start the X-43A combustor with a mixture of gaseous SiH_4/H_2 to ensure ignition during its short flight. Silane mixed with hydrogen can also reduce the pressure required for combustion in ground testing (Turner and Smart 2010).

A disadvantage of hydrocarbon fuels is that they ignite more slowly than hydrogen, which means these fuels require longer combustion length, thus increasing combustor weight and viscous drag. To overcome the slow ignition, a pilot mechanism can be used to decrease the reaction times of hydrocarbons to scales of the same order to those of hydrogen. Ethylene (C_2H_4) gas is used as ignition promoter in hydrocarbon-fueled scramjet combustors. One example of this is the X-51A's combustor, which ignited on a mixture of ethylene and JP-7 fuel. In ground-test experiments of the HIFiRE-2 scramjet, spark plugs were active to help initiate combustion because ignition with just the primary fuel could not be achieved (Cabell et al. 2011). The fuel was a mixture of ethylene and methane (64% ethylene–36% methane molar mixture), which was used as a surrogate for an endothermically cracked JP-7 fuel.

Prediction of chemical induction and ignition is highly sensitive to the chemical model used in numerical analysis. Since ignition delay time has significant consequence on simulation of high-speed reacting flows, it is important to select the correct or most appropriate chemical model. A review of reaction mechanisms is in Section 5.4.7.

5.4.6 Flameholding

There are different techniques for flameholding: (i) promote a recirculation area where the fuel jet and the airstream can mix partially at low velocities; (ii) use the interaction of a shock wave with partially or fully mixed fuel and air; and (iii) promote formation of coherent structures that may contain unmixed fuel and air in which a diffusion flame results as the gases are convected downstream. Staged injection of fuel from wall orifices can also serve as flameholder because the large region of separated flow that develops between the injectors results in significant reaction of fuel and air taking place in that region.

Cavity flameholders are recesses in the walls of supersonic combustion chambers that have shown to be excellent flameholding devices (Donohue 2012). Cavities are nonintrusive, resulting in reduced drag, lower total pressure losses, and minimal aerodynamic heating when compared with other means of piloting core combustion such as, for example, struts. Cavity flameholders may integrate fuel injectors or pilots as well, as illustrated in Figure 5.9. Characterized by their aspect ratio (L/D) and aft ramp angle (θ), cavity flameholders have been studied and their effectiveness has been tested. The key issues controlling their performance include entrainment rate, residence time, and distribution of fuel and air within the cavity (Hsu et al. 2000; Baurle 2017).

Cavity flameholders were used for the first time in a joint Russian/French dual-mode hydrogen-fueled scramjet flight test (Roudakov et al. 1993). The performance predictions indicated that these cavities would be effective as autoignition and flameholding devices. A flight test of the Russian Central Institute of Aviation Motors (CIAM) combustor in 1998 encouraged further investigation of cavities for scramjet application. Since then, research efforts were directed to establishing flameholding mechanisms and structures of cavity flames. Cavity-based flameholders were designed with various flush wall fuel injection schemes aimed at finding the most effective location for fuel injection that yields the best cavity/injector flameholder with high mixing efficiency for a given scramjet configuration.

An experimental study by Driscoll and Rasmussen (2005) (Mach 2 combustor entrance conditions) set out to establish the lean and rich blowout limits of rectangular and ramp cavities fueled with ethylene and methane directly injected from the aft wall and cavity floor. They concluded that better cavity performance is found near the lean blowout limit for injection from the aft wall, and they obtained a stable flame near the rich blowout limit for injection from the cavity floor. Regarding the effect of the fuel, this study found that ethylene yielded a broader range for stable cavity operation as compared with ethane. This is attributed to ethylene having greater reactivity, or a relatively shorter ignition delay time and higher flame speed.

Rasmussen and Driscoll (2008) studied direct fueling from inside a rectangular cavity using the aft wall and the cavity floor. They discovered that cavity fueling port placement can affect flameholding mechanisms as this changes the structure of the flames. For example, fueling from the cavity floor, the fuel jet creates a recirculation zone, which provides a hot zone for flame ignition. Then the shear layer distributes the ignited flame. See Yokev et al. (2021).

Other studies have focused on determining the optimal location for fuel injectors inside and near wall cavities. For example, when injecting fuel from the upstream side inside the cavity into a Mach 2 combustor flow field, Driscoll and Rasmussen (2005) found a uniform fuel/air distribution within the cavity, and this was relatively unaffected to changes occurring during ignition. However, when the combustor entrance flow was Mach 4, the flame was generated along the shear layer above the cavity rather than inside the cavity (Neely et al. 2005). Gruber (2004) also considered the injection angle for injectors positioned upstream of a cavity. He found that angled injection before a cavity reduces the stagnation pressure losses compared with normal injection and further contains the flame in the cavity.

The injector location and the angle of the fuel jet have a great effect on the flame structure, fuel/air mixing, and overall performance of the cavity, especially important for scramjet combustors with high Mach number entrance conditions.

More recently, Jeong et al. (2020) investigated the effect of three different fuel injection locations on cavity flameholding performance. Using the same experimental facility and test model used by Neely et al. (2005), this study considered angled injection upstream of the cavity (15°), parallel injection from the front face of the cavity, and upstream injection from the rear face of the cavity, at Mach 4 combustor inlet flow conditions. They concluded that angled fuel injection had greater fuel penetration into the airflow and enhanced gas diffusion. As the equivalence ratio increased, angled injection generated a weak bow shock in front of the injector and a recirculation zone to hold the flame. On the other hand, parallel fuel injection started the ignition at the midpoint in the cavity and reaction occurred only along the shear layer. For the high equivalence ratio condition, the supersonic flame developed along a single line according to the cavity geometry in the vertical section. More work is needed to establish the best configuration for the cavity/injection scheme at similar combustion conditions.

5.4.7 Combustion Chemical Kinetics Mechanisms

The complexity of the combustion process in the scramjet makes it challenging to model and simulate. The correct modeling of the supersonic and turbulent combustion processes is crucial for the development and design of a scramjet propulsion system. This requires simulation of the reaction, which is done via chemical models of the fuel and oxidizer. Reaction of fuel and air involves a sequence of many reactions and a number of different reactive species present in reactive mixture. In chemical kinetics, we use the terms mechanism or model to mean an assumed reaction scheme or a possible sequence of steps for a given reaction. The kinetic mechanism must be correct in order to predict combustion parameters of interest, such as heat release rates, flame temperature, and concentration of important principal species in the combustor. Chemical kinetics deals with how fast chemical reactions proceed (reaction rates), and what chemical reactions occurs in a chemical process (reaction mechanisms). Thus, chemical kinetics is the key to understand how a flame propagates in a combustor.

The rate of reaction of chemical reactions is defined with the Arrhenius law, which is a formula for the temperature dependence of reaction rates. A modified form of this law is usually employed when modeling supersonic combustion. The forward rate of each reaction j given by the modified Arrhenius law is expressed as:

$$k_j = A_j T_j^N \, exp\left(-\frac{E_j}{R\,T}\right) \tag{5.13}$$

where k_j is the rate constant for each reaction, T is the absolute temperature (K units), R is the universal gas constant, and E is the activation energy, which is the minimum amount of energy that must be provided for compounds to result in a chemical reaction. The constant A, power coefficient N, and values of E are determined for a number of reaction schemes.

There is a great deal of uncertainty on the value of the constants for many chemical reactions. Fuel oxidation involves many reaction species. This means that a large number of differential equations are required to develop the kinetic mechanism that represents the actual combustion process. These differential equations are usually numerically stiff and require special integration techniques. Another complication in this effort is that the rate constants of the elementary reactions may not be available or may be incorrect. Thus, researchers propose kinetic mechanisms that attempt to simplify the chemistry of a given fuel/air mixture.

Many reaction mechanisms have been proposed for the reaction between hydrogen and air. They range in complexity from a 2-step global model to a 33-step model proposed by Jachimowski (1988). The Jachimowski model uses 33 reactions to represent H_2-air combustion mechanism (the eight active species are: H_2, O_2, H, O, OH, H_2O, HO_2, and H_2O_2, and the ninth (N_2) is assumed inert). The reactions and Arrhenius coefficients are given by Jachimowski (1988) for the full $H_2 +$ air reaction mechanism.

For the analysis of a given supersonic combustion model, a trade-off analysis is usually required for the selection of the proper combustion kinetic mechanism. For example, due the short residence time in the scramjet combustor, chemical nonequilibrium effects may also need to be evaluated. However, since this would add computational cost, the analysis can be done with a reduced kinetic model that has sufficient accuracy to yield reliable computational results for the given case. In general, a two-step reduced mechanism is good for autoignition analysis, while a four-step global reduced mechanism is accurate for flames. The 9-species, 18-reaction model used by Drummond (1988), which is based on Jachimowski's H_2–air reaction mechanism, has been extensively applied to scramjet combustion analysis. This model is described in Table 5.2, where values for the Arrhenius constants *A*, *N*, and *E* are given for each reaction. Please note that this mechanism includes the formation of hydrogen peroxide (H_2O_2) and the perhydroxyl radical (HO_2), which is important for ignition delay of hydrogen/air mixtures. On the other hand, it lacks the species N, NO, and HNO, which may be important in hydrogen–air reaction systems for scramjets operating at high velocity conditions, $M_0 > 12$. Also, the calculated ignition delay times by Jachimowsky are very sensitive to rate coefficients assigned to reactions (5.2) and (5.8), so care must be taken when applying the model when accuracy of ignition times is important.

A novel hydrogen–air finite-rate chemistry mechanism was developed by Vincent-Randonnier et al. (2018) to model the combustion process in the LAPCAT-II dual-mode ramjet/scramjet combustor at ONERA-LAERTE facility. Called the Z22, this H_2–air reaction model employs the H_2–O_2 chemical structure with three additional fuel breakdown reactions, creating a reaction mechanism of 9 species and 22 irreversible reactions.

A goal of the Z22 development was to obtain a mechanism that could match the ignition delay times in the crossover region, which is an intermediate temperature region dominated by complex chemistry, and where many ramjets, scramjets, and dual-mode ram/scramjet engines operate. This was important for predicting ignition, flame anchoring, and subsequent flame stabilization. Z22 includes reactions important for the complete temperature spectrum, above and below the crossover region. For example, it includes the reaction $H + O_2 \rightarrow OH + O$, which effectively creates a pool of radicals decreasing the ignition time, whereas the hydroperoxyl (HO_2) producing reaction $H + O_2 + M \rightarrow HO_2 + M$ constrains the chain-branching nature of the chemistry, effectively increasing the ignition time.

Table 5.2 Hydrogen–air finite-rate chemistry model and rate coefficients for each reaction.

Reaction number	Reaction	A_i	N_i	E_i (cal/g-mole)
1	$H_2 + O_2 = OH + OH$	0.170×10^{14}	0	48 150
2	$H + O_2 = OH + O$	0.142×10^{15}	0	16 400
3	$OH + H_2 = H_2O + H$	0.316×10^{8}	1.80	3 030
3	$H_2 + OH = H + H_2O$	3.16×10^{7}	1.80	1 525
4	$O + H_2 = OH + H$	0.207×10^{15}	0	13 750
5	$OH + OH = H_2O + O$	0.550×10^{14}	0	7 000
6	$H + OH + M = H_2O + M$	0.221×10^{22}	−2.00	0
7	$H + H = H_2 + M$	0.653×10^{18}	−1.00	0
8	$H + O_2 + M = HO_2 + M$	0.320×10^{19}	−1.00	0
9	$HO_2 + OH = H_2O + O_2$	0.500×10^{14}	0	1 000
10	$HO_2 + H = H_2 + O_2$	0.253×10^{14}	0	700
11	$HO_2 + H = OH + OH$	0.199×10^{15}	0	1 800
12	$HO_2 + O = OH + O_2$	0.500×10^{14}	0	1 000
13	$HO_2 + HO_2 = H_2O_2 + O_2$	0.199×10^{13}	0	0
14	$HO_2 + H_2 = H_2O_2 + H$	0.301×10^{12}	0	18 700
15	$H_2O_2 + OH = HO_2 + H_2O$	0.102×10^{14}	0	1 900
16	$H_2O_2 + H = OH + H_2O$	0.500×10^{15}	0	10 000
17	$H_2O_2 + O = OH + HO_2$	0.199×10^{14}	0	5 900
18	$H_2O_2 + M = OH + OH$	0.121×10^{18}	0	45 500

Source: Data from Drummond (1988).

The competition between those two reactions, and the distribution of the fast radicals O, H and OH, and slow radical HO_2, are temperature dependent and produce three regions: a region of rapid ignition (chain-branching explosion at high temperatures), a region of slow ignition (thermal explosion at low temperatures), and a connecting region (Vincent-Randonnier et al. 2018).

Combustion of hydrocarbons and air adds further modeling challenges due to the more complex molecular form of the fuel. JP-7 fuel, for example, is a blend of many different hydrocarbon molecules, and there is a variation in the exact chemical composition used by different researchers. For example, a five-component surrogate blend for JP-7 was studied by Mawid et al. (2003). Its components were 30% decane ($n - C_{10}H_{22}$), 30% dodecane ($n - C_{12}H_{26}$), 30% butylcyclohexane ($C_{10}H_{20}$), 5% isooctane ($i - C_8H_{18}$), and 5% toluene (C_7H_8) by weight. As the fuel is cracked as it flows from the tank to the combustor, it will change its molecular structure. The chemistry model must take into account this process.

A JP-7 reaction mechanism, consisting of 205 species and 1438 reactions, was evaluated by Mawid et al. (2003) using a lean JP-7-air mixture over a temperature range of 900–1050 K at atmospheric pressure conditions. The study predicted autoignition delay times and compared them to the available experimental data for Jet-A fuel. At the higher temperatures, the predicted autoignition delay times were in satisfactory agreement with the Jet-A data. For the kinetics of the cracked JP-7 surrogate, consider the 32 species, 206 reaction ethylene mechanism of Luo et al. (2011).

5.4.8 Supersonic Turbulent Combustion Characterization

The analysis of supersonic combustion process requires characterization of the combustion field, which involves determining the combustion modes, i.e. premixed or nonpremixed. The Takeno flame index (TFI) is used as the metric indicative of flame activity, which can help to identify regions of premixed combustion. Introduced by Yamashita et al. (1996), TFI gives information on whether a flame regime is premixed or of the diffusion type. The TFI is typically defined as the dot product of the gradients of fuel and oxidizer mass fraction present in the combustion field:

$$TFI = \nabla Y_F \cdot \nabla Y_O$$

where ∇Y_F and ∇Y_O denote gradients of fuel and oxidizer mass fractions, respectively.

This metric assumes that in nonpremixed flames, $TFI = -1$, when the gradients of fuel and oxidizer species are oriented in opposite directions. In premixed flames, $TFI = +1$, and the gradients are oriented in the same direction. Therefore, a positive TFI value indicates that the angles of the vectors ∇Y_F and ∇Y_O are aligned, i.e. combustion is premixed when fuel and oxidizer species arrive from the same direction. A negative TFI value indicates ∇Y_F and ∇Y_O approach from opposite directions, a characteristic of nonpremixed combustion. These combustion modes are qualitatively illustrated in Figure 5.11.

In determining the applicability of flamelet models for their dual-mode scramjet numerical analysis, Quinlan et al. (2014) redefined this metric by taking the dot product of the gradients and normalizing. Thus, the normalized Takeno index becomes

$$\Lambda_T = \frac{\nabla \widetilde{Y}_F \cdot \nabla \widetilde{Y}_O}{\left|\nabla \widetilde{Y}_F \cdot \nabla \widetilde{Y}_O\right|}$$

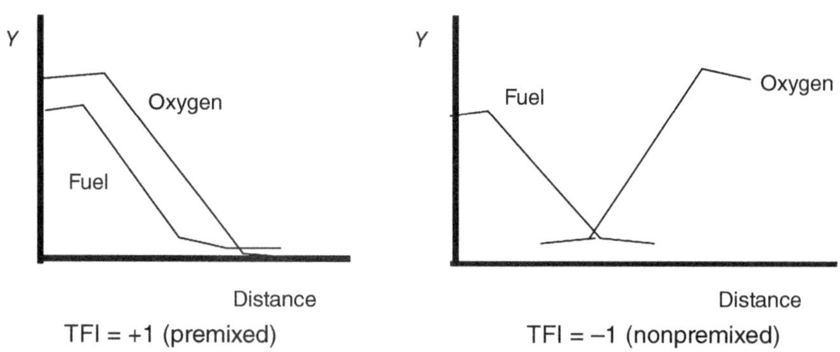

Figure 5.11 Flame index representation. +1 TFI when the reactants arrive from same direction; −1 TFI when reactants approach from opposite direction.

where $\nabla \widetilde{Y}_F$ and $\nabla \widetilde{Y}_O$ are the gradients of Favre-averaged fuel and oxidizer species mass fractions, respectively. Favre averaging is used in compressible flow to separate turbulent fluctuations from the mean flow.

By weighting the Takeno index with the flame index, Quinlan et al. defined a new index, Λ_f, which they named the *flame-weighted Takeno index*:

$$\Lambda_f = f\Lambda_T$$

Since the value of this new index is in the range $-1.0 < \Lambda_f < 1.0$, they used it to determine both the flame intensity and dominant combustion mode at each point in the flow field for both dual-mode combustion (Mach 5.84) and scramjet-mode combustion (Mach 8.0). The analysis then focused on identifying regions of nonpremixed and premixed combustion. They reported the following results:

For dual-mode case, combustion occurs primarily in a nonpremixed mode ($\Lambda_f < 0$); hence, the assumptions made for nonpremixed flamelet models are likely satisfied. Flamelet models, therefore, may sufficiently predict the dual-mode combustion physics governing the HIFiRE-2 combustor at low flight Mach number operation. However, for the scramjet mode, supersonic combustion was found split among both nonpremixed and premixed modes. The conclusion was that a suitable simulation of the scram-mode operation would likely require both premixed and nonpremixed flamelet models, and the assumptions made for these models may only be valid for limited regions of the combustion field (Quinlan et al. 2014).

In another study, the above assertion was confirmed. Gibbons (2019) explored the effects of turbulence/chemistry interaction (TCI) in a scramjet-like environment adopting a simplified planar domain with a double ramp, and a single hydrogen injector at the inlet section, based on an experimental model (M_0~10). The analysis used Improved Delayed Detached Eddy Simulation (IDDES) with detailed chemical kinetics (33 reactions and 13 species), enforced thermal equilibrium, and no explicit model for TCI (quasi-laminar chemistry).

Gibbons (2019) plotted the TFI in the experimental domain, using a combined view of three crossplanes cutting through the main combustion zones (arrow on top left indicates hydrogen injection), which we duplicate as Figure 5.12. Note the mixing structures at the entrance of the combustor, which produce a pair of coherent premixed vortex cores. The lightest contours visible inside the darkest boundary (negative TFI) indicate the mixture eventually burn out and leave the second half of the combustor as almost entirely nonpremixed flow. Most of the flow has a negative TFI, particularly at the start and end of the combustor, indicating that the fuel and oxygen gradients are opposed in direction as in nonpremixed combustion. Observe the bottom image where the positive TFI (lightest contour) in the centerplane indicates the recirculating flow has had sufficient time to mix the hydrogen/air reactants (Gibbons 2019).

While characterizing supersonic turbulent combustion in a generic hydrogen-fueled Mach-10 scramjet combustor, Moura et al. (2020) also concluded that the supersonic combustion process, driven by its flow nonuniformities, is multimode and multi-regime, with nonpremixed and premixed combustion both happening in the combustor, either alternating or, in some regions, simultaneously.

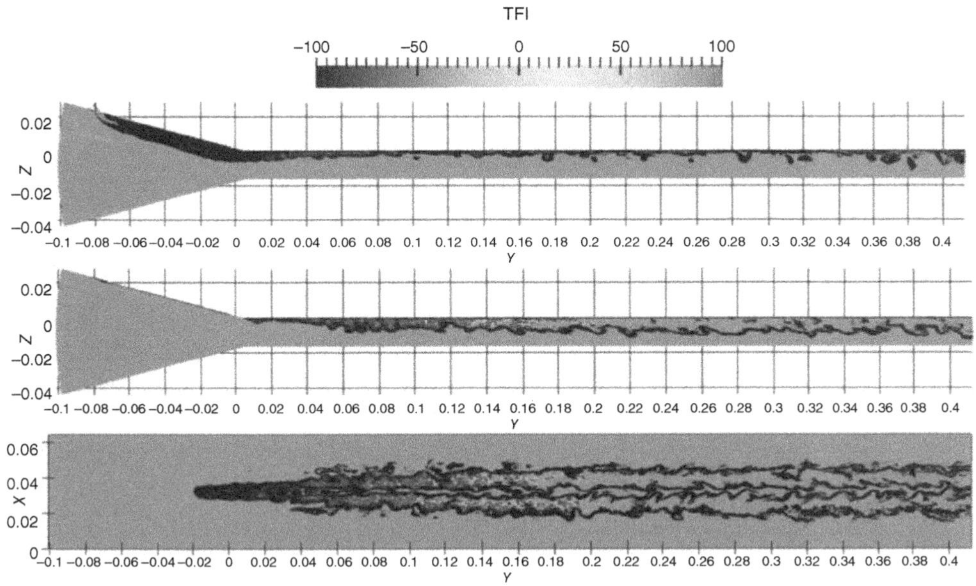

Figure 5.12 Takeno flame index (TFI). Top: $x = 32$ mm; middle: $x = 39$ mm; bottom: $z = 3.22$ mm. Source: From Gibbons (2019).

Moura et al. (2020) found that upstream of the combustor, the TFI, is almost completely negative around the fuel plume, an indication of nonpremixed combustion, although small regions close to the wall had positive TFI. As explained by these researchers, this is likely due to mixing occurring in the recirculation regions upstream and downstream of the injector, where fuel is entrained and mixed with air. In the combustor proper, a premixed region starts immediately downstream of the throat, close to the bottom wall, but then disappears downstream after the 150-mm marking. The counter-rotating vortex pair (CVP) was found to enhance mixing up to halfway through the combustor, retaining unburned fuel long enough to properly mix it with surrounding air, even as the premixed region decreases in the central plane. A layer of nonpremixed combustion enveloped the premixed region, persisting through the entire combustor and becoming more distorted as the length scale of the turbulence increases moving downstream.

Not surprisingly, Moura et al. (2020) noted that the region where combustion starts to become premixed coincides precisely with the region where there is a reduction in the production rate of OH. The production of H_2O was not affected. Overall, this study provided results indicating that both combustion modes are present simultaneously in the first half of the combustor, after which combustion reverts to nonpremixed only. The TFI values indicate that, in this scramjet, combustion seems to be mixing-limited.

The work by Gibbons (2019) and Moura et al. (2020) supports the conclusion that in a scramjet combustor multiple regimes of premixedness occur, manifested in separated layers controlled by the interaction between mixing and reaction rates. The analysis show that premixed regions occur where mixing is much more rapid than the chemical reactions. Away from these areas, nonpremixed combustion dominates. Gibbons suggested that for application to scramjet combustors, the most appropriate TCI model must be a hybrid model that can detect and switch between nonpremixed and premixed behavior, or be a generalized model that does not make simplifying assumptions about the gas composition.

Let us keep in mind that the above examples of supersonic turbulent flow characterization are not directly comparable, as they were done with different TCI models, and the analysis was carried out with different CFD codes, applied to much different scramjet combustion conditions. Quinlan et al. (2014) analyzed a JP-7-fueled Mach 8.0 scramjet combustor using the NASA VULCAN-CFD code, while Moura et al. (2020) analyzed a H_2-fueled Mach 10.0 scramjet combustor using the US3D code developed at the University of Minnesota. Nevertheless, they both arrived at the same conclusion: the turbulent combustion process in scramjets is multimode, where neither premixed nor nonpremixed combustion dominates, and both substantially contribute to heat release.

Read Chapter 12 for additional discussion related to simulation and modeling of the TCI in scramjet combustors.

5.5 Scramjet Combustor Geometry

The combustor section of a scramjet is designed to conform to the configuration of the required propulsion flowpath, including the transition from the inlet/isolator and the exhaust nozzle, which in turn depends on the integration of the engine with the vehicle airframe. Therefore, the geometry of the combustor can be 2-D/3-D rectangular, annular or axisymmetric, round, or round-to-elliptical 3-D.

The scramjet combustor design must account for many processes to properly occur in order for the combustor to operate in flight, e.g. fuel/air mixing, ignition, flameholding, and flame propagation. For dual-mode operation, the combustor configuration must consider thermal choking and fuel injection staging. Moreover, the integration with variable geometry propulsion components (inlet/isolator and nozzle) will impose further constraints on the ultimate design of the scramjet combustor. Since mechanical and thermal loads both increase with pressure, supersonic combustor structural design is very challenging, requiring advanced materials and effective cooling methods (see Chapter 9).

For the National Aerospace Plane (NASP) program conceived in the late 1980s to early 1990s, a two-dimensional inlet/rectangular combustor geometry was selected to maximize the air-breathing portion of the flight envelope for a Single-Stage to Orbit (SSTO) vehicle. For such ambitious flight mission, high propulsion efficiency over a very wide Mach number range, $0 < M_0 < 15$, required significant variable geometry in order to optimize engine inlet contraction ratio, dual-mode combustion efficiency, and nozzle expansion ratios. The two-dimensional planar design would allow movement of propulsion flowpath walls to accomplish the needed variable geometry in a vehicle-integrated configuration. The NASP vehicle concept became the baseline for which significant research, analytical tools, and technology development were performed.

The following are examples from scramjet engines that have been designed and tested.

5.5.1 NASA X-43A Vehicle with Rectangular Scramjet Geometry

After the NASP program was canceled in 1994, the two-dimensional propulsion architecture continued development under the NASA Hyper-X and the USAF HyTech programs. The Hyper-X conceived the X-43A vehicle, a scaled version derived from a 200-ft Global Reach concept, a Mach 5–10 airplane with high range potential of over 12 964 km (7000 nmi). Three X-43A autonomous research aircraft were designed and built intended for flight testing. Each of the 3.7 m-long, 1.5-m wide (12-ft long and 5-ft-wide) lifting body vehicles appear identical but incorporated slight vehicle/scramjet differences to simulate variable engine geometry (a function of Mach number). The first and second vehicles were designed to fly at Mach 7, and the third vehicle at Mach 10.

The X-43A airframe-integrated scramjet propulsion system was designed with a rectangular scramjet flowpath, predominantly two-dimensional, with a short external cowl, (76.2 cm long, about 20% of the total length of the vehicle), as depicted in Figure 2.1. Designing a small-scale scramjet as this one is challenging. The combustor size must be optimized to ensure the combustion conditions, including strategic positioning of fuel injectors and ignition promoters, to help the scramjet yield the demanded positive net thrust. The X-43A scramjet was started on gaseous hydrogen fuel mixed with silane (SiH_4). Although the hydrogen/air mixture was combustible, an external ignition source was deemed necessary to prevent flameout and ensure combustion stability. Since silane ignites on contact with oxygen, it is considered the best igniter for hydrogen combustion. Once the scramjet combustion had stabilized in flight, the silane gas flow was shut off.

In a small-scale scramjet, higher heat loads on a per-unit area basis make the thermal/structural design of the engine challenging. Hence, to protect the X-43A scramjet from the heat load during boost and engine operation, both the sharp-edged cowl and the vertical leading edges were water-cooled. In addition, zirconia coating was used on the forward section of the engine cowl and in key places throughout the engine for additional thermal protection, as described in Chapter 9. A nitrogen purge system was also used to cool several vehicle subsystems during the flight test phase.

5.5.2 NASA Hypersonic Research Engine with Conical Axisymmetric Geometry

The first scramjet concepts were conceived as extensions of ramjets for missile applications, or as podded engines that could be flight-tested on the underside of an aircraft such as the North American X-15 hypersonic rocket-powered airplane. Hence, the basic geometry of the scramjet was annular/axisymmetric, with a translating spike that could be moved to control the flow into the engine (Andrews and Mackley 1993).

The NASA axisymmetric Hypersonic Research Engine (HRE) depicted in Figure 5.13 was conceived for this purpose. It required inlet close-off to minimize the use of hydrogen coolant before and after the engine test portion of the X-15A-2 flight and to minimize foreign object damage to the engine during takeoff and landing. The design included staged fuel injectors to operate the engine at subsonic and supersonic combustion conditions.

Two full-scale engine models were fabricated and ground-tested: a structural model and a combustion/propulsion model, both with 18-in.-diameter cowls. A dummy version of the HRE was installed below the bottom ventral of the X-15 and tested in flight twice.

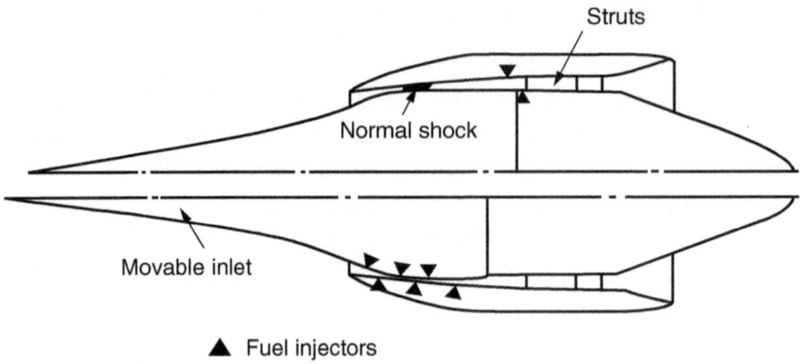

Figure 5.13 NASA axisymmetric dual-combustion HRE design Source: From Rubert and Lopez (1992)/NASA/Public Domain.

A complete water-cooled HRE engine, designated the Aerothermodynamic Integration Model (AIM), was ground-tested.

The HRE would operate as follows: at the onset of the engine test, the inlet spike would translate aft from inlet close-off to start the inlet. The spike would be in a fixed position for Mach 4–6 operation with the tip shock falling outside of the cowl lip. From Mach 6–8, the spike-tip shock would impinge on the cowl lip. This shock-on-lip condition would be maintained by translating the spike in a forward direction as the Mach number increased. Subsonic combustion was planned over the Mach range of 4–6, with transition from subsonic to supersonic combustion from Mach 5–6, and all supersonic combustion up to Mach 8. The locations of the staged fuel injectors, which were to be used to accomplish these mode changes, are shown in Figure 5.13. However, no operational HRE was ever flown because the X-15 program was canceled in 1968.

5.5.3 3-D Elliptical and Round Scramjet Geometries

In order to reduce propulsion structural weight and increase propulsion performance, airframe-integrated scramjets with rounded or elliptical combustors have been conceived. Such concepts take advantage of the inherent structural efficiency of rounded structures (Tam et al. 2007; Yao et al. 2018). Round and round-to-elliptic shape transition scramjet combustors avoid detrimental fluid dynamic effects of corner flows on the development of boundary layer and wave structures, improve the contact area between the transverse fuel jet (s) and the air crossflow, and contribute to the structural strength and weight reduction of the combustor. A nonrectangular combustor can ease the modular installation of the flowpath on a vehicle. Moreover, scramjets with elliptical combustor cross sections have a reduced wetted area compared with rectangular combustors of the same cross-sectional or flow area. This is an important characteristic for a scramjet flowpath, since a reduced wetted area lowers viscous drag and cooling requirements. Round geometries may improve the back pressure limits of the inlet/isolator or reduce isolator length requirements (Smart and Ruf 2006).

Researchers in Australia conceived a 3-D scramjet comprised of a rectangular-to-elliptical shape transition (REST) inlet and an elliptical combustor that could operate over a large

Mach number range with fixed geometry, which is an additional benefit for hypersonic flight at high Mach numbers. NASA tested a REST scramjet engine, which had a design point of Mach 7.1, and was intended to operate with fixed geometry between Mach 4.5 and 8.0 (Smart and Ruf 2006).

A new 3-D REST scramjet studied at the Centre for Hypersonics at the University of Queensland (Australia) has a design point of Mach 12. It incorporates different approaches for fuel injection, including inlet injection, and a step injection scheme in the combustor proper, designed to generate combustion in the boundary layer. A REST model flowpath (1980 mm long and maximum width of 180 mm) was built for ground testing with hydrogen fuel at conditions simulating flight at Mach 8.7. The test model consists of four components: a forebody plate, a REST inlet, an elliptical combustor, and a generic elliptical nozzle. During the experimental program, hydrogen fuel was injected at equivalence ratio values ranging between 0.30 and 1.23 using four distinct injection schemes. To support these studies, Moule et al. (2015) reported on Reynolds-averaged Navier–Stokes numerical simulations performed on a full REST engine at representative Mach 8.7 flight conditions. To characterize the performance of the scramjet engine, the global efficiency of the combustion chamber was determined by the product of thermal efficiency, chemical efficiency, and mixing efficiency.

Another interesting elliptically shaped combustor was conceived for the scramjet engine integrated in the LAPCAT-II Mach 8 cruise waverider concept. The constant cross section dual-mode ramjet/scramjet combustion chamber is intended to operate in a scramjet designed for flight Mach numbers from 4.5 up to 8.0. A small-scale engine model derived from the full-scale Mach 8 vehicle was designed and built for testing in the high-enthalpy facilities at the German Aerospace Center (DLR) (High Enthalpy Shock Tunnel Göttingen, HEG) and ONERA (F4) that simulate the vehicle flight conditions (Langener et al. 2013).

The combustor geometry was optimized in order to keep the pressure rise within the boundary layer separation limit and to avoid thermal choking. The 0.3-m-long elliptic combustor is fitted with staged fuel injectors, as depicted in Figure 5.14. The first upstream injection stage is comprised of two semi-strut injectors (each short strut with two injector nozzles), located on the geometrical transition from the inlet to the combustion chamber. The second stage (full strut with four injector nozzles) was placed in the center of the

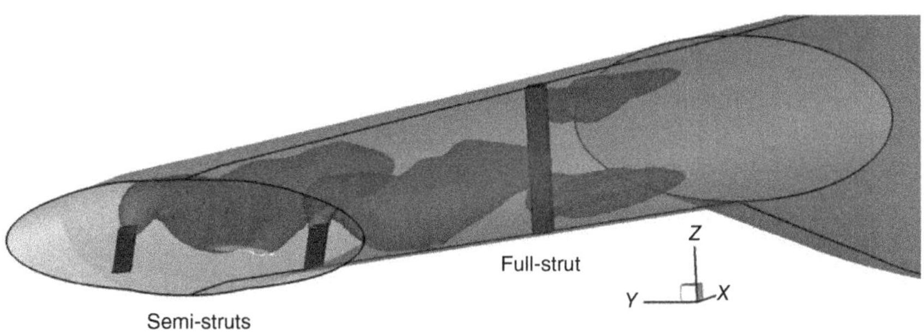

Figure 5.14 LAPCAT-II small-scale scramjet combustor layout with strut injectors. The brown isobars represent computed hydrogen distribution of 5% mass fraction. Source: Karl et al. (2020)/Springer Nature/CC BY.

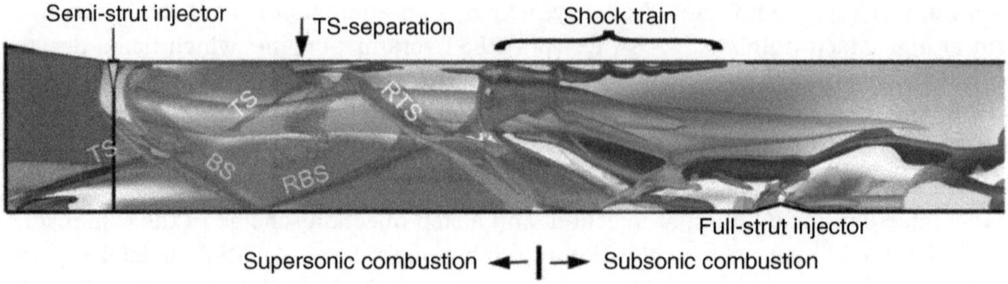

Figure 5.15 Top view of the flow structure in the LAPCAT-II small-scale scramjet elliptic combustor. It shows terminal shock (TS), reflected terminal shock (RTS), bow shock of semi-strut hydrogen injection (BS), and reflected bow shock (RBS). Source: Karl et al. (2020)/Springer Nature/CC BY.

chamber. This fuel injection scheme was found to provide the best configuration concerning combustion efficiency, flame anchoring, and separation control (Karl et al. 2020).

A recent numerical study was performed with the hybrid structured–unstructured DLR Navier–Stokes solver TAU (Langener et al. 2014). Figure 5.15 shows the top view of the combustor flow field structure, which is characterized by a complex interplay of turbulent flow, shock and expansion waves, and combustion. The researchers noted a strong dependency of the combustor inflow conditions on the boundary layer transition characteristics on the intake (engine inlet section). Thus, the inlet boundary layer can have a major effect on the flow structure developing in the combustor. They concluded that application of scale-resolving turbulence models together with improved prediction of boundary layer transition in the regions which are not accessible by ground experiment (e.g. upper part of the combustor) will lead to improved predictions in the future (Karl et al. 2020).

The Axisymmetric Cavity-Based Scramjet combustor depicted in Figure 5.16 is another example of round combustor. Its main feature is the recessed wall cavity spanning the entire circumference of the combustor entrance. The aft wall of the cavity is angled at 30° and 15°. To study the effect of fuel injection on combustor performance, four equi-spaced 1-mm-diameter injectors were placed at four different locations: injection into the cavity at a 45° angle (case 1); injection inside the cavity parallel to the main flow (case 2); transverse injection at the cavity bottom wall (case 3); and injection at a 30° angle upstream of the cavity (case 4). Using Reynolds Averaged Navier-Stokes (RANS) analysis, the effect of cavity-based injection was studied in a reacting combustion flow field. Scramjet performance was evaluated in terms of wall static pressure, hydrogen and H_2O mass fraction, and combustion efficiency, as well as total pressure loss (Relangi et al. 2021).

Results of the study are illustrated in Figure 5.16, which show enhanced jet penetration by the transverse injection upstream of the cavity (case 1), but penetration is rather low for parallel injection (case 2), as the fuel jet moves parallel to the mainstream. However, the H_2O mass fraction contour indicates that reaction has occurred inside the cavity, acting in effect to hold the flame there, especially in case 1. Normal injection from the cavity bottom wall (case 3) disturbs the core flow and H_2O fills the cavity, indicating that good mixing occurred and reaction took place inside the cavity. In the last case 4, injecting the fuel at an angle upstream of the cavity shows a very interesting flow pattern, as the bow shock generated by the fuel jet interacts with the shear layer as well as the separated boundary layer.

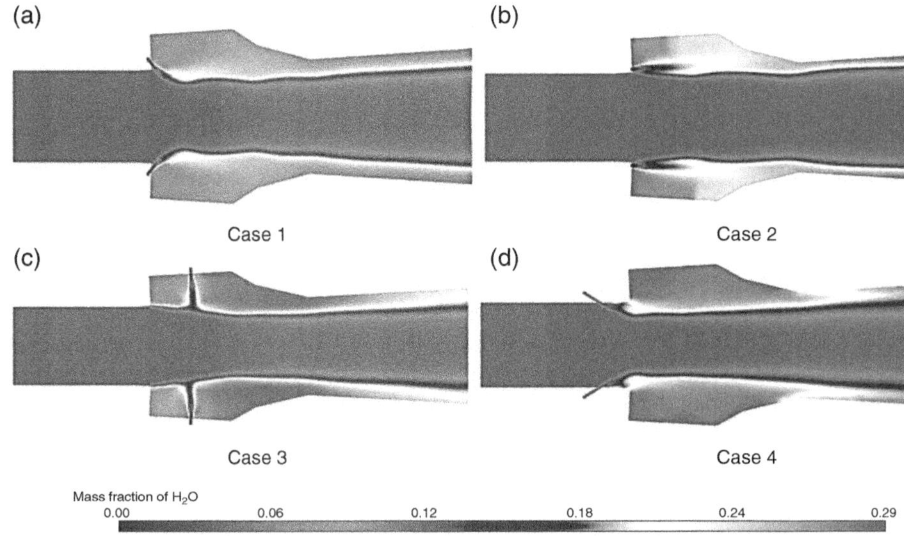

Figure 5.16 H_2O mass fraction contours for different fuel injection schemes in the Axisymmetric Cavity-Based Scramjet combustor. Source: Relangi et al. (2021)/MDPI/CC BY.

This interaction promotes better fuel/air mixing and reaction. Most remarkable in this case, the fuel jet penetrates further into the core flow. Results from the performance analysis of this combustor are reported by Relangi et al. (2021).

5.6 Scramjet Combustor Design Issues

The most important design requirement for the scramjet combustor is that mixing, ignition, and reaction must be very rapid. A short combustion length is required to minimize weight, cooling, and viscous drag. The combustor design must incorporate optimized injection schemes, ignitors, and special devices designed to help maintain continual and stable combustion. Such design will ensure that the combustion chamber has regions of higher temperature, higher pressure, or lower velocity, which serve as self-ignition points and flameholders.

Scramjet combustor designs all share the same technical challenges. A trade-off between competing system configurations is off-design performance, operability, and weight (including thermal protection system). Supersonic combustion challenges are typically addressed by flowpath variable geometry and multiple fuel injection stations. Fuel injection schemes become an issue at speeds greater than Mach 10, because the momentum of the fuel begins to have a strong impact on the achievable thrust.

Internal drag can reduce the performance of a scramjet engine significantly. Therefore, the combustor design must aim at reducing internal surface areas: combustor must be made as short as possible. Intrusive injectors are also a source of base and wave drag. Thus, hyper-mixers must be carefully assessed to determine their effective applicability.

If heat released by combustion becomes greater than a critical value, the flow becomes thermally choked. Choked flow causes a normal shock to form at the engine inlet, creating drag and reducing engine performance at high flight Mach numbers. Thus, the supersonic combustor area ratio must increase to prevent thermal choking and excessive pressure gradients and prevent unstart.

Scalability of the combustor cross section is an important design requirement. A high aspect ratio combustor allows for fuel penetration into the core airflow using reasonable fuel injection pressures from flush wall injectors. However, scalability of high aspect ratio combustors from smaller to larger engines is still a challenging issue. In general, a circular or elliptical combustor section is easier to scale.

By considering all aspects of the system design, the scramjet flowpath design driving factors may not be just fuel/air mixing optimization, and/or combustion efficiency. The combustor overall design must optimize and balance its full performance. For example, it may require to maximize net positive thrust at the expense of not achieving complete fuel/air mixing.

5.7 Closing Remarks

Supersonic turbulent combustion is still not completely understood. Using sophisticated CFD tools to model scramjet flow fields is part of hypersonic air-breathing propulsion technology development. It helps us predict effects of new geometries and is required to optimize fuel injection and combustion strategies (see Shenoy et al. 2018; Drozda et al. 2019). For example, Large Eddy Simulation (LES) in conjunction with appropriate Sub-Grid Scale (SGS) models have become increasingly useful. SGS modeling, capable of simulating injection and flameholding helps researchers develop practical scramjet combustors tools.

Before applying computational tools to analyze air-breathing hypersonic propulsion, it is imperative to assure that they properly represent the complex flow physics, especially in the turbulent supersonic combustor. CFD codes must include adequate chemical kinetics models to simulate hydrogen/air and by endothermic hydrocarbons/air combustion.

In addition, CFD tools that integrate all scramjet flow field phenomena (internal or flowpath and external) are needed to predict overall performance (thrust, fuel consumption, and drag). Finally, to evaluate the accuracy of the numerical algorithms and the physical models used, data from benchmark ground experiments is needed.

Today, a growing area of research involves modeling and simulating regeneratively cooled scramjets fueled by hydrocarbon endothermic fuels. Hydrocarbon-fueled, dual-mode scramjet combustors are extensively studied in ground-based facilities over a range of test conditions simulating the lower hypersonic flight regime. Investigations use simple fuels (e.g. ethylene) and more complex liquid hydrocarbons. Fuel cracking (converting the original substance into a mixture of smaller hydrocarbons in a heat exchanger/reactor) prior to combustion is required. As described in Chapter 6, the cracking process results in JP-7 fuel decomposing to smaller, lighter hydrocarbons such as C_2H_4 and CH_4. Such cracked fuel will have different ignition and/or reaction characteristics than the parent fuel (Puri et al. 2005). Hence, the combustor model must be appropriate to simulate the cracked fuel/air combustion.

Questions

1. What is the optimum entrance static temperature for the scramjet combustor?
2. How to inject fuel to maximize mixing and ignition in short residence time?
3. How much fuel a scramjet needs to completely burn with ingested air?
4. Why silane gas was added to the X-43A combustor?
5. What is the optimum combustor entrance static temperature?
6. In a Mach 6 scramjet, the air ingested enters the combustor at ~Mach 3. How do we inject fuel to promote mixing with the oxidizer in the hypervelocity airflow?
7. What are typical hypermixers most appropriate for scramjet combustors?
8. The turbulence/chemistry interaction in a scramjet combustor is a complex process. Do you think the temperature difference of reactants affect this interaction? How? Consider both hydrogen/air and endothermic hydrocarbon/air mixtures.

References

Anderson, G.Y., McClinton, C.R., and Weidner, J.P. (2000). Scramjet performance. In: *Scramjet Propulsion*, AIAA Progress in Astronautics and Aeronautics (ed. E.T. Curran and S.N.B. Murthy), Vol. 189, 2000.

Andrews, E.H. and Mackley, E.A. (1993). Review of NASA's hypersonic research engine project. *AIAA Paper 93-2323. 29th Joint Propulsion Conference and Exhibit*, Monterey, CA (28–30 June 1993).

Baurle, R.A. (2017). Hybrid reynolds-averaged/large-Eddy simulation of a scramjet cavity flameholder. *AIAA Journal* 55 (2): 544–560.

Cabell, K., Hass, N., Storch, A., and Gruber, M. (2011). HIFiRE Direct-Connect Rig (HDCR) phase I scramjet test results from the NASA Langley arc-heated scramjet test facility. *AIAA Paper 2011-2248. 17th AIAA International Space Planes and Hypersonic Systems and Technologies Conference*, San Francisco, CA (11–14 April 2011).

Cabell, K., Drozda, T.G., Axdahl, E.L., and Danehy, P.M. (2014). The enhanced injection and mixing project at NASA Langley. *JANNAF 46th CS/34th APS/34th EPSS/28th PSHS Joint Subcommittee Meeting*, Albuquerque, NM.

Cain, T. and Walton, C. (2006). Review of experiments on ignition and flameholding in supersonic flow. *Technologies for Propelled Hypersonic Flight*, Vol. 2. Subgroup 2: Scram Propulsion, RTO-TR-AVT-007-V2 (January 2006).

Di Stefano, M., Hosder, S., and Baurle, R.A. (2021). Effect of turbulence model uncertainty on scramjet strut injector flow field analysis. *Computers & Fluids* 229: 1–19.

Donohue, J.M. (2012). Dual-mode scramjet flameholding operability measurements. GRC-E-DAA-TN10262, 2012. *Journal of Propulsion and Power* 30 (3): 592–603.

Doster, J.C. (2008). Hypermixer pylon fuel injection for scramjet combustors. Ph.D. dissertation. Department of the Air Force Air University, Air Force Institute of Technology. http://www.dtic.mil/dtic/tr/fulltext/u2/a487085.pdf.

Driscoll, J.E. and Rasmussen, C.C. (2005). Correlation and analysis of blowout limits of flames in high-speed airflows. *Journal of Propulsion and Power* 21 (6): 1035–1044.

Drozda, T.G., Drummond, J.P., and Baurle, R.A. (2016). CFD analysis of mixing characteristics of several fuel injectors at hypervelocity flow conditions. *AIAA Paper AIAA 2016-4764. 52nd AIAA/SAE/ASEE Joint Propulsion Conference*, Salt Lake City, UT (25–27 July 2016).

Drozda, T.G., Shenoy, R.R., Axdahl, E.L., and Baurle, R.A. (2019). Numerical investigation and optimization of a flushwall injector for scramjet applications at hypervelocity flow conditions. *AIAA Paper 2019-4196. AIAA Propulsion and Energy 2019 Forum*, Indianapolis, IN (19–22 August 2019).

Drummond, J.P. (1988). *A Two-Dimensional Numerical Simulation of a Supersonic, Chemically Reacting Mixing Layer*, NASA TM 4055, 1–99. Hampton, Virginia: Langley Research Center.

Drummond, J.P. (2014). Methods for prediction of high-speed reacting flows in aerospace propulsion. *AIAA Journal* 52 (3): 465–485.

Falempin, F.H. (2000). Scramjet performance. In: *Scramjet Propulsion,* AIAA Progress in Astronautics and Aeronautics (ed. E.T. Curran and S.N.B. Murthy), Vol. 189, 2000.

Gibbons, N. (2019). Simulation and dynamics of hypersonic turbulent combustion. Ph.D. thesis. The University of Queensland, School of Mechanical and Mining Engineering, St Lucia, QLD.

Gruber, M.R. (2004). Mixing and combustion studies using cavity-based flameholders in a supersonic flow. *Journal of Propulsion and Power* 20 (5).

Heiser, W.H. and Pratt, D.T. (1994). *Hypersonic Airbreathing Propulsion*, AIAA Education Series. Washington, DC: AIAA.

Hsu, K.-Y., Carter, C., Crafton, J., et al. (2000). Fuel distribution about a cavity flameholder in supersonic flow. *AIAA-2000-3585. 36th AIAA/ASME/SAE/ASEE Joint Propulsion Conference and Exhibit*, Las Vegas, NV (24–28 July 2000).

Jachimowski, C.J. (1988). An analytical study of the hydrogen–air reaction mechanism with application to scramjet combustion. Technical Paper 2791, NASA.

Jeong, E., O'Bryne, S., Jeung, I.-S., and Houwing, A.F.P. (2020). The effect of fuel injection location on supersonic hydrogen combustion in a cavity-based model scramjet combustor. *Energies* 13 (193): 1–16. https://doi.org/10.3390/en13010193.

Karl, S., Martinez Schramm, J., and Hennemann, K. (2020). Post-test analysis of the LAPCAT-II subscale scramjet. *CEAS Space Journal* 12: 385–395. https://doi.org/10.1007/s12567-020-00307-7.

Koshi, M., Murakami, Y., and Matsui, H. (1991). Chemical kinetics of silane combustion. *Journal of Physics D: Applied Physics* 23 (1): 195–202.

Kouchi, K. Sasaya, J., Watanabe, H.S., and Masuya, G. (2010). Penetration characteristics of pulsed injection into supersonic crossflow. AIAA 2010-6645

Kutschenreuter, P. (2000). Supersonic flow combustors. In: *Scramjet Propulsion*, Progress in Astronautics and Aeronautics (ed. E.T. Curran and S.N.B. Murthy), 513–568. Reston, VA: AIAA.

Langener, T., Steelant, J., Karl, S., and Hannemann, K. (2013). Layout and design verification of a small scale scramjet combustion chamber. *ISABE-2013-1655. Proceedings of the XXI International Symposium on Air Breathing Engines (ISABE 2013)*, Busan, Korea (9–13 September 2013).

Langener, T., Erb, S., and Steelant, J. (2014). Trajectory simulation and optimization of the LAPCAT-MR2 hypersonic cruiser concept. *29th Congress of the Intl. Council of the Aeronautical Sciences* (September 2014).

Luo, Z., Lu, T., and Liu, J. (2011). A reduced mechanism for ethylene/methane mixtures with excessive NO enrichment. *Combustion and Flame* 158 (7): 1245–1254.

Mao, M., Riggins, D.W., and McClinton, C.R. (1991). Numerical simulation of transverse fuel injection. In: *Computational Fluid Dynamics Symposium on Aeropropulsion*, NASA-CP-3078, 635–667. NASA: Cleveland, OH.

Mawid, M.A., Park, T.W., Sekar, B., and Arana, C.A. (2003). Development and validation of a detailed JP-8 fuel chemistry mechanism. *ASME Int. Gas Turbine Inst. Publ. IGTI 2*, 613–624.

Morris, N.A., Morgan, R.G., Paull, A., and Stalker, R.J. (1987). Silane as an ignition aid in scramjets. *AIAA Paper 87-1636. 22nd Thermophysics Conference*, Honolulu, HI (8–10 June 1987).

Moule, Y., Sabelnikov, V., Mura, A., and Smart, M. (2015). Computational fluid dynamics investigation of a Mach 12 scramjet engine. *Journal of Propulsion and Power* 30 (2): 461–473.

Moura, A.F., Gibbons, N., Wheatley, V. et al. (2020). Characterisation of supersonic turbulent combustion in a Mach-10 scramjet combustor. *AIAA Journal* 58 (5): 2180–2196.

Murugappan, S., Gutmark, D., and Carter, C. (2005). Control of penetration and mixing of an excited supersonic jet into a supersonic cross sream. *Physics of Fluids* 17: 106101-1–106101-13. https://doi.org/10.1063/1.2099027.

Neely, A., Stotz, I., O'Bryne, S. et al. (2005). Flow studies on a hydrogen-fueled cavity flame-holder scramjet. *AIAA/CIRA 13th International Space Planes and Hypersonics Systems and Technologies Conference*, Capua, Italy (16–20 May 2005).

Puri, P., Ma, F., Choi, J.-Y., and Yang, V. (2005). Ignition characteristics of cracked JP-7 fuel. *Combustion and Flame* 142: 454–457.

Quinlan, J.R., McDaniel, J.C., Drozda, T.G. et al. (2014). A priori analysis of flamelet-based modeling for a dual-mode scramjet combustor. *AIAA 2014-3743. 50th AIAA/ASME/SAE/ASEE Joint Propulsion Conference*, Cleveland, OH (28–30 July 2014).

Rasmussen, C.C. and Driscoll, J.E. (2008). Blowout limits of flames in high-speed airflows: critical Damkohler number. *AIAA 2008-4571*. 44th AIAA/ASME/SAE/ASEE Joint Propulsion Conference & Exhibit, Hartford, CT (21–23 July 2008).

Relangi, N., Ingenito, A., and Jeyakumar, S. (2021). The implication of injection locations in an axisymmetric cavity-based scramjet combustor. *Energies* 14: 2626. pp. 1–13. https://doi.org/10.3390/en14092626.

Roudakov, A.S., Schikhmann, Y., Semenov, V. et al. (1993). Flight testing of an axisymmetric scramjet: Russian recent advances. *IAF Paper 93-S.4.485. 44th Congress of the International Astronautical Federation*, Graz, Austria (16–22 October 1993).

Rubert, K.F. and Lopez, H.J. (1992). The NASA hypersonic research engine program. *Rocket-Based Combined-Cycle (RBCC) Propulsion Technology Workshop*. Tutorial session, NASA, Lewis Research Center (January 1992).

Shenoy, R.R., Drozda, T.G., Norris, A.T., et al. (2018). Comparison of mixing characteristics for several fuel injectors at Mach 8, 12, and 15 hypervelocity flow conditions. *AIAA 2018-4540. 2018 Joint Propulsion Conference,* Cincinnati, Ohio (9–11 July 2018).

Smart, M.K. (2012). How much compression should a scramjet inlet do? *AIAA Journal* 50 (3): 610–619.

Smart, M. (2016). Comparison Between Hydrogen and Methane Fuels in a 3-D Scramjet at Mach 8. Final Report, AFRL-AFOSR-JP-TR-2016-0066.

Smart, M.K. and Ruf, E.G. (2006). Free-jet testing of a REST scramjet at off-design conditions. *AIAA 2006-2955. 25th AIAA Aerodynamic Measurement Technology and Ground Testing Conference*, San Francisco, CA (5–8 June 2006).

Tam, C.-J., Hsu, K.-Y., Gruber, M.R., and Raffoul, C.N. (2007). Aerodynamic Performance of an Injector Strut for a Round Scramjet Combustor. Technical Report, *AFRL-RZ-WP-TP-2008-2067. 43rd AIAA/ASME/SAE/ASEE Joint Propulsion Conference & Exhibit*, Cincinnati, OH (8–11 July 2007).

Turner, J.C. and Smart, M.K. (2010). Application of inlet injection to a three-dimensional scramjet at Mach 8. *AIAA Journal* 48 (4): 829–838.

Vincent-Randonnier, A., Ristori, A., Sabelnikov, V. et al. (2018). Combined experimental and computational study of the LAPCAT II supersonic combustor. *22nd AIAA International Space Planes and Hypersonics Systems and Technologies Conference*, Orlando, US (September 2018). HAL Id: hal-02420953.

Yamashita, H., Shimada, M., and Takeno, T. (1996). A numerical study on flame stability at the transition point of jet diffusion flames. *Symposium (International) on Combustion* 26: 27–34. Elsevier.

Yao, W., Yuan, Y., Li, X. et al. (2018). Comparative study of elliptic and round scramjet combustors fueled by RP-3. *Journal of Propulsion and Power* 34 (3): 584–594.

Yokev, N., Brod, H.E., Cao, D., and Michaels, D. (2021). Impact of fuel injection distribution on flame holding in a cavity-stabilized scramjet. *Journal of Propulsion and Power* 37 (4): 584–594.

6

Fuels for Hypersonic Air-Breathing Propulsion

Fuel is considered as an enabling technology for hypersonic air-breathing vehicles. Each class – expendables (missiles), reusable accelerators (launch vehicles), and hypersonic transports (sustained cruisers) – places very different constraints on the fuel. The key drivers and limits for the fuel characteristics and fuel system development arise from combustion, thermal management, mission requirements, ground infrastructure, and cost. Therefore, the fuel ultimately selected for a given hypersonic vehicle will be a compromise between cost, operability, and performance requirements.

Table 6.1 summarizes some of the advantages and disadvantages of both hydrogen and endothermic hydrocarbon fuels. Liquid hydrogen has superior combustion performance and heat sink capability. However, hydrogen has very low density, about 1/11th that of hydrocarbons, and high handling cost, and these disadvantages drive the incentive to develop alternative fuels for future hypersonic cruisers and expandable military vehicles, applications that are incompatible with large bodied (hydrogen) vehicles and related infrastructures.

Hydrocarbon fuels have a low heating value and their combustion is much more complex and slower, as compared with hydrogen. However, endothermic hydrocarbons have an inherent thermal stability and demonstrated the ability to provide a sizable endothermic heat sink capability. Owing to the long ignition delay associated with combustion of hydrocarbon fuels, a robust ignition and combustion-piloting source is required.

In the following sections, we review important characteristics of endothermic fuels and the potential they offer to fuel a next generation of hypersonic air-breathing vehicles. The main objective of this chapter is to provide an overview with sufficient detail to help us answer questions such as:

- Endothermic fuels are a critical technology to enable the use of hydrocarbons at hypersonic speeds. *Why? What characterizes endothermic fuels?*
- Onboard reforming and cracking are endothermic processes considered to enhance the quality and operability of fuels. *Are these processes viable for hypersonic propulsion?*
- *Which fuel is a better choice for a reusable hypersonic cruise vehicle? Which is the best fuel for an expandable Mach 7 missile?*

Scramjet Propulsion: A Practical Introduction, First Edition. Dora Musielak.

Table 6.1 Advantages and disadvantages of hydrogen and hydrocarbon fuels.

Hydrogen	Hydrocarbons
• Highest heating value	• Low heating value
• High cooling capacity	• Low cooling capacity
• Wide flammability limits	• Adequate flammability limits
• Low density	• High density
• Smallest ignition delay time • Green, environmentally friendly combustion exhaust	• Long residence time to complete chemical reactions • Non-clean combustion exhaust
• $0 < M_0 < 25$	• $0 < M_0 < 7.0$

6.1 Introduction

The primary purpose of a fuel is to provide propulsive energy to the engine. In high-speed applications, the fuel must burn rapidly and generate the thermal energy required for thrust production. The fuel choice is of crucial importance and must be made with careful consideration to the mission and the vehicle type. Within the range of application, the specific impulse of an air-breathing engine is always higher when it is fueled by hydrogen. Hydrogen is the fuel of choice for high Mach numbers, $M_0 \geq 10$. The NASA X-43A vehicle, the first mission to prove the viability of scramjet engines in hypersonic flight, was fueled with gaseous hydrogen. However, hydrocarbon fuels offer other benefits for the low range of hypersonic flight speeds, as demonstrated by the X-51A Waverider whose scramjet was fueled by JP-7.

The higher speeds of hypersonic flight have a direct impact on the fuel itself, as higher speeds mean higher air stagnation temperatures. This increases aerodynamic heating on the vehicle, which increases cooling requirements and prevents the use of air as a coolant. Thus, a second fundamental requirement for the fuel is to cool the hypersonic vehicle's hot surfaces, which will be exposed to severe thermal environments, particularly in the engine. At Mach 8, for example, uncooled combustor surfaces may surpass 3000 K (5000 °F), a temperature far above the limit of known structural material capability (see Chapter 9). For a hypersonic cruise aircraft, the time exposed to high temperatures (hot time) and stress will be many times greater than other aerospace vehicles experience. A fuel-cooled thermal management system is then essential to reduce the extreme temperatures in the hottest parts of the vehicle. As propulsion performance requirements and flight speeds increase, the needs for thermal management using the fuel onboard are augmented.

Figure 6.1 gives the properties of candidate fuels for scramjet application, normalized by hydrogen property values: specific energy = 120 MJ/kg, density (STP) = 71 kg/m^3, energy density = 8300 MJ/kg. These data indicate the much higher energy per unit mass of hydrogen; however, its smaller density results in a significantly lower energy per unit volume when compared to hydrocarbon fuels. As a result, a hydrogen-fueled scramjet vehicle will

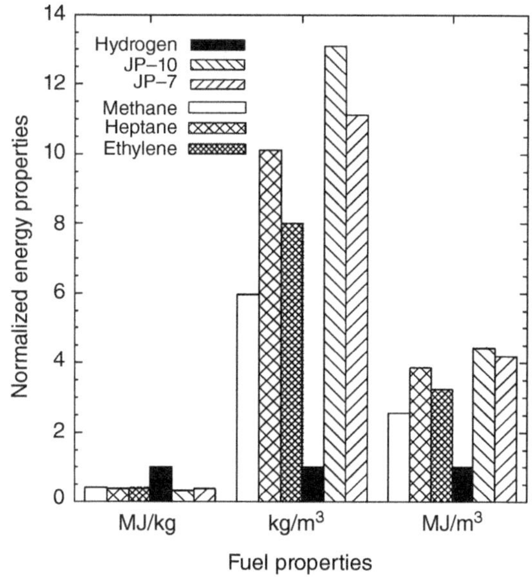

Figure 6.1 Normalized properties of candidate fuels for scramjet application. Source: From Denman (2017).

be comparatively larger than a hydrocarbon equivalent, as it must carry a greater volume of fuel. The low energy density of hydrogen shifted interest to hydrocarbon fuels that exhibit much higher energy densities. As shown in Figure 6.1, low-order hydrocarbons such as ethylene and methane offer attractive potential for air-breathing propulsion in terms of energy per volume characteristics. Ethylene is about eight times denser than hydrogen, with an energy density over three times greater. Methane is approximately six times denser than hydrogen, enough to make it interesting for air-breathing hypersonic flight applications.

Hydrogen is the fuel of choice for hypersonic flight due to its high energy content per unit mass, rapid burning rate, and excellent heat transfer properties for active cooling. However, we can also use the endothermic properties of hydrocarbon fuels to create an extra heat sink and cool the combustor walls and adjacent surfaces. A general consensus exists regarding fuel selection for a mission: hydrogen for flight speeds from Mach 10 and greater, while storable, endothermic JP-type hydrocarbon fuels can be used to speeds up to Mach 6–8, although the upper end of this range today poses a significant technical challenge. Vehicles fueled with JP-7 could cruise to about $M_0 = 8$, while those with methane fuel could fly to $M_0 = 10$. Endothermic hydrocarbon fuels are also attractive for use in the booster stage of reusable two-stage-to-orbit (TSTO) vehicles due to the higher density and logistical benefits that endothermic fuels offer.

6.1.1 Fuel Energy for Combustion

In addition to hydrogen, hydrocarbon fuels are important for hypersonic air-breathing propulsion. Hydrocarbons include methane, jet fuel, and military jet propellant or JP fuels. Methane (CH_4) is a promising fuel for hypersonic air-breathing propulsion. It is more stable than liquid hydrogen (most common rocket fuel) and can be stored at more manageable temperatures. Methane has an energy density of 50 MJ/kg.

Jet fuels are made from kerosene derived from petroleum. Kerosene results from the distillation of crude oil at atmospheric pressure or from catalytic, thermal, or steam cracking of heavier petroleum streams (cracked kerosene). Kerosene jet fuels primarily consist of C_9 to C_{16} hydrocarbons designed to maintain viscosity at low temperatures and boil in the range of 145–300 °C. Two important types of jet fuels exist for commercial aviation, Jet A and Jet A-1. Jet A is predominantly used in the United States, while Jet A-1 is used throughout the rest of the world. These fuels are nearly identical; however, Jet A-1 is refined to have a lower maximum freezing point (−47 °C) than Jet A (−40 °C). The lower freezing point makes Jet A-1 a better choice for international flights, especially on polar routes during the winter season.

Jet propellant (JP) fuels are hydrocarbon-based kerosenes used as aircraft fuels by military organizations around the world, which use a different classification system of JP numbers. The US military use JP-5, JP-8, JP-7, and JP-10. The fuel JP-5 was developed in 1952 for use in aircraft stationed aboard aircraft carriers, where the risk from fire is particularly great. JP-8 fuel is used as a "universal fuel" in both turbine-powered aircraft and diesel-powered ground vehicles. It was first introduced at NATO bases in 1978 and is projected to remain in use at least until 2025. JP-8 contains a corrosion inhibitor and anti-icing additive that is not required in the ASTM specification of Jet A-1. NASA considered fueling the Mach 4 Revolutionary Turbine Accelerator (RTA) with JP-8. The RTA is an advanced concept that integrates a turbofan fitted with a ramburner, the booster engine in a turbine based combined cycle (TBCC) propulsion concept (see Chapter 12). The primary difference between JP-5 and JP-8 fuels is that the flash point temperature for JP-5 is higher (60 °C) as compared to JP-8 (38 °C). The higher flash point for JP-5 makes it more suitable for safe handling and fueling practices aboard aircraft carriers, as the US Navy requires.

Military jet fuels contain additives to enhanced thermal stability. Jet fuels act as a heat sink for modern aircraft engines. They absorb heat from engine oil, hydraulic fluid, and air conditioning equipment. Jet fuels used for high-performance military aircraft engines have even greater need of thermal stability as compared to commercial aviation fuels. In the late 1990s, the US Air Force began development of an additive to increase the thermal stability of jet fuels. JP-8 fuel containing this additive package is usually referred to as JP-8 +100, because this additive increased the thermal stability by 37.78 °C (100 °F) without coking. Coking refers to fuel decomposition when it is heated to temperatures higher than its thermal limit, leaving a residue that can clog the fuel lines and injectors. Beginning in 2013, the US Air Force began using Jet A (plus additives) rather than JP-8 for continental flight usage in order to save on fuel costs.

JP-7 fuel was developed specifically for the Pratt & Whitney J58 (JT11D-20) afterburning turbojet engine that powered the Lockheed SR-71 Blackbird aircraft. During flight, the SR-71 could attain speeds in excess of Mach 3+, which was the most efficient cruising speed for the J58 engines. Since very high skin temperatures are generated at this speed due to aerodynamic friction, a new jet fuel was needed with a high flash point and high thermal stability. JP-7 fuel was developed for this purpose. Being a low-volatility fuel, JP-7 requires a chemical ignition system. That is why the J58 engine used triethylborane (TEB) as ignitor. The flash-resistant JP-7 fuel protected the SR-71 from

Table 6.2 Heats of reaction or heating values for typical hydrocarbon fuels reacting with air at the standard reference state; all reactants and products are gaseous.

Fuel	h_{PR} (BTU/lbm)	h_{PR} (kJ/kg)
Hydrogen, H_2	51 571	119 954
Methane, CH_4	21 502	50 010
Ethane, C_2H_6	20 416	47 484
Octane, C_8H_{18}	19 256	44 786
JP-5	18 300	42 566
JP-7	18 702	43 500
JP-8	18 585	43 228

the threat of inadvertent ignition as a result of the combination of high airframe and fuel temperatures during the hot portion of flight at high-speed cruising. In scramjet applications, the combustor can be ignited with a mixture of ethylene (C_2H_4) and JP-7 fuel, as it was done in the X-51A demonstration flight.

JP-10 is a high energy fuel used for air-breathing missiles, such as the ALCM (subsonic air-launched cruise missile) and the Tomahawk missile. JP-10 is essentially a pure component: exo-tetrahydrodicyclopentadiene (tricyclo[5.2.1.0^{2,6}] decane). This fluid has an appreciable density (0.94 g/cm^3) and produces a specific impulse of 297.4 seconds. These characteristics as well as its very low freezing point (−79 °C) have made JP-10 the only air-breathing missile fuel used by the United States at the present time.

A fuel is characterized by its energy content per unit mass. The heat of reaction or heating value of the fuel, denoted as h_{PR}, represents the ideal fuel energy density, i.e. fuel thermal energy per unit mass of fuel. The heating value varies relatively little over the normal ranges of initial pressures and temperatures of the reactants, and thus, it is evaluated at the standard reference state (25 °C, 1 atm). The heats of reaction of typical gaseous fuels reacting with air are found in Table 6.2. Hydrogen has low molecular mass and the highest heating value, giving the largest amount of heat release per unit mass of fuel, and this makes hydrogen the most attractive fuel for realizing the highest specific impulse air-breathing engines. In contrast, hydrocarbons have less than 50% that amount of heat release. Nonetheless, hydrocarbon fuels contain approximately 15–20% hydrogen.

Of course, not all available h_{PR} can be realized. Thus, analysis must account for the possibility that some of the chemical energy of the fuel is not released due to inadequate mixing or reaction time for complete combustion. The combustor efficiency represents this limitation:

$$\eta_b = \frac{h_{PR,actual}}{h_{PR,ideal}} \tag{6.1}$$

From a thermodynamic cycle analysis, the amount of heat addition to the combustor can be represented by the enthalpy change that occurs within. The rate at which the chemical reactions make energy available to the engine cycle is modeled by the product of fuel flow rate and the fuel's heat of reaction, $\dot{m}_f h_{PR}$. Hence, denoting heat addition as $\dot{q}_{in}$, we write

$$\dot{q}_{in} = (h_4 - h_3) = \frac{\eta_b \dot{m}_f h_{PR}}{\dot{m}_0} = \eta_b f h_{PR} \tag{6.2}$$

where f denotes the fuel-to-air ratio, and η_b is the combustion efficiency.

As shown in Chapter 5, the stoichiometric mixture in the engine is the ideal ratio of air to fuel that burns all fuel with no excess. Denoted by f_{st}, the stoichiometric fuel-to-air ratios with hydrogen and JP-7 fuels are $f_{st} = 0.0291$ and $f_{st} = 0.0664$, respectively. Assuming a combustion efficiency of 98%, heat addition values with stoichiometric fuel-to-air ratio are $\dot{q}_{in} = 3421$ kJ/kg (for H_2/air) and $\dot{q}_{in} = 2831$ kJ/kg (for JP-7/air).

Hydrogen is an attractive propellant for hypersonic systems both from energetic and vehicle coolant perspectives. Hydrogen can be physically stored as either a gas or a liquid. Storage as a gas typically requires high-pressure tanks: 5000–10 000 psi tank pressure; the higher the pressure, the more fuel a tank can hold. Storage of hydrogen as a liquid requires cryogenic temperatures because the boiling point of hydrogen at one atmosphere pressure is −252.8 °C (liquid hydrogen has a density of close to 71 kg/m^3).

The NASA X-43A aircraft was fueled by hydrogen gas to demonstrate hypersonic flight at Mach 7 and 10. Hydrogen fuel was selected because of its intrinsic cooling capability and its capability to achieve the highest specific impulse. The X43A scramjet was manufactured as a self-contained engine with a rectangular flowpath and fit with a silane igniter. Silane is a pyrophoric gas that ignites on contact with air. To provide sufficient flight time, the hydrogen and hydrogen/silane igniter mixture (80 : 20 by volume) were stored at 8500 and 4500 psi, respectively.

The U.S. Air Force/Boeing X-51A Waverider used JP-7 to fuel its Pratt & Whitney SJY61 scramjet engine. The endothermic fuel cooled the engine by absorbing the heat load from the interior spaces of the airframe. The hot fuel was then pumped through fuel injectors into the combustor. Ignition sequence began by burning ethylene, C_2H_4, to ensure supersonic combustion. After approximately 10 seconds, the scramjet engine transitioned to JP-7 fuel to accelerate the vehicle to its hypersonic speed. Ethylene is an easy-to-handle, high-energy fuel used extensively in scramjet engine ground tests to simulate fuel properties expected from endothermically cracked JP-7. Cracking refers to a reaction in which larger saturated hydrocarbon molecules are broken down into smaller, more useful hydrocarbon molecules. Cracked JP-7 is a fuel that is heated prior to combustion and yields different compositions, depending on the cracking method utilized.

6.2 Endothermic Fuels

The limit of conventional thermal management systems is set by the physical heat sink capacity of fuels. In the late 1960s and early 1970s, researchers found that heat sink in a fuel could be obtained from two sources: the traditional sensible heat sink derived from heating of the fuel, and also from endothermic reactions such as cracking, a class of chemical reactions that are made to occur in the fuel system prior to ignition. This led to the development of endothermic hydrocarbon fuels. Heat-absorbing chemical reactions were first promoted by energy extracted from the engine heated air (Lander and Nixon 1971).

Since that time, studies of endothermic fuel reactions were pursued in order to develop endothermic fuels that can fuel air-breathing propulsion for high-speed applications without resorting to cryogenic propellants. In fact, the development of JP-7 endothermic fuel developed specifically for the Mach 3.5 SR-71 reconnaissance aircraft began a path to "high-temperature fuels" development. In that application, thermal cracking reactions generated relatively small molecules when the JP-7 fuel was used as a coolant prior to combustion in the two afterburning turbojet engines that propelled the aircraft.

We define an endothermic fuel as that characterized by an additional heat sink capability over that of conventional fuels. The endothermic state is obtained by undergoing chemical reactions through heat absorption, reforming, and thermal cracking prior to injecting the fuel into the combustor. Endothermic processes such as cracking yield a substance that is much lighter than the heavy hydrocarbon fuel. The process is called cracking because the original long-chain hydrocarbon molecules are broken into lighter molecules that absorb heat (an endothermic process).

JP-7 is a low-volatility, highly refined kerosene with low sulfur and aromatics, consisting mainly of alkanes with an average carbon number of 12, and it has a thermal stability limit of 560 K (550 °F) (Croswell and Biddle 1992), giving it a flight Mach number limit of about 4.6 (Wiese 1992) at a typical operating flight altitude. Cracked JP-7 has an additional source of heat capacity, which allows to extend the flight operational velocity to perhaps Mach 7–8.

Two endothermic fuel reactions are catalytic dehydrogenation (Carbon–Hydrogen C—H bond breaking), and thermal cracking (Carbon–Carbon [C—C] bond breaking). The cracking reaction will naturally occur at higher temperatures than dehydrogenation, or it may be enhanced through the use of catalysts or chemical initiators. Studies confirm that the breaking of C—C bonds over a catalyst, or catalytic cracking, is an endothermic reaction that can provide a substantial heat sink for hydrocarbons.

Consider the cracking of JP-7 for cooling purposes. The fuel circulates through a heat exchanger and absorbs heat before it is injected into the combustor. In the process, the fuel temperature and heating value are increased (due to cracking reactions), permitting a reduction in the fuel flow rate. The cracking process results in JP-7 decomposing to smaller, lighter hydrocarbons (e.g. C_2H_4 and CH_4). The resulting mixture will have better ignition and/or reaction characteristics than the parent fuel. Such endothermic fuel meets the operational demands for supersonic and hypersonic aircraft, as it was successfully demonstrated by the X-51A Waverider that flew in 2013 to Mach 5.1 for 240 seconds. Endothermic JP-7 has a greater heat sink capacity and less carbon formation, or coking, as compared with other hydrocarbon fuels.

Hydrocarbon-fueled scramjet combustors are extensively studied in ground-based facilities over a range of test conditions simulating flight from Mach 4–7. Investigations have used simple fuels (e.g. ethylene) and more complex liquid hydrocarbon fuels. For example, Kay et al. (1992) conducted hydrocarbon-fueled combustor experiments at Mach 5.6 conditions using ethylene and JP-5. Studies demonstrated viability of combustor operating in dual mode.

Steam reforming of liquid hydrocarbon fuels is used in the Russian "AJAX" concept (Gurijanov and Harsha 1996). This approach utilizes a two-stage steam reforming process, with individual optimization of catalysts and temperatures for each stage. The first stage is steam cracking of the kerosene fuel to form primarily methane; the second stage is the highly endothermic steam reforming of methane to form CO and H_2. Steam reforming offers a very large heat sink potential; however, this process needs appreciable quantities of steam. Methanol reforming is considered a promising alternative to steam reforming.

6.3 Heat Sink Capacity of Hydrogen and Endothermic Fuels

The fuel selected for high-speed propulsion must have not only the required heat of reaction for combustion, but it must be sufficiently attractive as a coolant. The fuel must be thermally stable and provide a heat sink capability through sensible heating and latent phase change. The heat sink capacity due to its sensible heat is proportional to the maximum temperature the fuel can achieve, and the magnitude of its specific heat capacity (i.e. $c_p\Delta T$).

Hydrogen has a much higher specific heat than hydrocarbon fuels, and for the same ΔT, it has a higher heat capacity than hydrocarbon fuels. The constant pressure specific heat (c_p) of typical jet fuels at 344 K (160 °F) lies in the range from 2.15 to 2.30 kJ/kgK (0.51–0.55 BTU/lbm °F). By contrast, the constant pressure specific heat of hydrogen at 343 K (158 °F) is 14.38 kJ/kgK (3.44 BTU/lbm °F). This explain why, to obtain a given amount of heat release, less hydrogen mass is required; hence, the hydrogen fuel flow rate to a combustor is lower than that of a hydrocarbon fuel.

However, it is not the specific heat that gives hydrogen fuel better heat capacity than hydrocarbon fuels have. Rather, it is the ability of hydrogen to sustain higher temperatures and thus to increase the ΔT part of its heat capacity; hydrogen can sustain temperatures up to the operating temperature of the scramjet. On the other hand, conventional hydrocarbon fuels cannot be driven to as high a temperature as hydrogen.

The hydrocarbons in the fuel systems of hypersonic vehicles will be subject to supercritical pressures and temperatures to ensure adequate flow of fuel to the engine under all operating conditions. For example, the JP-8 fuel used in the fuel system of the McDonnell Douglas F-15 fighter aircraft (top speed Mach 2.5+) is subject to pressures up to 27.6 MPa and temperatures as high as 450 K for residence times up to 20 minutes (Heneghan et al. 1996). These pressures are above the critical pressures of typical commercial and military jet fuels (JP-7, JP-8, and JP-10), which are given in Table 6.3. The thermal

Table 6.3 Characteristics and properties of military jet fuels.

	JP-7	JP-8	JP-10
Approximate formula	$C_{12}H_{25}$	$C_{11}H_{21}$	$C_{10}H_{16}$
H/C ratio	2.1	1.9	1.6
Critical pressure (MPa)	2.1	2.3	3.7
Critical temperature (K)	672	683	698

Source: Adapted from Sobel and Spadaccini (1997).

loads of hypersonic vehicles are much greater, and thus, the fuel may withstand temperatures on the order of 813 K or higher, which is greater than the critical temperature of current military jet fuels (Edwards 1993).

The heat sink capability of the endothermic fuel is made up of its sensible heat plus any net endothermic capacity derived from high fuel dissociation reactions. It is important to enhance the endothermic capability of a hydrocarbon without degrading its exothermic capability. Moreover, hydrocarbon decomposition reactions are accompanied by carbon formation, or coking. Coking is undesirable as carbon deposits on heat transfer surfaces degrade thermal management efficiency. Hence, the upper limit of a hydrocarbon fuel heat sink capability must consider both the maximum temperature achievable without coking up and the endothermic capacity of the cracking reactions that can occur.

Lander and Nixon (1971) developed a relationship between fuel temperature and heat sink for thermally cracking (pyrolytic) of a kerosene-class hydrocarbon. As shown in Figure 6.2, typical kerosene-class fuels yield a total sensible heat sink of approximately 2300 kJ/kg (1000 BTU/lb) for fuels heated from ambient to 1000 K (1340 °F). By contrast, the heat sink for hydrogen under comparable conditions is 10 000 kJ/kg (4300 BTU/lb).

Catalytic cracking of endothermic fuels has the potential to improve the fuel's ignition properties. Solid acid catalysts, such as zeolites, are known to be effective for catalyzing the reactions needed for endothermic fuel cracking. In 2002, Huang, et al. reported on the cracking of JP-7 over a zeolite catalyst under supercritical conditions (978 K and 4.1 MPa). They determined a physical heat sink of 388 kJ/mol and a chemical heat sink of 178 kJ/mol, assuming an average molecular weight of 166 g/mol for JP-7. This fuel cracking approach was successfully implemented in the test flights of the US Air Force/Boeing X-51A Waverider, which reached flight speeds up to Mach 5.1.

JP-10 has superior thermal stability compared to conventional JP-8, making it an attractive endothermic fuel. Its thermal cracking, or pyrolysis, has been studied extensively. Huang et al. (2002) compared the gas-phase product distribution, heat sink capacity, and coke deposition results for the cracking of JP-7, JP-8, and JP-10 over a zeolite catalyst under supercritical conditions using a benchtop reactor setup with a residence time of 1.2 seconds. They reported that JP-10 had a lower chemical heat sink (90 kJ/mol) than JP-7 (178 kJ/mol) and JP-8 (137 kJ/mol).

More recently, Huang (2017) compared the endothermic cooling potential of pyrolytic and catalytic cracking of JP-10. He found that catalytic cracking of JP-10 did not increase

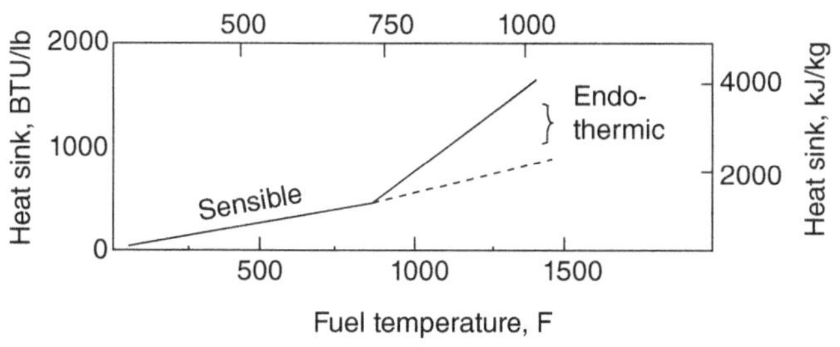

Figure 6.2 Fuel heat sink as a function of temperature. Source: Adapted from Lander and Nixon (1971).

endothermic cooling capacities at low conversions and exhibited a high tendency for coke formation at high conversions. These characteristics suggest that H-Y/JP-10 is not a good catalyst/fuel pairing for endothermic fuels.

6.4 Fuel Heat Sink Requirements

The cooling capacity or heat sink of a substance is known as *chemical heat absorption*. A number of system trade studies can be done to determine the heat sink requirements of a fuel for a given vehicle and mission. The cooling capacity requirement increases with flight Mach number. For example, the cooling requirements of current supersonic military aircraft (such as the McDonnell Douglas F-15 Eagle, with a top speed of Mach 2.5), exhibit a maximum heat load of about 12 660 kJ/min. It is predicted that aircraft with flight speeds greater than Mach 5 will require to absorb heat loads greater than 38 000 kJ/min (Harrison 1990).

Lander and Nixon (1971) published a plot of typical fuel heat sink requirements as a function of flight Mach number, M_0, for both crewed and uncrewed vehicles, which we duplicate in Figure 6.3. As shown, at $M_0 = 5$, the range for fuel sink required is in the range of 620–820 BTU/lbm, and it increases to 1450–2000 BTU/lbm at $M_0 = 8$. At higher speeds, the fuel sink required exceeds 1810 BTU/lbm. This is a dramatic increase in thermal load with Mach number. The cooling capacity of fuels must accommodate those requirements.

Consider the amount of thermal energy that a fuel can absorb (by unit mass of fuel), which we denote by h_{fc}. Then the quantity required for fuel cooling (to balance total engine surface heat flux as a function of freestream Mach number and dynamic pressure) is given by the product of stoichiometric fuel-to-air ratio, f_{st}, and the absorption capacity: $f_{st}h_{fc}$. The physical absorption capacities of hydrogen, methane, and JP-7 fuels are given by Heiser and Pratt (1994). They present latent plus sensible or physical heat absorbed by the liquid fuels going from tank conditions and heated to 1800 °R (1000 K), achieving a vapor state. Hydrogen has the highest heat absorption capacity with

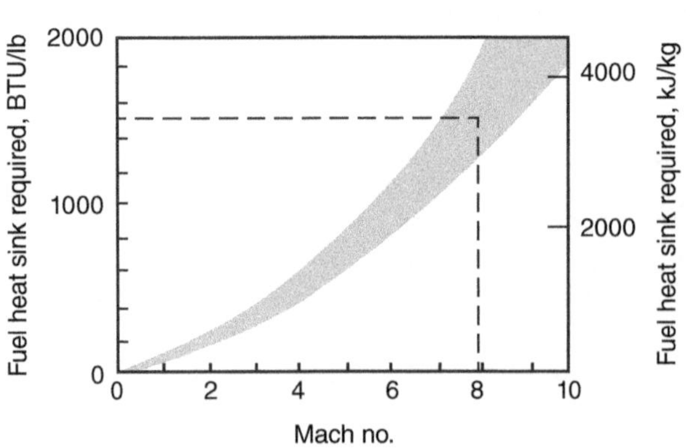

Figure 6.3 Heat sink required as a function of flight Mach number. Lower bound: missiles; upper bound: aircraft. Source: Adapted from Lander and Nixon (1971).

440 kJ/kg (189 BTU/lbm), while methane has less than half of that. With a heat absorption capacity 135 kJ/kg (58 BTU/lbm), JP-7 has the lowest value. Note that since JP-7 is a blend of many different hydrocarbon molecules and there is a variation in the exact chemical composition, the fuel is represented by the average molecular ratio of carbon to hydrogen, i.e. $C_{12.5}H_{26}$.

If we wish to know whether the existing fuel flow rate $\dot{m}_f$ is sufficient to satisfy the total cooling requirements of an entire system, Heiser and Pratt (1994) suggest to begin with the total rate at which the fuel can remove the thermal energy from the vehicle surfaces:

$$\dot{Q}_f = \dot{m}_f h_{fc} \tag{6.3}$$

Assuming that the majority of the heating is due to turbulent flow, the total rate at which heat must be removed is $\dot{Q}_r = \dot{q}_w A_w$, where A_w is the area to be cooled by the fuel, and $\dot{q}_w$ is the surface area-averaged heat flux. The required total heat to be removed can be approximated as:

$$\dot{Q}_r \propto \frac{\rho_0 V_0 V_0^2 A_w}{(\rho_0 V_0)^{1/5}} = \dot{m}_0 \frac{V_0^{11/5}}{q_0^{1/5}} \cdot \frac{A_w}{A_0} \tag{6.4}$$

Then, the ratio of the *available* fuel cooling to the *required* fuel cooling is

$$\frac{\dot{Q}_f}{\dot{Q}_r} = \frac{f_{st} h_{fc}}{K} \cdot \frac{q_0^{1/5}}{V_0^{11/5}} \cdot \frac{A_0}{A_w} \tag{6.5}$$

where K is a constant, f_{st} denotes the stoichiometric fuel-to-air ratio, and the product $f_{st}h_{fc}$ represents the quantity required for fuel cooling to balance the total engine surface heat flux as a function of flight velocity V_0 and dynamic pressure q_0 (see Heiser and Pratt 1994).

The value of $f_{st}h_{fc}$ that balances the heat load, $\dot{Q}_f = \dot{Q}_r$, is

$$f_{st} h_{fc} = K \cdot \frac{V_0^{11/5}}{q_0^{1/5}} \cdot \frac{A_w}{A_0} \tag{6.6}$$

The constant K can be obtained for a specific configuration from a consistent set of design data at any reference point. An example given by Heiser and Pratt (1994) is for a Mach 10 vehicle flying at a dynamic pressure $q_0 = 47.88\ \text{kN/m}^2$ with a surface area-averaged heat flux $\dot{q}_w = 163.5\ \text{W/cm}^2$; the constant is expressed as

$$K = \frac{\dot{q}_w}{\rho_0 V_0} \cdot \frac{q_0^{1/5}}{V_0^{11/5}} = 9.67 \times 10^{-6} \frac{\text{kJ}}{\text{kg}} \frac{(\text{N/m}^2)^{1/5}}{(\text{m/s})^{11/5}}$$

As noted by Heiser and Pratt, the value of $f_{st}h_{fc}$ increases rapidly with V_0 and decreases slowly with q_0, the latter being due to the fact that the fuel flow rate increases slightly more rapidly than the heat flux with q_0 at a given V_0. Clearly, the fuel heat sink required will depend on the engine design and size, cooling scheme design, structural materials selected, and cooling required by the airframe and other subsystems (such as avionics and flight controls).

6.5 Ignition Characteristics of Fuels

A fundamental requirement for scramjet combustors is that the mixing, ignition, and chemical reaction be very rapid, so that combustion occurs in the shortest length scale possible. A great abundance of research dealing with the problem of ignition and reaction of various fuel-to-air or fuel-to-oxygen mixtures at supersonic velocities and the associated higher reactant temperatures, show that hydrogen fuel is much more desirable than hydrocarbon fuels, since it exhibits fast reaction rates and high heat release per unit mass. Early scramjet studies focused on the ignition characteristics of hydrogen/air mixtures, finding that although the reaction rates for hydrogen combustion are very fast, a rather high temperature is required for ignition, and for typical supersonic combustor entrance flow conditions the static temperatures are too low to achieve self-ignition in a short flow length, even though the total temperature of the flow may be high enough (Huber et al. 1979).

Thus, much effort was devoted to define the combinations of gas conditions and configuration parameters which allow for self-ignition of supersonic hydrogen–air mixtures in combustors. Self-ignition (and combustion) occurs in a flowing combustible mixture when static temperature, static pressure, fuel–air mixture, and the residence time are at the correct conditions. In general, temperature has a strong effect on ignition and combustion. Huber et al. (1979) used the following expressions to determine ignition time τ_i and reaction time τ_r (in seconds):

$$\tau_i = \frac{8 \times 10^{-9}\, e^{9600/T}}{p}, \quad \tau_r = \frac{1.05 \times 10^{-4}\, e^{-1.12T/1000}}{p^{1.7}}$$

They considered ignition to be accomplished when the temperature rise in the combustor reaches 5% of the complete reaction temperature rise. Thus, sufficient free radicals, or chain carriers (OH, H, and O species used over and over to maintain the ignition process), are formed to initiate the reaction system, but no appreciable heat is released (formation of the chain carriers does not involve a significant exothermicity) (Huber et al. 1979).

The ignition delay time is the time interval between the creation of a combustible mixture and the onset of a flame. As a measurable quantity, this time interval is a function of the initial temperature, pressure, and composition of the reactant mixture. Based on the hydrogen–air combustible mixtures, reaction time τ_r is defined as that time required from ignition until 95% of the heat is released (due to formation of the product, H_2O). From the expressions above, note that the ignition time τ_i has a strong exponential dependence on temperature, but the reaction time τ_r is dependent on temperature to a much lesser extent. Over the range from 1000 K to 2000 K, the researchers found that τ_i varies by about a factor of 100, while τ_r varies by only about a factor of 3 at constant pressure.

Hydrocarbon fuels have relatively long ignition delay times compared with hydrogen fuel. This was an obstacle to the development of hydrocarbon-fueled scramjets. For example, for a hydrogen/air mixture at 1200 K, the ignition delay time is about 10^{-1} ms, while for a methane/air mixture at the same temperature, the ignition delay time is almost 10^2 ms.

The long ignition time of hydrocarbon fuels became one of the most challenging problems in hypersonics due to the extremely short flow residence time in a scramjet operating at high Mach numbers. Significant research on ignition characteristics of important JP fuels has

Table 6.4 Ignition delay coefficients in Eq. (6.7).

Fuel	A	E (kcal/mol)	a	b
Methane	2.21×10^{-14}	45 000	−1.05	0.33
Ethylene	2.82×10^{-17}	35 000	−1.20	0.00
JP-10	7.63×10^{-16}	46 834	−1.20	0.40
JP-7 (cracked)	8.32×10^{-10}	39 166	−0.3573	−0.3455
Hydrogen	1.60×10^{-14}	19 700	−1.00	0.00

Source: Adapted from Colket and Spadaccini (2001) and Puri et al. (2005).

been conducted to date. This is especially important for the development of endothermic fuels, as researchers seek to determine the autoignition characteristics.

In 2001, Colket and Spadaccini used shock tube experiments to measure and compare the ignition delay times of several fuel candidates for scramjet propulsion. In particular, they assessed the effect of fuel cracking on the autoignition of endothermic fuel/product mixtures. In particular, they measured ignition delays of ethylene, heptane, and JP-10 in dilute mixtures behind reflected shock waves at temperatures in the range 1100–1500 K, pressures of 3–8 atm, and equivalence ratios of 0.5–1.5. They compared the experimental results to other published data for heptane and ethylene and determined new ignition delay correlations for all fuels, comparing those correlations with those available for methane and hydrogen. The ignition delay time for hydrocarbon fuels is expressed in the following Arrhenius form:

$$\tau_i = A\, exp\left(\frac{E}{RT}\right)(C_xH_y)^a(O_2)^b(Ar)^c \tag{6.7}$$

where T is the mixture temperature at point of ignition, R is the universal gas constant, E is a parameter equivalent to a global activation energy, and A is an empirically determined constant. The terms C_xH_y, O_2, and Ar are the molar concentrations (mol/cc) of hydrocarbon, oxygen, and argon, respectively, in the combustible mixture. The empirical exponents a and b express power dependencies of ignition delay time on fuel and oxidizer concentrations. Values of these constants are given in Table 6.4.

Correlating experimental data for typical hydrocarbons using these equations show that ignition delay time decreases for both increasing temperature and pressure, although temperature changes provide a much greater effect (Colket and Spadaccini 2001). For example, ignition delay time for methane decreased from 16.8 to 14.6 when the temperature increased from 1300 K to 1400 K (constant pressure 7 atm). This research also concluded that hydrocarbon fuel cracking enhances ignition.

Results for JP-8 and JP-10 are given in Colket and Spadaccini (2001), and Mawid et al. (2003), respectively. Emphasis was on determining ignition delay times for typical endothermic fuel product mixtures having different degrees of cracking of the parent fuel. The results confirm that fuel cracking enhances ignition and support the validity of published reaction mechanisms for methane and heptane. Interestingly, autoignition of the endothermic reaction product mixture was not driven entirely by the constituent with the shortest ignition delay (i.e. ethylene or hydrogen). Small changes in concentrations of the individual component species are not likely to make dramatic changes in the ignition delay times of the fuel (Colket and Spadaccini 2001).

Puri et al. (2005) studied the ignition of cracked JP-7 fuel with both oxygen and air over a wide range of pressures (1–20 atm), temperatures (1200–2000 K), and equivalence ratios (0.5–1.5). They obtained correlations of ignition delay times, of the form used by Colket and Spadicinni, Eq. 6.7 (using the Chemkin-II package and least-squares analysis). To determine the influence of fuel composition on ignition delay, researchers considered two different compositions of cracked JP-7 fuel. One cracked composition contains methane, ethane, ethene, propane, and propene and includes C_3 hydrocarbons. The ignition delay time for this fuel composition is given by:

$$\tau_i = 3.47 \times 10^{-9}\, exp\left(\frac{37\,388}{RT}\right)(F)^{0.1886}(O_2)^{-0.9087} \tag{6.8}$$

The second cracked JP-7 fuel consists of hydrogen, methane, ethane, and ethene but no C_3 hydrocarbons. The ignition delay time for this simpler fuel is given by:

$$\tau_i = 4.69 \times 10^{-9}\, exp\left(\frac{34\,775}{RT}\right)(F)^{0.1357}(O_2)^{-0.7607} \tag{6.9}$$

Note the ignition delay for the first cracked fuel is 7.5% higher and the preexponential factor is lower. The conclusion reached is that the presence of C_3 species tends to reduce the ignition delay time over the temperatures of concern, but the difference appears to be limited.

Denman (2017) collected the ignition delay time correlations and plotted them to easily identify the order of magnitude differences for all fuels. Figure 6.4 gives ignition delay time as a function of temperature of methane, ethylene, liquid kerosene, and hydrogen. The data

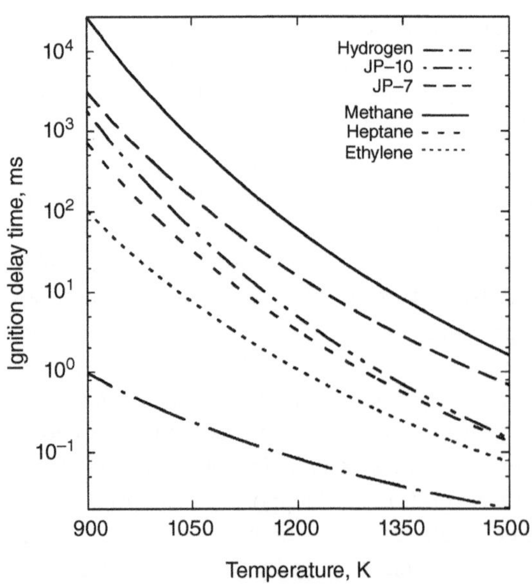

Figure 6.4 Ignition delay times of fuels of interest for high-speed propulsion application, determined at $\phi = 1.0$ and $p = 50$ kPa. Source: Data compiled by Denman (2017) from Colket and Spadaccini (2001) and Puri et al. (2005).

shows, for example, that the delay time for methane decreases several orders of magnitude as temperature increases from 900 to 1500 K.

6.6 Mixing Characteristics of Cracked Hydrocarbon Fuels

Endothermic processes, such as cracking, yield a substance that is much lighter than the original hydrocarbon fuel. Combustion with light hydrocarbon fuels is expected to be more efficient, as the fuel enters the combustion chamber in a hot, vaporized state, helping with faster mixing, faster ignition, and lower soot production (Maurice et al. 2006).

Researchers in Australia investigated mixing characteristics of uncracked fuel and cracked fuel compositions using numerical methods (Ravindran et al. 2018). They studied flow features and mixing process after injecting fuels into a scramjet combustor at a given stagnation pressure and temperature, using several performance parameters such as streamwise circulation, mixing efficiency, jet penetration, and stagnation pressure losses. The main results were as follows: (i) mixing rates and flow structures change with fuel compositions; (ii) as cracking increases, mixing and streamwise circulation increase, but jet penetration and stagnation pressure losses decrease; and (iii) jet penetration is found to decrease as cracking increases. Figure 6.5 shows that 100% cracked fuel has the highest mixing efficiency.

Since streamwise circulation produced by cracked fuel is higher than that of uncracked fuel, a stronger streamwise circulation appears to be the main reason for mixing

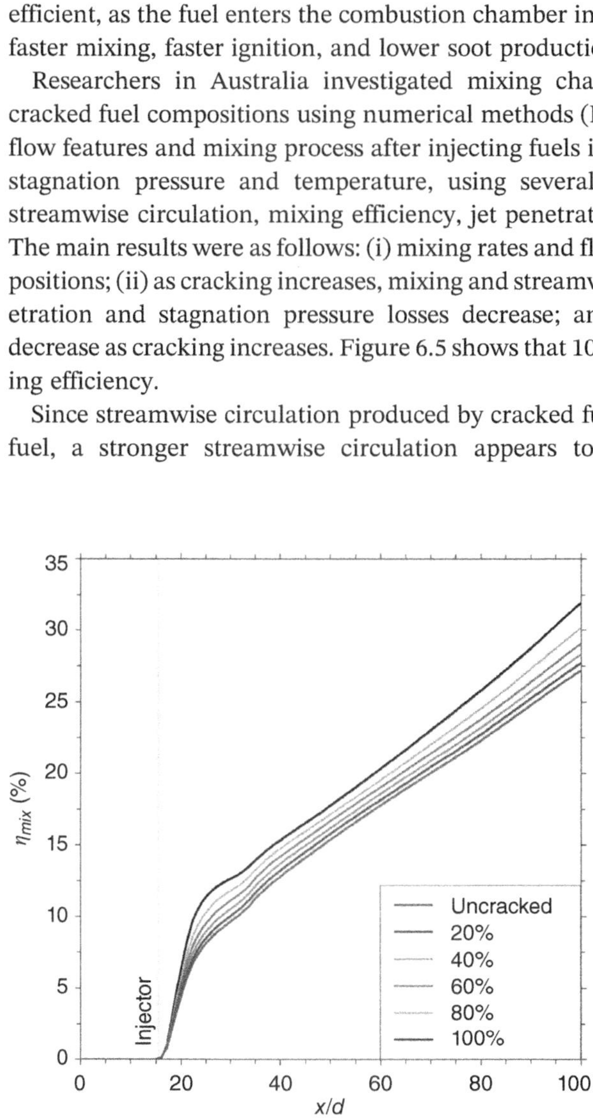

Figure 6.5 Mixing efficiency of cracked and uncracked fuels. Source: From Ravindran et al. (2018)/ With Permission of Elsevier.

enhancement. The overall jet penetration was found to decrease with cracked fuels as flow moved further downstream. The main conclusion of this computational study is that "there are mixing benefits to be gained by injecting cracked hydrocarbon fuels compared to heavy uncracked fuels in scramjets" (Ravindran et al. 2018).

6.7 Structural and Heat Transfer Considerations

High combustion temperatures and long operation durations require the use of cooling and thermal management techniques as integral part of the hypersonic vehicle design. The entire cooling capacity of the fuel will probably be needed for the engine alone (Heiser and Pratt 1994). Practical application of storable endothermic fuels for scramjets relies on the cooling capability demonstrated primarily in single-tube heat exchanger tests, and such data is then extended to the overall fuel-cooled structure.

In a hypersonic vehicle powered by a scramjet, the fuel-cooled combustor may resemble regeneratively cooled liquid rocket engines, which are designed with coolant passages around the combustor and nozzle walls and through which the propellant flows before it is injected in the combustion chamber. Convective cooling is accomplished by circulating the coolant fuel through passages in the structure to remove heat from the hot walls by fuel absorption. If the heat is transferred to the fuel before it is burned, this is referred to a regenerative cooling system (see Chapter 9). With regenerative cooling, the pressure in the cooling channels is greater than the chamber pressure. The inner liner is under compression, while the outer wall of the engine is under significant hoop stresses.

Using a numerical analysis, Ulas and Boysan (2013) studied the effects of geometry and number of cooling channels on the maximum temperatures of a rocket thrust chamber wall and coolant, modeled to operate on a liquid oxygen (LOX)/kerosene mixture. They investigated the pressure drop in cooling channel to determine the effect of different aspect ratios and number of cooling channels on gas-side wall and coolant temperatures. One of their findings was that increasing the aspect ratio with constant number of cooling channels will increase the cooling efficiency up to an optimum level, and then efficiency will decrease because of decreasing heat transfer area.

In hypersonic air-breathing propulsion, the fabrication of fuel-cooled structures, including manifolds, is expected to be a greater challenge. At high fuel temperatures and heat fluxes, heat transfer instabilities may be encountered. The mechanism of these instabilities is still unclear. Spadaccini and colleagues at United Technologies patented a method of cooling high-speed vehicles. They determined that using an endothermic fuel, which provides a high total heat sink yields products with superior combustion characteristics, it does not require precious metal catalysts and has handling and storage characteristics similar to those of conventional aircraft turbine fuels (Spadaccini et al. 1993).

To extend the Mach number capability of hydrocarbon-fueled hypersonic vehicles, it requires to improve the fuel heat sink capability and reduce the heat load from the vehicle into the fuel. High-temperature ceramic engine structures may be capable of significantly reducing the required scramjet engine cooling (see Chapter 9).

The thermal protection system (TPS) for a given vehicle will require a sophisticated approach that utilizes a combination of materials and various modes of heat transfer designed in concert to deal with acreage external surfaces (which are a substantial fraction of the vehicle's empty mass) and local regions of intense heating, such as leading edges and sharply concave corners.

6.8 Fuel System Integration and Control

For a hydrocarbon-fueled hypersonic vehicle, the fuel system is one of the key integration challenges for the vehicle designer, as the fuel must serve as a coolant as well as the source of energy for combustion. This challenge is compounded, as the fuel requirements change during the different phases of a mission. The most thermally challenging part of the mission is at the highest Mach number (cruise phase), and this is where the combustor may operate for longer times at the lowest equivalence ratio, which means the fuel flow rate is lower, and therefore less coolant is available.

For example, in a typical Mach 8 mission, in addition to the cruise phase, the other challenging flight point is the transition from boost to ramjet operation. At ramjet takeover speed (about Mach 4), the fuel is a liquid, and the result is a fuel/air mixture with relatively poor combustion properties (Maurice et al. 2000).

A high-temperature fuel has significantly different fuel properties from a cold fuel. The fuel density and viscosity change significantly with temperature, especially near the fuel critical temperature (Edwards 1993). As we noted before, the fuel for hypersonic aircraft may withstand temperatures greater than the critical temperature of current military jet fuels, which are almost 700 K (see Table 6.3). Hence, the fuel control system must deal with a fuel with properties that vary significantly over its operating time. The simple fuel pressure controls used in current aircraft will not work properly (Maurice et al. 2000). The mass flow through a simple valve will be quite different for a 311 K fuel and a 980 K fuel at a given pressure. Similar concerns apply to fixed-orifice injectors.

The NASA X-43A scramjet engine and fuel system were built as separate units and assembled to the vehicle. The fuel system, which delivered gaseous hydrogen and pyrophoric silane for ignition, was designed for a limited hypersonic velocity operation and a very short time. However, the control and operability requirements for the fuel system of future hypersonic vehicles at all flight Mach numbers are to ensure that cooling and combustion are both satisfied in a controlled manner for optimum operation in a long duration flight. Understanding and characterization of the fuel properties during the different phases of a mission is a crucial element of the design of such system.

6.9 Combustion Technical Challenges with Hydrocarbon Fuels

The typical combustion process with liquid fuels involves the following steps: fuel injection $\rightarrow$ atomization $\rightarrow$ vaporization $\rightarrow$ pyrolysis $\rightarrow$ mixing $\rightarrow$ oxidation. In a typical jet engine combustor, this process is carried out in a low subsonic flow field. However, in a scramjet, such combustion must occur in a supersonic flow field where fuel injection, mixing, kinetics, piloting, and heat-release phenomena are much more complex. Combustion in dual-mode scramjet, therefore, becomes even more challenging, as these processes will transition from subsonic to supersonic flow conditions. The effect of the fuel preprocessing as a function of Mach number in the fuel system before combustion is an important issue to consider. This processing can range from preheating to partial reaction. Hence, the typical liquid fuel combustion process in the dual-mode scramjet is modified to the extent that all but the mixing and oxidation steps might occur in the fuel system (Maurice et al. 2000). In addition, the ramjet to scramjet transition operation requires more research to evaluate combustion stability during transition.

A key combustion challenge is achieving ignition in the short residence times available in supersonic flow combustors. As noted in Chapter 5, the scramjet combustor is limited to residence times of the order of fractions of a millisecond, values which are commensurate with a broader range of combustion reaction timescales (1 μs–10 ms). For example, the residence time for operation at Mach 8 is usually on the order of 0.4 ms. During this time, fuel injection, fuel/air mixing, ignition, and combustion must all occur efficiently. Such conditions require a fuel with short ignition delay time, and at the same time the fuel must have high energy per unit mass in order to maximize specific impulse (Lewis 2001). This explains why gaseous hydrogen is the fuel of choice for scramjets operating at speeds greater than Mach 7.

At lower Mach numbers, the residence times are still very short, making this much more challenging if burning hydrocarbon fuels. Thus, the ignition delays of endothermic hydrocarbon fuels must be reduced sufficiently to achieve optimum supersonic combustion.

Experimental research has demonstrated that cavity flameholder promotes the ignition of gaseous ethylene fuel at a flight Mach number of 7.3, with fuel–air equivalence ratios ranging from 0.60 to 0.70 (Smart 2016). These results suggest that the passive entrainment of ethylene and air into the cavity sufficiently increases the effective residence time, thus enhancing the combustion of the hydrocarbon fuel.

Hydrocarbon fuels ignite much slower than hydrogen and must be vaporized in order to react. "Barbotage" fuel injection, fuel additives, ignition schemes, and piloting are options to improve endothermic fuel combustion and ignition performance. Barbotage refers to effervescent atomization in which gas is introduced into a liquid with a very low velocity, leading to turbulent two-phase flow that can improve penetration and vaporization of the fuel jet spray. Russian researchers demonstrated supersonic combustion of kerosene, using barbotage in aeroramp injectors (gas wedges) to inject the liquid hydrocarbon fuel aerated by air for the enhancement of the supersonic mixing and combustion. In another study (Vinogradov et al. 1992), the fuel vaporization was enhanced by the addition of hydrogen gas bubbles to the fuel before it was injected. The bubbles rapidly expand after the fuel is injected into the lower pressure air flow thereby breaking the liquid into tiny droplets that vaporize quickly. Barbotage injection may improve mixing of hydrocarbon fuels with the supersonic airstream to enhance combustion.

There are other issues associated with endothermic fuel combustion such as the requirement to increase the flame speeds and broaden the blowout limits of conventional hydrocarbons that require further investigation.

6.10 Impact of Fuel Selection on Hypersonic Vehicle Design

The characteristics of the fuel must be considered not only with respect to propulsion performance and cooling capacity, but also in terms of the impact a fuel poses on the design of the airframe itself, including its storage tanks, and fuel control system. Although hydrogen fuel provides the largest amount of heat release per unit mass of fuel (see Table 6.2), its high specific volume is a major concern. For example, if we consider a requirement to store liquid hydrogen, the cryogenic temperatures range between 18 and 25 K, and the density of the liquid will vary from 70.8 to 89.9 kg/m^3. This cryogenic state requires very large, insulated tanks, which in turn results in a large and heavy vehicle structure and increased drag. Liquid methane is three to four times as dense as liquid hydrogen, while kerosene is about 10–12 times denser, and so the tanks can be smaller.

The impact of fuel selection can be evaluated in terms of a vehicle's cruise capability, which is typically characterized by the range or distance flown by a vehicle during a given time. The Breguet formula for range puts in perspective several aspects of a hypersonic air-breathing vehicle performance, calling attention to important propulsion and vehicle design issues. Disregarding takeoff, climb and approach and landing distances, the flight range can be expressed as:

$$R = \frac{\eta_0 h_{PR}}{g_0} \cdot \frac{L}{D} \cdot ln\left(W_i/W_f\right) \tag{6.10}$$

assuming unaccelerated level flight, constant lift-to-drag ratio, and engine overall efficiency.

According to Eq. (6.10), the range of an aircraft depends on the following parameters:

- **Overall propulsion efficiency** η_0. This represents the ability of the propulsion system to convert the fuel chemical energy into useful thrust.
- The **fuel efficiency** is represented by the fuel heat of reaction h_{PR}. Hydrocarbon fuels have much lower energy content per unit mass as compared with hydrogen.
- **Aerodynamic efficiency** is given by the lift to drag ratio L/D. This represents the ability of the vehicle's aerodynamic shape to produce lift without generating excessive drag. A high value of L/D increases the range. The design of hypersonic vehicles with reasonable values of L/D is a design challenge.
- **Structural efficiency** is represented by the vehicle weight ratio W_i/W_f, where W_i and W_f are the initial and the final values of the vehicle weight.

The propulsion efficiency is given by the ratio of thrust power (FV_0) and chemical energy rate ($\dot{m}_f h_{PR}$):

$$\eta_o = \frac{FV_0}{\dot{m}_f h_{PR}} = \frac{I_{sp} V_0}{h_{PR}} \tag{6.11}$$

It has been demonstrated that light hydrocarbons yield higher performance (better I_{sp}) than typical jet fuels. Hence, the propulsion efficiency is enhanced with endothermic fuels.

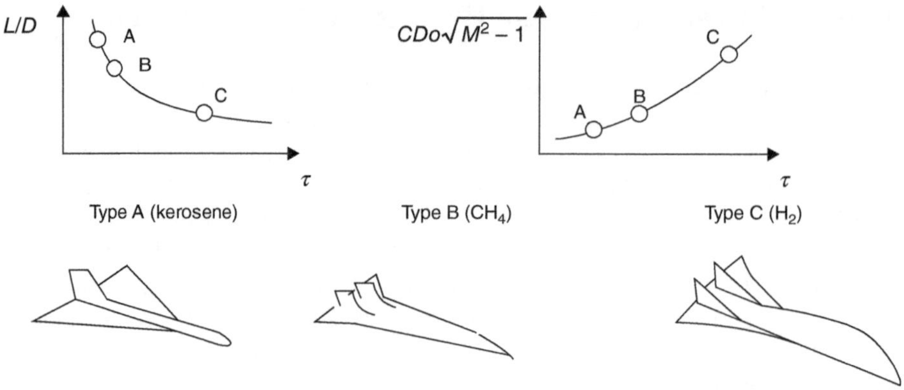

Figure 6.6 Fuel density effects on *L/D* and zero-lift drag coefficient. Source: Maurice et al. (2006)/ Research and Technology Organisation. The original version of this figure was published by the Research and Technology Organization, North Atlantic Treaty Organization (NATO RTO) in the publication number RTO-TR-AVT-007-V2.

The fuel density impacts the *L*/*D* factor as it influences largely the incompressible zero-lift drag coefficient, $C_{D,\,0}$, a dimensionless parameter which relates an aircraft's zero-lift drag force to its size, speed, and flying altitude. In Figure 6.6, Maurice et al. (2006) expressed the impact of this parameter as $C_{D,0}\sqrt{M^2-1}$, shown plotted as a function of the Küchemann parameter τ. The Küchemann parameter gives the ratio between vehicle volume and surface area raised to the 1.5 power, and it was to establish to set the *L*/*D* limit for aircraft aerodynamic design (Küchemann 1978). Clearly, the vehicle fueled by hydrogen (Type C in Figure 6.6) has a much lower *L*/*D* compared with vehicle A fueled by kerosene. Maurice et al. also proposed a possible law for air-breathing aerospace planes reaching hypersonic cruise speed in air-breathing mode: $L/D = 0.998\left[C_{D,0}\sqrt{M^2-1}\right]^{-0.564}$.

It is important to emphasize that air-breathing hypersonic flight vehicles are extremely drag-sensitive. The smaller the drag, the smaller the required installed engine performance, the fuel consumption and, depending on the mission, the mass and volume of fuel (size of fuel tank). Engine-airframe integrated hypersonic vehicles exhibit lower *L*/*D* ratios as compared with classical aerodynamic geometries. With low-density fuels such as hydrogen, matching lift to vehicle weight results in configurations that cannot take advantage of high-lift aerodynamics (Lewis 2001), especially since vehicles with air-breathing propulsion integrated geometries must operate in high dynamic pressures for both cruise and acceleration missions. Also, optimization studies conclude that hydrocarbon-fueled engine integrated vehicles may have superior cruise range performance in comparison with cryogenic-fueled designs (Lewis 2001).

For missile propulsion, a scramjet engine is highly attractive because it can yield greater range than rocket engines for the same propellant weight. However, attaining high aerodynamic and structural efficiencies is extremely challenging for vehicles powered by air-breathing propulsion in sustained hypersonic cruise flight. For a reusable hypersonic vehicle, the choice of fuel can affect the structural efficiency, and the addition of thermal management system can reduce it significantly. For a Mach 5–6 missile, it may be possible to use a composite structure that is not actively cooled, which could result in perhaps a 25% weight saving and, thereby, longer range, better acceleration, and lower cost.

For the air-breathing trans-atmospheric flight envelope, hydrogen-fueled vehicles exhibit the highest specific impulse. However, hydrocarbon fuels, despite having substantially lower heating values and much longer burn times, have demonstrated to fulfill mission requirements, especially for the lower hypersonic flight Mach number regime. For example, for cruise vehicles operating below Mach 10, or the first-stage vehicle of a TSTO space launcher (with staging at about Mach 10), hydrocarbons offer better packaging (e.g. distribution of fuel tanks within fuselage) and handling benefits, which may result in better overall flight performance. This is because hydrogen has very low density, and it has boil-off issues that challenge packaging and increase the structural mass and required overall volume for a given mission (Lewis 2001). Having up to 11 times the storage density, hydrocarbons have over 3.5 times more energy content per unit volume.

Slush hydrogen has been proposed as a fuel in place of liquid hydrogen in order to use smaller fuel tanks and thus reduce the dry weight of the vehicle, especially the gross takeoff weight of space transportation systems. Slush hydrogen is a mixture of liquid and solid hydrogen, with a 16% higher density and 18% higher heat capacity than liquid hydrogen. Slush hydrogen was selected for the NASP project in the 1990s. The production costs of slush hydrogen now may be considerable.

Although vehicle aerodynamic design is outside the scope of this chapter, this section is meant to call attention to the importance of fuel selection in the design of hypersonic vehicles powered by air-breathing propulsion systems.

6.11 Fuels Research for Hypersonic Air-Breathing Propulsion

Efforts must continue to develop methods to increase the cooling or heat sink capacity of hydrocarbons fuels. Hypersonic flight vehicles demand greater total cooling capacity from the fuel initially in their tanks, since they are propelled by higher temperature engine cycles.

In addition to increasing the heat sink and thermal stability of storable hydrocarbon fuels, efforts to develop highly reactive fuels, and an understanding of combustion chemistry, are critical components of the development of future hypersonic propulsion systems.

A key challenge of supersonic combustion is achieving ignition in very short residence times. Design and development of endothermic fueled engines require an understanding of the autoignition characteristics of cracked product mixtures, and this requires both experimental and computational research. In addition, we must devote efforts toward the development of modeling and simulation rules and tools for the effective use of the improved fuel technology.

The following are some areas that require attention for research and development of fuels.

1) Improve heat sink and coke reduction for conventional endothermic fuels. This effort must address the development of improved catalysts, additives, and coke-resistant designs for the *n*-octane/JP-7 kerosene class of hydrocarbons.
2) Develop endothermic fuel and catalyst pairings that provide higher heat sink capacity to increase the hypersonic flight speed limit for hydrocarbon fuels.

3) Establish a robust program to test supersonic ignition and combustion of hydrocarbon fuels under realistic conditions. This effort is required to improve the combustion behavior of endothermic fuels, especially at the lower Mach numbers (including ignition).
4) Improve endothermic hydrocarbon physical properties. The physical properties of these fuels under high-temperature and high-pressure conditions must be better understood. Studies must be conducted to develop correlations for the ignition delay time of cracked JP-7 fuel and other endothermic fuels of interest.
5) Develop and design reduced heat rejection vehicle structures. Two key issues need to be addressed: (i) the structural manufacturability and durability of ceramic structures, and (ii) exposure of fuel to high-temperature surfaces and the impact on coking.
6) Evaluate unconventional endothermic fuels and use of steam reforming for increased heat sink. This effort requires to investigate other approaches to obtaining dramatically higher heat sinks than available with conventional endothermic fuels, including exotic alternatives, such as liquid metal decomposition and high-energy-density fuels.
7) Combustion and regenerative cooling research is required to develop effective thermal protection management systems.
8) Studies must continue to firmly establish the impact of fuel selection on vehicle designs, especially those that require reusability and commonality with propulsion and airframe integration.

In closing this chapter, it is important to affirm that development of reusable hypersonic flight vehicles relies on the ability to cool the combustor using the fuel. The cooling capacity of kerosene-based fuels is low even with endothermic cracking reactions. Therefore, there is a strong need to develop new endothermic fuels and reactors that can deliver substantially higher heat sink capacities.

Questions

1. The choice of fuel is crucial for hypersonic air breathing propulsion, why? Give all possible reasons, technical and otherwise, and select the best fuel for future hypersonic transports.

2. What differentiates endothermic fuels from conventional jet fuels?

3. What is fuel cracking, and why this process is important for hypersonic air-breathing propulsion?

4. How would you model and simulate the supersonic combustion process with a new form of endothermic fuel?

5. What are your views regarding green fuels for hypersonic cruise vehicles?

References

Colket, M.B. and Spadaccini, L.J. (2001). Scramjet fuels autoignition study. *Journal of Propulsion and Power* 17 (2): 315–323.

Croswell, B.M. and Biddle, T.B. (1992). High temperature fuel requirements and payoffs. In: *Aviation Fuel: Thermal Stability Requirements* (ed. P.W. Kirklin and P. David). Philadelphia, PA: American Society for Testing and Materials (ASTM STP 1138).

Denman, Z.J. (2017). Supersonic combustion of hydrocarbon fuels in a three-dimensional Mach 8 scramjet. PhD thesis. The University of Queensland, Australia.

Edwards, T. (1993). USAF supercritical hydrocarbon fuels interests. *AIAA-93-0807 Paper. AIAA 31st Aerospace Sciences Meeting*, Reno, NV (11–14 January 1993).

Gurijanov, E.P. and Harsha, P.T. (1996). Ajax: new directions in hypersonic technology. *AIAA 96-4609. Space Plane and Hypersonic Systems and Technology Conference*, Norfolk, VA (18–22 November 1996).

Harrison, W.E. (1990). Aircraft Thermal Management. Report of the Joint WRDC/ASD Thermal Management Working Group. WRDC-TR-90-2021. Ilcpoii No. TR-90-2021. Wright-Patterson Air Force Base, Ohio.

Heiser, W.H. and Pratt, D.T. (1994). *Hypersonic Airbreathing Propulsion*, AIAA Education Series. Washington, DC: American Institute of Aeronautics and Astronautics (AIAA).

Heneghan, S.P., Zabarnick, S., Ballal, D.R., and Harrison, W.E. (1996). JP-8+100: the development of high thermal stability jet fuel. *The Journal of Energy Resources Technology* 118 (3): 170–179.

Huang, B. (2017). Solid acid catalysts for endothermic fuels. Ph.D. thesis. University of Virginia.

Huang, H., Sobel, D., and Spadaccini, L. (2002). Endothermic heat-sink of hydrocarbon fuels for scramjet cooling. *AIAA-2002-3871. 38th AIAA/ASME/SAE/ASEE Joint Propulsion Conference & Exhibit*, Indianapolis, IN (7–10 July 2002).

Huber, P.W., Schexnayder, C.J., and McClinton, C.R. (1979). Criteria for self-ignition of supersonic hydrogen–air mixtures. NASA Tech Paper 1457 (August 1979).

Kay, I.W., Peschke, W.T., and Guille, R.N. (1992). Hydrocarbon-fueled scramjet combustor investigation. *Journal of Propulsion and Power* 8 (2): 507–512.

Küchemann, D. (1978). *The Aerodynamic Design of Aircraft*, AIAA Education Series. London, UK: Pergamon Press, American Institute of Aeronautics & Astronautics (21 September 2012).

Lander, H. and Nixon, A.C. (1971). Endothermic fuels for hypersonic vehicles. *Journal of Aircraft* 8 (4): 200–207.

Lewis, M.J. (2001). Significance of fuel selection for hypersonic vehicle range. *Journal of Propulsion and Power* 17 (6): 1214–1221.

Maurice, L.Q., Edwards, J.T., and Griffiths, J.F. (2000). Liquid hydrocarbon fuels for hypersonic propulsion. In: *Scramjet Propulsion* (ed. E.T. Curran and S.N.B. Murthy). American Institute of Aeronautics and Astronautics.

Maurice, L.Q., Edwards, J.T., Cuoco, F., et al. (2006). Chapter 2: Fuels, in Technologies for Propelled Hypersonic Flight, Volume 2 – Subgroup 2: Scram Propulsion. Report RTO-TR-AVT-007-V2, January 2006. Research and Technology Organisation North Atlantic Treaty Organisation BP 25, F-92201 Neuilly-sur-Seine Cedex, France.

Mawid, M.A., Park, T.W., Sekar, B., and Arana, C.A. (2003). Development and validation of a detailed JP-8 fuel chemistry mechanism. *ASME International Gas Turbine Institute to Publication (IGTI)* 2: 613–624.

Puri, P., Ma, F., Choi, J.-Y., and Yang, V. (2005). Ignition characteristics of cracked JP-7 fuel. *Combustion and Flame* 142: 454–457.

Ravindran, M.R., Bricalli, M., Pudsey, A., and Ogawa, H. (2018). Mixing characteristics of cracked gaseous hydrocarbon fuels in scramjets. *Paper AIAA 2018-5260. 22nd AIAA International Space Planes and Hypersonics Systems and Technologies Conference*, Orlando, FL (17–19 September 2018).

Smart, M.K. (2016). Comparison Between Hydrogen and Methane Fuels in a 3-D Scramjet at Mach 8. Final Report AFRL-AFOSR-JP-TR-2016-0066, June 2016. Air Force Research Laboratory, Arlington, VA.

Sobel, D.R. and Spadaccini, L.J. (1997). Hydrocarbon fuel cooling technologies for advanced propulsion. *Journal of Engineering for Gas Turbines and Power* 119: 344.

Spadaccini, L.J., Marteney, P.J., Colket, M.B. III, Stiles, A.B. (1993). Method of cooling with an endothermic fuel. US Patent 5,176,814. https://patents.google.com/patent/US5176814.

Ulas, A. and Boysan, E. (2013). Numerical analysis of regenerative cooling in liquid propellant rocket engines. *Aerospace Science and Technology* 24: 187–197.

Vinogradov, V., Kobigsky, S., and Petrov, M. (1992). Experimental investigation of liquid carbohydrogen fuel combustion in channel at supersonic velocities. *AIAA Paper 92-3429. AIAA/SAE/ASME/ASEE 28th Joint Propulsion Conference and Exhibit*, Nashville, TN (6–8 July 1992).

Wiese, D.E. (1992). Thermal management of hypersonic aircraft using noncryogenic fuels. *SAE Transactions* 100 (1): 1313.

7

Dual-Mode Combustion Scramjet

7.1 Introduction

For the flight regime $3 < M_0 < 10$, a scramjet must be able to operate also as a ramjet engine. As depicted in Figure 7.1, the pure ramjet engine requires two physical throats: (i) at the inlet diffuser exit – the throat is required to stabilize the final, normal shock wave in the flowpath expansion region downstream in order to deliver subsonic flow to the combustor, and (ii) downstream of the combustor – a choked throat ($M = 1$) is required to accelerate the subsonic combustion flow to supersonic velocity in the engine's expansion nozzle.

A pure scramjet engine depicted in Figure 7.2 has no physical throat, as the flow is not required to decelerate to subsonic conditions. *How can the scramjet operate in ramjet mode?* At low flight Mach numbers M_0, the scramjet must deliver subsonic flow to the combustor without physical throats because having area constriction would limit the mass flow rate required at higher M_0 when operation switches to supersonic combustion.

For high-speed flow ($M_0 > 8.0$), the combustor flow field is dominated by supersonic combustion. The pressure increase due to fuel–air chemical reaction is confined to the combustion chamber, thus influencing only the downstream flow. However, as the flight Mach number is lowered, subsonic combustion occurs, and significant effects drastically change the flow character in the scramjet flowpath, and thus its performance and operability. For example, the ratio of heating rate (due to fuel–air combustion) to total enthalpy rate of the flow entering the combustor becomes large. In addition, the combustor experiences a significant pressure rise due to heat release, which is coupled with the decrease of the Mach number in the combustor, and this can lead to thermal choking ($M \to 1$). We require, therefore, to design the flowpath in such a way that the scramjet engine can operate in dual mode, transitioning from subsonic to supersonic combustion, and vice versa.

Dual-mode combustion is characterized by extensive interaction with the flow upstream of the combustor entrance, developing as an oblique shock train with associated low velocity or large recirculation flow adjacent to the walls. A high-speed scramjet operated in such mode will experience a condition where these phenomena immediately propagate into the inlet. Hence, an isolator duct section must be added between the inlet and the combustor to "isolate" them and to prevent inlet/engine unstart. Having sufficient

Scramjet Propulsion: A Practical Introduction, First Edition. Dora Musielak.

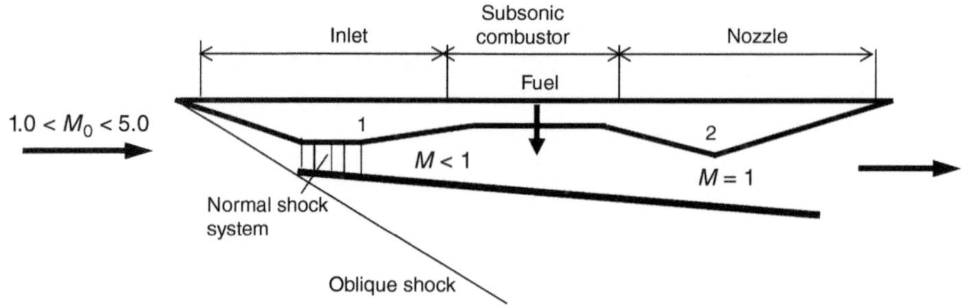

Figure 7.1 Pure ramjet engine concept.

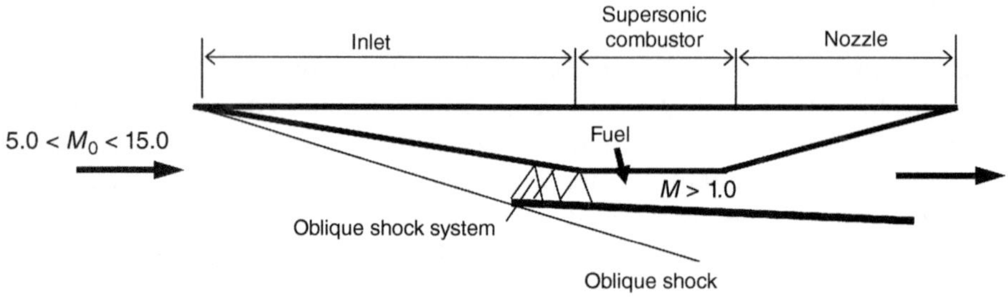

Figure 7.2 Pure scramjet engine concept.

length, the isolator will stabilize the oblique shock system and associated pattern of recirculation regions.

As we found in Chapter 4, the Inlet Isolator is a constant area diffuser duct designed to contain the static pressure rise created by the scramjet combustor to allow the inlet to operate and prevent inlet unstarting. The role of the isolator is to prevent engine unstart from either thermal choking or from flow separation due to heat release.

The transition process from subsonic to supersonic combustion is characterized by the dissolution of a precombustion shock train (PCST) in the isolator, and if the heat addition exceeds a critical value the flow will become thermally choked, and the flow will transition back to subsonic combustion ramjet mode, or cause engine unstart.

Therefore, a dual-mode scramjet flowpath must incorporate other features to control the flow field. For example, a diverging combustor geometry can relieve the combustion-generated pressure rise, delay the development of thermal choke, and facilitate reestablishment of supersonic flow subsequent to thermal choking. Staged fuel injection is also used to distribute the heat release according to the varying inflow conditions, aided with recessed cavity flameholders to maintain the flame in the different modes, and other devices to control the transition from subsonic to supersonic combustion (Micka and Driscoll 2009). Backward facing steps at the combustor entrance or at several locations along the combustor can also be added in order to isolate or stabilize flow recirculation and slow the development of flow interactions.

7.2 Phenomenological Description of Dual-Mode Scramjet

The dual-mode scramjet engine is a unique propulsion system that combines the benefits of subsonic combustion at the lower flight speeds (pure ramjet) with supersonic combustion in the hypersonic regime (pure scramjet). The main components of the dual-mode scramjet are: Inlet + Isolator + diverging Combustor with thermal throat + Thrust Nozzle.

Consider a scramjet engine required to operate in the range $3 < M_0 < 10$. At the lowest flight Mach number, the engine operates in ramjet mode, decelerating the incoming air for subsonic combustion process, Figure 7.3a. Ramjet mode operation is characterized by a strong PCST, no boundary layer separation in the subsonic combustor, and no thermal choking. The fuel is injected into the combustor, which has no physical throat. Thermal choking occurs in the combustor as a result of the heat addition. Controlling the choke drives a shock wave upstream to a stable position in the inlet throat. Fuel injection, mixing, and combustion now occur in a subsonic region between the shock wave and the choke point in the combustor. Downstream of the choke point, supersonic expansion of the gases takes place in the same manner as in a conventional ramjet. The heat release from the combustion process drives the supersonic inflow to sonic conditions, via thermal choking, and forming a PCST in the isolator. The shock train aids flame stabilization by increasing pressure and temperature while at the same time decelerating the flow. The thermal throat in the diverging combustor section allows the exhaust flow to reaccelerate to supersonic speed, as required for propulsion.

As the operating speed (M_0) and inlet total temperature (T_{t0}) increase, the temperature ratio associated with combustion (T_{t4}/T_{t2}) decreases, and the shock wave in the inlet throat begins to move downstream. The shock wave can be maintained at the inlet throat by

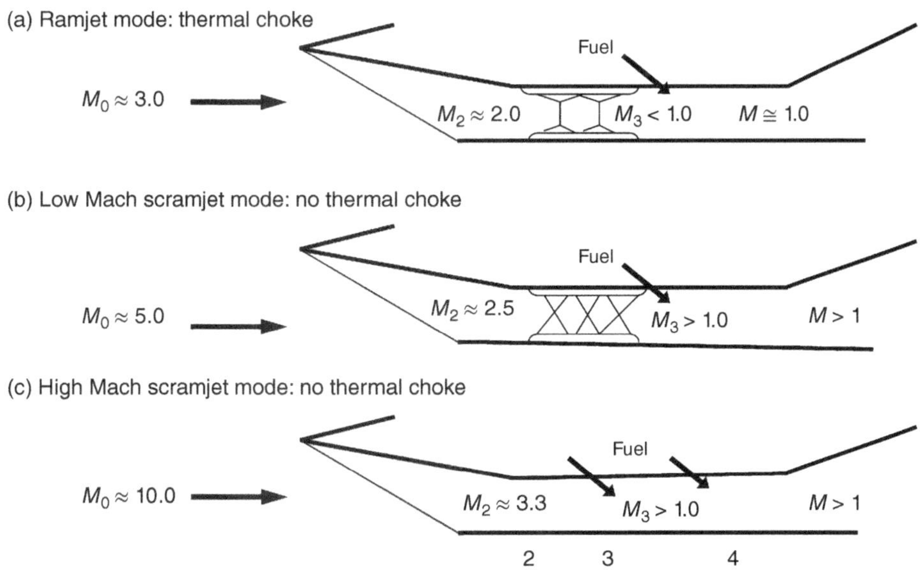

Figure 7.3 Combustion modes for dual-mode scramjet.

moving the fuel injection and associated choke point upstream to a smaller combustor flowpath flow area (Figure 7.3b). The low Mach scramjet mode operation is characterized by a weak PCST, separation of the boundary layer in the combustor, and no thermal choking. Critical inlet operation can be maintained by varying the fuel injection stations as a function of the engine operating speed.

At high hypersonic flight Mach numbers, the engine operates in pure scramjet mode. Due to the increase in operating speed, the diffuser shock wave moves downstream past the fuel injectors, and supersonic combustion occurs as the available heat release no longer chokes the combustor flowpath (Figure 7.3c). At the high speed, the PCST weakens until the isolator has an exit Mach > 1 and scramjet mode is achieved. There is no PCST in pure scramjet mode operation and no thermal choking.

The complex physics of dual-mode combustion poses unique problems for research, design, and analysis. For example, fueling control, fuel injection, and flameholding schemes are more important for dual-mode combustor design. Transition from subsonic to supersonic combustion is achieved by controlling the heat release to prevent thermal choking (see, for example, Fotia 2015; Laurence et al. 2015). Once heat release is reduced, the flow in the combustor remains supersonic. Pressure increase due to combustion is confined to the combustor, thus influencing only the downstream flow going into the nozzle.

The flowpath geometry of the dual-mode combustor must be carefully designed to support the inherent complex aerothermodynamic processes and operate efficiently in three regimes: supersonic, subsonic, and transition, as needed.

The combustor shape and/or fuel injection location must vary with Mach number to assure optimum combustor operation (Aguilera et al. 2017). Combustion must start close to the combustor entrance for very high Mach operation. For example, the scramjet combustor can be designed with two fuel injection stages in order to operate in dual mode. The first stage of injectors is sized for supersonic combustion and is located at the entrance of the combustor. The second stage of injectors for subsonic combustion mode is typically located near the combustor end in order to add thermal energy and promote a thermal throat before the overall combusted flow enters the engine nozzle for expansion.

The isolator and the thermal throat are key elements of mode transition. The isolator must be designed to contain the complex wave structure or pseudo-shock, which occurs during the deceleration of the supersonic flow from the engine inlet to the lower required speed at the combustor entrance. The precombustion shock structure provides the flow boundary condition upstream of the region of heat addition. As we described in Chapter 4, the shock structure itself changes with the variation of the combustion heat release.

7.3 Heat Addition to Flow in Constant Area Duct

Only a limited amount of heat can be added to the flow in constant area ducts. This limiting amount is given by the ratio of heat addition per unit mass, q, and the enthalpy of the gas, h, which is expressed as:

$$\left(\frac{q}{c_p T}\right)_{limit} = \frac{\left(M^2 - 1\right)^2}{2(\gamma + 1)M^2} \tag{7.1}$$

where c_p is the specific heat at constant pressure, γ is the ratio of specific heats, T is the static temperature of the inlet gas, and M is the axial Mach number at the entrance of the duct.

Equation (7.1) implies that, in a constant area combustor, as the entry Mach number approaches unity (low supersonic flight speeds), the amount of heat that can be added decreases to a very small value. Thus, very little thrust can be produced.

Another representation of heat addition in constant area ducts is obtained from a general differential control volume analysis for a constant area heating (Rayleigh flow). It yields a relation between Mach number and the gas total temperature T_t, which represents the heat added to the flow (or rate of combustion heat release):

$$\frac{dM}{M} = \frac{(1 + \gamma M^2)(1 + ((\gamma - 1)/2)M^2)}{2(1 - M^2)} \frac{dT_t}{T_t} \tag{7.2}$$

This differential equation states that heat addition to a flow in a duct always drives the Mach number toward 1, and thus to thermal choking. Equation (7.2) has no solution at $M = 1$ because of the singularity in the denominator. The numerator shows that the rate at which the Mach number approaches 1 is greater for supersonic flow than for subsonic flow. This is an important result to remember for controlling the heat addition in the dual-mode combustor.

Moreover, adding heat to a constant area flow always decreases the total pressure, a phenomenon known as Rayleigh heating loss. The total pressure decrease is greater as the Mach number increases, as represented by the following expression:

$$\frac{dp_t}{p_t} = -\frac{\gamma M^2}{2} \bullet \frac{dT_t}{T_t} \tag{7.3}$$

Thus, the constant area combustor requires a strict control of Rayleigh heating loss. It is important to note that, since the heat addition forms a shock train upstream of the combustor, if too much heat is added, then the shock train moves out of the inlet, and this leads to engine unstart. One solution to alleviate the restriction on heat addition is to open up the flowpath by increasing the combustor cross-sectional area.

7.4 Divergent Combustor and Heat Release

Most practical scramjet engines include combustor designs with area relief. This can be done by step increases in area, or by use of diverging duct segments. Thus, since the combustor area changes, the longitudinal heat release must be carefully controlled in a given duct geometry to ensure combustion stability and prevent engine unstart. Heiser and Pratt (1994) give an excellent treatment of this topic.

To assess the interaction between the axial variation of cross-sectional area $A(x)$ and of total temperature $T_t(x)$, consider axial variation of Mach number along the combustor:

$$\frac{dM}{dx} = M\left(\frac{1 + ((\gamma - 1)/2)M^2}{1 - M^2}\right)\left[-\frac{1}{A}\frac{dA}{dx} + \frac{(1 + \gamma M^2)}{2}\frac{1}{T_t}\frac{dT_t}{dx}\right] \tag{7.4}$$

This differential equation can be used to determine how much increase of $A(x)$ is required to accommodate increases of $T_t(x)$ in order to control the axial variation of Mach number, pressure, and other thermodynamic properties and flow variables. The total temperature $T_t(x)$ depends on the rate of combustion heat release, which in turn is determined by the finite-rate processes of fuel–air mixing and chemical reaction in the combustor.

When $M = 1$, the quantity in the square bracket in Eq. (7.4) must equal zero. Hence, there must be a balance between the area change $A(x)$ and the heat release (dT_t/dx). This condition yields a thermal throat.

When $M > 1$, occluding the flow (either by decreasing A or by increasing T_t) causes M to decrease in the axial direction. On the other hand, relieving the flow (either by increasing A or decreasing T_t) will cause M to increase.

In dual-mode combustion, $A(x)$ will not decrease with x, as there is no physical throat, and $T_t(x)$ will not decrease, because exothermic combustion can only add heat to the flow. Choking must be provided within the combustor by means of a choked thermal throat, which can be done by choosing the right combination of area distribution $A(x)$, and fuel–air mixing and combustion. Therefore, the primary technical challenges in dual-mode combustors are modulation of the thermal throat location, fuel distribution, and ignition and flameholding in the large cross section.

The thermal throat is a crucial design requirement for practical application of the dual-mode scramjet. If the heat released by combustion becomes greater than a critical value, the flow becomes thermally choked. Choked flow causes a normal shock to form at the engine inlet (unstart), creating drag and reducing engine performance at high flight Mach numbers. The combustor area ratio is increased to prevent thermal choking and excessive pressure gradients. The cross-sectional area at which the thermal throat must form increases as flight Mach number decreases, unless the fuel-to-air ratio is reduced. For a given duct, this effect determines the minimum flight Mach number for thermally choked operation. At Mach 3, the required thermal throat area approaches that of the inlet capture area, and it could be surmised that combustion extends into the nozzle expansion region. Some researchers suggest that in-stream devices must be retractable or expendable so as not to inhibit supersonic combustion operation.

7.4.1 Heat Addition and Thermal Choke

Consider a diverging combustor flowpath depicted in Figure 7.4. From the previous differential equation, we determine that the axial gradient dT_t/dx has its greatest value near the combustor entry (station 3) and decreases monotonically to its least value at the exit (station 4), so that the ratio $T_t(x)/T_{t2}$ can be represented by a function τ:

$$\tau(x) = \frac{T_t(x)}{T_{t2}} = 1 + (\tau_b - 1)\left[\frac{\theta\chi}{1 + (\theta - 1)\chi}\right] \tag{7.5}$$

where θ is a constant that depends on the method of fuel injection and of fuel–air mixing, taking values $\theta \geq 1$; τ_b denotes the total temperature ratio T_{t4}/T_{t2} within the combustor, and χ is the nondimensional distance along the combustor axis:

$$\chi \equiv \frac{x - x_i}{x_4 - x_i} \tag{7.6}$$

which is defined downstream of the point x_i at which combustion (heat addition) begins.

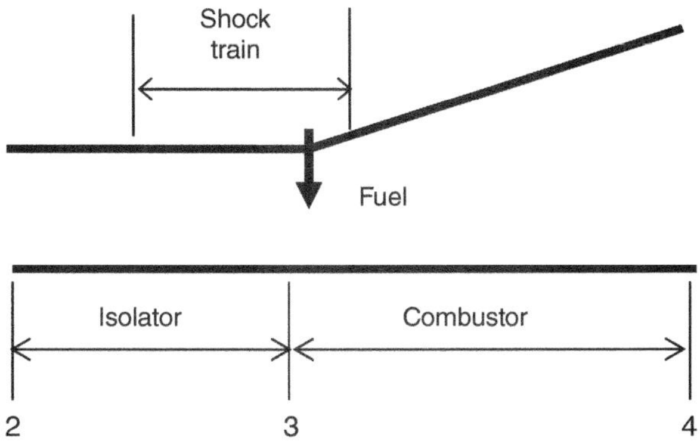

Figure 7.4 Diverging combustor station numbering.

For preliminary design, $T_t(x)$ can be varied within realistic limits to explore the effects of various fuel injection and mixing strategies on scramjet mode combustor performance. Heiser and Pratt (1994) gave the following illustrative example for a dual-mode scramjet:

Supersonic combustion mode with shock-free isolator, $M_2 = M_3 = 1.5$, and $\tau_b = 1.4$. At combustor entry, where heat addition rate dT_t/dx is greatest, the process path is close to a desirable constant pressure trajectory because pressure rise due to heat addition is largely counteracted by pressure relief due to area increase. However, as dT_t/dx tapers off toward the exit, the relieving effect of increasing area now dominates occlusion due to heat addition, so pressure starts to decrease, and the combustor process approaches a constant Mach process, then a constant static temperature, and finally a constant T_t (adiabatic) process near the exit.

In supersonic-to-subsonic combustion mode, the Mach number passes through a minimum $M_s = 1.33$ (no choke). By reducing the combustor exit area, it yields thermal throat Mach numbers of 1.19 and 1.03, respectively. These increase pressure, but the combustor remains in scramjet mode of operation. Further reduction of combustor exit area causes the flow to reach Mach 1 and choke (thermal throat). The flow is now controlled at the choking location and must be subsonic at the entry, which also causes the entry Mach to be lower there, allowing more room for heat addition. In such state, the combustor is in ramjet mode of operation. See Heiser and Pratt (1994) for further discussion.

7.4.2 Dual-Mode Scramjet Isolator

The shock train and the flow field developed in the isolator are illustrated in Figure 7.5 indicating separated flow. The shocks produce a static pressure rise in the constant area diffuser. The back pressure may be due to chemical energy release in the combustor upstream or due to choking of a downstream area; it may also be caused by obstructions in the flowpath caused by fuel injectors, wall mounted ramps, and fuel jets.

A qualitative description of the two types of shock train that develop in the isolator is given in Chapter 4. A normal shock train configuration tends to occur at lower inlet Mach numbers and with thick inlet boundary layers. An oblique shock train structure tends to occur at higher inlet Mach numbers and thinner inlet boundary layers. The maximum static

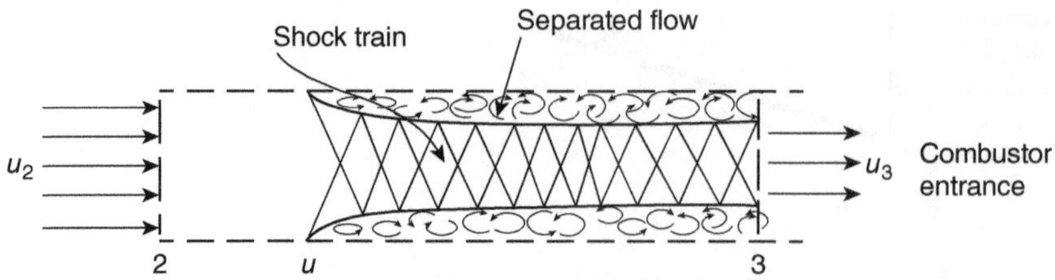

Figure 7.5 Shock train showing a separated core area.

pressure rise generated is exactly the same as a simple normal shock discontinuity would have generated, provided that the shock train is complete, i.e. the flow properties are uniform across the exit plane, and total wall friction is negligible (Vu et al. 2016).

7.4.3 Axial Location of Choked Thermal Throat

The thermal throat must occur at a unique location in the scramjet flowpath. In the dual-mode combustor, a choked thermal throat is achieved by the right combination of area distribution $A(x)$ and fuel–air mixing and heat addition by chemical reaction. Referring to the differential equation (7.2), observe that the axial location for thermal choke should occur where this equality holds

$$\frac{1}{A}\frac{dA}{dx} - \frac{1+\gamma}{2}\left(\frac{1}{T_t}\frac{dT_t}{dx}\right) = 0 \tag{7.7}$$

If Eq. (7.7) is satisfied, choked flow may exist within the combustor. If Eq. (7.7) is not satisfied, the combustor would not be capable of ramjet mode operation. Axial derivatives dM/dx at the choking point can also be obtained from algebraic relationships given by special conditions at the sonic point. If $A(x)$ and $T_t(x)$ do admit a choking point, Eq. (7.2) is numerically integrated upstream from there to the combustor entry, starting with two calculated dM/dx values. These limiting processes determine combustor entry conditions outside of which either all subsonic ($M_3 < M_{3+}$), or all supersonic ($M_3 > M_{3-}$) combustion is possible.

Using Eq. (7.7), one can determine the value of x where a critical point can occur in a flowpath between stations 3 and 4, for any given pair of the functions $A(x)$ and $T_t(x)$. This equation gives sufficient conditions for a critical point to exist at x, but it does not give the necessary conditions for the critical point to occur (Heiser and Pratt 1994). Moreover, if there is no solution for a particular set of $A(x)$ and $T_t(x)$ that satisfies Eq. (7.7), then the combustor cannot operate in ramjet mode.

The isolator accommodates supersonic combustor entry flow to the conditions imposed by thermal choking. For frictionless flow without mass or heat addition where the exit flow is assumed to be confined to a separated core area A_c less than the total throughflow area A_2,

$$\frac{A_c}{A_2} = \left[\left(1+\gamma M_2^2\right)\frac{M_3}{M_2}\sqrt{\frac{1+((\gamma-1/2))M_3^2}{1+((\gamma-1/2))M_2^2}} - \gamma M_3^2\right]^{-1} \tag{7.8}$$

where M_2 is the Mach number at the entrance of the isolator, and M_3 is the combustor entry Mach number, which can be found from the analysis of the core flow in the isolator.

Use the impulse function to estimate the velocity of the core flow: $I_3 = p_3A_2 + \dot{m}u_3$. This expression must be equal to the impulse function at the entrance of the isolator, while the isolator gas temperature T_3 can be estimated from an energy balance equation (see Chapter 2).

Finally, for a given isolator entry Mach number M_2 and back pressure ratio p_3/p_2 imposed by the combustor, the isolator exit Mach number M_3 can be determined from the following expression (see Heiser and Pratt for details):

$$M_3 = \left[\frac{\gamma^2 M_2^2(1 + ((\gamma - 1/2))M_2^2)}{\left(1 + \gamma M_2^2 - \frac{p_3}{p_2}\right)^2} - \left(\frac{\gamma - 1}{2}\right)\right]^{-1/2} \tag{7.9}$$

Note that M_3 decreases approximately linearly with increasing the back pressure p_3, from $M_3 = M_2$ when $p_3/p_2 = 1$ (scramjet mode with shock-free isolator) to the lowest, subsonic value, which is the post-normal shock Mach number corresponding to the normal shock pressure ratio. The maximum and minimum possible isolator exit values of M_3 correspond both to $A_c = A_2$ and to the jump conditions across a normal shock wave.

7.4.4 Combustion-Induced Pressure Rise and Flow Separation

Combustion-induced pressure rise in a supersonic combustor is a well-known phenomenon, extensively discussed in the literature. Frost et al. (2009) concluded the combustion-induced pressure gradient is the main factor responsible for initiating scramjet unstart by inducing boundary layer separation on the combustor walls. They considered the Korkegi criterion to predict scramjet unstart. Korkegi (1975) developed a relation to predict boundary layer separation induced by an impinging oblique shock.

The Korkegi criterion relates the minimum pressure ratio necessary for separation p_2/p_1 to the incoming Mach number M_1:

$$\frac{p_2}{p_1} = 1 + 0.3M_1^2; \quad \text{for}\, M_1 \leq 4.5 \tag{7.10}$$

where p_1 and p_2 are the pressure upstream and downstream of the shock impingement location, respectively.

Researchers agree that scramjet unstart and boundary layer separation are closely correlated. Smart (2010) also found that, as fueling increased, the pressure rise in the isolator could lead to flow separation. Using Korkegi's separation criterion to determine if separation occurs, Smart determined separation right at the intersection of the isolator and the combustor (for a combustor with equivalence ratio $\phi = 0.72$). In this study, the core flow began diffusing at this point at a rate dictated by Ortwerth's equation (Ortwerth 2001), and it reached its minimum area, $A_c/A_2 = 0.822$. He noted that combustion pushed the flow toward reattachment, at an axial location downstream of fuel injection, where the Mach number is 2.027. After re-attachment, the Mach number continued to drop reaching a minimum of 1.087, and the pressure continued to rise to a maximum, after which the flow accelerated under the action of increasing area, and it left the combustor at supersonic conditions (Smart 2010).

7.5 Combustor Mode Transition Studies

The prediction of scramjet operation in the low hypersonic regime ($4 < M_0 < 7$) is challenging. This is because, as we saw earlier, in this flight regime the combustor operates in dual mode. Much effort has been devoted to understanding the inevitable transition regime where mixed characteristics of both subsonic and supersonic combustion exist within the same combustor. The physics of dual-mode transition flows are characterized by strong shock trains, complex shock-boundary layer interactions, significant regions of separated flows, and mixing-limited chemical reactions.

The heat release caused by combustion pushes the PCST upstream of the combustor. This means that the shock system causes the flow to transition from supersonic to subsonic conditions within the isolator. The combustion heat release eventually expands the flow back to sonic or thermally choked conditions, while the combustor area divergence downstream further expands the flow to supersonic conditions. The reaction mechanism developing between the PCST (isolator) and downstream combustion process can be easily disturbed. Modeling this complex dual-mode combustion process requires turbulence and combustion models capable of capturing the mixed supersonic and subsonic flow regions in the combustor.

Moreover, because of computational expense, it is impractical to simulate mode transition as it would occur in flight. Hence, from the solution of the dual-mode simulation, the computation imposes a sudden change in the boundary conditions (at the entrance to the isolator and the fuel injector stages) to match operational conditions in scramjet mode, using data from ground tests (Bermejo-Moreno et al. 2013).

In the past two decades, research was focused on advancing the methodologies to deal with dual-mode combustors. A number of studies applied RANS, using both finite-rate chemistry and flamelet models applied to different dual-mode combustors (Baurle and Eklund 2002; Dessornes and Scherrer 2005; Vyas et al. 2010; Quinlan et al. 2014). Not surprisingly, the results of those studies indicate a strong sensitivity of the predictions to the numerical methods, computational grids, flow thermal nonequilibrium, reaction mechanism as well as the combustion model, and particularly the turbulence model (Fiévet et al. 2015). To better capture the unsteady nature of the flow fields, other studies abide by the power of hybrid Reynolds-Averaged Simulation (RAS)/Large Eddy Simulation (LES) methodology (Fulton et al. 2014; Bermejo-Moreno et al. 2013; Lacaze et al. 2016; Huang and Yan 2016) to study different dual-mode ramjet combustors for which extensive ground test data exists to validate their models.

To support flight test programs, a number of dual-mode scramjet engines were successfully ground-tested at discrete Mach numbers. These programs include the HyShot and the HIFiRE flight experiments. The following sections review examples representative of the research, design, and development efforts related to dual-mode scramjet engines.

7.5.1 HIFiRE-2 Dual-Mode Combustor

The hydrocarbon-fueled HIFiRE-2 dual-mode scramjet was designed to accelerate from Mach 5.5 to 8.5 through a nearly constant dynamic pressure flight corridor. During its flight, the scramjet has to demonstrate dual-mode transition. (The flight payload is shown in

Figure 1.17). The two-dimensional scramjet flowpath is comprised of two opposing forebody ramps that extend back to the sidewalls of the rectangular inward turning Inlet, followed by the Isolator. The combustor was designed with a recessed wall cavity for flameholding. Aft of the combustor, the flow is split with a bifurcating nozzle that vents the flow outward and overboard.

A ground test rig was developed in parallel with the flight program to verify the HIFiRE-2 dual-mode scramjet performance and operability. The ground tests provided required data for Computational Fluid Dynamics (CFD) validation and comparison with flight data. Designated the HIFiRE Direct-Connect Rig (HDCR), the full-scale, heat-sink, test article was designed for direct-connect ground tests that duplicated both the flowpath lines and most of the instrumentation layout of the isolator and combustor portion of the flight test hardware (Cabell et al. 2011).

Figure 7.6 is a schematic of the HIFiRE-2 isolator/combustor flowpath design used in the direct-connect test (flow direction is from left to right). This flowpath includes a constant area isolator section, flush wall fuel injection, primary (PI) and secondary (SI) injector stages, opposed cavity-based flameholders with spark plugs and injectors (CI) for ignition, and a combustor section with a constant divergence angle (total included angle = 2.6°). The cross section at the isolator entrance is 25.4 mm × 101.6 mm (1 in × 4 in), and the overall length of isolator/combustor section is 711.3 mm (just over 28 in).

As shown in Figure 7.6, the two opposed cavity flameholders separate the fuel injection stages, each with eight injection ports (four on the body side and four on the cowl side, equispaced spanwise). The PIs upstream of the cavity are angled at a 15° inclination from the wall, with a circular diameter of 3.175 mm. The SIs downstream of the cavity are perpendicular to the wall with a 2.38-mm circular diameter. Cavity injectors were also included in the design but were not used (Cabell et al. 2011). Equivalence ratios for the primary and SIs were denoted as ϕ_{PI} and ϕ_{SI}, respectively, and ϕ_{CI} for injectors at the cavity.

Cavity injectors were used for ignition only at Mach 8 enthalpy test conditions. For all cases considered, the combustor was fueled by a JP-7 surrogate consisting of 64% ethylene and 36% methane by volume. The fuel total temperature for all cases was 540 °R. Fueling began two seconds after the facility reached the target test conditions, and then a small amount of fuel ($\phi_{CI} = 0.1$) was injected from the cavity. Simultaneously, fuel was injected from the primary injection site at $\phi_{PI} = 0.7$. Cavity injection was then reduced to zero, and the primary and secondary fuel levels were programmed to their target schedules: $\phi_{PI}/\phi_{SI} = 0.4/0$, 0.4/0.3,

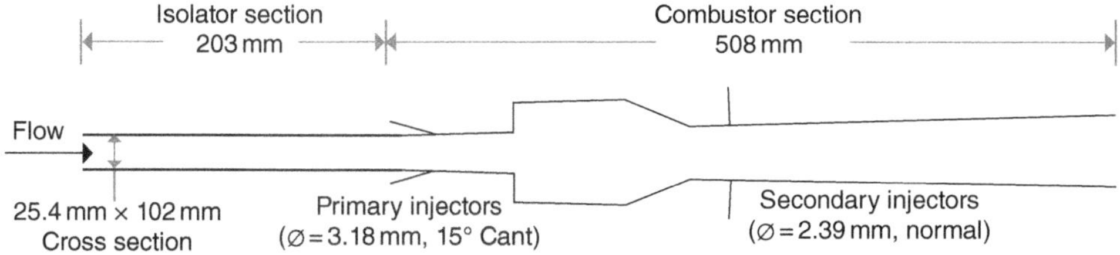

Figure 7.6 Side view and key dimensions of the HDCR combustor flowpath (HIFIRE-2 scramjet). Fuel injectors internal diameter denoted as ∅. Source: Quinlan et al. (2014)/American Institute of Aeronautics and Astronautics, Inc.

and 0.4/0.4 after the cavity fuel was turned off. Subsequent splits tested were $\phi_{PI}/\phi_{SI} = 0.4/0.5$, 0.4/0.6, and 0.4/0.7 (Cabell et al. 2011).

In flight, the HIFiRE-2 combustor transitions from dual mode to scramjet mode at some point in the Mach 6–8+ trajectory. In ground tests, it is not possible to provide continuous acceleration, so no transition between the two modes can be simulated in the HDCR facility. Hence, a different facility nozzle needs to be used for each operating regime. In this manner, researchers determined dual-mode operation at the low Mach numbers (5.84 and 6.5) and scramjet mode operation at the high Mach numbers (7.5 and 8.0). In the HDCR Mach 6 test series, a fuel split of $\phi_{PI}/\phi_{SI} = 0.4/0.6$ was identified as giving acceptable dual-mode operation at Mach 6.5 enthalpy. Thus, at Mach 8, this split was used, giving a fixed ratio of primary to secondary fuel level from dual mode to scramjet mode, thus simplifying the HIFiRE-2 flight fuel system configuration.

In a separate study, Storch et al. (2011) carried out three-dimensional CFD simulations, using the VULCAN-CFD and CFD++ codes. They used measurements obtained from ground testing to specify inflow conditions and used combustor data as benchmarks from four representative tests at simulated flight enthalpies of Mach 5.84, 6.5, 7.5, and 8.0. This analysis demonstrated dual-mode operation at Mach 6.0 through 7.0. Moreover, it needed to verify that the HIFiRE-2 combustor operated in scramjet mode at higher Mach numbers, which means there is no significant pressure rise ahead of the PI, and one-dimensional (1D) supersonic Mach number throughout the flowpath. These two requirements were confirmed.

Figure 7.7 shows the computed 1-D Mach number distributions through the isolator/combustor flowpath for Mach 8.0. The flow Mach number drops quickly downstream of the primary fuel injector (PI), reaching a minimum just downstream of the cavity ramp (never reaching sonic condition), and then the flow begins a gradual acceleration to exit the combustor (Storch et al. 2011).

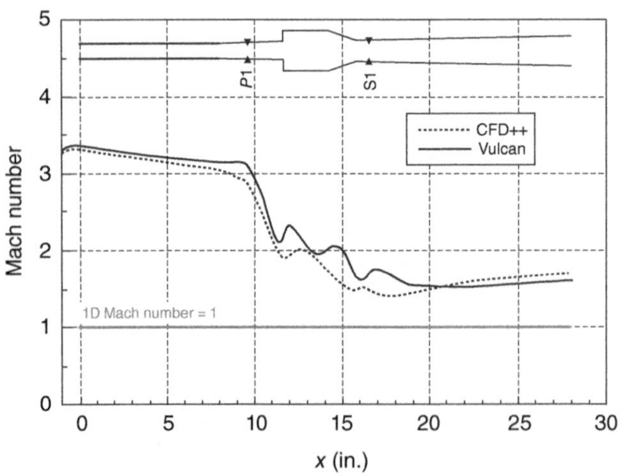

Figure 7.7 One-dimensional Mach number distribution from CFD simulations for HIFiRE-2 at Mach 8.0, scramjet mode operation. Source: Storch et al. (2011)/American Institute of Aeronautics and Astronautics, Inc.

Quinlan et al. (2014) conducted an a priori investigation of the applicability of flamelet-based combustion models to dual-mode scramjet combustion utilizing RAS via the NASA VULCAN-CFD code. They chose the HDCR flowpath, operating in dual and scramjet mode to study the possibility of replacing finite-rate kinetics with flamelet models, and thus the flow field is characterized using flame and combustion mode indices. To model the chemistry of the JP-7 fuel surrogate, a 22-species, 18-step chemical reaction mechanism was used. Simulation results were compared to experimentally obtained, time-averaged, wall pressure measurements to validate the RAS solutions. Researchers selected dual-mode cases tested at a flight Mach number of 5.84 and a total equivalence ratio of 0.65, and the scramjet-mode cases where those tested at a flight Mach number of 8.0 and a total equivalence ratio of 1.0. The total equivalence ratio was split between the primary and SIs, respectively, as 0.15 and 0.5 for the dual-mode cases and as 0.4 and 0.6 for the scramjet-mode cases.

The RAS simulation of dual-mode and scramjet-mode operation provided much insight into the HIFiRE-2 combustor flow field, showing differences in flow structure and flame location, as illustrated in Figure 7.8. The following are some interesting results:

1) **Dual-Mode Flowfield – Mach Number**. Simulated conditions: $M_0 = 5.84$, $\phi_{PI} = 0.15$, $\phi_{SI} = 0.50$. The Mach number contours in Figure 7.8a clearly show upstream of the Primary Injector (PI) the leading oblique shock (caused by combustor pressure rise), and a series of shock reflections all the way through the flowpath. The black sonic line was added to identify the regions of subsonic and supersonic flow.
2) **Supersonic Mode Flowfield – Mach Number (Figure 7.8b)**. Simulated conditions: $M_0 = 8.0$, $\phi_{PI} = 0.40$, $\phi_{SI} = 0.60$. The core flow remains highly supersonic throughout the flowpath, but the flow inside the cavities is subsonic.
3) **Dual-Mode Flowfield – Chemical Heat Release (Figure 7.8c)**. Simulated conditions: $M_0 = 5.84$, $\phi_{PI} = 0.15$, $\phi_{SI} = 0.50$. The chemical release contours show a strong thin flame anchored directly over the PIs, upstream from the cavities. Moreover, the SIs flames release considerably more heat than the flames produced by the PIs. Farther downstream, the flow expands out of the combustor.
4) **Supersonic Mode Flowfield – Chemical Heat Release (Figure 7.8d)**. Simulated conditions: $M_0 = 8.0$, $\phi_{PI} = 0.40$, $\phi_{SI} = 0.60$. As indicated by the chemical release contours, combustion downstream from the PIs appears more uniform, as compared with the dual-mode case. These primary flames are formed behind the leading oblique shock and above the cavity region. Keep in mind that the PIs have a much lower equivalence ratio in the dual mode as compared with that of the scramjet-mode case. Combustion that results from the SIs appear to be of a similar nature to those of the dual-mode case (Quinlan et al. 2014). The differences in flame structure obtained with the flamelet model are given in Chapter 12.

In an independent study, Bermejo-Moreno et al. (2013) carried out analysis with LES aimed at resolving the complex flow fields of the HIFiRE-2 scramjet at two operating conditions: Mach 6.5 (dual mode) and Mach 8 (scramjet mode). They used the flamelet progress variable approach (FPVA) to model combustion. The FPVA model reduces the otherwise computationally intractable complexity of a hydrocarbon fuel chemical mechanism (owing to the large number of species and reactions involved) to a flamelet look-up table, precomputed for a set of flame boundary conditions.

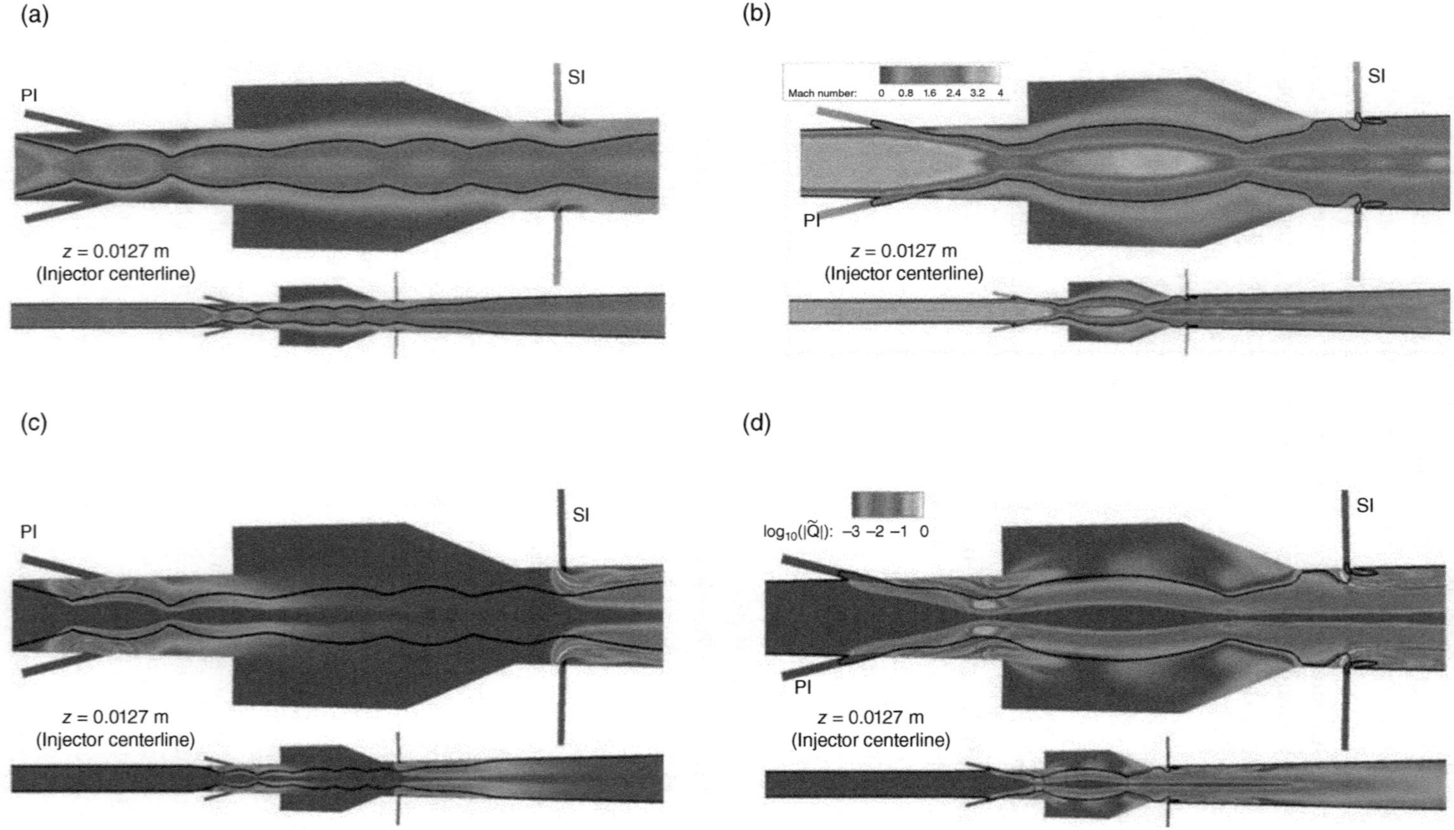

Figure 7.8 RAS simulation of HIFiRE-2 HDCR flowpath with JP surrogate fuel. Mach number contours and logarithm of chemical heat release normalized by its global maximum for indicated operation modes. (a) Dual-mode flow-field – Mach number. (b) Scramjet-mode flow-field – Mach number. (c) Dual-mode flow-field – chemical heat release. (d) Scramjet-mode flow-field – chemical heat release. Source: Quinlan et al. (2014)/American Institute of Aeronautics and Astronautics, Inc./Public Domain.

In addition to presenting scramjet flow features from different snapshots of the simulations, the authors compared their time-averaged results with available experimental data from the ground tests. Figure 7.9 shows the simulated time-averaged wall pressure profiles compared with the ground test experimental data at the spanwise center plane for the two operation conditions considered, including both reacting and nonreaction flow fields.

As shown in Figure 7.9a, for the dual-mode operation, the simulation results (dash-dotted) compare well with the pressure profile obtained in ground testing (hollow dots)

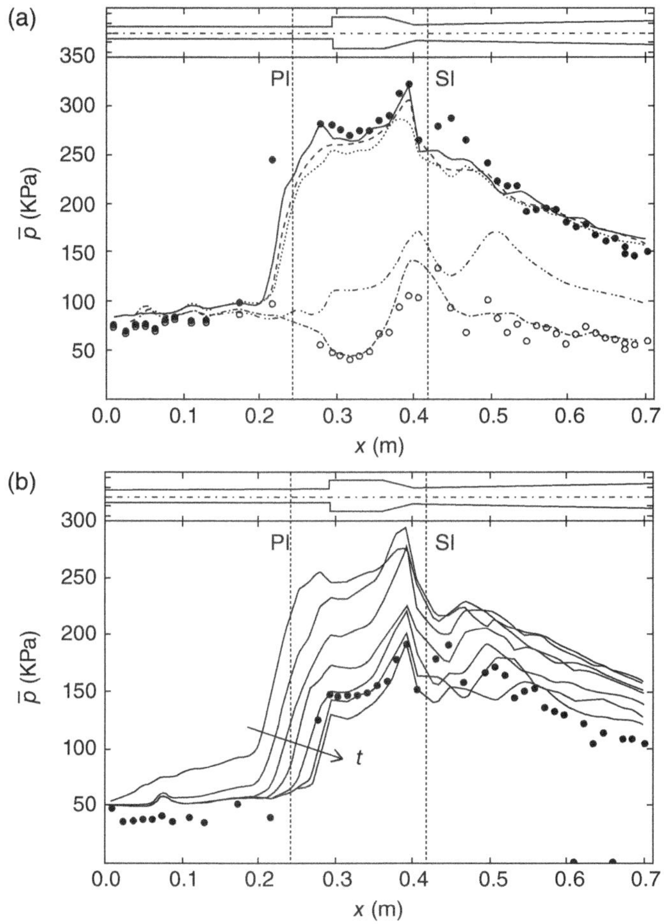

Figure 7.9 Time-averaged wall pressure profiles at the center plane. (a) Experimental data represented by hollow dots (nonreacting flow) and solid dots (reacting flow); lines correspond to simulation results: dash-dotted for nonreacting; dash–dot–dot for reacting flow with original FPVA formulation; dotted, dashed, and solid for reacting with flamelet-based fast chemistry combustion model on coarse-, medium-, and fine-resolution meshes, respectively. (b) Transition from dual to scramjet mode: experimental data given by solid dots; Wall-Model LES (WMLES) results with flamelet-based fast chemistry combustion model on medium-resolution mesh shown with each line corresponding to the simulation data averaged over consecutive time intervals of 0.3 ms, evolving from the sudden change of conditions from dual to scramjet mode. Source: Bermejo-Moreno et al. (2013)/Stanford University.

for nonreacting (no fuel injection) flow conditions at Mach 6.5. However, the simulation predicts higher pressure recovery along the cavity ramp. The authors attributed this to possible differences in the reattachment location obtained with the wall model as compared with the actual experiment.

In the reacting flow case, the simulation using the FPVA model, dash-dot-dotted line in Figure 7.9a, does not reproduce the pressure rise observed in the experiments (solid black dots), which begins upstream of the PI. The model predicts pressure rise farther downstream, inside the cavity, and then decrease to pressure levels much lower than measured in the experiments. This is indicative of insufficient combustion. Past the cavity and downstream of the SI, the experiment shows a rise in pressure, but it eventually decays. After the pressure peak, the simulation shows a trend similar to the experimental profile.

Using a flamelet-based fast chemistry combustion model, researchers considered three-grid resolutions and found good agreement with the experimental data. As shown in Figure 7.9a, the simulation recovered the pressure levels inside the cavity and downstream, and it captured the pressure rise occurring upstream of the PI; however, the prediction shows this downstream of the corresponding experimental location, which is observed to penetrate farther into the isolator. It is also important to note that the pressure level in the combustor (particularly inside the cavity) increases slightly with mesh resolution, approaching the experimental data. The authors concluded that the predictions may not have been grid-converged.

For the scramjet mode, the simulation started from the solution of the dual-mode simulations at Mach 6.5, imposing a sudden change in the boundary conditions at the entrance to the isolator and the fuel injectors in order to match the scramjet operation at Mach 8 in the ground tests. The flow then evolves through a transient, as reflected in Figure 7.9b, which shows center line wall pressure profiles averaged over consecutive time intervals of 0.3 ms. Within the first 2 ms, the pressure levels throughout the scramjet decrease to the experimental values. The pressure level reached in the cavity at the last time interval shown in Figure 7.9b is below the experimental data. The study concluded that the combustion model might not be able to capture the supersonic combustion in scramjet operating mode. Overall, their attempts to apply FPVA combustion modeling strategy to the HIFiRE-2 simulations failed for both operating regimes under consideration. The model yielded insufficient burning, particularly upstream of the cavity flameholder (Bermejo-Moreno et al. 2013).

7.5.2 LAPCAT II Dual-Mode Combustor

An interesting study combining experimental testing and computational LES analysis is carried out at ONERA (*Office National d'Etudes et Recherches Aérospatiales*). The study focuses on the LAPCAT-II dual-mode ramjet/scramjet combustor tested at ONERA-LAERTE facility, which is operated in the blow-down mode with the combustor acting as a heat sink. The 1257-mm-long combustor comprises four sections with gradually increasing cross-sectional area: the first section has a constant cross section, the following sections have 1°, 3°, and 1° of diverging half-angles, respectively, to prevent/stunt thermal choking. The combustor is fueled with hydrogen, which is injected at sonic velocity from the upper and lower walls (in the first section) through two flush-mounted 2.0-mm porthole injectors. Experiments are conducted in a Mach 2 vitiated air flow with 0.40 MPa total pressure (p_0) and total temperature (T_0) in the range 1414–1707 K, with fuel injection at a fixed equivalence ratio of $\phi = 0.15$. Transition from supersonic to subsonic flow occurs at the beginning of the fourth combustor section (Vincent-Randonnier et al. 2018).

From the combined OH$*$ chemiluminescence and Schlieren images taken during the experiments, researchers noted that the shock structure and boundary layer separation are very sensitive to T_0 and dominate the ignition and flame stabilization mechanisms. In particular, at low T_0 combustion occurs in the low-speed region (third combustor section), at intermediate T_0 there occurs either unsteady combustion with the combustion fronts oscillating between two locations, or thermal choking, and finally shock-induced combustion occurs at high T_0.

The computational effort is based on finite-rate chemistry LES combined with skeletal H_2–air combustion chemistry, and the filtered reaction rates are modeled using the Partially Stirred Reactor (PaSR) model. The LES-PaSR model equations are solved using a fully explicit finite volume code based on the OpenFOAM C++ library. One characteristic of this model that is different from others is the compressible finite-rate chemistry LES model using a novel 22 step H_2–air reaction mechanism called Z22 (9 species and 22 irreversible reactions) which is suitable for the low-temperature chemistry of ignition; the Z22 is the only investigated mechanism that reproduces accurately the experimental data at the lowest temperatures (Vincent-Randonnier et al. 2018).

Figure 7.10, adapted from several figures in Vincent-Randonnier et al. (2018), is used here to illustrate the predicted temperature (T) and heat release (Q) distributions which were compared with measured reference values for all cases investigated, keeping constant the equivalence ratio at $\phi = 0.15$. Consider the following results:

Nonreacting flow, Case 0 (air $T_0 = 1414$ K): Just downstream of the transition from 1° to 3° combustor sections, at $0.70 < x < 0.80$ m, the nonreacting flow separates. Researchers noted that low-speed zones are formed along the upper and lower walls, with separation occurring either along the upper or the lower wall, i.e. the separation line oscillated back and forth, resulting in an asymmetric flow field. Clearly, this case without fuel injection has just a slight temperature increase, and no heat release.

Subsonic Combustion, Case 1 (air $T_0 = 1414$ K): The predicted distributions of temperature, T, and heat release, Q, are found to be consistent with experimental results: there is a significant increase in T after the intense combustion heat release that occurs in the upper and lower wall boundary layers, just after flow separation. The researchers analyzed the modified Takeno Flame Index (TFI) (TFI is the product of the gradient of the fuel mass fraction and the gradient of the oxygen mass fraction, giving information on whether a flame regime is premixed or of the diffusion type) and concluded that these two flame branches are premixed. This means that the injected H_2 and the vitiated air had sufficient time to mix prior to the ignition taking place just downstream of the flow separation.

Unsteady Supersonic Combustion, Case 3a (air $T_0 = 1505$ K): In this case, both simulated and experimental combustion images indicated that supersonic combustion started between the first and second combustor sections, occurring much earlier as compared to subsonic combustion (compare images in Figure 7.10 for Cases 1 and 3a). The experiments showed an abrupt combustion front oscillating between $0.30 < x < 0.45$ m, while LES predicted a more gradual combustion front, oscillating between $0.28 < x < 0.40$ m. Observe the

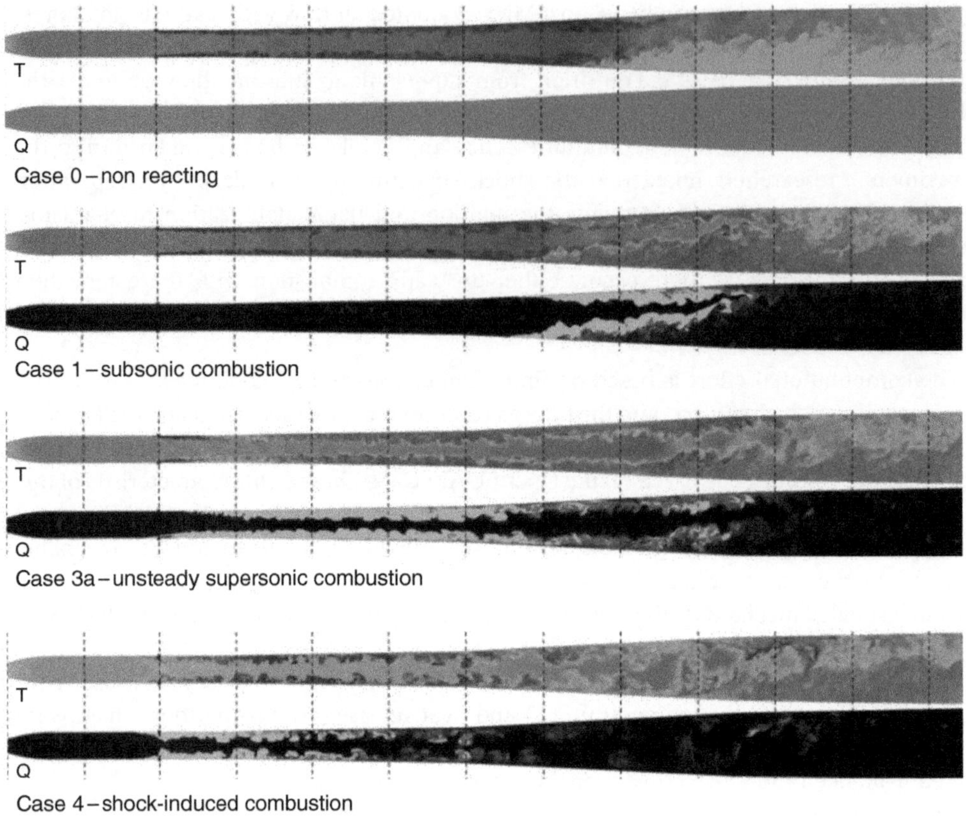

Figure 7.10 Computational images of temperature (*T*) and heat release (*Q*) for cases investigated in LAERTE-LAPCAT-II combustor with H_2 injection at x = 0.20 m. Source: Adapted from Vincent-Randonnier et al. (2018).

high-temperature and heat release distributions in the first two sections of the combustor. Downstream of this location, Q decreases as the H_2 is consumed, but the temperature remains high. According to researchers, the modified TFI imply that non-premixed combustion dominates this case; the numerical simulation predicts that the two (upper and lower) flame branches are relatively loosely coupled and hence behave rather independently.

Shock-Induced Combustion, Case 4 (air $T_0 = 1697$ K): As indicated by the bottom two images of Figure 7.10, in this case combustion starts even earlier than previous cases. Experimental images indicated that combustion starts abruptly at $x \approx 0.26$ m caused by shock-induced ignition (fuel is injected at $x = 0.20$ m). However, LES predicted intermittent combustion beginning even earlier, as shown by T and Q images, with intermittent combustion occurring also in front of the fuel injectors, beneath the bow shocks and partially in the horseshoe vortices surrounding the H_2-rich jets. According to researchers, all numerical simulations indicated that the upper and lower flame branches alternated from being attached to the upper and lower walls, respectively, with a frequency of approximately 50 Hz (Vincent-Randonnier et al. 2018).

Overall, this study found excellent agreement of flow fields predicted by LES as compared with experiments for all cases investigated. Both LES and experiments revealed a strong sensitivity of the combustor to variations in air total temperature T_0. The work at ONERA is ongoing, investigating the dual-mode characteristics of the LAERTE-LAPCAT-II combustor. For updates on this effort, the interested reader is referred to Vincent-Randonnier et al. 2019.

7.5.3 JHU APL Axisymmetric Dual-Combustor Engine (DCE)

The Dual-Combustor Ramjet (DCR) shown schematically in Figure 7.11 was designed and tested at the Johns Hopkins University Applied Physics Laboratory, (JHU APL). In the DCR, a subsonic combustion ramjet is used as the pilot to a scramjet engine, enabling efficient operation over a wider range of supersonic and hypersonic Mach numbers using hydrocarbon JP fuels (Van Wie et al. 2005).

As indicated in Figure 7.11, the DCR is an axisymmetric dual-mode propulsion system in which the forebody serves as the initial compression surface of the supersonic inlet, and the incoming air flow enters the engine via eight openings at the cowl lip. Four smaller inlets supply air to a subsonic combustor, operating supercritically (swallowing the normal shock) to avoid the interaction of the normal shock with the flow entering the larger inlets that feed the supersonic combustor. In addition, in order to provide stable combustor operation over a wide range of flight Mach numbers, the flow passages to the subsonic combustor have an increasing cross-sectional area in the streamwise direction. The major portion of the air is captured by the four larger inlets and the external cowl compression surface, turned supersonically inward toward the engine axis. Ingested air flow is spread radially to form an annulus of supersonic flow that surrounds the outlet of the dump combustor.

In order for the DCR to operate as a dual-mode propulsion system, the aft sections of the supply ducts have slightly diverging flow passages in the streamwise direction, which effectively act as the combustor inlet isolator. Thus, when the DCR operates at high equivalence ratio and/or at low flight Mach number, the isolator can operate with a shock train causing

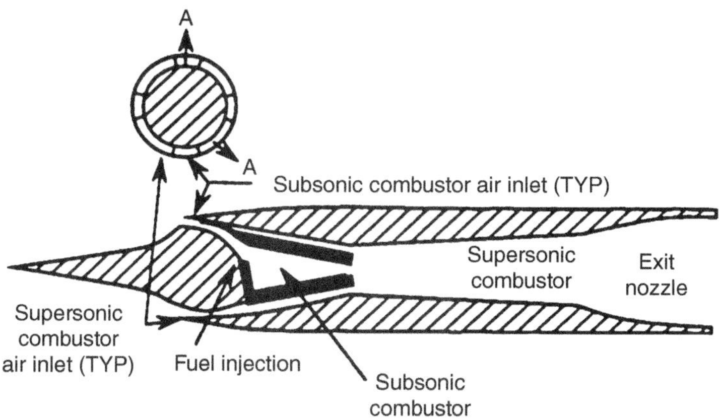

Figure 7.11 Schematic of the JHU APL dual-combustor ramjet engine (DCR). Source: Van Wie et al. (2005)/The Johns Hopkins University Applied Physics Laboratory.

a pressure rise equivalent to that of normal shock, and thus the Mach number at the combustor entrance is subsonic, while the mean Mach number at the combustor exit is either sonic or supersonic. On the other hand, when the DCR operates at lower engine equivalence ratios and/or higher flight Mach numbers, the isolator shock train pressure rise is equivalent to that of an oblique wave structure. In such case, the mean flow Mach number throughout is supersonic.

Waltrup (1992) compared the performance of the DCR engine to both a scramjet and ramjet, showing that its performance is comparable to a subsonic combustion ramjet at Mach 3 but is better at Mach 6. This is to be expected since the DCR operates more like a scramjet at the higher flight speeds. Waltrup also found that the DCR performs better than the scramjet at Mach 4 but has poorer performance at Mach 8, again supporting the fact that the DCR operates much like a ramjet at the lower flight speeds.

7.5.4 Free-Jet Dual-Mode Combustor

Researchers at NASA conceived the Free-Jet Dual-Mode Combustor (Figure 7.12), a novel concept that can operate in both subsonic and supersonic combustion modes.

As depicted in Figure 7.12a, supersonic combustion occurs in an unconfined free jet that traverses the larger subsonic combustion chamber, and the high-energy stream exhausts

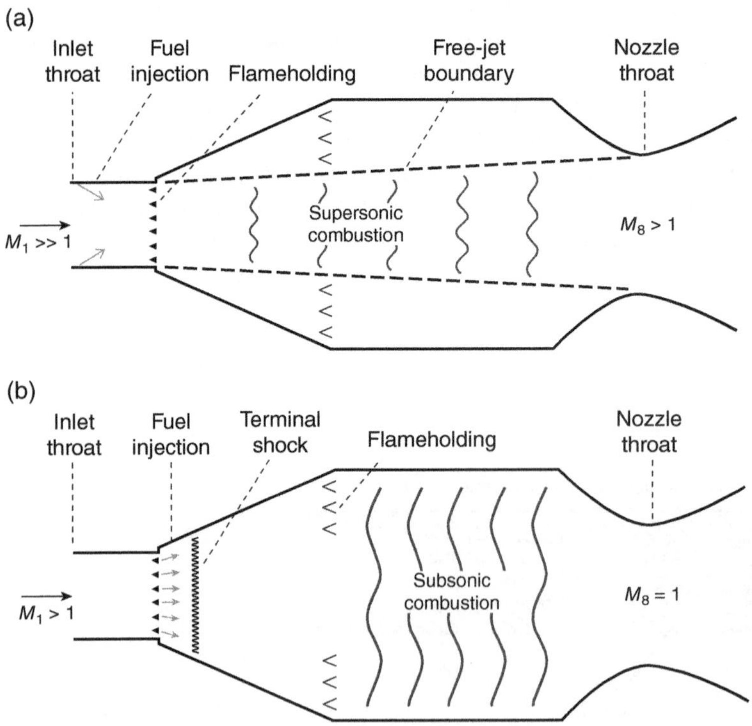

Figure 7.12 Free-jet dual-mode combustor. (a) Supersonic combustion free-jet mode. (b) Subsonic combustion ramjet mode. Source: Adapted from Trefny and Dippold (2010).

directly through the throat plane of a variable area nozzle. In this mode, the combustion (propulsive) stream is not in contact with the combustor walls, and it equilibrates to the combustion pressure. In subsonic combustion mode, it operates as a conventional combustor, with the entire flow field filling the combustion chamber and then exhausting through the converging–diverging nozzle, as shown in Figure 7.12b. At the desired flight condition, transition to free-jet mode occurs by increasing the nozzle throat area, and by inducing flow separation at the diffuser inlet via carefully scheduled fuel injection (Trefny et al. 2017).

Numerical studies demonstrated the feasibility of supersonic free-jet combustion at flight Mach numbers of 5, 8, and 12. Researchers concluded that a fixed geometry propulsion system could operate over a range of flight Mach numbers.

7.6 Closing Remarks

The complexity of reacting flow in the scramjet combustor operating in dual-mode makes both analysis and design extraordinarily difficult. Such combustion flows are strongly coupled and are driven by complex turbulence and chemical kinetics mechanisms, representing the ultimate challenge for analysis, numerical simulation, and experimental research.

Choking, for example, must occur within the combustor flowpath by means of a thermal throat. This can be done by selection of an optimum combination of flowpath area distribution $A(x)$, and heat addition, which means having optimum fuel injection stagging, effective fuel–air mixing, strategic flameholding, and efficient combustion. If chemical energy release in the combustor increases above a threshold (e.g. excessive fuel), inlet unstart will occur caused by back pressure increases. Dual-mode scramjet design must include start/restart capability.

The required cross-sectional area for the thermal throat increases as flight Mach number decreases, unless the fuel–air ratio is reduced. Hence, the technical challenges in practical application of the dual-mode scramjet are modulation of the thermal throat location, fuel distribution, and ignition and flameholding in the large cross section of the flowpath. Any in-stream devices (e.g. strut injectors, ramps, and pylons) must be retractable or expendable so as not to inhibit supersonic combustion operation.

Maintaining combustion stability through the various modes and mode transitions needs further investigation. Most dual-mode scramjet designs use two injection stages strategically located in the combustor flowpath. The first stage is mainly used to drive supersonic combustion, whereas the second stage promotes subsonic combustion with a thermal throat located near the combustor exit. The effects of staged versus simultaneous mixing must be investigated. Thermal management for the dual-mode combustor needs to be addressed as well.

Questions

1. What engine parameter controls transition from subsonic to supersonic combustion?

2. What is the function of the Isolator section in a scramjet engine?

3. How does one control shock train position and make it reside within a shorter Isolator?

4. A pure scramjet engine has no physical throat; how can it operate in ramjet mode?

5. In a dual-mode scramjet, where and when will thermal choke occur?

6. What duct length is required to "isolate" the engine Inlet from influences propagating upstream from the combustor?

7. What is the back pressure effect on shock train location in the scramjet Isolator?

References

Aguilera, C. and Yu, K.H. (2017). Scramjet to ramjet transition in a dual-mode combustor with fin-guided injection. *Proceedings of the Combustion Institute* 36: 2911.

Baurle, R.A. and Eklund, D.R. (2002). Analysis of dual-mode hydrocarbon scramjet operation at Mach 4–6.5. *Journal of Propulsion and Power* 18: 990–1002. (RANS study).

Bermejo-Moreno I., Larsson J., Bodart J., and Vicquelin R. (2013).Wall-Modeled Large-Eddy Simulations of the HIFiRE-2 Scramjet. CTR Annual Research Briefs (LES study).

Cabell, K., Hass, N., Storch A., and Gruber, M. (2011). HIFiRE direct-connect rig (HDCR) phase I scramjet test results from the NASA Langley arc-heated scramjet test facility. *AIAA 2011-2248. 17th AIAA International Space Planes and Hypersonic Systems and Technologies Conference*, San Francisco, CA (11–14 April 2011).

Dessornes, O. and Scherrer, D. (2005). Tests of the JAPHAR dual mode ramjet engine. *Aerospace Science and Technology* 9: 211–221.

Fiévet, R., Koo, H., and Raman, V. (2015). Numerical simulation of a scramjet isolator with thermodynamic nonequilibrium. *AIAA Aviation (AIAA 2015–3418). 22nd AIAA Computational Fluid Dynamics Conference*, Dallas, TX (22–26 June 2015).

Fotia, M.L. (2015). Mechanics of combustion mode transition in a direct-connect ramjet–scramjet experiment. *Journal of Propulsion and Power* 31: 69.

Frost, M.A., Gangurde, D.Y., Paull, A., and Mee, D.J. (2009). Boundary-layer separation due to combustion-induced pressure rise in supersonic flow. *AIAA Journal* 47 (4): 1050–1053.

Fulton, J.A., Edwards, J.R., Hassan, H.A. et al. (2014). Large-Eddy/Reynolds-Averaged Navier–Stokes simulations of reactive flow in dual-mode scramjet combustor. *Journal of Propulsion and Power* 30: 558–575. (LES study).

Heiser, W.H. and Pratt, D.T. (1994). *Hypersonic Airbreathing Propulsion*, AIAA Education Series. Washington, DC: American Institute of Aeronautics and Astronautics (AIAA).

Huang, W. and Yan, L. (2016). Numerical investigation on the ram–scram transition mechanism in a strut-based dual-mode scramjet combustor. *International Journal of Hydrogen Energy* 41 (8): 4799–4807. (LES study).

Korkegi, R.H. (1975). Comparison of shock induced two- and three-dimensional incipient turbulent separation. *AIAA Journal* 13 (4): 534–535.

Lacaze, G., Vane, Z.P., and Oefelein, J.C. (2016). Large Eddy simulation of the HIFiRE direct connect rig scramjet combustor. *SAND2016-12566C (LES study). 55th AIAA Aerospace Sciences Meeting*, Grapevine, TX (9–13 January 2017).

Laurence, S.J., Lieber, D., Schramm, J.M. et al. (2015). Incipient thermal choking and stable shock-train formation in the heat-release zone of a scramjet combustor. Part I: shock-tunnel experiments. *Combustion and Flame* 162 (4): 921–931.

Micka, D.J. and Driscoll, J.F. (2009). Combustion characteristics of a dual-mode scramjet combustor with cavity flameholder. *Proceedings of the Combustion Institute* 32: 2397–2404.

Ortwerth, P.J. (2001). *Scramjet Vehicle Integration, Scramjet Propulsion, Progress in Astronautics and Aeronautics*. Washington, DC: AIAA Chapter 17.

Quinlan, J., McDaniel, J.C., Drozda, T.G. et al. (2014). A priori analysis of flamelet-based modeling for a dual-mode scramjet combustor. *AIAA 2014-3743 (RANS study). 50th AIAA/ASME/SAE/ASEE Joint Propulsion Conference*, Cleveland, OH (28–30 July 2014).

Smart, M. (2010). Scramjet isolators. RTO-EN-AVT-185 10. Presented at the AVT-185 RTO AVT/VKI Lecture Series held at the von Karman Institute, Rhode St. Genèse, Belgium (13–16 September 2010). http://www.dtic.mil/dtic/tr/fulltext/u2/a581901.pdf.

Storch, A., Bynum, M., Liu, J. et al. (2011). Combustor operability and performance verification for HIFiRE flight 2. *17th AIAA International Space Planes and Hypersonic Systems and Technologies Conference*, San Francisco, CA. (11–14 April 2011).

Trefny, C. J. and Dippold, V.F. III. (2010). Supersonic free-jet combustion in a ramjet burner. NASA/TM-2010-216932, NASA Glenn Research Center, Cleveland, OH (November 2010).

Trefny, C.J., Dippold, V.F. III. and Yungster, S. (2017). Dual mode free-jet combustor. *International Symposium on Air Breathing Engines (ISABE)*, Manchester, England (3–8 September 2017).

Van Wie, D.M., D'Alessio, S.M., and White, M.E. (2005). Hypersonic Airbreathing Propulsion. *Johns Hopkins APL Technical Digest* 26 (4): 430–437.

Vincent-Randonnier, A., Ristori, A., Sabelnikov, V., Zettervall, N. and Fureby, C. (2018). A combined experimental and computational study of the LAPCAT II supersonic combustor. *Paper AIAA-2018-5208. 22nd AIAA International Space Planes and Hypersonic Systems and Technologies Conference*, Orlando, FL (17–19 September 2018).

Vincent-Randonnier, A., Sabelnikov, V., Ristori, A. et al. (2019). An experimental and computational study of hydrogen–air combustion in the LAPCAT II supersonic combustor. *Proceedings of the Combustion Institute* 37: 3703–3711.

Vu, N.L., Quan, L.H., Nguyen P. H., Le, D. Q., "Analysis of a dual-mode scramjet engine isolator operating from Mach 3.5 to Mach 6." *International Journal of Mechanical Engineering and Applications*, Volume 4, Issue 5, October 2016, Pages: 189–198.

Vyas, M.A., Engblom, V.A., Georgiadis, N.J. et al. (2010). Numerical simulation of vitiation effects on a hydrogen-fueled dual-mode scramjet. *AIAA-2010-1127 (RANS study). 48th AIAA Aerospace Sciences Meeting Including the New Horizons Forum and Aerospace Exposition*, Orlando, FL (4–7 January 2010).

Waltrup, P.J. (1992). The dual combustor ramjet: a versatile propulsion system for hypersonic tactical missile applications. *AGARD-CP-526, Paper No. 8 (September 1992). AGARD Conference Proceedings*, Issue 526, (September 1992).

8

Scramjet Nozzle/Aftbody

8.1 Introduction

The development of a hypersonic air-breathing aircraft relies heavily on the optimal integration of the engine and the airframe. This is necessary in order to reduce excessive drag and weight, optimizing the integrated design to extract thrust from the high-pressure, high-temperature flow generated by the supersonic flow combustor at the high Mach numbers at which the aircraft will be traveling. In such airframe-propulsion configuration (Figure 8.1), the entire aft end of the vehicle acts as a single (one sided) expansion ramp nozzle. Such aftbody/nozzle geometry provides a very large exit-to-throat area ratio that can be best utilized at the high nozzle pressure ratios associated with hypersonic flight Mach numbers at high altitudes. This type of exhaust engine nozzle is known as Single Expansion Ramp Nozzle (SERN).

In such asymmetric nozzle, with only an upper solid surface, the lower region of the scramjet exhaust flow interacts with the hypersonic airflow around the vehicle. The overall propulsive efficiency of this nozzle is determined, to a large extent, by the exhaust plume flow over the afterbody section. Scramjet cycle performance requires that the SERN exhaust cross section be about 30% larger than the inlet capture cross-sectional area. In principle, the large afterbody nozzle area can provide very efficient expansion of the hot gas exhaust with minimal aerodynamic drag. The large ramp exhaust surface can also be used advantageously to increase favorable lift for cruise aircraft.

The tight coupling between the engine and the vehicle requires a combined analysis of internal and external flows. The hypersonic freestream and the supersonic exhaust flow mix through a shear layer, where mass, momentum, and energy transfers occur. The interference of the exhaust on the control surfaces of the airframe can have adverse effects on the stability of the vehicle. Therefore, a method of simulating this type of flow is required to properly design the nozzle and the afterbody section.

In addition to the SERN, there are other engine exhaust nozzle designs (Section 8.2). Their geometries depend on the type of vehicle where they will be integrated. For hypersonic missile applications, for example, the typical design incorporates axisymmetric engines, and thus the nozzle will be contained within a flowpath that require an axisymmetric shape similar to that found in podded afterburning turbojet engines.

Scramjet Propulsion: A Practical Introduction, First Edition. Dora Musielak.

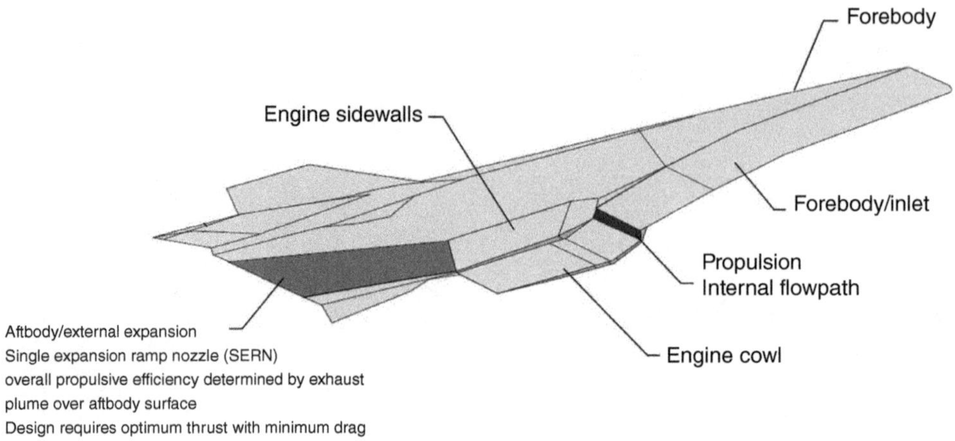

Figure 8.1 Hypersonic air-breathing integrated airframe-propulsion aircraft.

In all hypersonic flight applications, the supersonic nozzle must be designed to optimize thrust, while at the same time aiming to reduce airframe weight at large expansion ratios, reduce drag, increase controllability with thrust vectoring, and offer high operational reliability.

8.1.1 Nozzle Function

The primary function of the engine nozzle is to provide flow acceleration from combustor exit static pressure to local atmospheric or freestream static pressure over the entire range of the powered vehicle operation. This acceleration must occur in a controllable and reliable manner with maximum performance, i.e. maximum expansion efficiency (minimum losses) and exhausting flow vector aimed in the desired direction to achieve optimum thrust.

In the flowpath of a propulsion-airframe hypersonic vehicle, the hot combustor-exit flow expands in the internal nozzle and along the vehicle afterbody where the bulk of the work potential that was added in the combustor is finally realized by increased wall pressure and yielding the generation of vehicle thrust. The nozzle is required to accelerate the flow over a wide range of supersonic conditions from the combustor exit.

A 1-D control volume analysis of the ideal hypersonic engine yields an expression for the (uninstalled) thrust, representing the change in momentum flux from entry to exhaust. Using the nomenclature and engine reference stations we adopted in Chapter 2, the thrust force is expressed as

$$F = \left(\dot{m}_0 + \dot{m}_f\right)V_{10} - \dot{m}_0 V_0 + (p_{10} - p_0)A_{10} = \dot{m}_0[(1 + f)V_{10} - V_0] + (p_{10} - p_0)A_{10} \tag{8.1a}$$

or equivalently

$$F = [\dot{m}V + pA]_{10} - [\dot{m}V + pA]_0 - p_0(A_{10} - A_0) \tag{8.1b}$$

From this expression is clear that the nozzle exhaust system design must maximize the nozzle exit flow velocity V_{10}, especially as the flight Mach number increases. We expect the exhaust velocity to be a function of temperature, $V_{10} = M_{10}\sqrt{\gamma RT_{10}}$ with $M_{10} > 1$. The 1-D steady energy equation yields an expression for the expanding flow velocity:

$$V_{10} = \sqrt{V_4^2 + 2c_p(T_4 - T_{10})} \tag{8.2}$$

Although the exhaust velocity must exceed the flight velocity in order to obtain positive thrust, for the scramjet engine the velocity ratio V_{10}/V_0 is only slightly greater than 1.

Hence, we must optimize the nozzle design to maximize nozzle thrust (accounting effect of exhaust and external flow interaction). Moreover, since the rear portion of the airframe can produce large force and moment contributions that significantly affect overall vehicle stability and trim, the nozzle-aftbody must be designed for maximum stability and low trim drag. This requires careful contouring of the external nozzle surface (aftbody expansion ramp in Figure 8.1) in order to distribute these effects effectively.

For a typical airframe-integrated scramjet, the optimum exhaust area A_{10}, inclusive of both the internal nozzle and the vehicle afterbody, is approximately 1.5–2.0 times the freestream engine capture area A_0. Moreover, the engine will require a very large nozzle area ratio A_{10}/A_4. Thus, the design is constrained by the physical size of the airframe afterbody. As a consequence, the scramjet will operate underexpanded as the nozzle may not be large enough to fully expand the combustor flow to the ambient pressure.

For a hypersonic vehicle such as the NASA X-43A aircraft, the asymmetric two-dimensional (2-D) SERN was an essential component of the scramjet engine. Such design produced the required level of thrust with minimum weight and frictional drag. The 2-D SERN appears to be geometrically simple. However, the design and analysis for this asymmetric nozzle-afterbody section is very challenging, as there are many parameters that must be considered, in addition to those encountered in a conventional supersonic flow nozzle.

8.1.2 Expansion Process

At hypersonic speeds, with the engine operating as a scramjet (supersonic combustion), the nozzle inlet Mach number is supersonic; the nozzle would operate with a minimum length shock-free condition, expanding to freestream ambient pressure. The SERN depicted in Figure 8.2 is perfectly expanded ($p_{10} = p_0$), and the limiting characteristic in the exhaust flow emanating from the trailing edge of the cowl intersects the expansion surface at the trailing edge.

The scramjet nozzle flowfield is predominantly supersonic at high Mach numbers, except in thin subsonic boundary layers adjacent to the nozzle surfaces. Initially separated by the engine cowl, the internal hot gases and external air flow join in and mix through shear layers in the exhaust flow. The cowl and shear layer of the SERN can be thought of as the symmetry plane of a symmetric 2-D nozzle. In addition to the usual fluid and thermodynamic properties which vary across the shear layer, there is a large pressure gradient between the two flows, which causes the shear layer to bend inward or outward, depending on the sign of the pressure gradient between the internal and freestream flows. We must note that, since the underside of the aft body serves as a free expansion surface, the flow

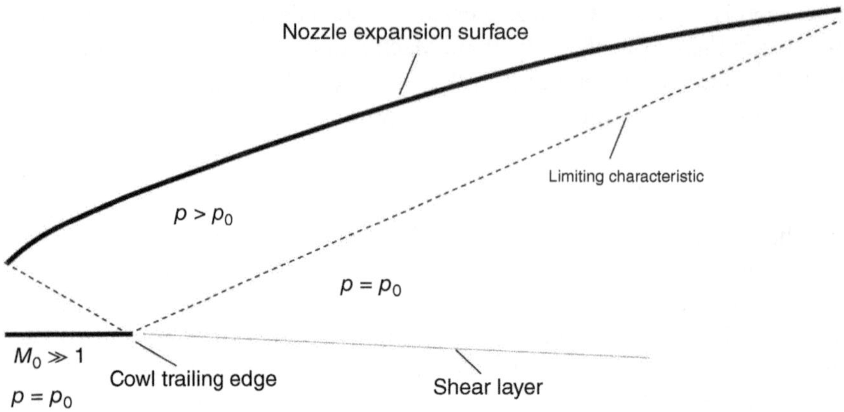

Figure 8.2 SERN ideal expansion at high-speed flight conditions.

is expected to become 3-D in the external part of the expansion process. With perfect expansion ($p_{10} = p_0$), the SERN would give optimum performance at high-speed flight conditions.

Nozzle off-design conditions are characterized by either overexpanded or underexpanded flow at the nozzle exit plane, according to conventional nozzle terminology. In the internal region, the upstream flow at the nozzle throat feeds downstream toward the afterbody surface and undergoes a centered expansion as it reaches the corner of the single ramp. Another expansion occurs as the flow encounters the corner of the nozzle cowl. The wall boundary layers along the cowl surfaces become an expanding shear layer as the flow moves downstream past the nozzle exit plane.

In the asymmetric nozzle with overexpanded flow, as the flow reaches the nozzle exit plane, the static pressure is lower than that of the external freestream pressure ($p_{10} < p_0$), which results in a shock emanating from the tip of the nozzle cowl. Leaving the internal nozzle, the exhaust stream must adjust to the local base pressure through a shock diamond wave structure. The shock wave impinging on the aftbody surface of the airframe makes it a rather undesirable flow condition that would have an adverse effect on the stability and trim of the vehicle.

In the asymmetric nozzle with underexpanded flow, the internal flow continues to expand beyond the nozzle cowl, and the static pressure is higher than the external freestream pressure ($p_{10} > p_0$). In this condition, the upstream internal flow goes through the same two centered expansions at the ramp corner and at the cowl corner, but when the flow reaches the nozzle exit plane, instead of encountering a shock, the flow continues expanding to match the freestream conditions. The interaction between the expanding jet and the external flow produces a lip shock and a contact discontinuity.

It is unlikely that a scramjet nozzle would achieve near perfect expansion at high altitude operation where the local ambient pressure is extremely low. Also, a three-dimensional (3-D) flow field will develop when the internal nozzle hot flow mixes with the freestream cold flow in the lateral direction, and the same type of shocks and expansion waves would occur in the spanwise direction.

The numerical analysis of the flow structure developed by a scramjet asymmetric nozzle requires to consider the interaction between the internal flow and the external freestream.

Mixing shear layers emanating from the engine cowl and the nozzle sidewalls, plus the expansion fans, compression or shock waves (caused by the deflection angles of the shear layers) will affect the overall nozzle flowfield. Numerical simulation should consider a 3-D computational domain that contains the external freestream surrounding the nozzle, in addition to the internal, expanding flow. Moreover, it is important to consider the state of the expanding combustion gases. If the gas pressure drops very rapidly, it will freeze the reaction chemistry before the desired amount of heat has been released. Whether the nozzle is over- or underexpanded, thrust will be less than it would be achieved with the ideal flow expansion.

8.2 Nozzle Geometric Configurations

Engine exhaust systems or thrust nozzles can be classified according to their geometry or according to their application. Considering the wide regime from subsonic to hypersonic flight, we can distinguish nozzles designed for a turboramjet, for a pure ramjet engine, those for dual-mode scramjet propulsion, and nozzles designed for combined cycle propulsion, such as the double nozzle for TBBC engines.

The following are common nozzle geometric configurations for ramjet and scramjet engines. As depicted in Figure 8.3, nozzles for conventional ramjet engines are converging–diverging. This is because the flow entering the nozzle is subsonic, having a Mach number of 0.3 (typical subsonic combustion condition), and thus the flow chokes at the throat (minimum area). Then the flow expands and accelerates to supersonic speed through the divergent section. The physical throat of such nozzle provides backpressure to the ramjet combustor.

The nozzle for a pure scramjet engine (or dual-mode scramjet) does not require a physical choking throat since the flow entering the nozzle is supersonic. Leaving the supersonic combustor, the flow continues to accelerate through the divergent nozzle. The definition of combustor exit/nozzle entrance can vary with operating condition. The flow is not choked in the expansion process. The nozzle for a basic scramjet engine is often configured as SERN.

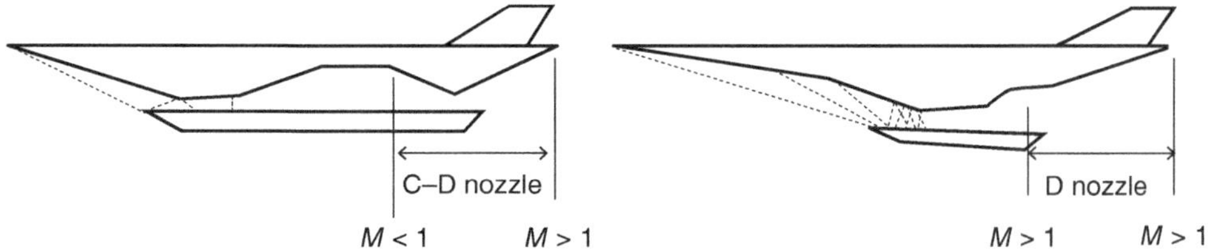

Figure 8.3 Geometric differences between the ramjet and scramjet nozzles. In the pure ramjet, the combustor flow expands through a convergent–divergent nozzle to produce thrust. In the pure scramjet, the combustion flow expands and accelerates through a divergent nozzle.

8.2.1 Two- and Three-Dimensional Nozzles

The expanding flowfield can be 2-D or 3-D, depending on whether the nozzle is integrated with an airframe having flat or round cross section, or perhaps shaped as a wave-rider. Rectangular 2-D nozzles are geometrically simple, integrate best on planar aftbodies, and are amenable to variable geometry. A planar 2-D flowfield can be easily analyzed. Axisymmetric nozzles are widely used in hypersonic missile applications.

The rectangular 2-D nozzles (Figure 8.4a) can be longer than axisymmetric or 3-D nozzles, thus adding weight and skin friction, and may experience performance issues due to regions of vortical flow. Nevertheless, 2-D geometries can adapt to different flight conditions with less effort than 3-D nozzles. The 2-D nozzle in the NASA X-43A vehicle demonstrated successful scramjet propulsion in flight with the rectangular scramjet engine mounted under the vehicle.

A 3-D expanding flowfield is more complicated, but 3-D nozzles can be shorter by their use of lateral expansion as well as medial expansion, leading to shorter length for a given area-ratio A_{10}/A_4, resulting in high expansion efficiency. They integrate well into a wide array of vehicle shapes. However, 3-D nozzles are not amenable to variable geometry, and their design and analysis require more sophisticated tools for performance optimization.

Converging–diverging (C–D) axisymmetric nozzles (Figure 8.4b) combine the simplicity of the 2-D flowfield with the gains of lateral flow expansion. Unlike a 2-D rectangular nozzle, an axisymmetric nozzle does not have corners, thus impeding the formation of vortices, which results in a higher quality expanded flow. Axisymmetric nozzles are best for physically throated ramjet engines. Although they provide high efficiency in the shortest length, these nozzles are not amenable to variable geometry configurations.

Baidya et al. (2018) reported on a method to optimize ramjet exhaust expansion to produce sufficient thrust to propel the vehicle into altitudes and Mach regimes where scramjet operation can be initiated. Using a CFD parametric analysis, three candidate C–D designs

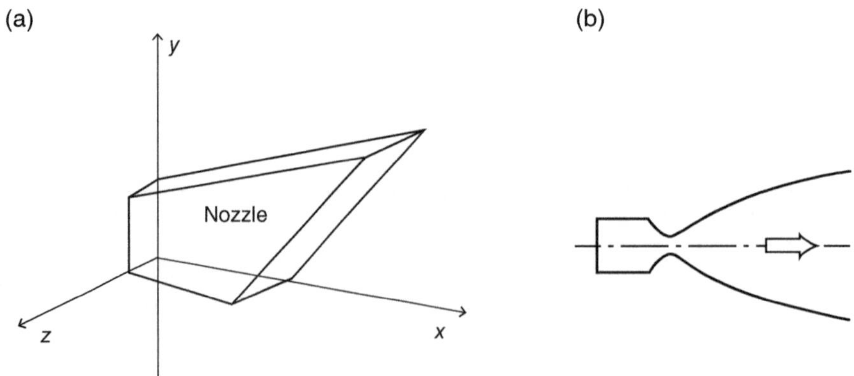

Figure 8.4 Rectangular (a), and C–D axisymmetric (b) nozzle geometries.

were evaluated: conical, bell, and dual-bell nozzles. Of these, the dual-bell nozzle design produced the highest thrust when compared to the other nozzle geometries.

8.2.2 Single Expansion Ramp Nozzle (SERN)

The SERN is a variable, asymmetric nozzle with only an upper solid surface upon which the engine combustor flow expands. As shown in Figure 8.5, the SERN is composed of the following pieces: an internal upstream plug, an expansion ramp, two sidewalls, and a cowl. The upstream plug, in conjunction with the cowl and sidewalls, provided a specific internal nozzle contour (convergent or converging–diverging). The expansion ramp surface provides a specific internal-to-external nozzle contour. Geometrically, the expansion ramp can be straight or curved (with initial angle β and trailing angle θ), while the cowl extending past the nozzle throat can be short, or long, its trailing edge straight or upturned. The sidewalls may terminate at the cowl trailing edge, or extend along the sides of the expansion ramp. The geometry and dimensions of those four sections determine the nozzle internal expansion ratio (A_{ex}/A_t) and the performance of the nozzle at a given set of flight conditions.

Internal expansion process occurs between nozzle throat and trailing edge of the cowl, whereas external expansion process occurs along the vehicle's afterbody surface or expansion ramp. Thus, the SERN maximum propulsive efficiency depends mainly on nozzle pressure ratio and nozzle expansion ratio. Properly designed, the high-performing SERN can be a short section that results in lower weight and less skin friction drag compared with axisymmetric nozzles. And just as in designing traditional symmetric supersonic nozzles, if the shape of the walls of a SERN is not just right, oblique shock waves can occur inside. The proper contour for a supersonic nozzle can be determined from the method of characteristics (MOC).

The SERN adapts well for operation at off-design conditions by managing the changing static pressure over a range of Mach numbers. Its major disadvantage is the poor performance at transonic and supersonic speeds where low-pressure ratios result in a highly overexpanded nozzle, as the exhaust mass flow is insufficient to fill the large area-ratio nozzle designed for high Mach number operation. The shear layer and shock wave system adjust and stabilize such that the pressure in the nozzle plume and external flow equalize. If the freestream flow turns around the aerodynamic boundary formed by the cowl and the shear

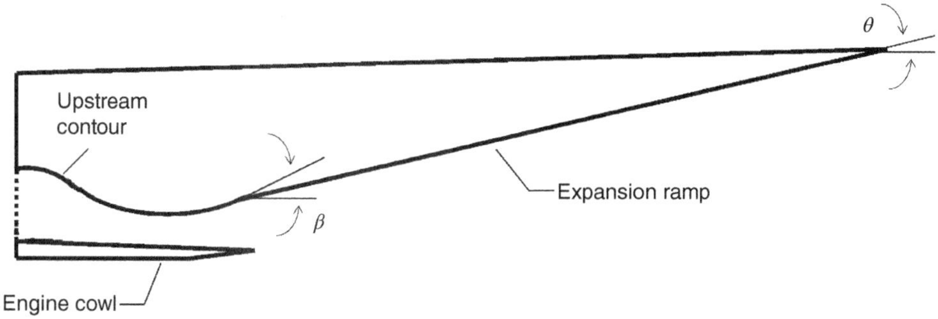

Figure 8.5 Single Expansion Ramp Nozzle.

layer, the resulting pressure acting on the nozzle expansion surface will be less than the atmospheric pressure. This overexpanded flow will yield low thrust. Hence, methods to improve SERN performance must be sought if the design requires fixed geometry for operation over a wide range of flight Mach numbers that includes transonic speeds (see Ridgway et al. 2018; Xu et al. 2017).

8.2.3 Three-Dimensional Elliptical to Rectangular Shape Transitioning Nozzles

Based on the REST inlet design approach by Smart (1999), Mo et al. (2014) developed a circular to rectangular shape transition nozzle, which yielded good thrust and lift characteristics. Then in 2018, Kunze et al. described the design procedure for 3-D elliptical to rectangular shape transition nozzles. Their method, derived from the REST inlet design, captures properties of a modified axisymmetric thrust optimized nozzle in a complex 3-D shape via streamline tracing. As we emphasized in Chapter 5, REST inlets are perfectly adaptable to elliptical combustors and are fully integrated into the body of the vehicle.

According to the nozzle design methodology developed by Kunze et al. (2018), an annular parent flowfield with a uniform inflow and outflow is first designed. Then the desired inflow and outflow cross sections are constructed and discretized through the same number of points, respectively. Each of these points is streamline traced through the parent flowfield to receive two streamline traced nozzles of different geometries. Finally, the two streamline traced nozzles are blended to create the inviscid shape transitioning nozzle. The expanding flowfield is determined by the flow conditions at the exit of the combustor and the area ratio A_{10}/A_4. The area ratio is preset per the geometry of the vehicle or the area of the capture shape and the exit area of the combustor A_4, depending on the vehicle configuration. Clearly, any nozzle should be designed to use the maximum available area for thrust production, without unreasonably increasing the vehicle's drag (Kunze et al. 2018).

8.2.4 Nozzles for Combined Cycle Propulsion Systems

The thrust nozzle for combined cycle propulsion systems will encounter a number of design issues to meet more complex performance requirements, especially during inlet mode transition. For example, for the first-stage vehicle with a TBCC propulsion system, the turbine and ramjet/scramjet flowpaths are in parallel in an over/under configuration. At Mach 3, the air-breathing propulsion transitions from gas turbine to ramjet/scramjet mode (Slater and Saunders 2010).

The low-speed nozzle design depicted in Figure 8.6 is highly integrated within the ramjet/scramjet SERN optimized for high-speed performance. Each turbine engine has its own individual nozzle, which starts with a circular to rectangular transition section. The upper nozzle surface is fixed and contoured. The rotating lower body flap gives a 30% variation for the nozzle throat over its expected Mach 0–3.5 speed range. The short nozzle limits the surface area subjected to the hot turbine exhaust flow, but there is still sufficient length to achieve good flow, variable geometry, and performance qualities (Snyder and Espinosa 2013).

Analyses reported indicated a significant installation penalty from overexpansion losses on the low-speed nozzle exhaust flow, including high base drag from the SERN. The study considered using high-speed inlet and nozzle ramps along with the rest of the high-speed

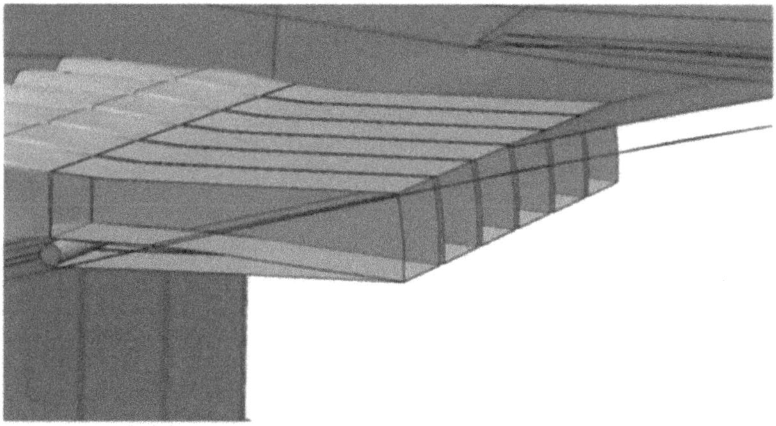

Figure 8.6 Schematic of low-speed (Mach 3.5) nozzle for TBCC air-breathing propulsion system. Source: Snyder and Espinosa (2013)/NASA/Public Domain.

flow path to improve performance. For example, by opening the high-speed inlet midway, introducing additional flow into the aft region could relieve the adverse pressure gradients on the aft surface. This could also change the magnitude and direction of the force on the high-speed inlet ramp. However, the high-speed propulsion design has limited internal flow capability, and opening the high-speed inlet further actually penalizes the forebody without significant aft body force and moment improvements (Snyder and Espinosa 2013).

External burning was considered to reduce adverse forces and for thrust improvement. Results of CFD analysis indicated that overexpansion losses could indeed be mitigated and the aft body pressures increased to near or even above ambient pressure levels. Greatest improvement with external burning with "reasonable" hydrogen flow rates was best at transonic conditions, where overexpansion losses are most severe (Snyder and Espinosa 2013).

8.2.5 Issues Related to Nozzle for Dual-Mode Scramjet

For a dual-mode scramjet engine, the nozzle must operate over a wide range of flow conditions, as the engine transitions from subsonic combustion (ramjet) to supersonic combustion (scramjet), which can be achieved using the no-throat geometry of the scramjet flowpath. Let us recall the ordinary differential equation for 1-D axial variation of Mach number along a frictionless channel, Eq. (8.3). This equation is used to assess the interaction between axial variation of cross-sectional area $A(x)$ and of total temperature $T_t(x)$:

$$\frac{dM}{dx} = M\left(\frac{1 + ((\gamma - 1)/2)M^2}{1 - M^2}\right)\left[-\frac{1}{A}\frac{dA}{dx} + \frac{(1 + \gamma M^2)}{2}\frac{1}{T_t}\frac{dT_t}{dx}\right] \tag{8.3}$$

When $M = 1$, the quantity in square bracket must equal zero; hence, there must be a balance between the area change $A(x)$ and the heat release (dT_t/dx). This condition yields a thermal throat.

When $M > 1$, occluding the flow, either by decreasing A or by increasing T_t, causes M to decrease in the axial direction; relieving flow, either by increasing A or decreasing T_t, will cause M to increase.

In the dual-mode combustion engine, $A(x)$ will not decrease with x, as there is no physical throat, and $T_t(x)$ will not decrease because exothermic combustion can only add heat to the flow. Because very few closed form, integral solutions are known to exist, approximate methods are required to solve Eq. (8.3). For very simple algebraic forms for $A(x)$ and $T_t(x)$, we can use this equation to determine how much area increase $A(x)$ is required to accommodate increases of total temperature $T_t(x)$ in order to control the axial variation of Mach number, pressure, and other thermodynamic properties and flow variables. $T_t(x)$ depends on the rate of combustion heat release, which in turn is determined by the finite-rate processes of mixing and chemical reaction in the combustor.

At transonic speeds, when the engine operates as a ramjet with a choked throat, the flowfield in the nozzle changes. Critical is the condition when the nozzle pressure ratio and freestream Mach number are reduced at the transonic flight condition. The internal nozzle section (enclosed by the cowl) now has a significant area ratio and, at low nozzle pressure ratios, is itself overexpanded to less than freestream pressure ($p_{10} < p_0$). In such condition, the local base pressure (to which the exhaust flow eventually equalizes) would be lower than freestream because the exhaust stream fills only a small part of the large nozzle exit area. Nozzle drag results from flow overexpanding in the internal nozzle and because the exhaust stream cannot fully recompress to freestream pressure due to the low overall vehicle surface base pressure.

Hence, the engine nozzle for a vehicle intended to fly from subsonic to hypersonic velocities will encounter large variations in back pressure over its flight regime. Typically, back pressure variations are handled with a variable area nozzle that adjusts the exit area to changes in back pressure. However, for hypersonic vehicles where nozzle pressure ratios will reach very high values, variable geometry nozzles are impractical; mechanical difficulties limit extreme variations in expansion ratio.

8.3 Nozzle Performance Parameters

To describe the efficiency of the nozzle expansion process, the state variables at the nozzle entrance and exit stations can be combined to obtain several different performance parameters. The definitions of the parameters can be easily visualized using the Mollier diagram shown in Figure 8.7, which indicates the specific static enthalpy and static entropy at each of the endpoints of the thermodynamic processes (or nozzle reference stations). A nozzle pressure ratio ($NPR \equiv p_{t4}/p_0$) of 3000 is representative of Mach 10 flight conditions. An average exit-plane static temperature of approximately 1895 K (3410 °R) is typical of actual scramjet exit temperatures (Tatum and Huebner 1995).

8.3.1 Adiabatic Expansion Efficiency

The adiabatic efficiency of expansion η_n assumes that the exhaust flow is perfectly expanded to the surrounding atmospheric pressure (from the burner static pressure $p_3 = p_4$ to the

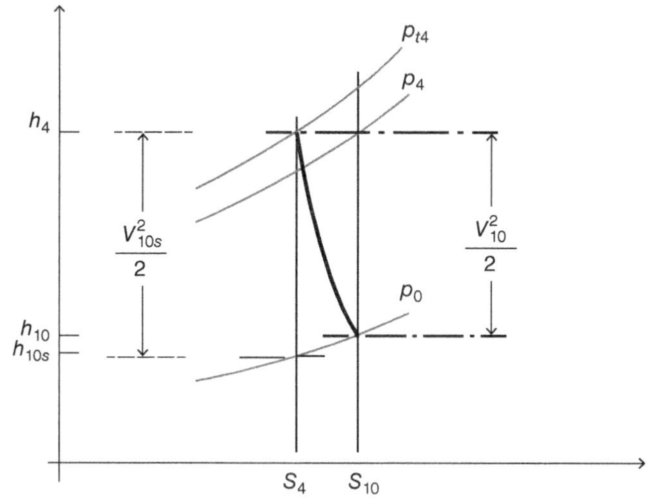

Figure 8.7 Nozzle expansion process represented by enthalpy–entropy diagram.

freestream static pressure $p_{10} = p_0$) and that the expansion process is adiabatic. Although the latter assumption may not be satisfied for the scramjet nozzle of interest, we use η_n to get a measure of the goodness of the expansion process in terms of the actual change in static enthalpy (or temperature) as referenced to the ideal or isentropic change in static enthalpy (or temperature) that would accompany the same change in static pressure. By definition,

$$\eta_n = \frac{h_4 - h_{10}}{h_4 - h_{10s}} \cong \frac{c_{p4}(T_4 - T_{10})}{c_{p4}(T_4 - T_{10s})} = \frac{1 - (T_{10}/T_4)}{1 - (T_{10s}/T_4)} \tag{8.4}$$

where $T_{10s}/T_4 = (p_{10s}/p_4)^{(\gamma_n - 1)/\gamma_n}$. Incorporating this relation and the nozzle static pressure ratio p_0/p_4 helps us rewrite η_n as

$$\eta_n = \frac{1 - [(1/\pi_n) \bullet (p_0/p_4)]^{(\gamma_n - 1)/\gamma_n}}{1 - [p_0/p_4]^{(\gamma_n - 1)/\gamma_n}} \tag{8.5}$$

Here π_n is the total pressure ratio across the nozzle, $\pi_n = p_{t10}/p_{t4}$, or

$$\pi_n = \frac{p_0}{p_4}\left[\frac{1 + ((\gamma_n - 1)/2)M_{10}^2}{1 + ((\gamma_n - 1)/2)M_4^2}\right]^{\gamma_n/(\gamma_n - 1)} = \frac{p_0}{p_4}\left(\frac{T_4}{T_{10}}\right)^{\gamma_n/(\gamma_n - 1)}$$

Ideally, combustion takes place at constant static pressure so that $p_4 = p_3$; thus, π_n is afected by the cycle compression pressure ratio (see Chapter 2):

$$\frac{p_3}{p_0} = \left[\frac{\psi}{\psi(1 - \eta_c) + \eta_c}\right]^{\gamma_c/(\gamma_c - 1)}$$

A typical range for the pressure ratio is 100–1000. For $\psi = 7.0$, $\eta_c = 0.90$, and $\gamma_c = 1.36$, obtain $p_3/p_0 = 264$. This compression pressure ratio would be higher if the combustion takes place in a constant area combustor.

8.3.2 Nozzle Velocity Coefficient

The velocity coefficient is defined as the *actual* average total velocity at the exit of the nozzle (V_{10}) divided by the *ideal* total velocity, V_{10s}, or the velocity that would develop if the expansion process were isentropic. This velocity ratio is a function of the adiabatic expansion efficiency η_n and the velocity ratio that would exist across the ideal nozzle, V_4/V_{10s}:

$$C_{ev} = \frac{V_{10}}{V_{10s}} = \sqrt{\eta_n + (1-\eta_n)\left(\frac{V_4}{V_{10s}}\right)^2} \tag{8.6}$$

where the velocity ratio V_4/V_{10s} is a function of the freestream Mach number.

From a simple scramjet cycle calculation using nominal component efficiency values, Heiser and Pratt (1994) obtained $V_4/V_{10s} = 0.55$ for $M_0 = 8.0$ and $V_4/V_{10s} = 0.70$ for $M_0 = 10.0$, values that are rather conservative but provide a reference for an estimate of the nozzle velocity coefficient.

In all cases, the nozzle velocity coefficient is both sensitive to the flight conditions, and especially to the adiabatic efficiency of the nozzle, i.e. $C_{ev} \to 1$, as $\eta_n \to 1$.

The effect of friction in performance analyses may be expressed by a group parameter that modifies the velocity ratio, e.g.

$$\frac{V_4}{V_{10}} \approx 1 + \frac{c_f}{2}\frac{A_w}{A_{10}}$$

where c_f is a friction coefficient, and A_w/A_{10} is the surface-to-throughflow area ratio. The expansion fan in the nozzle reduces the magnitude of A_w/A_{10} for the entire expansion process.

8.3.3 Entropy Increase

The expansion process in the nozzle is irreversible, even if we assume it is adiabatic. Losses such as skin friction in the nozzle, flow dissociation, and shock waves that develop during the expansion cause the entropy of the process to increase from the combustor exit value s_4–s_{10} at the end of the expansion. Since entropy increase is the natural indicator of the effects of flow losses, it is customary to give an entropy increase term as a nozzle performance measure. Heiser and Pratt obtain the following expression for a dimensionless entropy increase (entropy change across the nozzle divided by the specific heat of the air at constant pressure) in terms of the adiabatic expansion efficiency:

$$\frac{s_{10} - s_4}{c_{p_n}} = ln\left[\eta_n + (1-\eta_n)\left(\frac{p_4}{p_0}\right)^{(\gamma_n - 1)/\gamma_n}\right] \tag{8.7}$$

where typical values of the static pressure ratio p_4/p_0 can range from 100 to 1000.

Since both the dimensionless entropy increase and the adiabatic expansion efficiency (performance measures that we use interchangeably) are referenced to freestream properties, their results are insensitive to freestream Mach number. However, both parameters are indicative of the irreversibilities of the expansion process, losses that the nozzle designer must aim to minimize in order to optimize the overall propulsion system (see Section 8.4).

8.3.4 Nozzle Efficiency or Gross Thrust Coefficient

The nozzle cannot realize the full isentropic potential because friction and other losses use some of the internal energy that might have gone into increasing the kinetic energy of the gas (velocity). Both underexpansion and overexpansion reduce the nozzle's ability to produce the maximum or theoretical amount of thrust. Therefore, the efficiency of the nozzle can also be determined by the *gross thrust coefficient*, C_{fg}. This performance parameter is given by the ratio of the actual thrust produced by the nozzle to the theoretical thrust that would be produced if the flow through the nozzle were expanded isentropically to the ambient pressure. This parameter gives a measure of how well a real exhaust nozzle performs as related to an ideal isentropic nozzle.

Using the notation of the nozzle sketched in Figure 8.8, we define the thrust coefficient as

$$C_{fg} = \frac{(\dot{m}v + pA)_{in} + \int pdA - p_0 A_{ex}}{(\dot{m}v + pA)_{in} + \int p_0 dA_0 - p_0 A_0} \tag{8.8}$$

where $\int pdA$ denotes the axial pressure force generated by the nozzle, and $\int p_0 dA_0$ stands for the theoretical axial pressure force generated if the nozzle flow were expanded isentropically to ambient pressure p_0 (Goel et al. 1990).

The value of the denominator can be simplified as follows. Since

$$\int p_0 dA_0 = (\dot{m}_0 v_0 + p_0 A_0) - (\dot{m}v + pA)_{in}$$

the denominator becomes

$$(\dot{m}v + pA)_{in} + \int p_0 dA_0 - p_0 A_0 = \dot{m}_0 v_0 = \rho_0 v_0^2 A_0.$$

Furthermore, since

$$v_0^2 = M_0^2 a_0^2 = \left(M_0 \sqrt{\frac{\gamma p_0}{\rho_0}} \right)^2$$

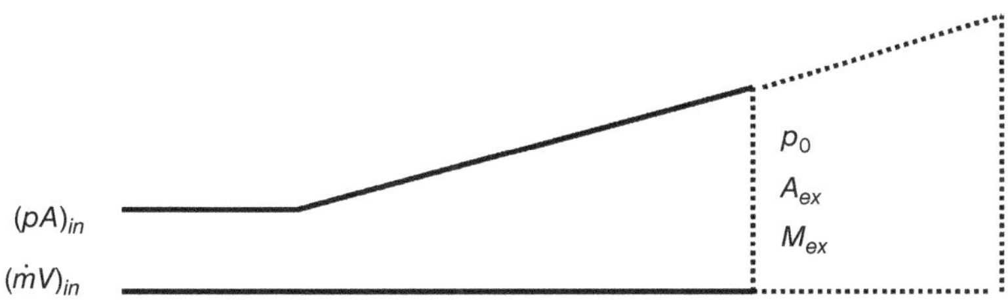

Figure 8.8 Notation for nozzle efficiency.

the denominator is written as

$$\rho_0 v_0^2 A_0 = M_0^2 A_0 \gamma p_0$$

Therefore, the nozzle efficiency is given in the form of the gross thrust coefficient as

$$C_{fg} = \frac{(\dot{m}v + pA)_{in} + \int pdA - p_0 A_{ex}}{M_0^2 A_0 \gamma p_0} \tag{8.9}$$

where the value of the numerator is obtained directly from computation.

A well-designed nozzle would have a value of C_{fg} of about 0.96, and 0.97 represents a practical upper limit (Anderson et al. 2000). This nozzle efficiency is also known as gross thrust coefficient, which is consistent with terminology used for low speed, conventional air-breathing propulsion systems.

Another parameter used to evaluate the performance of a scramjet nozzle is the net thrust coefficient. While developing hypersonic technologies at the University of Queensland in Australia, Stalker et al. (2005) published an expression for the net thrust coefficient C_{TN} for a scramjet, equating the combustion energy release ΔQ to the kinetic energy gain of the flow through the engine, using a momentum balance that neglected the pressure difference between capture and exit. Their original expression for the thrust coefficient is

$$C_{TN} = \frac{2T_N}{\rho_0 V_0^2 A_0} = \left(\frac{2\Delta Q}{V_0^2}\right)\left(1 - \frac{0.5\Delta Q}{V_0^2}\right) - C_D \tag{8.10}$$

where V_0 is the freestream velocity (m/s), ΔQ is theheat released by fuel combustion in air (=3.45 MJ/kg for stoichiometric hydrogen), and C_D is the drag coefficient $= 2D/\left(\rho_0 V_0^2 A_0\right)$.

During acceleration, the net thrust coefficient can be viewed as the product with the flight velocity, which is proportional to the net fuel specific impulse, and thus C_{TN} is a measure of the efficiency with which the engine uses fuel to acquire vehicle velocity; C_{TN} is equal to zero for the vehicle in the cruise phase. Equation (8.10) implies that the drag force must be minimum in order to maximize the thrust. This requirement becomes more difficult as the flight velocity increases.

At hypersonic speeds, the efficient generation of thrust and the aerodynamic balance of the vehicle are critical factors because the nozzle must work on the total flow through the engine, and the net propulsive thrust is just a small difference between two large numbers, i.e. the nozzle gross thrust and the ram drag of the stream tube entering the inlet. The size and weight of the nozzle/aftbody must be kept to a reasonable size. These and other design requirements yield a nozzle that is not large enough to fully expand the flow to ambient pressure. The result is underexpansion losses that reduce the nozzle gross thrust values.

Moreover, in the integrated hypersonic vehicle, the flow expansion is done over a large portion of the vehicle afterbody, and the shape of the afterbody gives a direction to the propulsion flow, which in turn establishes the angle of the gross thrust vector relative to the vehicle's flight direction. Hence, the efficient generation of thrust and the aerodynamic balance of the vehicle must be given primary consideration in the design of the hypersonic nozzle. The nozzle thrust direction dominates the trim of the vehicle at hypersonic speeds. The impact of these factors on the design of the combined vehicle and propulsion system is crucial for achieving optimal sustained hypersonic flight.

8.4 Nozzle Flow Losses

In conventional nozzles, the forces and moments generated can be determined by analyzing the flow up to the nozzle exit plane only. However, in the asymmetric SERN with an internal and an external expansion portion, the analysis must extend further downstream since the lower aft portion of the vehicle forms the external portion of the nozzle. Moreover, the flow characteristics over this afterbody region have a dramatic effect on the thrust vector and pitching moment generated by the scramjet nozzle.

The expansion process in hypersonic air-breathing nozzles involves a number of important viscous effects that require accurate prediction of the wall skin friction and boundary-layer thickness, shape, transition, and separation. The analysis is complicated by a number of factors such as continuing chemical reaction in the expansion section, flow separation, heat transfer, and flow unsteadiness. As we have noted, the nozzle has to expand the hot, high pressure gas mixture leaving the combustor into a high velocity exhaust with greater momentum than the captured airflow in order to generate net thrust.

Other losses inherent in the expansion process that we must consider in analysis include the flow profile at the nozzle entrance, low recombination of dissociated species, skin friction, and flow divergence. Exit flow divergence result when the streamlines at the nozzle exit plane are at different angles relative to that of the flight path. Internal flow divergence is a result of the nozzle shape and 3-D flow features.

Ideally, the nozzle entrance flow is uniform. In reality, the flowfield developed at the entrance may have large asymmetric components normal to the axial direction. Such flow profiles may be the result of boundary-layer growth from the body side and cowl surfaces, shock waves that were not canceled, and fuel injection, mixing, flameholding, and combustion process. Researchers have considered altering the flow profile with speed by changing the way fuel injection is proportioned between body side and cowl side injectors, aiming to change the gross thrust vector angle and ultimately improve the aerodynamic balance of the vehicle over a range of Mach number.

In preliminary, parametric performance analysis of a nozzle, we typically assume exhaust flow to maintain equilibrium to the nozzle exit station. In reality, freezing of the hot gas exhaust flow tends to occur just downstream of the combustor exit plane, and this results in significant performance losses. Ideally, an equilibrium nozzle expansion process allows the dissociated species in the hot combustion gas to recombine and recover the dissociation energy. However, in reality, the dissociation energy is lost from the propulsion cycle, and, consequently, the net thrust is reduced. This is mainly due to the requirement for rapid nozzle area expansion in the initial nozzle contour, which generates a correspondingly rapid decrease in static pressure that results in a frozen chemical process (Anderson et al. 2000).

This explain why dissociation is not a significant problem at low speeds (below Mach 4) since dissociation within the combustor is a strong function of static temperature and pressure, and, at low speeds, pressure tends to be high and static temperature is relatively low (compared with the gas temperature in supersonic combustion). Therefore, design features and performance levels within the air-breathing engine that affect static temperature at high Mach numbers will affect dissociation and the potential for losses through freezing the flow in the nozzle expansion process.

Moreover, dissociation losses are maximum when the combustor burns fuel–air at stoichiometric conditions ($\phi = 1$), which correspond to the highest combustor-exit temperatures. To minimize such loss, it is necessary to limit the static temperature, deviate from stoichiometric conditions to reduce dissociation at the combustor exit, which can be done through combustor divergence, for example, to drive the process to equilibrium.

In addition, the losses are very sensitive to the engine inlet design parameters. For example, a high contraction ratio is desirable at high Mach numbers to increase static pressure and temperature for rapid ignition and combustion. However, such high contraction leads to more kinetic freezing effects. Such losses result in a reduction in specific impulse. Hence, the design of the inlet must aim for an optimized contraction ratio and increased inlet kinetic efficiency.

Wall friction within the nozzle affects its gross thrust coefficient C_{fg}. It is clear that the larger amount of friction is generated in the first half of the nozzle where pressures are higher. This is where the behavior of the boundary layer must be well characterized since it controls friction and heat transfer, i.e. a laminar boundary layer significantly reduces friction drag and heat transfer. Therefore, efforts should be made to relaminarize the boundary layer just downstream of the nozzle throat. This flow laminarization could be achieved by cooling the nozzle walls.

Anderson et al. (2000) considered the potential for losses in the gross thrust coefficient, C_{fg} caused by three effects: dissociated flow, inlet contraction ratio, and inlet kinetic energy efficiency for a Mach 14 flight condition and, as they noted, the losses are particularly sensitive to the engine inlet design parameters. Although a high contraction ratio is desirable at high Mach numbers (to increase static pressure and temperature for rapid ignition and combustion), the contraction must be limited to avoid high losses caused by kinetic freezing affects. Based on analysis of finite-rate chemical kinetics of the expansion process leads to the conclusion that the flow is very close to a frozen process. Therefore, efforts to minimize this loss must rely on revising component design parameters to limit static temperature and dissociation at the combustor exit. Another approach is to catalyze the reactions in the nozzle flow and drive the expansion process further toward equilibrium (Anderson et al. 2000).

8.5 SERN Design Approach

It is essential to have accurate definition of entry flow for designing the exhaust nozzle during preliminary and final design of hypersonic propulsion systems. Nozzle design is mainly controlled by thrust and stability requirements. While designing the nozzle section, propulsion parameters across the entire aircraft flight envelope must be considered.

Traditionally, a nozzle contour can be created with the MOC, which is a numerical procedure appropriate for solving, among other things, 2-D compressible flow problems. By using this technique, the nozzle flow properties, such as direction and velocity, can be calculated at distinct points throughout the flow field.

In the late 1950s, Rao developed the maximum thrust theory to obtain the optimal thrust of a rocket nozzle with a fixed length and mass flow rate. This solution (Rao 1958), which he sought to increase the performance of the axisymmetric nozzle, has widely been used to design rocket nozzles. Rao's approach can also be used as a reference for the design of 2-D SERNs.

The nozzle must be designed for optimal system performance across the entire operating range of the propulsion system. Flowline generation usually involves MOC or Rao contours and boundary layer (BL) theory. The MOC provides a technique for properly designing the contour of a supersonic nozzle for shock free, isentropic flow, taking into account the multidimensional flow inside the duct.

In traditional nozzle design using MOC, the uniform, parallel flow at the desired exit Mach number maximizes the resulting thrust (Yu et al. 2019). However, the resulting nozzle length may be unacceptable for a hypersonic propulsion system. To decrease the length of the scramjet nozzle, two approaches have been considered: using a minimum length nozzle and introducing an asymmetry factor to increase the expansion on the cowl. Although these methods may achieve promising results, the lengths may still be unacceptable for highly integrated hypersonic vehicles.

Other studies have attempted to reduce the length of a SERN by compressing or truncating a perfect nozzle. Hoffman (1987) investigated the performance of linearly compressed perfect nozzles and obtained an adequate propulsive nozzle. Shyne and Keith (1990) also considered a truncation method based on the balance of thrust and friction to reduce the length and weight of the nozzle. Other studies have been done to optimize SERN configurations to match specific geometric constraints (Gruhn et al. 2002; Huang et al. 2013).

In 2011, Riehmer and Gülhan presented a method to find the optimal shape of a 3-D supersonic nozzle for a rectangular scramjet combustion chamber with rounded edges, taking into account the skin friction effects. The nozzle inlet conditions were mean values of the combustor exit conditions. The shape of the nozzle was generated with a streamline tracing technique applied to an axisymmetric flow field determined by MOC. The skin friction was calculated using the reference temperature method (RTM). The RTM was developed by Rubesin and Johnson (1956) to determine the heat flux and skin friction on an isothermal wall with a simple engineering approximation for hypersonic flow over a flat plate. According to Riehmer and Gülhan (2011), integrating RTM into a streamline tracing tool leads to a very fast method for calculating thrust and heat flux for different nozzle contours. Their results yielded differences between the design method and the 3D numerical calculations.

Several 3-D asymmetric scramjet nozzles were developed using streamline tracing (Lu et al. 2009; Mo et al. 2014) and hybrid optimization strategies (Xing and Damodaran 2004). Compared to the 2-D SERN, the 3-D asymmetric nozzle appears to be better in some aspects, but the complex contour of these 3-D nozzles may not be amenable to variable geometry, such as that required for the exhaust system of turbine-based combined cycle (TBCC) engine with the wide range of operating flight Mach numbers, or for hypersonic flight demonstrator aircraft (e.g. X-43A), for which the 2-D SERN is much more attractive.

Previous efforts to trim ideal nozzles via compressing and truncating methods did not consider the performance of the SERN simultaneously. Optimization methods can yield optimal SERN performance under geometric constraints, but they may require long computational time.

In 2017, Lv et al. presented a new method based on maximum thrust theory to design a 2-D SERN with geometric constraints. To generate the contour of the nozzle, they calculated the inviscid flowfield with MOC and applied the reference temperature method to correct the boundary-layer thickness. The nozzle designed by this method showed increases in the axial thrust coefficient, lift, and pitching moment at the design point (Lv et al. 2017).

Today, computational fluid dynamics (CFD) provides the most powerful tools that researchers and industry are undertaking. CFD codes provide robust high-fidelity methods for performance analysis that can take into account (i) nonuniform entrance conditions, (ii) vehicle/nozzle interactions, (iii) nonequilibrium thermochemical effects, and (iv) base drag in under- and overexpanded flowfields.

The preliminary design of the scramjet nozzle/afterbody section using CFD analysis is economical and effective. A number of CFD codes have capability of handling arbitrary free stream conditions, so that effects of different flight conditions can be considered. However, since there is very little experimental data for such type of expanding flows at the flight conditions of interest, calibrating the CFD codes and validating their results remain challenging.

Assuming uniform, supersonic nozzle entry flow is useful for preliminary performance analysis and conducting nozzle conceptual design. However, the flow is nonuniform. The nozzle designer must know the state of the flow at the exit plane of the combustor, which can be obtained either by CFD analysis or by measurements in a wind tunnel, as nonuniformities will affect nozzle performance.

From the design perspective, the nozzle and engine forebody must be designed to minimize the external drag of the vehicle in order to produce net thrust. This requires to minimize the size of the external surfaces. From the performance perspective, the actual nozzle performance must be cast as a sum (or product) of losses: expansion, viscous, divergence, and chemistry. The designer must understand the relative magnitudes of losses to focus the design and technology efforts on the highest priorities for the overall program.

8.6 Nozzle Ground Testing Issues

Ground testing a scramjet nozzle-afterbody section using the actual engine combustion products is challenging if not impossible in a conventional wind tunnel. This is because the actual chemistry composition of the exhaust gas and its high total enthalpy values are very difficult to duplicate and match in a small-scale test. Hence, alternative approaches to ground testing the design of scramjet nozzles have been developed. One such approach uses a simulant gas to substitute for the actual combustion products and pass the simulant gas through scale models of the actual nozzle/forebody section.

Oman et al. (1975) developed a method to use a noncombusting, thermally perfect gas at cold temperatures to simulate the gas dynamic effects caused by air/hydrogen combustion products in a hot exhaust external nozzle flow field. Since the cold simulant gas has a temperature much less than that of a hot combustion gas, the test gas requires to match the ratio of specific heat γ, the Mach number, and the static pressure ratio throughout the exhaust flow field. Oman et al. used a mixture of Freon-12 and argon (Ar) to achieve a value for γ close to that obtained from actual combustion.

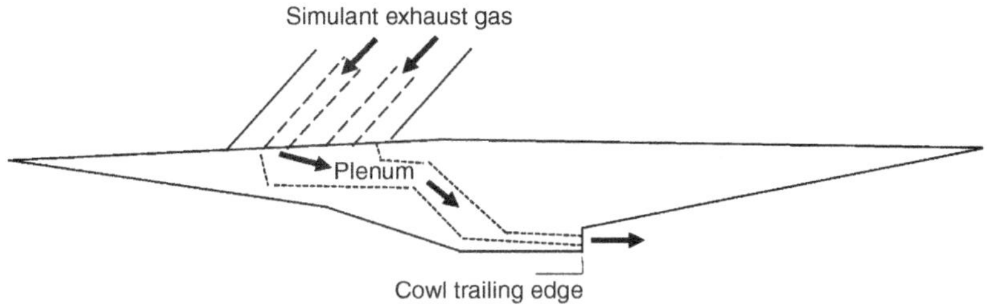

Figure 8.9 Representation of ground test/small-scale model showing exhaust simulation flowpath.

In the 1990s, NASA researchers developed the test technique demonstrator (TTD), a generic NASP-like configuration designed primarily for investigating wind tunnel test techniques for hypersonic air-breathing vehicles. Two small-scale wind tunnel models approximately 1.5 and 3 ft (0.457 and 0.9144 m) long were fabricated. As illustrated in Figure 8.9, the simulant exhaust gas was routed to the vehicle model through the support strut into a plenum chamber and exhausted out through an internal nozzle designed to yield specified flow conditions at the exit. Testing at NASA LaRC was carried with cold air and with a 70/30% mole fraction mixture of tetrafluoromethane (CF_4) and Ar at cold temperatures. The internal nozzle was contained within the engine module as part of the forebody, and the nozzle exit plane (engine cowl trailing edge) defined the downstream end of the forebody.

Computational analysis of this configuration with simulated scramjet exhaust was also performed for CFD code calibration and validation and to get more insight into the flow physics associated with the nozzle/aftbody expansion process. Using the NASA General Aerodynamic Simulation Program (GASP), the CFD analysis also studied powered effects caused by changing angle of attack and effects of inlet representation on aftbody flows. It also examined the differences in aftbody performance and plume flow field definition resulting from various degrees of complexity of exhaust gas modeling using four exhaust chemistry models.

Since each exhaust gas chemistry model had different flow expansion characteristics, four different internal nozzles were required to provide average cowl exit-plane Mach numbers of approximately 3.4 for the appropriate exhaust gas, with an average exit-plane static temperature of about 3410 °R (1895 K), typical of actual scramjet exit temperatures. The exit plane geometry was held fixed (nozzle length was allowed to vary only 0.2 in). To impede recirculation on the cowl, the nozzle walls had sharp trailing edges, thus allowing supersonic space-marching CFD solutions (Tatum and Huebner 1995).

This research concluded that a cold CF_4-Ar mixture provides a much better representation of the hot combustion products exhaust as compared with cold or hot air simulant. This is because the ratio of specific heats γ of the cold CF_4-Ar mixture at the nozzle exit more closely matched that of hot scramjet exhaust. The lift, thrust, and pitching moment components were simulated within approximately 5–6% accuracy by the cold CF_4–Ar exhaust, while the air exhaust accuracy was inconsistent: good for lift, poor for thrust, and moderate for pitching moment.

8.7 Special Topics for Further Research

Regardless of its configuration or application, the exhaust or thrust nozzle of the scramjet engine controls the expansion of the exhaust gas from combustor-exit static pressure p_4 to ambient pressure p_0, requiring optimal performance through the operation of the propulsion system. Integrated airframe/propulsion aircraft requires that the nozzle use the vehicle after-body as the expansion surface. In turn, the expansion surface also helps to stabilize the vehicle by generating lift and strong nose-down pitching moments over all flight conditions.

Optimization of the nozzle/afterbody surfaces is required to simultaneously achieve the highest net thrust with proper vehicle trim characteristics and low exhaust system weight. Hence, optimization methods will continue to be refined as there are numerous trades to optimize the overall propulsion system performance that account for diverse type of losses. There are many issues that remain to be addressed, or topics that are not well understood regarding the aerothermodynamics, performance, and operation of thrust nozzles. The following are some of the topics that deserve further consideration.

8.7.1 Flow Separation

The nozzle may experience flow separation when the overexpanded nozzle wall static pressure at the wall p_w is sufficiently below ambient pressure ($p_w < p_0$). Flow separation occurs when the momentum of the expanding flow within a boundary-layer region becomes too small to overcome a positive (adverse) pressure gradient. This causes the flow to break away from the boundary layer, which in turn results in a backflow or reversed flow region. In 2-D flow, the point of separation can be defined as the point where the velocity gradient near the surface in the direction normal to the boundary becomes zero. The nozzle contour will affect the pressure ratio p_w/p_0 where separation occurs. External burning (Section 8.7.5) can be used to relieve overexpansion.

Flow separation in scramjet nozzles can be investigated via CFD. However, numerical simulation of flow separation is difficult because turbulence models may be unable to predict correctly flow separating away from the aftbody walls, especially when considering the state of the gas as it continues to reach thermochemical equilibrium.

8.7.2 Relaminarization

Studies on laminarization, or reverse transition of turbulent flow, must be refined to assess the potential of reducing nozzle wall heating and friction. Laminarization or relaminarization is the result of a strong flow acceleration on a turbulent boundary layer, transitioning to a boundary layer that contributes to a transport of heat. Laminarization can be described by a greatly increased laminar sublayer region with a significant decrease in fully turbulent flow in the wall region. A rule of thumb for laminarization of accelerating flows is that it happens when the streamwise acceleration parameter K exceeds 10^{-5}, that is,

$$K = \frac{\mu}{\rho V^2} \bullet \frac{\Delta V}{\Delta L} > 10^{-5}$$

where μ is the flow dynamic viscosity, ΔL is the maximum allowable acceleration length, and V is the velocity.

Rapid flow acceleration at nozzle corners can cause laminarization or relaminarization of the boundary layer, which will reduce the friction coefficient C_f of a nozzle surface. Experiments and computational studies have been conducted to evaluate the possibility of the boundary layer relaminarizing just downstream of the nozzle throat (Baker et al. 1993; Jentink 1993). A combined experimental and computational effort with cooled and uncooled nozzles that includes heat transfer measurements will reveal the complex flowfield and its interrelationship with the aerothermodynamic environment when laminarization occurs. Such concentrated research will establish the levels of heat transfer and friction reductions that are possible to achieve on the large aftbody/nozzle section of the hypersonic vehicle.

8.7.3 Aft-Body Performance at Transonic Speeds

High transonic drag is another issue related to the large aftbody/nozzle performance that requires further study. As we noted earlier, the nozzle must be designed for a very high-pressure ratio to operate at hypersonic flight conditions, but this results in a highly over-expanded nozzle at transonic speeds. NASA conducted extensive research to characterize SERN configurations. Anderson et al. (2000) reported results from nozzle tests conducted in the 16-foot transonic tunnel at LaRC using aftbody models with different expansion ramp angles. Data showed pressure dropping below atmospheric pressure immediately downstream of the cowl exit, and then it recovered to atmospheric pressure at about half of the nozzle length. The downstream pressure recovery results from the overexpansion shock originating from the cowl lip as well as outside airflow filling in the large base area. As shown in Figure 8.10, the measured drag coefficient maximizes at Mach 1.2, and the value depends on nozzle-afterbody expansion angle (Anderson et al. 2000). External burning, a process which is used for thrust augmentation, has shown promise for reducing transonic base drag. This is another topic that deserves further study.

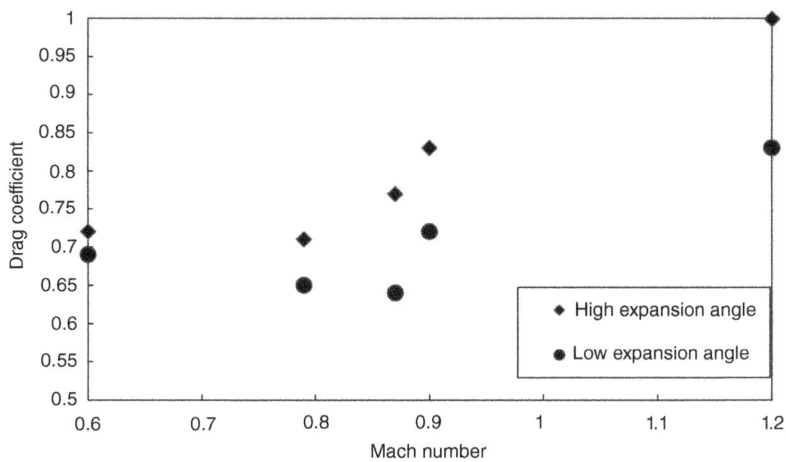

Figure 8.10 Aft-body performance at transonic speeds. Source: Adapted from Anderson et al. (2000).

8.7.4 Variable Area Nozzle

The type of nozzle and its geometry must be tailored to the specific vehicle integration and mission requirements. All hypersonic vehicles tested to date have achieved satisfactory performance with a fixed geometry at the on-design conditions. However, it is crucial for future hypersonic air-breathing vehicle design to determine the best nozzle geometry that can perform optimally at off-design conditions as well.

Although the SERN designed for a fixed design point (fixed expansion ratio) performs very well over a range of conditions as a result of internal and external expansion process, the SERN (like most fixed geometry nozzles) experiences significant performance losses at far off-design conditions due to the changing expansion ratio requirements. As we found earlier, for maximum performance at high speeds and altitudes, a SERN is designed with a large expansion ratio. However, at low flight conditions, the expansion ratio is too large to maintain attached, fully expanded flow along the entire length of the expansion ramp. Thus, the hot gas flow overexpands and separates from the expansion ramp surface. As a result of boundary-layer separation, vortical flow may roll over the sidewalls, creating low-pressure regions along the expansion ramp. Such flow conditions in the nozzle result in decreased thrust, increased afterbody pressure drag, and increased vehicle trim requirements to lessen large moments that may result along the vehicle's afterbody surfaces.

Methods to improve SERN off-design performance are sought. For example, using a translating throat concept was considered by NASA researchers (Deere and Asbury 1999). By translating the throat axial location (with actuated doors integrated into the afterbody), the result is a nozzle with a variable expansion ratio with a changing exit area. This design allows for a more optimum exhaust expansion at various flight conditions.

8.7.5 External Burning

External burning is a method pursued for thrust augmentation. In this concept, an additional amount of oxidant gas is injected into the expanding exhaust gas in the nozzle that contains a significant amount of unburned fuel. The idea of afterburning or external burning has the potential to significantly increase the thrust produced by the nozzle while also maintaining an ideal nozzle expansion ratio ($p_{10} = p_0$) by decreasing the injection pressure of liquid oxygen as flight altitude increases (Candon et al. 2015). After burning is illustrated in Figure 8.11, where a jet of liquid oxygen is injected via the nozzle wall.

The researchers (Candon et al. 2015) conducted a numerical study of Mach 8 scramjet after burning through the injection of liquid oxygen into the axisymmetric nozzle and found a maximum thrust augmentation of 300%. Their computed flowfields for Mach number and hydrogen mass fraction are presented in Figure 8.12. The Mach number contours show considerable effects of the injected oxygen on the crossflow. As the authors indicated, a prominent bow shock forms due to interactions between the fuel jet and crossflow, followed by expansion in the separated region downstream. Shock reflection of the fuel jet on the symmetry axis is also noted at approximately $x = 1$ m followed by an impingement on the nozzle wall between $x = 1.1$ and 1.2 m causing further expansion.

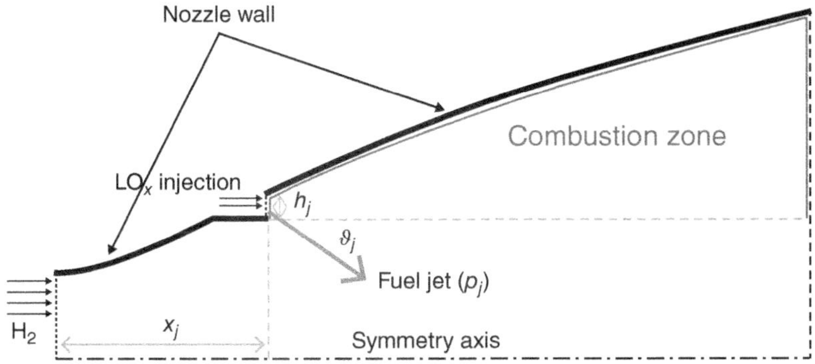

Figure 8.11 External burning by injecting liquid oxygen into expanding flow for thrust augmentation. Source: Candon et al. (2015)/Korea Institute of Science and Technology Information.

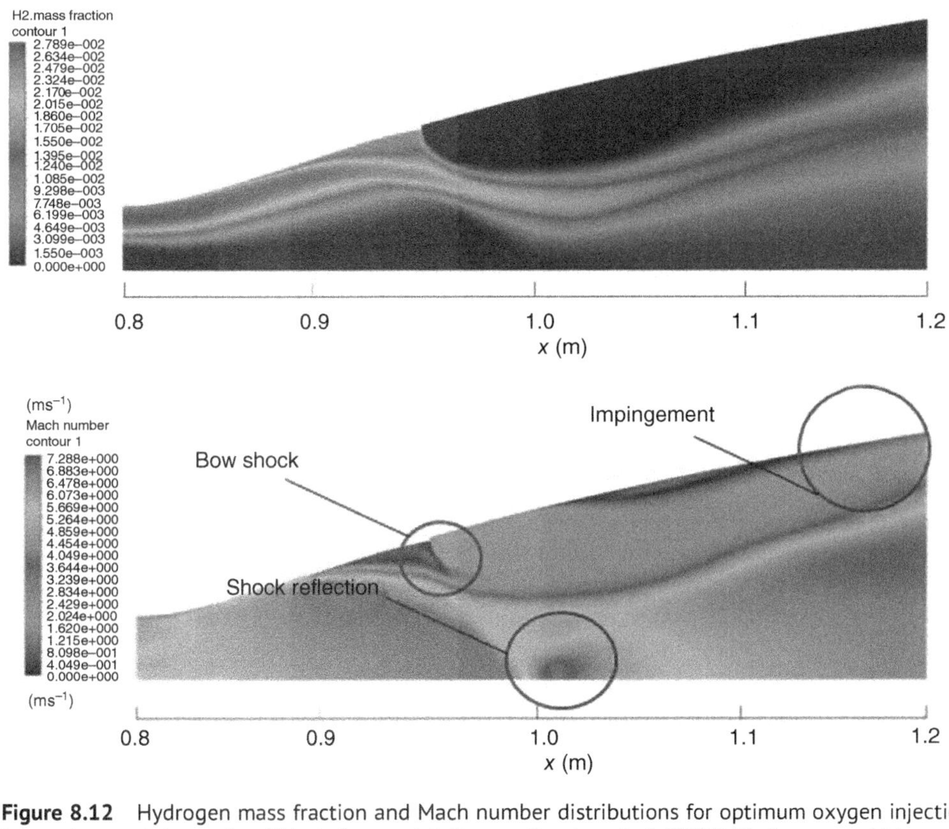

Figure 8.12 Hydrogen mass fraction and Mach number distributions for optimum oxygen injection into axisymmetric nozzle of Mach 8 scramjet. Source: Candon et al. (2015)/Techno-press Ltd.

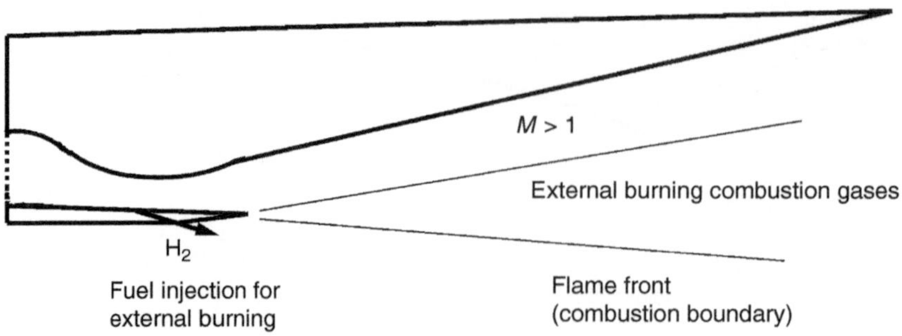

Figure 8.13 External burning with hydrogen fuel injected into expanding flow for improving SERN performance at transonic speeds.

External burning has also been studied as effective method for improving SERN performance at transonic speeds. This requires to inject fuel into the external nozzle flow via the engine cowl, as depicted in Figure 8.13. The result of mixing the fuel with the hot exhaust gases is additional combustion, which in turn pressurizes the entire expansion ramp surface and the cowl trailing edge. The effectiveness of external burning for improving the performance of the SERN depends on the controlled combustion of fuel and air at the freestream conditions.

The CFD study of the afterburning SERN by Yungster and Trefny (1994) showed that the external burning process create a large subsonic flow region that permits higher back pressure levels to be imposed along the expansion ramp. They also noted that localized high-pressure regions are created by the external burning process. Since adverse pressure gradients can force the boundary layer to separate, such interaction may be a major design issue and requires further study. More recent RANS analysis by Ju et al. (2017) suggests that scramjet nozzle performance can benefit from the application of energy addition.

8.8 Closing Remarks

The main function of the engine nozzle is to efficiently turn thrust potential of the hot gas flow leaving the combustor into usable thrust for the hypersonic vehicle. The nozzle is also a key contributor to vehicle pitching moments. Hence, its design and configuration must be tailored to specific vehicle integration/mission requirements, as those constraints determine whether the nozzle should be 2-D or 3-D, or have fixed or variable geometry. MOC methods are often used to design nozzle contour, and CFD is used for analysis, especially for complex 3-D shapes.

The performance of the nozzle can be characterized with coefficients that facilitate engine cycle analyses (Marathe and Thiagaraian 2005). Such performance must consider many losses (expansion, divergence, viscous, and chemistry). It is important to note that the standards for rocket nozzles are applicable to some hypersonic air-breathing engine nozzles. However, ramjet/scramjet performance analysis approach developed during the NASP program established important baseline for analysis of future vehicle configurations. In the end, we must understand the relative magnitudes of nozzle losses to focus design and technology efforts on performance and features with highest priorities. It is important to characterize losses to correct performance that does not include full physics – i.e. frozen chemistry CFD.

Questions

1. How can overexpansion/underexpansion flow characteristics be controlled in a fixed geometry SERN?

2. What are some practical approaches for designing a nozzle to maximize thrust for a scramjet propelling a hypersonic cruise aircraft?

3. Explain the laminarization process of a hot, high velocity, turbulent combustion gas; how does it affect the scramjet nozzle?

4. What operational/design issues do you think external burning presents, and how would you resolve them?

5. What type of materials/TPS are more appropriate for a Mach 12 scramjet nozzle?

References

Anderson, G.Y., McClinton, C.R., and Weidner, J.P. (2000). Scramjet performance. In: *Scramjet Propulsion*, AIAA Progress in Astronautics and Aeronautics (ed. E.T. Curran and S.N.B. Murthy). American Institute of Aeronautics and Astronautics.

Baidya, R., Pesyridis, A., and Cooper, M., Ramjet nozzle analysis for transport aircraft configuration for sustained hypersonic flight, *Applied Sciences* 2018, 8, 574; pp. 1–12. doi: https://doi.org/10.3390/app8040574.

Baker, N.R., Northam, G.B., Stouffer, S.D., and Capriotti, D.P. (1993). Evaluation of scramjet nozzle configurations and film cooling for reduction of wall heating. *AIAA Paper 93-0744. 31st Aerospace Sciences Meeting*, Reno, NV (11–14 January 1993).

Candon, M.J., Ogawa, H., and Dorrington, G.E. (2015). Thrust augmentation through afterburning in scramjet nozzles. *Advances in Aircraft and Spacecraft Science* 2: 183–198.

Deere, K.A. and Asbury, S.C. (1999). Experimental and computational investigation of a translating-throat, single-expansion-ramp nozzle. NASA/TP-1999-209138. NASA LaRC (May 1999).

Goel, P., Barson, S.L., and Halloran, S.D. (1990). The Effect of Combustor Flow Nonuniformity on the Performance of Hypersonic Nozzles. Hypersonic Combined Cycle Propulsion, AGARD CP-479, 1990. NATO Advisory Group for Aerospace Research and Development (AGARD) Scientific Publications.

Gruhn, P., Henckels, A., and Sieberger, G. (2002). Improvement of the SERN nozzle performance by aerodynamic flap design. *Aerosp. Sci. Technol.* 6: 395–405.
Heiser, W.H. and Pratt, D.T. (1994). *Hypersonic Airbreathing Propulsion*. Washington, DC: AIAA Education Series.
Hoffman, J.D. (1987). Design of compressed truncated perfect nozzles. *Journal of Propulsion* 3 (2): 150–156.
Huang, W., Wang, Z., Ingham, D.B. et al. (2013). Design exploration for a single expansion ramp nozzle (SERN) using data mining. *Acta Astron.* 83: 10–17.
Jentink, T.N. (1993). An evaluation of nozzle relaminarization using low reynolds number κ–e turbulence models. *AIAA Paper 93-0610. AIAA, Aerospace Sciences Meeting and Exhibit, 31st*, Reno, NV (11–14 January 1993).
Ju, S., Yan, C., Wang, X. et al. (2017). Effect of energy addition parameters upon scramjet nozzle performances based on the variance analysis method. *Aerospace Science and Technology* 70: 511–529.
Kunze, J., Smart, M.K., and Gollan, R.L. (2018). A design method for shape transition nozzles for hypersonic vehicles. *AIAA 2018-5318. 22nd AIAA International Space Planes and Hypersonics Systems and Technologies Conference*, Orlando, FL (17–19 September 2018).
Lu, X., Yue, L.J., Xiao, Y.B. et al. (2009). Design of scramjet nozzle em-ploying streamline tracing technique. AIAA Paper 2009-7248.
Lv, Z., Xu, J., Yu, Y., and Mo, J. (2017). A new design method of single expansion ramp nozzles under geometric constraints for scramjets. *Aerospace Science and Technology* 66: 129–139.
Marathe, A.G. and Thiagaraian, V. (2005). Effect of geometric parameters on the performance of single expansion ramp nozzle. *Paper AIAA 2005-4429. 41st AIAA/ASME/SAE/ASEE J. Propulsion Conference*, Tucson, AZ, USA (10–13 July 2005).
Mo, J.W., Xu, J.L., Gu, R., and Fan, Z.P. (2014). Design of an asymmetric scramjet nozzle with circular to rectangular shape transition. *J. Propuls. Power* 30 (3): 812–819.
Oman, R.A., Foreman, K.M., Leng, J., and Hopkins, H.B. (1975). Simulation of hypersonic scramjet exhaust. NASA CR-2494 (March 1975).
Rao, G.V.R. (1958). Exhaust nozzle contour for optimum thrust. *Jet Propulsion* 28 (3): 377–382.
Ridgway, A., Sam, A.A., and Pesyridis, A. (2018). Modelling a hypersonic single expansion ramp nozzle of a hypersonic aircraft through parametric studies. *Energies* 11: 3449. https://doi.org/10.3390/en11123449.
Riehmer, J. and Gülhan, A. (2011). Design of a scramjet nozzle with streamline tracing technique and reference temperature methode. *ESA SP-692. 7th European Symposium on Aerothermodynamics*, Brugge, Belgium (9–12 May 2011). ISBN 978-92-9221-256-8., p. 29.
Rubesin, M.W. and Johnson, H.A. (1956). A critical review of skin-friction and heat-transfer solutions of the laminar boundary layer of a flat plate. *Transactions of the American Society of Mechanical Engineers* 78 (6): 1273.
Shyne, R. and Keith, T. (1990). Analysis and design of optimized truncated scarfed nozzles subject to external flow effects. *AIAA Paper 90-2222. 26th Joint Propulsion Conference,* Orlando, FL (16–18 July 1990).
Slater, J.W. and Saunders, J.D. (2010). Computational fluid dynamics (CFD) simulation of hypersonic turbine-based combined-cycle (TBCC) inlet mode transition. NASA/TM-2010-216362.

Smart, M. (1999). Design of three-dimensional hypersonic inlets with rectangular-to-elliptical shape transition. *J. of Prop. & Power* 15 (3): 408–416.

Snyder, C.A. and Espinoza, A.M. (2013). Lessons learned during TBCC design for the NASA-AFRL joint system study. *NASA/TM – 2013-218100. JANNAF Propulsion Meeting*, Arlington, VA (December 2013).

Stalker, R.J., Paull, A., Mee, D.J. et al. (2005). Scramjets and shock tunnels: the Queensland experience. *Progress in Aerospace Sciences* 41: 471–453.

Tatum, K.E. and Huebner, L.D. (1995). Exhaust gas modeling effects on hypersonic powered simulation at Mach 10. *AIAA Paper 95-6068. International Aerospace Planes and Hypersonics Technologies*, Chattanooga, TN (3–7 April 1995).

Xing, X.Q. and Damodaran, M. (2004). Design of three-dimensional nozzle shape using NURBS, CFD and hybrid optimization strategies. *AIAA Paper 2004-4368. 10th AIAA/ISSMO Multidisciplinary Analysis and Optimization Conference*, Albany, New York (30 August to 1 September 2004).

Xu, B., Xu, J., Wang, X. et al. (2017). Flowfield and performance analysis of a three-dimensional TBCC exhaust nozzle. *Journal of Engineering for Gas Turbine and Power* 139: 112602-1–112602-9.

Yu, K.K., Xu, J.L., Lv, Z., and Song, G.T. (2019). Inverse design methodology on a single expansion ramp nozzle for scramjets. *Aerospace Science and Technology* 92: 9–19.

Yungster, S. and Trefny, C.J. (1994). Computational study of single-expansion-ramp nozzles with external burning. *NASA TM 106550. 32nd Aerospace Sciences Meeting sponsored by the American Institute of Aeronautics and Astronautics*, Reno, Nevada (10–13 January 1994).

9

Materials, Structures, and Thermal Management

Reusable long-duration hypersonic vehicles require advanced high-temperature materials to build the airframe, propellant tanks, and other parts of the overall structure. For a cruise vehicle, the structure will operate at high dynamic pressure and high temperature for long period of time (hours) and the heat will conduct throughout the entire vehicle. For a spaceplane, the structure will be subject to launch, hypersonic acceleration, and the very harsh ascent/reentry conditions. Figure 9.1 illustrates an air-breathing spaceplane, showing typical ascent leading-edge heat flux data published by Glass (2008). Note the extreme heat transfer rates which are orders of magnitude greater than the heating rates experienced by the Space Shuttle Orbiter on entry. For example, the high heat rate on the engine cowl of the spaceplane (50 000 Btu/ft^2·s with shock-on-lip) results from a combination of factors such as aerodynamic heating, the small radius of the cowl lip, and the shock–shock interactions.

In addition to utilization of high-temperature materials, the design of these vehicles requires advanced options for thermal management, including passive, semi-passive, and actively cooled techniques. Hypersonic speeds will subject the vehicle to aerodynamic heating and extreme thermomechanical loads on the structure for the service life of the aircraft.

Material and structural solutions for rocket-propelled vehicles are well characterized. However, hypersonic vehicles powered by air-breathing propulsion systems are still under development. Their trajectories, their missions, and the length of time they will fly at hypersonic velocities are very diverse. For example, the air-breathing cruiser will fly in a relatively narrow Mach number/altitude corridor corresponding to flight dynamic pressures, between 0.24 and 0.95 atm, the exact altitude optimized by conflicting requirements. An optimized flight trajectory will ensure the vehicle operates below its structural heating and load limits while delivering enough airflow to its propulsion system to ensure an optimum level of net thrust for acceleration. The material requirements for each component of the air-breathing engine vary with flight conditions, whether or not the surfaces are actively cooled, whether the engine is fueled by hydrogen or hydrocarbons, and whether the intended mission requires an expendable or reusable vehicle.

Although many high-temperature materials are common with rocket-powered vehicles, their design and use in hypersonic air-breathing vehicles are very different. Air breathers must accelerate and cruise in Earth's atmosphere, and thus are slender and have sharp leading edges. Some vehicles are configured for horizontal launch, some will be part of

Scramjet Propulsion: A Practical Introduction, First Edition. Dora Musielak.

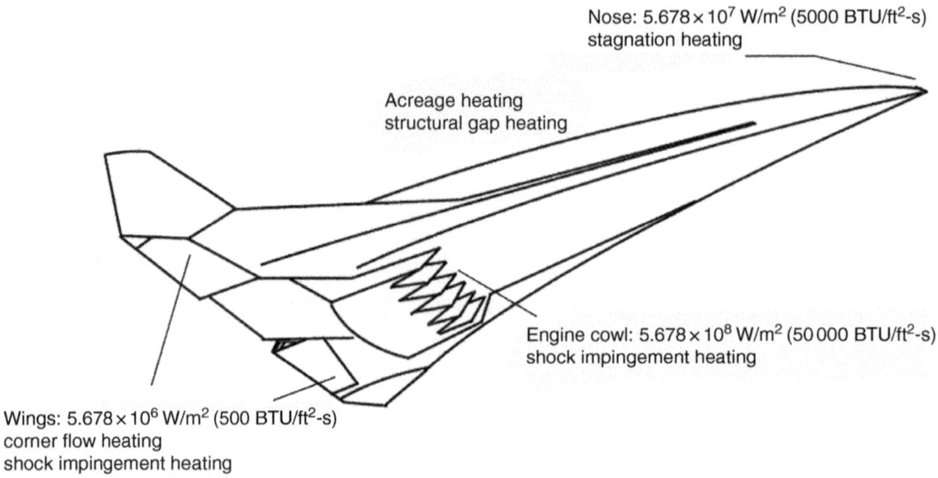

Figure 9.1 Ascent leading-edge heat flux estimates for an air-breathing spaceplane. Source: NASA.

multistage configurations with complex geometries, and all will fly in high dynamic pressure trajectories.

The hot structure of the NASA X-43A expendable research aircraft that flew for 10 seconds at Mach 6.83 was designed to demonstrate the viability of hydrogen-fueled scramjet propulsion. On its third flight, the X-43A vehicle flew at Mach 9.68 for 11 seconds, experiencing airframe temperatures of about 1982 °C (3600 °F). This vehicle flew in a thermodynamic environment more than 538 °C (1000 °F) harsher than that faced by the Mach 6.7 vehicle (Hallion 2008). The selected materials were just adequate for the short flight test of this subscale, prototype vehicle. However, for sustained hypersonic flight of full-scale vehicles, those material solutions may not be sufficient.

In this chapter, we wish to address questions such as:

- *Where are the most severe heating rates in a hypersonic aircraft?*
- *What material has capability to sustain the thermal loads found in the combustion chamber of a hypersonic air-breathing propulsion system?*
- *What materials are best for the nose of a Mach 10 missile?*
- *What are the most appropriate solutions for thermal management of reusable hypersonic cruisers?*

9.1 Hypersonic Flight Mission Characteristics

The three main hypersonic vehicle categories – expendables (missiles), accelerators (launch vehicles), and reusable aircraft – have different requirements for advanced materials and thermal management technologies. In general, a body in a very high-speed flow heats due to compression and friction within the boundary layer. As shown in Chapter 3, heat flux grows exponentially with flight velocity. An access to space vehicle propelled by air-breathing engines, for example, will experience major heat loads in the ascent phase of its mission and in hypersonic acceleration and cruise phases of flight.

For a two-stage-to-orbit (TSTO) vehicle concept, the first stage is an aircraft propelled by an air-breathing propulsion system. Both the Booster and the Orbiter vehicles demand advanced high-temperature materials and complex thermal management systems to survive the different harsh aerothermodynamic environments they will encounter in their respective mission trajectories. A lightweight structure is crucial for spaceplanes, since relatively small changes in vehicle weight will impact whether or not orbit is reached.

For a cruise vehicle, the structure will operate at high dynamic pressure and high temperature for periods of time long enough so that heat will conduct throughout the entire vehicle. The temperature will be extreme at the stagnation regions, i.e. the leading edges, the engine cowl, and the nose of the fuselage. These sections of the structure will be exposed to extreme temperatures for hours for airline-like hypersonic vehicles. Moreover, the ascent flight profile will be constrained by sonic boom overpressure, dynamic pressure, inlet engine pressure, and heating rate limits on the structure. Hence, reusable human-rated vehicles have an extremely demanding set of materials and thermal management requirements.

All vehicles will experience significant aerodynamic heating when traveling at hypersonic speeds through the Earth's atmosphere. The heating is severe and varies over the surface of a vehicle requiring several different types of thermal protection systems (TPS) on the same vehicle. Boundary layer transition drives the design of the TPS, which increases weight (i.e. less payload). Hence, understanding of the boundary layer transition is a crucial phenomenon that reflects on the requirements we impose on thermal management in the different parts of the vehicle. For all air-breathing hypersonic flight vehicles, the materials and thermal management requirements include

- Low weight,
- Long-life structural concepts,
- Optimized thermal management and maximized volume utilization,
- Maintainability,
- Reusability.

This chapter will highlight technologies that support development of hypersonic air-breathing propulsion. In general, the flight trajectory and the requirement for reusability have a major influence on the selection of airframe structural arrangement, structural concepts, materials, insulation approaches, and cooling techniques. Moreover, the total thermal load causes the structure to heat up and degrade the properties of the structural materials. Temperature gradients introduce thermal stresses into structural elements, which are as important as over stresses due to aerodynamic loads, thrust loads, and landing loads. All these effects impose stringent design requirements for the TPS.

9.2 Aerodynamic Heating

In designing a hypersonic vehicle, we must determine the aerodynamic heating environment it will encounter as well as the aerodynamic forces and moments it will experience. Designers have relied on wind-tunnel-based empirical correlations complemented by analytical solutions to obtain reasonable estimates of the actual flight environment. The method employed to determine the flow field depends to a large extent on the shape of the vehicle, Mach number, Reynolds number, and Knudsen number (for very high-altitude operation). This requires appropriate gas models and appropriate analysis of multidimensional flow fields. Special attention is given to the stagnation point regions.

Aerodynamic heating, that is the heat transfer from the hot boundary layer to the cooler surface, dominates the design of the vehicle and its propulsion system. It restricts the flight envelope in which the vehicle can operate. It also affects the design of its TPS and, hence, the vehicle weight. The heat transfer rate into a vehicle's skin from the external environment includes the convective heat transfer from the air into the skin, and the radiative heat transfer from the skin to its colder surroundings. The convective heat transfer rate is a function of the geometry, flight conditions, boundary layer state, and surface temperature. The radiative heat transfer rate is a function of the material emissivity and surface temperature. The difference between these heat transfer modes determines the heat flux into the structure.

9.2.1 Stagnation Temperature

In high-speed flow, the effects of viscous dissipation in the boundary layer become important. For flow over an adiabatic surface (perfectly insulated wall), the high temperature at the surface is the combined result of the heating due to **viscous dissipation** and the temperature rise of the fluid as the kinetic energy of the flow is converted to internal energy while the flow decelerates through the boundary layer. The shape of the temperature profile inside the boundary layer depends on the relation between the rate at which viscous shear work increases the internal energy of the fluid and the rate at which heat is conducted toward the freestream.

Although the processes in a high-speed boundary layer are not adiabatic, they can be related to adiabatic processes. The conversion of kinetic energy in a gas being slowed down adiabatically to zero velocity is described by the energy equation:

$$h_{t0} = h_0 + \frac{V_0^2}{2} \tag{9.1}$$

where h_{t0} is the total enthalpy in the inviscid flow outside the boundary layer, h_0 is the enthalpy of the gas in the freestream, and V_0 is its velocity. This expression shows that the total enthalpy for hypersonic velocity flows can become very large (the static enthalpy of the freestream flow in the atmosphere remains fixed).

Thus, the temperature rise associated with slowing the flow near a surface is very high. The stagnation temperature must be determined from the exact set of equations of state. Then, to find the adiabatic wall temperature, we use the recovery factor, defined to account for the irreversibility in the boundary layer flow. In a real boundary layer, where the real fluid has a finite thermal conductivity, some of the thermal energy is conducted away from the higher temperature region; the fluid is not brought to rest reversibly because the viscous shearing process is thermodynamically irreversible.

Aerodynamic heating: It is a phenomenon due to viscous dissipation. When a gas flows over a surface, the gas in contact is brought to rest as a result of viscosity. When the velocity near the surface decreases, the temperature increases. Aerodynamic heating is aggravated by skin friction and shock wave heating.

9.2.2 Wall Temperature Estimation for TPS

The objective of the TPS is to minimize heat conduction into the vehicle to reduce operational risk, and to maximize performance without adding unnecessary mass to the system. This requires aerothermodynamics modeling to obtain an accurate and conservative prediction of the heating environment encountered by the vehicle. Aerothermodynamics is the study of the energy balance at the surface of the material. Aerothermal modeling is tightly coupled with the design of the TPS for the vehicle to withstand the predicted environment with risk-appropriate margin. The heat flux (with pressure and shear) is used to select TPS materials, since the heat load determines the required thickness of the TPS. Using a heat transfer model to represent the sum of the incident aeroheating (conduction and radiation) and material response (Figure 9.2), we can estimate wall temperatures.

In order to determine the temperature-related boundary condition at a solid wall, consider the energy balance for the wall element of thickness Δ shown in Figure 9.2. In a one-dimensional model like this, we can neglect conduction through the (end) surfaces perpendicular to the exposed surface. That is, we assume that the changes in the flow field and in the wall temperature in directions tangent to the vehicle surface are relatively small. We use the flux per unit area, since the heat flux vector is one-dimensional (pointed in the negative y-direction).

The incident convective heat flux $\dot{q}$ is given by:

$$\dot{q} = h(T_r - T_w) \tag{9.2}$$

where T_r denotes the recovery or adiabatic temperature, T_w is the wall temperature, and h is the heat convection coefficient.

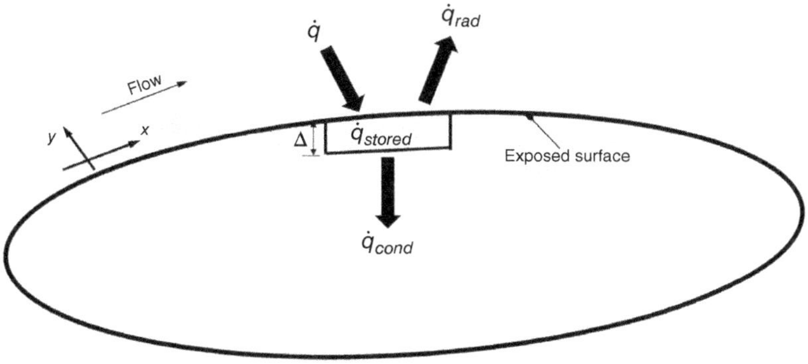

Figure 9.2 Energy balance for a surface element of the vehicle TPS.

At the surface of the TPS element, we must have an energy balance with the contribution of conduction and radiation heat fluxes, and a term to represent the rate at which energy is stored in the element ($\dot{q}_{\text{stor}}$), as depicted in Figure 9.2:

$$\dot{q} = \dot{q}_{stor} + \dot{q}_{cond} + \dot{q}_{rad} \tag{9.3}$$

The rate at which energy is stored in the TPS element is given by:

$$\dot{q}_{stor} = \rho_w c_w \Delta \frac{dT_w}{dt} \tag{9.4}$$

which is zero for steady state heat transfer calculations.

The rate at which energy is conducted through the backface TPS element is

$$\dot{q}_{cond} = k_w \frac{dT}{dy} \tag{9.5}$$

The rate at which energy is radiated from the exposed surface of emissivity ε is

$$\dot{q}_{rad} = \varepsilon \sigma T_w^4 \tag{9.6}$$

where σ is the Stefan–Boltzmann constant (derived from other known physical constants), $\sigma = 5.670\,374 \times 10^{-8}$ W/(m^2 • K^4) $= 4.76 \times 10^{-13}$ BTU/(ft^{2}s • R^4).

Assuming that the energy stored in the element is negligible, we can consider three types of thermal boundary conditions at the wall:

1) *The wall temperature T_w is prescribed.* Knowing T_w, it fixes the incident convective heat flux, Eq. (9.3), and the rate at which heat is radiated from the surface, Eq. (9.6). We can then determine the rate at which heat is conducted through the backface of the TPS element into the adjacent material.
2) *The incident convective heat flux $\dot{q}$ is prescribed.* Hence, we can determine the wall temperature from the energy flux balance, Eq. (9.3).
3) *The wall is adiabatic.* This means that the wall is perfectly insulated, so $\dot{q}_{cond} = 0$. If $\dot{q}_{rad}$ is significant, we can determine the wall temperature from the energy flux balance. On the other hand, if $\dot{q}_{rad}$ is negligible, the energy balance requires that

$$T_r = T_w$$

i.e. the wall temperature is equal to the recovery or adiabatic temperature. The recovery temperature is related to the total and static temperatures at the edge of the boundary layer by the recovery factor *r*, as:

$$r = \frac{T_r - T_e}{T_{te} - T_e} \tag{9.7}$$

The recovery factor is a function of the Prandtl number (*Pr*), a nondimensional number that correlates the viscosity of a fluid with its thermal conductivity.

For a laminar boundary layer, the recovery factor is given by:

$$r = Pr^{1/2} \tag{9.8a}$$

and for a turbulent boundary layer:

$$r = Pr^{1/3} \tag{9.8b}$$

The Prandtl number is the ratio of momentum diffusivity to thermal diffusivity given by Eq. (3.18), where μ is the gas viscosity and κ is its thermal conductivity, Eq. (3.2) and Eq. (3.22), respectively.

For a perfect gas, the recovery temperature and the total temperature are related as:

$$\frac{T_r}{T_{te}} = r + \frac{1 - r}{\left(1 + ((\gamma - 1)/2)M_e^2\right)}. \tag{9.9}$$

While evaluating theories for predicting turbulent skin friction and heat transfer in flat plates, Hopkins and Inouye (1971) noted: "At supersonic speeds (M_e~3), the surface temperature is essentially the adiabatic wall temperature. At hypersonic speeds (M_e~7), however, the external surface temperatures generally will be a fraction (0.3–0.5) of the adiabatic wall temperatures as a result of considerable radiative cooling and internal heat transfer."

Let us consider a simple example of a hypersonic vehicle with a 0.5 m nose radius, flying at a given velocity at an altitude where the air density is known. We can easily estimate the stagnation point heating rate and the wall temperature, assuming a TPS material with emissivity of 0.8. As a first approximation, we can neglect the energy stored in the element and the heat conducted through the backface TPS element. Using Eq. (3.26) to estimate the incident convective heating, then the wall temperature is found from the energy balance $\dot{q} = \dot{q}_{rad} = \varepsilon\sigma T_w^4$.

I must emphasize that the primary function of the TPS is to minimize heat conduction (must be good insulator) through the vehicle wall. Neglecting material response, we can assume that $\dot{q} = \dot{q}_{cond} + \dot{q}_{rad}$, and we can readily solve for the wall temperature. Thus, for hypersonic flight, the thermal design of the vehicle structure and the material selection depend on the heating rates. The thermomechanical design point of a scramjet must be established by the maximum flight Mach number, as this condition is where the maximum heat transfer will be reached.

9.3 Hypersonic Integrated Structures

We consider structural concepts applicable to hypersonic vehicles. Such vehicles may fall into the following categories: ballistic missiles, reentry vehicles, space access vehicles, interceptor missiles, hypersonic cruise missiles, and hypersonic cruise aircraft, which can be easily divided into single-use expendable and reusable systems. Although our principal emphasis is on applications to vehicles powered by air-breathing propulsion systems, some structural configurations equally apply to other types of hypersonic vehicles.

9.3.1 Hot, Cooled, and Warm Structures

Hypersonic air-breathing airframe-integrated propulsion vehicles require development of a unique structure that fully integrate its fuselage/engine/tank/TPS system. Such structure may incorporate hot, cold, and warm structures, including those that use ablative heat shields (Thornton 1996). Ablators are attractive for short exposure to very high temperatures and can also be used for flight demonstration systems; however, they are not suitable for reusable hypersonic sustained cruise vehicles. Hence, structure concepts which employ ablative heat shield materials are not included in this overview.

9.3.1.1 Hot Structures

A hot structure is one that does not employ a TPS. The external surface of a hot structure acts as the primary load-bearing vehicle structure, and the external surface absorbs all the heat. This is possible because the external surface is allowed to attain a radiation equilibrium temperature, which of course depends on the aerodynamic heating environment and material limitations. Hot structures are easy to maintain and inspect but since they are subjected to high thermal stresses, this type of structure requires materials with the highest temperature and strength capability.

Hot structures are recommended for moderate heat fluxes over a period of time to allow structures to reach steady state condition. Although hot structures are likely to be heavy, they are typically employed in supersonic aircraft as they are very durable, and many structural configurations are possible (e.g. corrugated, honeycomb sandwich, truss core sandwich, and integral stiffened), which give designers more options. Insulation for hot structures is optional, but in some parts of the vehicle it may be required such as in fuel tanks, and to protect internal equipment (hydraulics and avionics).

Hot structures technology involves lighter-weight materials and requires less maintenance than warm or cold structures that use parasitic TPS materials that attach to warm or cool substructure. Development of hot structures requires a thorough understanding of material performance in an extreme environment, boundary conditions and load interactions, structural joint performance, and thermal and mechanical performance of integrated structural systems that operate at temperatures ranging from 1500 to 3000 °C (2732–5432 °F) (Rivers and Glass 2006).

The DARPA's Hypersonic Flight Demonstration (HyFly) program used ground testing to demonstrate the ability of an uncooled, high-temperature composite engine to withstand the extreme engine environment experienced at Mach 6. This achievement gives credence to the recommendation to use hot structures for high-speed tactical missiles.

A hot structure is any part of airframe that can operate efficiently up to 3300 K without active cooling or insulation.

9.3.1.2 Cold Structures

A cold structure uses an external metallic heat shield and insulation material to protect the primary vehicle structure from aerodynamic heating. The metallic shield absorbs the heat, and the insulation layer protects the load-bearing structure from the heat load. Hence, a cold structure, also known as a heat shield structure, in fact decouples the load-bearing and heat-absorbing parts of the vehicle. Hence, this type of structure can meet a wide range of flight conditions by varying the heat shield materials and the insulation beneath the shield and can use active and semi-passive cooling. Cold structures are used for moderate and high heat fluxes over long periods of time. These structures are durable and reusable, and they can potentially be lighter than hot structures, but the load-bearing structure is difficult to access and inspect. However, with careful design, it is possible to inspect and maintain it.

Cold structures present several disadvantages as their designs must restrict boundary air leakage and also provide for thermal expansion. Differences in thermal expansion between the hot metal shield and cold structure can result in integrity issues. Moreover, surface roughness can be present, which will degrade aerodynamic performance if adjacent shields overlap. Since cold structure designs are more complicated, and inadvertent heat shorts can occur to the primary structure, a trade-off study will establish whether the advantages are greater and support vehicle performance and mission requirements.

9.3.1.3 Warm or Externally Insulated Structure

In this concept, the primary vehicle structure is protected by bonding an insulation material on the external surface; the structure can be primary or a carrier panel. In this type of structure, the external insulation material is a key to keep the structure cool. Warm structures are considered for moderate heat fluxes over relatively short period of time. The former Space Shuttle Orbiter used this type of structure with reusable insulation (Throckmorton and Hartung 1993). Some of the disadvantages are that the surface insulation has rather limited reusability, mission turnaround time is increased if insulation must be replaced, and inspection of the structure is difficult. Current efforts strive to develop reusable surface insulation materials that are more durable.

9.3.1.4 Actively Cooled Structure

This structure concept uses a coolant fluid to flow-through passages adjacent to the hot structure. The coolant is typically the fuel, especially when the vehicle carries a large volume of cryogenic or endothermic fuel aboard. There are different approaches to design the ducting system to circulate the coolant fuel through the hottest parts of the structure. The liquid cold fuel is pumped from the tanks and, after absorbing heat, it is then delivered to the fuel injectors in the combustion chamber.

Actively cooled structures are considered for moderate and high heat fluxes over long periods of time. However, it requires to match airframe heat loads with available fuel sink. Although adding active cooling may result in a lightweight structure, it presents design and fabrication complications and added weight of a ducting/heat exchanger (HEX) system to circulate the coolant. The X-43A vehicle used water for active cooling of its cowl and sidewall leading edges. The X-51A scramjet engine used its endothermic JP-7 fuel to actively cool its Inconel 625 metal structure.

9.3.2 Vehicle Nose and Leading Edges

While rocket space vehicles have blunt noses and leading edges to minimize drag and heat flux during reentry, the air-breathing hypersonic vehicle, whether cruise aircraft or spaceplane, must be slender (relatively low drag). Thus, while the entire vehicle body is subjected to atmospheric heating due to drag, the thin, sharp leading edges – including the nose region, tail edges, and scramjet forebody/inlet – are exposed to intense, highly localized stagnation point heating, heating due to interfering shocks, and a severe aerothermal

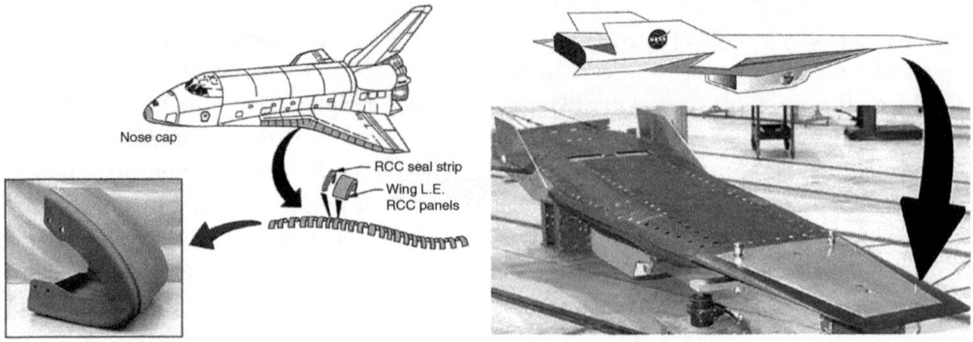

Figure 9.3 Comparison of blunt leading edges (Space Shuttle Orbiter) and sharp leading edges (X-43A aircraft) (vehicles sizes not to scale). Source: Jacobson (2014)/NASA/Public Domain.

environment in the high dynamic pressure altitudes it flies. The leading-edge radii must be sufficiently small so that the drag is not excessive, but large enough to maintain the heating rates within tolerable limits.

To get a perspective, compare the blunt leading edge of the Space Shuttle Orbiter and the sharp flat nose of the X-43A vehicle (Figure 9.3). The Orbiter fuselage nose radius was 60 cm (23.622 in) and had relatively blunt leading edges, and so the shock waves spread out, dissipating much of the heat away from the vehicle. Upon reentry, the Orbiter nose cap and wing leading edges took most of the heat – up to 1600 °C for short periods (~five minutes). In contrast, the Mach 7 X-43A vehicle had a nose tip radius of 0.0762 cm (0.030 in), and the Mach 10 vehicle had a nose tip radius of 0.127 cm (0.050 in). Even for full-scale-sized vehicles, the nose tip radii will be relatively small and the shock waves will be closer to the surfaces. Hence, for future hypersonic air-breathing cruisers with flat noses, specialized structural approaches are required to maintain the integrity of the thin leading edges, which are the hottest parts of these vehicles.

9.3.2.1 Hot Structure

Using a hot structure for the vehicle's nose and leading edges requires high-temperature materials such as refractory metals and ceramics. However, some of those materials are heavy and lead to large weight penalties for the overall vehicle. Hence, we seek lighter materials such as carbon–carbon composites to use as hot structure for nose, leading edge, and control surface designs. Carbon–carbon composites require effective oxidation protection. For example, reinforced carbon–carbon (RCC) was used to protect the nose cap and the wing leading edges of the Space Shuttle Orbiter, parts where the surface temperature was expected to exceed 1530 K (2294 °F). RCC is an example of a hot structure subtype of a passive TPS concept.

For low hypersonic Mach number vehicles, metal alternatives such as refractory metals (Niobium), superalloys (Inconel), and intermetallics (TiAl) could be used for the nose and leading edges, since these high-temperature metals also exhibit desirable characteristics such as strength/stiffness and high melt/oxidation temperature thresholds. However, for

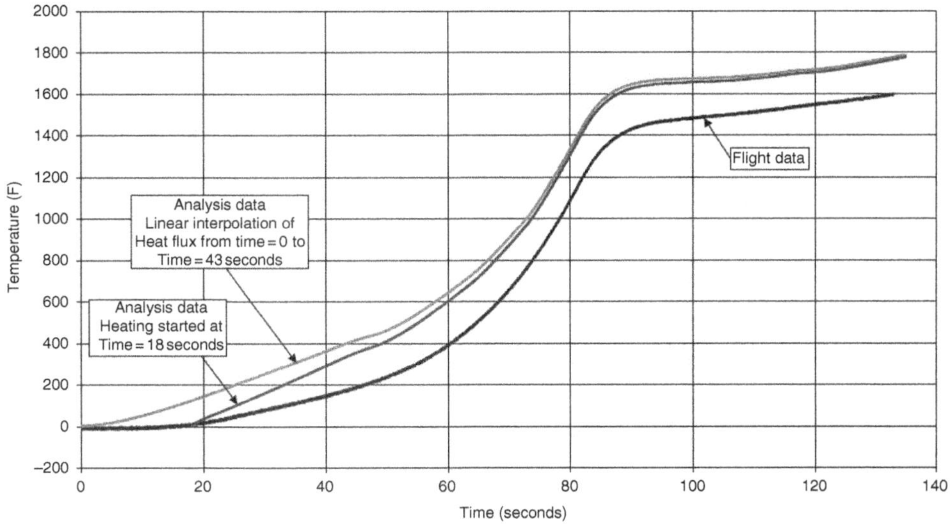

Figure 9.4 Nose temperature comparison of flight and analysis data. Source: Amundsen et al. (2004)/ NASA/Public Domain.

higher-speed applications, the component mass penalty associated with increased density may reduce their applicability.

The former Space Shuttle Orbiter used carbon–carbon for both the nose cap and leading edges, which performed reasonably well during the Orbiter's mission. However, for hypersonic cruise aircraft or for orbital vehicles powered by air-breathing engines, the total heat load and/or maximum heating will be much higher than that experienced by the Orbiter. For example, the heating to the cowl lip of the National Aero-Space Plane (NASP) on its ascent flight was estimated to be 6.24×10^8 W/m^2 (55 000 BTU/ft^2 • s), which is orders of magnitude greater than the heating rates experienced by the SS Orbiter on its entry trajectory (Bertin 1994). Hence, alternate structure design approaches will be required to protect the stagnation regions of future space vehicles intended to reach low Earth orbit, and for reusable vehicles intended for sustained hypersonic flight.

The NASA X-43A experimental vehicle was equipped with thermocouples in the hot structures (nose, horizontal tail, and vertical tail) to record the flight thermal response of these components. One thermocouple was installed in the carbon–carbon leading edge at a location 0.5 in (1.27 cm) aft of the nose stagnation point. After the Mach 7 flight (March 2004), researchers were able to compare the flight data and the thermal analysis results, which we reproduce in Figure 9.4 taken from Amundsen et al. (2004). As shown, very little temperature increase is observed during the first 18 seconds of the flight (expected). The data from the original analysis (upper curve) matches the flight data trend quite well. However, the magnitude of the predicted temperature is higher than the actual flight temperature. This was expected as a result of the early heating caused by linearly interpolating the prediction between time 0 and time 43 seconds (Amundsen et al. 2004).

9.3.2.2 Externally Insulated Structure

In this approach, the vehicle nose and leading-edge structure are covered with an external layer of insulation. The main disadvantage of this design is finding materials which are sufficiently durable and reusable in the high-temperature environment.

9.3.2.3 Active Cooled Structure

We can also consider using active cooling for the nose and leading-edge structures. A coolant fluid would be flown through passages in panels adjacent to the leading edges and subsequently routed to a HEX or simply rejected by radiation. Another approach to use in the vehicle's nose and leading edges is one that incorporates heat pipes to distribute heat evenly over the structure. This design is especially appropriate for applications with very high heat loads at the stagnation points. The cooled leading-edge structure can use high specific strength carbon–carbon to accommodate thermal/structural loads and very thin refractory metal D-shaped heat pipes embedded within the carbon–carbon structure to transport stagnation heat aft where it can be rejected by radiation (McComb et al. 1990).

9.3.3 Passive and Active Cooling Methods

Selection of thermal protection for a vehicle depends on the mission, e.g. thermal environment, flight duration, velocity, and the reusability, operation, and maintenance requirements. Possible cooling methods for hot structures in hypersonic vehicles include active, passive, and semi-active methods. Messe (2017) published an overview of these techniques and summarized them in Table 9.1.

Active Cooling. Active cooling methods involve the use of a fluid from an interior supply container and then regenerative cooling or injecting the fluid through porous skin or channels. Active cooling is required to manage very high heat fluxes and for long-time exposure to the thermal environment.

Regenerative cooling is a method of cooling in which a coolant fluid – a heat sink fluid – is passed through passages in the structure and thereby takes heat from the surroundings hot walls. In high heat flux areas, heat sink fluids are flown to hot sections via a cooling loop. A heat sink refers to a high thermal conductivity material that absorbs heat and distributes it quickly and uniformly away from parts of the vehicle's structure it is designed to protect. If the coolant is the fuel itself and the heat is transferred to the fuel before it is burned in the combustor, the system is called a regenerative cooling system.

This cooling method adds a parasitic weight to the system due to fluid distribution structure. Potential cooling fluids include water glycol, liquid metals (sodium or potassium), hydrogen, and endothermic fuels. While traditional aircraft fuels use only the sensible heat ($c_p \Delta T$) for cooling, an endothermic hydrocarbon fuel (EHF) provides cooling through the absorption of the sensible heat and also via endothermic reactions. Due to its excellent heat sink capacity, EHF is considered to be the most effective in regenerative on-board cooling for hypersonic vehicles flying at Mach numbers up to about 8. For the scramjet combustor, regenerative cooling using the fuel itself may be the best option (Section 9.3.4).

Transpiration cooling involves fluid ejection through a porous skin into the boundary layer over a surface, thus reducing the adiabatic wall temperature. The coolant decreases the heat flux to structure, such as in a combustor, which is subject to very high heat loads.

Table 9.1 Active and passive cooling methods for hot structures.

	Active cooling methods			
	Regenerative	**Wall jet cooling**	**Effusion**	**Transpiration**
	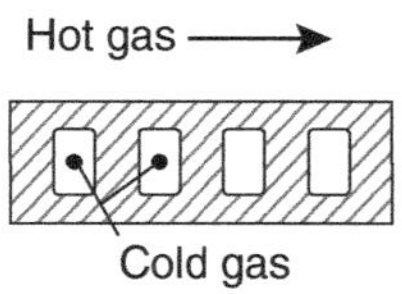	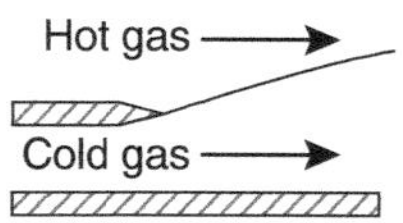	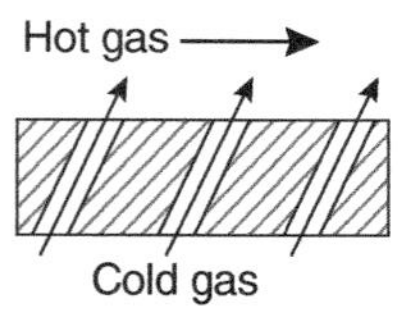	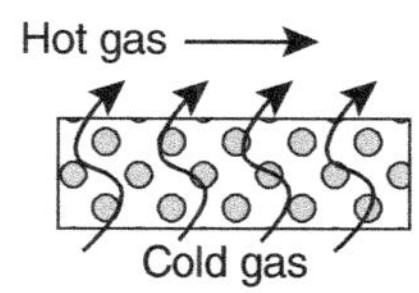
Cooling effect	High	High	High	Very high
System complexity	High	Medium	Medium	Medium
	+Field-tested in rocket engines +Can use fuel as coolant	+Field-tested in gas turbines +Can use fuel as coolant	+Field-tested in gas turbines +Can use fuel as coolant	+More efficient than effusion −Requires porous surface coating
	−Requires high channel pressure −Complex pump system required	−Requires high coolant mass flux	−Requires high coolant mass flux	−Requires high coolant mass flux

(*Continued*)

Table 9.1 (Continued)

	Passive cooling methods		Other cooling methods	
	Radiative	**Heat pipe**	**Core burning**	**Precooling**
	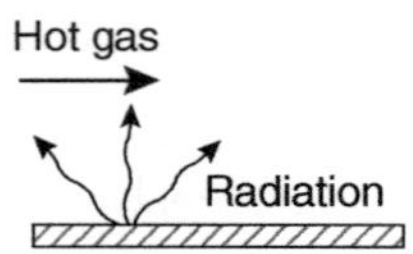	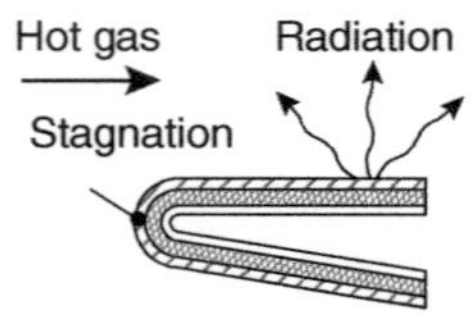	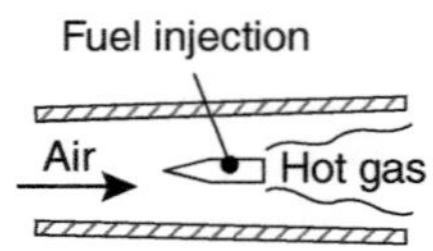	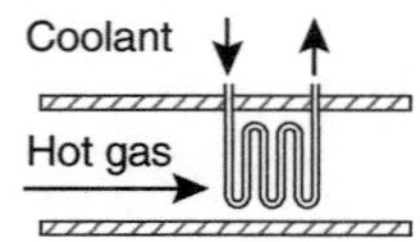
Cooling effect	Medium	Low	Low	Very high
System complexity	Low	Medium	Low	Very high
	+Very robust	+Robust	+Field-tested in gas turbines	−Requires high coolant mass flux
	−Requires heat sink (e.g. environment)	−Requires heat sink	+Combinable with other methods	−Reduces engine performance
	−Works mainly on external surfaces	−Limited operating range	−Low hypersonic regime only	−Heavily increases system mass
		−Works mainly on leading edges		

Source: Messe (2017)/Universität Stuttgart.

An example of a combustion chamber made of carbon–carbon material with transpiration cooling was designed at NASA (Glass 2008).

It is important to note that expending any amount of fuel for transpiration cooling will reduce engine performance (reduce I_{sp}). Water is a suitable transpiration and film coolant and may be effective convective coolant in some parts of the structure.

Passive Cooling. Passive cooling may be by radiation, or it may involve an ablating material underneath a porous skin that upon heating sublimes to produce a cooling gas. With radiation cooling, much of the heat flux is reflected back toward space (black body) by a high emissivity coating on the protected substrate. This method is very effective in orbit. For moderate heat fluxes and relatively short-time exposure, insulated structure can be used. If high heat fluxes persist for long times, the use of heat pipe (semi-passive method) is also recommended.

9.3.4 Fuels for Regenerative Cooling

Without active cooling, wall temperatures in a scramjet could reach 2760 °C (5000 °F), high enough to melt any metal. Hence, we can use regenerative cooling with the fuel itself. To illustrate how a fuel could serve as coolant, we can use Heiser and Pratt (1994) approach to estimate the total rate at which fuel can remove thermal energy ($\dot{Q}_f$) as:

$$\dot{Q}_f = \dot{m}_f h_{fc} \tag{9.10}$$

where $\dot{m}_f$ is the fuel flow rate and h_{fc} is the amount of thermal energy that can be absorbed by unit mass of fuel.

Assuming that most heat is due to turbulent flow, the total rate at which heat must be removed from engine surfaces (required cooling $\dot{Q}_r$) is expressed as:

$$\dot{Q}_r = \dot{q}_w A_w \propto \frac{\rho_0 V_0 \bullet V_0^2}{(\rho_0 V_0)^{1/5}} A_w \propto \dot{m}_0 \frac{V_0^{11/5}}{q_0^{1/5}} \frac{A_w}{A_0} \tag{9.11}$$

Thus, the ratio of available fuel cooling to required fuel cooling is

$$\frac{\dot{Q}_f}{\dot{Q}_r} = \frac{f\, h_{fc}}{K} \frac{q_0^{1/5}}{V_0^{11/5}} \frac{A_0}{A_w}$$

For a balanced system, we require

$$f\, h_{fc} = K \frac{V_0^{11/5}}{q_0^{1/5}} \frac{A_w}{A_0} \tag{9.12}$$

The physical heat absorption capacities of aerospace fuels are given in Table 9.2, where the heat absorbed refers to the fluid going from fuel tank conditions to 1000 K (1800 °R). In all cases, the fuel is stored in liquid state and heated to vapor state before injecting it into the engine flowpath as a gas.

Hydrogen is the best engine coolant overall, followed by methane fuel and then JP-7. In the process of cooling a structural component, the HC fuel is vaporized and eventually is partially cracked. Cracking is the process by which heavy hydrocarbon molecules are

Table 9.2 Fuels heat absorption characteristics.

Fuel	Liquid density kg/m³ (lbm/ft³)	f_{st}	Physical heat absorption, h_{fc} kJ/kg (BTU/lbm)	$f_{st}h_{fc}$ kJ/kg (BTU/lbm)
Hydrogen, H_2	75.9 (4.74)	0.0291	15 100 (6500)	440 (189)
Methane, CH_4	300 (18.7)	0.0583	3 400 (1460)	198 (85.1)
JP-7, $C_{12.5}H_{26}$	793 (49.5)	0.0675	2 000 (860)	135 (58)

broken up into lighter molecules by means of heat and usually pressure. Thermal cracking is accompanied and limited by the formation of undesirable carbon deposits (coking). The highly reactive fuel vapor is then directed to any or all of the fueling sites in the engine via flow control valves capable of handling the high-temperature gas. The practical limit of this process is the coking limit of the fuel. Since coking within the cooling channels must be avoided at all costs, temperatures are controlled to values safely below the coking temperature. This limits the hydrocarbon-fueled scramjet engine to approximately Mach 8. Comparing JP-type fuels, it is found that JP-7 fuel has better heat absorption than JP-8 and produces more of gaseous products that reduce ignition times. The X-51A vehicle successfully demonstrated a thermally balanced design using endothermic JP-7 fuel for active cooling.

9.3.5 Integral and Nonintegral Fuel Tanks

The fuel tanks of the hypersonic vehicle can be designed as integral or nonintegral structures.

- A nonintegral tank is separated from the airframe structure and does not carry body shears and bending moments; thus, it must support fuel inertial loads and internal pressure loads. The primary vehicle load is carried in the fuselage shell. Nonintegral tanks can be removed for inspection or maintenance and repair. Also, differential thermal expansion between the cryogenic tankage and the primary structure can be accommodated easily to reduce thermal stresses. In addition, nonintegral tanks can be configured in circular shapes (spherical, cylindrical, and conical) independent of the external mold line of the vehicle, ensuring that they carry the pressure loads efficiently.
- An integral tank carries all the primary vehicle load in the center fuselage and redistributes loads from all the vehicle. The tank walls are the primary structural paths for the airframe and carry body shears and bending moments as well as internal pressure and slosh, inertial, and gravity loads. Integral tank systems may be simpler and lighter overall than the nonintegral systems because mechanical and thermal loads are carried by the minimum number of structural elements, so there is less redundancy in the integral systems.

An annotated review of the state of the art on aircraft cryogenic tank structures was published by Mital et al. (2006). They consider materials, structural designs, and insulation systems, intending to develop a lightweight and long-term storage system for liquid hydrogen.

They reviewed classes of insulation systems such as foams (including advanced aerogels) and multilayer insulation (MLI) systems with vacuum. The structural configurations evaluated included single- and double-wall constructions. For long-duration flight applications, a double-wall construction with a vacuum-based insulation system may be optimum design (Mital et al. 2006). Among the potential wall material considered are monolithic metals, polymer matrix composites, and discontinuously reinforced metal matrix composites (MMCs). This review provided trends in liner material development (to minimize/eliminate loss of hydrogen fuel through permeation) of interest today.

The selection for the most satisfactory approach to design the fuel tanks for a hypersonic air-breathing vehicle must take into account all aspects of the vehicle design, construction, and operation. For example, to maximize available space for fuel, the X-51A used an integral fuel tank wet-bay approach, where the fuel is not contained in a separate tank, but rather the skin and bulkheads of the vehicle form the walls of the tank (Hank et al. 2008). During and following boost acceleration, the JP-7 fuel tank was pressurized and fuel flow initiated through the engine. The Interstage had flow-through ducts designed to allow the scramjet inlet to start during boost, preheating the hydrocarbon fuel prior to scramjet ignition (Rondeau and Jorris 2013).

9.4 High-Temperature Materials Requirements and Properties

Hypersonic airframes and propulsion systems have an extremely demanding set of materials requirements. These include

- High-temperature operating environments,
- High dynamic pressure trajectories,
- Aggressive weight requirements for successful missions,
- Multiple missions for reusable vehicles,
- Damage tolerance for human-rated hypersonic flight.

Materials and structures are technical challenges for sustained flight of hypersonic air-breathing-propelled vehicles. Advanced materials are being explored which retain strength and stiffness at high temperatures. In particular, low density, high-temperature materials with high strength, high stiffness, and corrosion-resistant are critical enablers for successful vehicle designs.

9.4.1 Design Drivers and Material Properties

The total thermal load on the vehicle causes the structure to heat up and may degrade the properties of the materials used to fabricate the structure. The design drivers for the hypersonic structure include

- Maximum heating rate,
- Duration of heating,
- Total heat load,
- Flight envelope and propulsion system,

- Mission life requirements,
- Vehicle design,
- Flight environment,
- Containment of cryogenic propellant.

The design requirements define the properties of the materials selected:

- Melting temperature and operating temperature limit,
- Density,
- Elastic Modulus,
- Strength (yield, ultimate, and fatigue),
- Hot corrosion and oxidation durability,
- Creep resistance,
- Fracture toughness,
- Thermal expansion,
- Processing and fabrication characteristics.

The required material characteristics for hypersonic air-breathing propulsion vehicles are high-temperature capability in the range 1093–2204 °C (2000–4000 °F), high strength at those elevated temperatures, high toughness, lightweight, and durability.

9.5 Selected Materials for Hypersonics

A number of materials are candidates to meet the structural requirements of a hypersonic vehicle. As shown in Figure 9.5, for environments with temperature less than 4000 °C (7232 °F), several material options exist (depending on operational requirements). However, no material developed to date is capable of operating above 4000 °C, and thermal management must be used at these high-temperature environments. In the following sections,

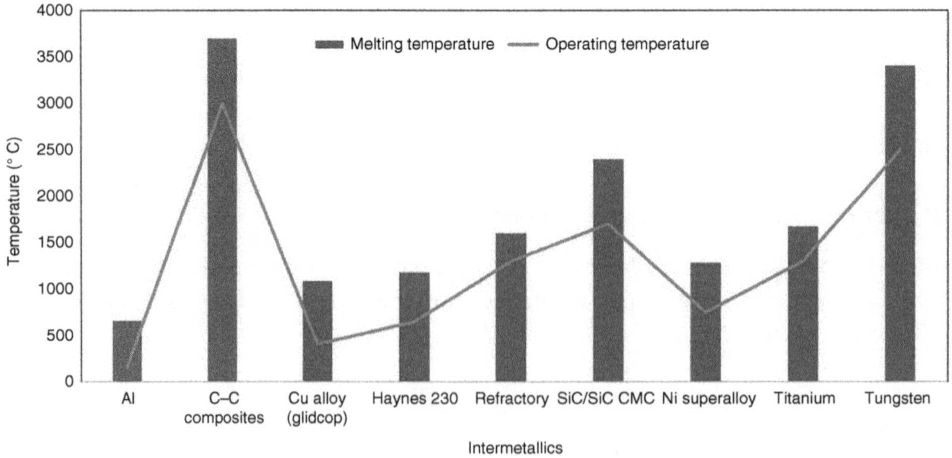

Figure 9.5 Operating and melting temperature of selected materials.

we review the characteristics of the most important high-temperature materials considered for hypersonic flight applications.

Please note that the material data in this chapter are given for illustrative purposes rather than as recommended design guidelines. For accurate property values, the designer is urged to consult materials property databases and incorporate those materials according to their particular design requirements. Reports on materials research for hypersonic aircraft are given by Marshall et al. (2014), and Glass (2008).

Technology goal for materials in hypersonic flight applications is to develop high specific strength and low-density materials that can operate effectively at elevated temperatures during the duration of the mission.

Low-density, high-temperature materials with high strength, high stiffness, and corrosion-resistant are critical enablers for reusable hypersonic air-breathing vehicle designs.

9.5.1 Superalloys: High-Temperature Metals

High-temperature resistance is essential for hypersonic flight since vehicles must operate in high heat environments, exposed to extreme oxidation, and cycling. Hypersonic materials must be lightweight with an extremely high Young's modulus (mechanical property that measures the stiffness of a solid material) in order to withstand the enormous stresses of hypersonic flight maneuvering. In addition, these materials must have long fatigue life and high fracture toughness, including creep and stress rupture resistance at temperatures up to 950–1300 °C (1223–1573 K).

Figure 9.6 provides the specific strength of the most important superalloys developed to date. Superalloys refer to a group of nickel, iron–nickel, and cobalt alloys used in aircraft structures due to their high strength, oxidation, and corrosion resistance, and in particular excellent heat-resistant properties.

The following superalloy materials are considered for hypersonic flight applications:

- **Nickel-based superalloys.** These are materials that can be used at temperatures greater than 550 °C (>823 K) in aerospace applications. This type of superalloys includes

HAYNES® 230® (UNS N06230): A nickel–chromium–tungsten–molybdenum alloy has excellent high-temperature strength and outstanding resistance to oxidizing environments up to 1149 °C (2100 °F).

HASTELLOY® Alloy X (HX): A nickel–chromium–iron–molybdenum superalloy has outstanding high-temperature strength, oxidation resistance, and fabricability, excellent high-temperature strength and oxidation resistance to 1204 °C (2200 °F), excellent forming and welding characteristics, and excellent resistance to oxidizing, reducing, and neutral atmospheres.

INCONEL® Alloy 625: This alloy has high strength and toughness from cryogenic temperatures up to 982.22 °C (1800 °F), good oxidation resistance, exceptional fatigue strength, and good corrosion resistance.

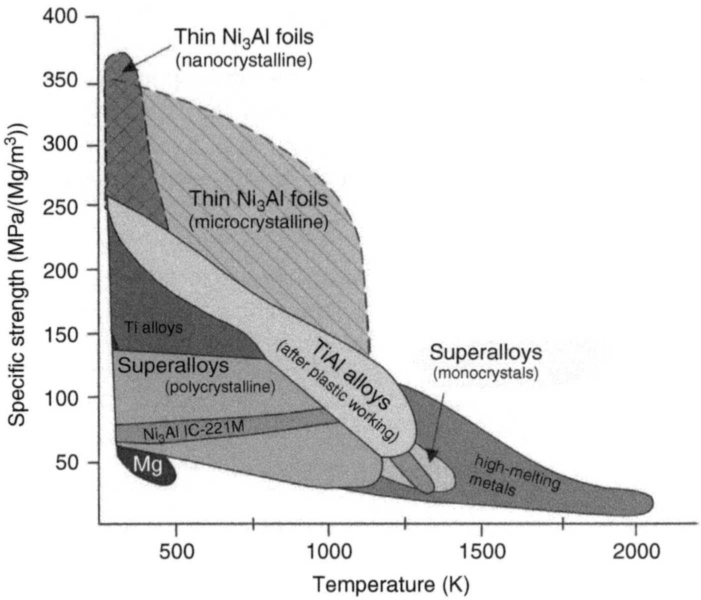

Figure 9.6 Diagram of specific strength as function of temperature of structural materials. Source: Jozwick et al. (2015)/MDPI/CC BY 4.0.

INCONEL® Alloy 718: This alloy has excellent strength from −253 to 704 °C (−423–1300 °F), is age hardenable, may be welded in fully aged condition, and has excellent oxidation resistance up to 982 °C (1800 °F).

- **Titanium (Ti) Alloys**. These alloys are used in airframe and turbine engine applications due to having a good mix of properties, including moderate density, high static strength and long fatigue life, and excellent resistance to oxidation and corrosion. Titanium alloys may be used for some hypersonic airframes due to having good mechanical performance at temperatures up to 550–600 °C (823–873 K). Titanium Ti-6Al-4V (UNS designation R56400), also sometimes called TC4, Ti64, or ASTM Grade 5, is an alpha–beta titanium alloy with a high strength-to-weight ratio and excellent corrosion resistance. Emphasis is directed to extend the temperature at which titanium is useful to above 816 °C (1500 °F).
- **Titanium aluminide Intermetallic Alloys (Ti-Al)**. Intermetallics refers to a group of materials composed of two (or more) types of metal (or metal and nonmetal) atoms, which exist as solid compounds and differ in structure from that of the constituent components. In comparison with conventional metals and alloys, intermetallic-based alloys exhibit several specific features. For example, titanium aluminide (Ti-Al) alloys have much greater percentage of aluminum (25–50%) than conventional Ti alloys, yielding desirable characteristics for hypersonics. They have similar or lower density than titanium alloys and higher elastic modulus, and higher oxidation resistance and strength retention at temperatures up to 750 °C (1023 K).

Das et al. (2004) and Draper et al. (2007) identified titanium aluminide as a potential back-structure material for maintainable composite panel HEXs in the inlet, combustor, and nozzle sections of a Turbine-Based Combined-Cycle (TBCC) propelled vehicle, where weight reduction is crucial. Their design trade-off studies showed that a Ti-Al structure, utilizing a high-strength, high-temperature Ti-Al alloy called Gamma MET PX™, reduces weight by 41–48% in comparison with baseline Inconel 718 configuration for the TBCC propulsion system.

Jozwick et al. (2015) published an overview of current and prospective applications of Ni_3Al-based intermetallic alloys. These are modern engineering materials with special properties that are potentially useful for both structural and functional purposes. They also reported development of composite materials with Ni_3Al-based alloys as a matrix hardened by TiC, ZrO_2, tungsten carbide (WC), SiC, and graphene.

Aluminum alloys are useful to moderately elevated temperatures (177 °C and 350 °F) only. However, they can be used in hypersonic airframes. Aluminum–lithium-based alloys and high-temperature aluminum alloys could improve performance of hypersonic vehicles. Several new aluminum alloy compositions have been developed which retain their mechanical properties to temperatures as high as 288 °C (550 °F).

9.5.2 Refractory Metals and Ceramic Matrix Composites

Refractory metals are a class of metals that are extraordinarily resistant to heat and wear. To be defined as refractory, a metal has to have a melting point above 2200 °C (4000 °F). Refractory metals include niobium, molybdenum, tantalum, tungsten, and rhenium.

Tungsten has the highest melting point of all metals, at 3410 °C (6170 °F). Its high melting point makes tungsten a good material for application in rocket nozzles. Tungsten and its alloys are often used in applications where high temperatures are present but still a high strength is necessary and the high density is not troublesome. In fact, the X-43A vehicle flat nose was made of solid tungsten, Densalloy SD 180, which was surrounded by carbon–carbon leading edges. The X-51A waverider was also designed with a tungsten nose cap, covered with silicon dioxide thermal barrier coating (TBC), which protected the nose tip from the high heat load and also served as ballast for the vehicle longitudinal stability (see Section 9.6).

Refractory alloys used in the aerospace industry are primarily columbium and tantalum alloys. Columbium alloys have a maximum use temperature of 1316 °C (2400 °F), and tantalum alloys have a maximum use temperature of 1538 °C (2800 °F). Columbium alloys have about half the density of tantalum alloys.

Refractory intermetallics are refractory metal-based alloys capable of operating at high temperatures. Seven intermetallic compounds, Nb_3Al, Cr_3Si, Co_2Nb, Cr_2Nb, $MoSi_2$, Mo_5Si_3, and Nb_2Al, with melting temperatures above 1600 °C, are candidate materials for high-temperature structural applications in advanced aero-turbines. These intermetallic materials were evaluated by Anton and Shah (1990), considering their creep strength, ultimate tensile strength and oxidation resistance for that application.

Ceramics have desirable properties for high-temperature applications, including high melting temperature, low thermal conductivity, high modulus and compressive strength, and creep resistance. Although ceramics have high compressive but low tensile strength, they have application in hypersonics due to their environmental resistance without dependence on coatings. Ceramics can be used in monolithic form as airframe nose and leading-edge inserts.

Composite materials have two parts: the reinforcement, which provides special mechanical properties such as stiffness or strength, and the matrix material, which holds everything together. Ceramic matrix composites (CMCs) are a special type of composite material in which both the reinforcement (refractory fibers) and matrix material are ceramics. In some cases, the same kind of ceramic is used for both parts of the structure, and additional secondary fibers may also be included. Because of this, CMCs are considered a subgroup of both composite materials and ceramics.

9.5.3 Carbon–Carbon Composites

One of the most advanced and promising engineering materials is the carbon fiber-reinforced carbon matrix composite (C—C composites). Carbon (fiber)–carbon (matrix) composites are among the strongest and lightest high-temperature engineered material in the world. Compared to other materials such as graphite, ceramics, metal, and plastic, carbon–carbon (C—C) is lightweight and strong and can withstand temperatures over 2000 °C without any loss in performance. C—C composites have excellent high-temperature thermal and mechanical properties such as:

- Mechanical properties are stable to temperatures approaching 3000 °C (5432 °F);
- Good resistance to thermal shock, corrosion, and creep;
- Specific strength and specific modulus are relatively high, and coefficient of thermal expansion (CTE) is relatively low;
- Stiffness, strength, and thermal conductivity are highest in fiber direction and lowest in direction transverse to fibers.

A disadvantage of C—C composites is that they exhibit severe high-temperature oxidation, which causes rapid deterioration and erosion. Carbon becomes susceptible to oxidation above 350 °C, and oxidation rate increases rapidly with temperature. Hence, C—C composites require oxidation-resistant surface coatings.

The thin, wide nose of the NASA X-43A vehicle was designed with C—C at the leading edges. For the Mach 7 vehicle, the C—C composite was tailored to reduce nose temperature by orienting the K321 fibers perpendicular to the leading edge. This enabled conduction heat transfer away from leading edges (Ohlhorst et al. 2005). Figure 9.7 shows the 3 : 1 weave architecture of the C—C leading edge. The side chine made of 3-D needled C—C polyacrylonitrile (PAN)-based fiber is shown in Figure 9.7. PAN is a carbon fiber polymer.

Carbon (fiber)–carbon (matrix) composites have the highest thermal conductivity per unit density among materials suitable for thermal management applications. The thermal conductivity of structural ceramics is plotted in Figure 9.8 as a function of temperature capacity. This is an Ashby-type map of temperature capability and nominal thermal conductivity of Ni-based superalloys, CMCs, Ultra-High Temperature Ceramic (UHTCs) and C—C composites (along high-conductivity orientation). The dashed curves are design lines for leading edges (inset) of different dimensions (t); arrows in the inset show conductive ($\dot{q}_{Cond}$) and radiative ($\dot{q}_{Rad}$) heat fluxes (Padture 2016). For thermal conductivity database of various structural carbon–carbon composites, the reader is directed to Ohlhorst et al. (1997).

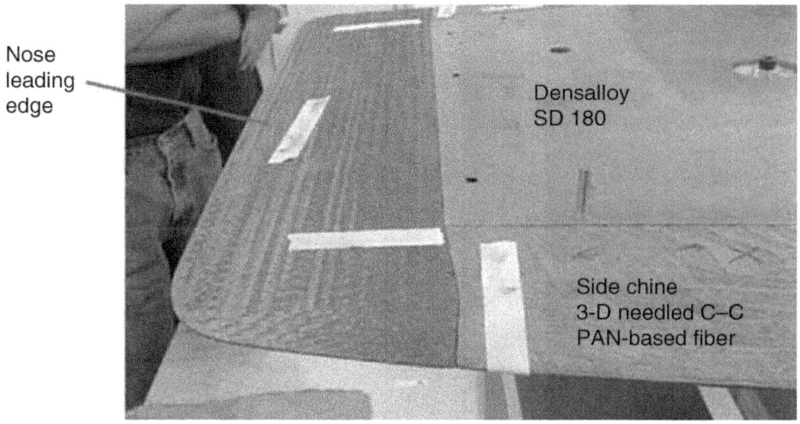

Figure 9.7 NASA X-43A Mach 7 vehicle nose leading edge made of high thermal conductivity carbon fibers woven in a 5 : 1 unbalanced weave. Source: Ohlhorst et al. (2005), NASA, Public domain.

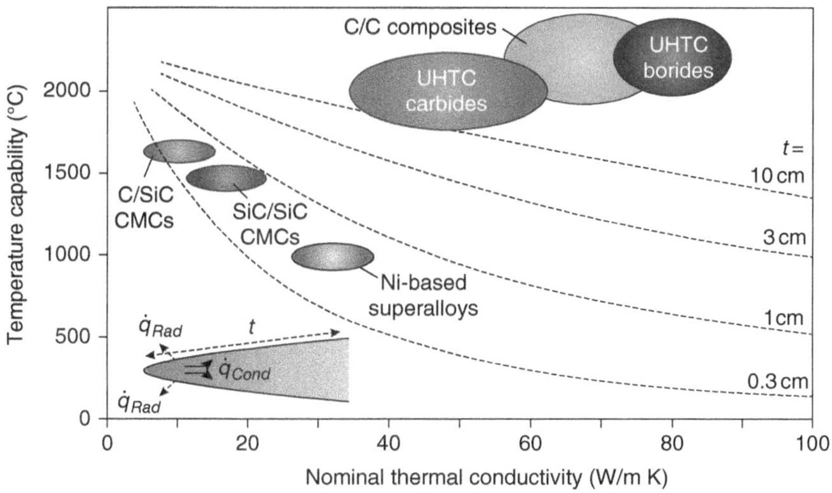

Figure 9.8 Thermal conductivity of CMC and Ni-based superalloy materials. Source: Padture (2016)/ Springer Nature.

9.5.4 Ceramic Matrix Composites (CMCs) and Metal Matrix Composites (MMCs)

CMCs are a subgroup of composite materials as well as a subgroup of ceramics. They consist of ceramic fibers embedded in a ceramic matrix. Both the matrix and the fibers can consist of any ceramic material, whereby carbon and carbon fibers can also be considered a ceramic material. Conventional ceramics are not appropriate for hypersonics due to their brittle failure and low fracture toughness and limited thermal shock resistance. However, CMCs have high-temperature strength, high strength-to-weight ratio, and outstanding environmental resistance without dependence on coatings. CMCs are normally identified by fiber/matrix. For example, C/SiC would be a carbon fiber in a silicon carbide matrix. The most commonly used CMCs include carbon–carbon (C/C), carbon–silicon carbide (C/SiC), SiC/SiC, and alumina silica (Al_2O_3-SiO_2). These CMCs can be used in airframe, control surfaces, and engine structures. These materials also exhibit resistance to hot hydrogen.

CMCs were developed to improve tensile strength and toughness while retaining compressive strength and thermal stability of pure ceramics. CMCs consist of a ceramic matrix phase reinforced by ceramic fibers or whiskers (typically SiC). CMCs can survive for long periods of time at very high temperatures, retaining their stiffness, strength, and toughness properties at temperatures that can exceed 3000 °C (5432 °F). CMCs are used for TPS in extreme environments, as they enable efficient structure materials (e.g. titanium or high-temperature polymer matrix composites) to carry structural loads.

SiC–SiC matrix composite is a particular type of CMC that can be used as an alternative to metallic alloys. A SiC/SiC composite is made by having a SiC (silicon carbide) matrix phase and a fiber phase incorporated together by different processing methods. SiC/SiC composites exhibit high thermal, mechanical, and chemical stability while also providing high strength-to-weight ratio.

Carbon fiber-reinforced silicon carbide matrix composites (C/SiC) and silicon carbide fiber-reinforced/silicon carbide matrix composites (SiC/SiC) are considered reusable materials because silicon carbide is a hard material with low erosion, and it forms a silica glass layer during oxidation which prevents further oxidation of inner materials. Above a certain temperature (it depends on environmental conditions of oxygen partial pressure), it starts the active oxidation of silicon carbide matrix to gaseous silicon monoxide, consequently loss of protection from further oxidation, which leads the material to an uncontrolled and fast erosion. For this reason, C/SiC and SiC/SiC are used in the temperature range between 1200 and 1400 °C.

Ultra-high-temperature ceramic matrix composites (UHTCMCs) or ultra-high-temperature ceramic composites (UHTCCs) are a class of refractory CMCs, which may overcome the limits associated with currently used CMCs (C/C and C/SiC) in TPS and in rocket nozzles. Carbon fiber-reinforced carbon matrix (C/C) can be used up to 3000 °C because carbon is the element with the highest melting point.

In the 1990s, NASA developed very high-temperature zirconium-based and hafnium-based CMCs. Tests have revealed that diborides of zirconium (ZrB_2) and of hafnium (HfB_2) were the most oxidation-resistant of the high-temperature materials studied. Arc jet tests of these materials demonstrated that they could survive heating rates over two times the capacity of RCC specimens and could survive temperatures in excess of 2500 K (4040 °F) (Rasky 1992).

CMCs can be used in applications requiring reliability at high temperatures (beyond the capability of metals) and resistance to corrosion and wear. CMCs are found in heat shield systems for rocket-propelled space vehicles, which need protection during the re-entry phase when the vehicle experiences higher temperatures, thermal shock conditions, and heavy vibration loads. CMCs are also used in gas turbine combustion chambers, stator vanes, and turbine blades. According to Glass (2008), CMCs provide the hypersonic vehicle the required strength at elevated temperatures.

Metal matrix composites (MMCs) are composite materials with at least two constituent parts, one is a metal and the other material may be a different metal or a ceramic or organic compound. When at least three materials are present, it is called a hybrid composite (Meysam et al. 2018).

MMCs are classified into different categories, depending upon their matrix materials. Some examples of most commonly used metallic matrix configurations are:

- Aluminum-based composites; aluminum as matrix can be either cast alloy or wrought alloy (i.e. AlMgSi, AlMg, AlCuSiMn, AlZnMgCu, AlCu, AlSiCuMg).
- Magnesium-based composites.
- Titanium-based composites.
- Copper-based composites.
- Superalloy-based composites.

Aluminum is the most commonly used matrix material in MMCs. As shown in Figure 9.6, MMCs, superalloys, and titanium all have good specific strength for temperatures up to 1093 °C (2000 °F). However, CMCs provide high strength at higher temperatures and thus are considered excellent materials for air-breathing vehicles. Ultimately, hypersonic materials must have high specific strength at elevated temperatures (Glass 2008).

Whereas silicon-based or oxide CMCs are aimed at applications that require longer life (100 000 hours of cycle time), non-oxide ultra-high-heat UHH CMCs are considered where temperatures are very high but the mission durations are very short.

9.5.5 Material for Scramjet Combustors

Combustor materials must maintain both adequate strength and resistance to corrosion/oxidation at the high-temperature conditions they will encounter. Combustor temperatures in excess of 2204 °C (4000 °F) are characteristic of engines designed to operate at hypersonic speeds. For example, considering a cruise Mach range 5–8, Van Wie et al. (2004) estimated steady-state temperatures for combustors to range from 4200 to 5500 °F (2316–3038 °C). These levels of temperature are excessive for any material.

Due to the enormous heat loads scramjet combustors must endure, they can be made of actively cooled metallic structures. An expandable missile may operate well with an uncooled combustor made of copper with sprayed zirconia TBC. However, CMCs are also considered, due to their high-temperature capabilities. A number of studies have estimated significant weight savings by making the combustor of CMCs (active and passive) such as carbon/carbon (C/C) and carbon/carbon-silicon carbide (C/C-SiC) materials. The design and operational advantages of this approach will depend on the specific requirements of the scramjet engine.

CMCs (fabricated by German Aerospace Center DLR) were ground-tested to establish their viability as a passively cooled combustor design for the Australia/US AFRL HIFiRE 8 Program, expected to be flight-tested at Mach 7 for about 30 seconds. Flat panels of DLR C/C and C/C-SiC materials were tested in the NASA Langley Direct Connect Supersonic Combustion Test Facility (DCSCTF) using the Durable Combustor Rig (DCR) test article. Simulating flight at Mach 5 and Mach 6, the estimated heat load on the CMC panels was estimated to be $\dot{q} \cong 1.9\ \text{MW/m}^2$ for several minutes (at a wall temperature $T_w = 300$ K). The results of these tests were very encouraging. The C/C-SiC panel survived the high-temperature scramjet combustor environment with very little erosion. The C/C panel experienced slightly more erosion, but still only a small amount, and for both exposed surfaces SEM analysis indicated very little oxidation (Glass et al. 2014).

9.5.6 Reusable Thermal Protection Materials

Reusable TPSs consist of materials that are mechanically or chemically unchanged by flight missions and can be safely flown a number of times (with or without servicing). NASA has developed many reusable TPS materials such as Silica-based Reusable Tiles, Ames Insulation Material (AIM), Fibrous Refractory Composite Insulation (FRCI), and Alumina Enhanced Thermal Barrier (AETB) (Myers et al. 2000). Typical applications include acreage areas for moderate heat fluxes. AETB is a reusable, higher temperature alumino-borosilicate ceramic tile that extends the temperature capability of insulating material. AETB-8 has a density of 0.13–0.15 g/cc while AETB-12 density is 0.17–0.21 g/cc. AETB consists of Nextel® fibers, alumina fibers, silica fibers, and silicon carbide, and is adaptable to complex geometries. AETB can be used for hypersonic vehicles operating with low-to-moderate heat flux rates. For the NASA X-43A, the windward (upper) surface was covered with black AETB with Toughened Uni-piece Fibrous Insulation (TUFI) coating.

9.5.7 Coatings

Materials and structural components for hypersonic vehicles require coatings for temperature control and for environmental protection. Coating for temperature control can be designed to have high emissivity and to be noncatalytic to the recombination of the dissociated gases present in the hypersonic airflow adjacent to the vehicle. Having such coating characteristics can lead to reductions in surface temperatures of several hundred degrees.

The TBC is a thin layer of a low thermal conductivity material (typically a ceramic) bounded to a material hot side to allow material to be used above its failure point. The coating has little structural strength. The wall material provides structural strength but if it cannot withstand high temperatures, and a TBC will allow this wall to operate at higher temperature. A bound layer between material and TBC allows for differential expansion.

Coatings for environmental protection are those that protect the hypersonic materials from the hot oxidizing atmosphere the airframe and engine flow path are exposed. For example, C—C composites suffer severe high-temperature oxidation, which causes rapid deterioration and erosion. Carbon becomes susceptible to oxidation above 350 °C, and oxidation rate increases rapidly with temperature. Hence, C—C composites require oxidation-resistant surface coatings. Titanium aluminides also require oxidation protection.

Moreover, when any part of the vehicle structure uses hydrogen fuel for cooling, the hydrogen readily diffuses through most materials and can form brittle compounds within the structure. Hence, hydrogen barrier coating is required. However, this poses a challenge because the coating must be thin, lightweight, resistant to damage, and easily applicable to complicated shapes, including intricate internal cooling passages.

For some missile applications, scramjet combustors may be designed with a liner from uncooled coated carbon–carbon of sufficient thickness to insure adequate strength after material losses from corrosion/oxidation. For example, C/C-SiC material (which has a maximum service temperature of 2200 K) can be protected from thermal oxidation if it is coated with yttrium silicate (YSi). The coating itself has been tested to up to 1800 K in ground and flight experiments. One drawback of this approach is that the complex heterogeneous microstructure of C/C-SiC makes prediction of material aging and failure very difficult (Glass et al. 2014).

Solid silicon carbide (SiC) and silicon nitride (Si_3N_4) are used as coating material on many refractory composite structures. They provide a diffusion-resistant barrier to oxygen at high temperatures. For example, the C—C tiles on the Space Shuttle Orbiter were coated with SiC to provide oxidation protection to about 1200 °C. The SiC coating is chemically compatible with C—C composite substrate material (used on the X-43A Mach 7 vehicle). However, SiC coating is susceptible to cracking due to CTE mismatch with the C—C composite material.

Research is now focused on improved coating concepts better suited to reusable hypersonic vehicles. Zirconia-based ceramic coatings, usually stabilized with dopants like yttria, have demonstrated superior performance in providing the best combination of thermal insulation, wear, oxidation, and corrosion resistance. Gas turbine blades and combustors and some rocket nozzles may use thin layer coatings of yttria-stabilized zirconia (YSZ). Zirconia coating was used on the forward section of the X-43A scramjet engine cowl and in key places throughout the engine for additional thermal protection (see Figure 9.9).

The European LAPCAT II supersonic combustor tested on the wind tunnel is made of a copper alloy. Its inner walls include a 0.3-mm-thick YSZ TBC. According to researchers (Vincent-Radonnier et al. 2014), the surface finish of this coating is quite rough (almost like sand paper) and can induce the thickening of the boundary layer in the combustor. Such material–boundary layer interaction is carefully examined to determine effects on overall combustor performance.

To protect C/C-SiC material from thermal oxidation, YSI coating has been tested to up to 1527 °C (1800 K) in ground and flight experiments and is chemically stable up to 1927 °C (2200 K). YSI coatings are promising complements to SiC coatings for protecting C/C—Si—SiC composites against oxidation owing to the pronounced chemical and mechanical properties of this material (low Young's modulus, low thermal expansion coefficient, low evaporation rate and oxygen permeability, and good erosion resistance).

Figure 9.9 X-43A vehicle hot structure: nose, horizontal tail, and vertical tail. Source: NASA.

Coatings based on SiC, such as those used on the Space Shuttle Orbiter leading edges, operate effectively at temperatures up to about 1649 °C (3000 °F). On certain hypersonic air-breathing propulsion vehicles, the temperatures on the thin leading edges may exceed 1649 °C. Thus, a different class of material coatings will be required, e.g. carbides, oxides, and diborides of Hf and Zr, and they could be used as part of a composite matrix. Research at NASA found that materials such as iridium (Ir) are more appropriate as a coating for air-breathing hypersonic applications (Glass 2008).

Ceramic–metal functionally graded materials (FGMs) have attracted attention as TBCs for aerospace structures working under super high temperatures and thermal gradients (Ho et al. 2007). The FGM technique involves tailoring the internal microstructure of composite materials to specific applications, producing a microstructure with continuously varying thermal and mechanical properties at the continuum or bulk level.

9.6 Examples of Vehicle Development Structure and Materials

9.6.1 X-43A Lifting Body Vehicle: LH_2 Fueled, Mach 7, Mach 10 Scramjet Flight Demonstrator

Vehicle Aerodynamic Design and Leading Edges: The X-43A is a lifting body aircraft with aft-mounted control surfaces which include all-moving horizontal tails and twin vertical fins and rudders. To reduce drag, all the leading edges are sharp, and the control surfaces are relatively thin. The windward (lower) surface of the forebody is contoured to provide the external air compression ramps for the scramjet inlet, and the lower afterbody section is also contoured to provide a single-expansion nozzle surface. The leeward (upper) surface of the fuselage is slightly curved, and its forward section is made of a massive solid wedge of tungsten, weighing about 392 kg (865 lbs). The tungsten section serves as ballast to provide a vehicle center of gravity sufficiently far forward to yield nearly neutral longitudinal stability characteristics.

Since the main goal of the NASA Hyper-X program was to test the hydrogen-fueled scramjet engine at Mach 7 and Mach 10, two slightly different vehicles were built. The more demanding flight conditions for the Mach 10 mission required some changes to the vehicle hardware, which impacted the outer mold line, the engine flowpath, the vehicle mass properties, and the TPS.

The leading-edge hardware for the Mach 10 flight vehicle consisted of 11 pieces: a nose flat tip, two forward chines, two aft chines, two horizontal tail pieces, two upper vertical tail pieces, and two lower vertical tail pieces. For the Mach 7 flight vehicles, only seven of the leading-edge pieces were fabricated of C/C since thermal analysis indicated that the four vertical tail pieces would not be subjected to high enough temperatures to require C/C and thus could be fabricated from a Haynes alloy (Ohlhorst et al. 2005).

Hot Structure and TPS: The materials for the X-43A structure were selected on the basis of vehicle stiffness requirements and thermal management requirements. These materials include steel, titanium, and aluminum longerons and bulkheads with steel and aluminum skins.

For its short flight demonstration test, the vehicle had a hot structure consisting of nose, horizontal tail, and vertical tail (Figure 9.9). These components are considered hot structure because they do not employ TPS. For example, the horizontal and vertical tails were made of Haynes 230, a nickel-steel alloy with a melting temperature range of 1302–1371 °C (2375–2500 °F) and relatively low thermal expansion characteristic. For the Mach 7 vehicle, only the leading edge of the tail horizontal surfaces were protected by carbon–carbon inserts, whereas for the Mach 10 vehicle, both horizontal and vertical surfaces leading edges were made of carbon–carbon (C—C) material.

The nose was fabricated of Tungsten Densalloy SD 180. The sharp leading edges around the tungsten nose were made of carbon–carbon (C/C) coated with a three-layer coating composed of silicon carbide (SiC), followed by a chemical vapor deposition (CVD) layer of SiC, followed by a thin CVD layer of hafnium carbide (HfC) (Rivers and Glass 2006).

The fuselage was thermally protected. As shown in Figure 9.9, the upper surface of the airframe (acreage) was covered with AETB tile with TUFI coating. For the Mach 10 vehicle, the C—C pieces on the leading edges of the nose, wings, and vertical tails were also treated with a HfC coating in addition to the SiC coating used on the Mach 7 vehicle.

Scramjet: The hydrogen-fueled engine was a self-contained unit (Figure 9.10) with a rectangular flowpath. It was an airframe-integrated design, i.e. the vehicle forebody and afterbody provided external compression and expansion surfaces for the flow through the engine. The engine proper had an actuated inlet cowl door, which remained closed during boost and was opened for scramjet operation; the door closed again during the vehicle descent after the scramjet completed its test in flight.

The engine was made of heat sink copper alloy (Glidcop) and incorporated active water cooling of its cowl and sidewall leading edges (Bakos 2008). Glidcop is a copper-based MMC alloy mixed primarily with small amounts of aluminum oxide ceramic particles. The Glidcop section was attached to a stainless steel strongback, which in turn connected it to the airframe. The heat sink design for the combustor (wall solid metal absorbed the intense heat of combustion) was deemed acceptable for the short-duration flight. The combustor walls would ultimately have melted if the engine had continued operating for more than a few seconds.

Figure 9.10 Top-down internal view of X-43 scramjet engine. Source: NASA.

Both the sharp-edged cowl and the vertical leading edges of the engine were water-cooled for thermal protection during boost and engine operation. The forward section of the engine cowl had zirconia coating, which was also used in other key places throughout the engine for thermal protection. In addition, the vehicle also incorporated a nitrogen purge system, which provided cooling of certain vehicle subsystems during the flight test phase (Grindle 2011).

Since the scramjet engine would only operate in flight for about 10 seconds, a simple cooling system using a water–glycol mixture was considered sufficient during the boost phase. This resulted in a far simpler design than would be required for a more advanced flight-ready, actively cooled engine. The mixture of water and ethylene glycol is a kind of traditional heat transfer fluid commonly used in many energy systems to increase the cooling capability of water. A water-only coolant system provided active cooling for the engine leading edges during free flight. The GH_2 HEXs, igniter tank, and GN_2 tank used thermostatically controlled resistive heating elements to maintain their temperatures at 150 °F (65.6 °C). The stored nitrogen gas was also used as a pressurant to move the coolant to the engine leading edges (Marshall et al. 2005).

Materials considered for vehicle nose and leading edges. For the Mach 10 vehicle with a 0.030-in. (0.762 mm) nose radius, preliminary thermal analysis predicted the temperature at the nose leading edge would reach temperatures near 2204 °C (4000 °F). Hence, typical high-temperature materials would not survive that extreme temperature for even a short-duration single flight. Therefore, a ground testing campaign was carried out to identify a suitable material for the nose leading edge. Testing was performed on 13 leading edge segments fabricated from as many different materials to evaluate their performance in a simulated flight environment. Table 9.3 shows the leading-edge design requirements, indicating the major concern of each (Ohlhorst et al. 2005). For the nose, major concerns were tip temperature and high thermal gradients leading to high compressive stresses. Testing was done to evaluate coatings for single use on a C/C substrate at Mach 10 heating conditions for 130 seconds.

Results from these tests showed ablation of the C—C nose leading edge at heating conditions and durations that were more severe than the planned flight trajectory. Even though these conditions were not likely to be experienced in flight, the Hyper-X team decided to design and machine a new nosepiece with a larger leading edge radius, reducing the likelihood of material ablation (Marshall et al. 2005). The nose leading edge was redesigned to

Table 9.3 X-43A leading edge design and requirements.

Part	Nominal size, cm (in.)	Max. temperature, °C (°F)	Major concerns
Nose	45.72 × 12.7 × 1.524 (18 × 5 × 0.6)	2093 °C (3800 °F)	High thermal gradient
Chine	45.72 × 10.16 × 7.62 (18 × 4 × 3)	704 °C (1300 °F)	Thickness
Horizontal tail	83.82 × 12.7 × 1.524 (33 × 5 × 0.6)	1760 °C (3200 °F)	High temperature at root
Vertical tail	35.56 × 12.7 × 1.778 (14 × 5 × 0.7)	1538 °C (2800 °F)	Fixed clearance

Source: Adapted from Ohlhorst et al. (2005).

have a 0.050-in. (1.27 mm) radius. This modification also required an alteration to the upper outer mold line to prevent a change to the nose planform.

After this testing, the NASA team concluded that the C—C parts were acceptable for flight. This composite material was required due to both the severe stagnation point heating on the leading edges and the shock-on-shock heating caused by the vehicle bow shock and the control surface shocks. Thus, all the leading edges for the X-43A flight vehicles were made of C—C composites, being lightweight materials with great strength and stiffness, even at very high temperatures. The thermal-resistant coating for the C—C had to be lightweight and thin, as the vehicle was designed to very precise specifications and could not afford to have a bulky coating.

Tungsten was selected for the vehicle nose because it has high strength, impact resistance, erosion and corrosion resistance, and resistance to thermal shock. Titanium (Ti-6A1-4V) has high strength-to-weight ratio compared to tungsten. However, the strength of titanium at high temperatures deteriorates much more quickly than tungsten.

For the leading edge of the nose, reinforced C—C composite was selected because it has low density, high strength-to-weight ratio, and ability to withstand high temperatures. It retains its properties up to 2273 K. Due to its low impact resistance, it was not used in the entire nose section.

Before the Mach 10 flight, engineers observed delamination of a section of the right forward C—C chine around the tungsten nose. Delamination occurs when the layers of fibers peel out and get detached from its adjacent layers, posing severe threat to the strength of the material. This defect required a replacement chine and, like all other C—C pieces, it underwent heat treatment cycles during the manufacturing process and was subjected to tap testing and thermographic inspection to ensure all pieces were intact before flight (Marshall et al. 2005).

Delamination is a critical failure mode in fiber-reinforced composites. It may lead directly to through-thickness failure owing to interlaminar stresses caused by out-of-plane loading, curved or tapered geometry, or discontinuities owing to cracks, ply drops, or free edges.

9.6.2 X-51A Waverider: JP-7-Fueled, Mach 6 Scramjet Flight Demonstrator

Conventional metal materials were used to construct the X-51A waverider structure. For thermal protection, the vehicle was covered with lightweight foam and tiles (Figure 9.11). The internal airframe and bulkheads were made from machined aluminum. The skin of the cruiser, the interstage, and the four movable fins on the booster were also made of aluminum. Titanium was used for only the interstage flow-through structure and the booster boat tail (Hank et al. 2008).

The waverider had four moveable fins made of Inconel with C—C composite leading edges. The nose cap was made of 150-lb (68 kg) solid tungsten, serving as ballast for the vehicle's longitudinal stability, and it rested on an Inconel nose cap adapter to prevent thermal conduction to the rest of the vehicle.

The tungsten nose cap was covered with silicon dioxide (SiO_2) TBC to protect the nose from the 1482 °C (2700 °F) temperature it experienced (Hank et al. 2008). This type of coating is a transparent to gray, odorless, crystalline, or amorphous solid. Its melting and boiling point are 1600 and 2230 °C, respectively. Refractory elements were chosen for control fin

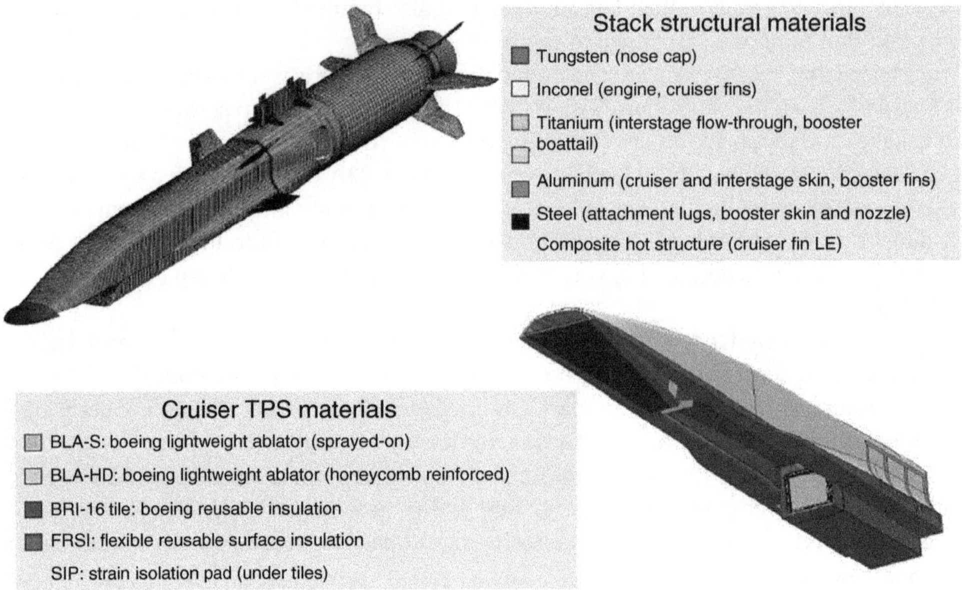

Figure 9.11 X-51A materials. Source: Mutzman et al. (2011)/American Institute of Aeronautics and Astronautics/Public Domain.

leading edges and nose cap. A composite hot structure (CHS) was utilized for leading edges due to its high specific strength and stiffness at high temperature. A refractory metallic was chosen for the nose cap due to its high-temperature capability and synergy with the ballast required to support flying qualities. Both materials were coated to control oxidation.

The external vehicle skin and exhaust nozzle wall temperatures were expected to range from about 816 to 1927 °C (1500–3500 °F) (Lane 2007). The TPS on the X-51 cruise vehicle included tiles, insulation, and ablator materials. Boeing Reusable Insulation (BRI-16) tiles were used in the chine leading edge and in the inlet ramp. The tiles were glued to a strain isolation pad (SIP) to absorb any strains caused by the different expansion rates of the tiles and the aluminum skin underneath. BRI-16 tiles are similar to those used on portions of the Space Shuttle Orbiter underside.

The interior walls of the scramjet engine were covered with Flexible Reusable Surface Insulation (FRSI) in order to protect the walls from the primarily radiative heat loads. A honeycomb-reinforced version of ablator material, identified as Boeing Lightweight Ablator (BLA-HD), was used on the interior of the vehicle nozzle, which was exposed to the hot combustion products from the scramjet engine.

The leeward (upper) surface of the cruiser was protected by FRSI, and this insulation was covered with Boeing Lightweight Ablator (BLA-S) foam of varying thicknesses. BLA-S can be sprayed to cover a large area quickly and then sanded down or carved into the exact outer mold line desired. This characteristic of the ablator foam is an advantage, as any seams or gaps around access panels used to service the vehicle immediately prior to flight can be quickly filled and shaped.

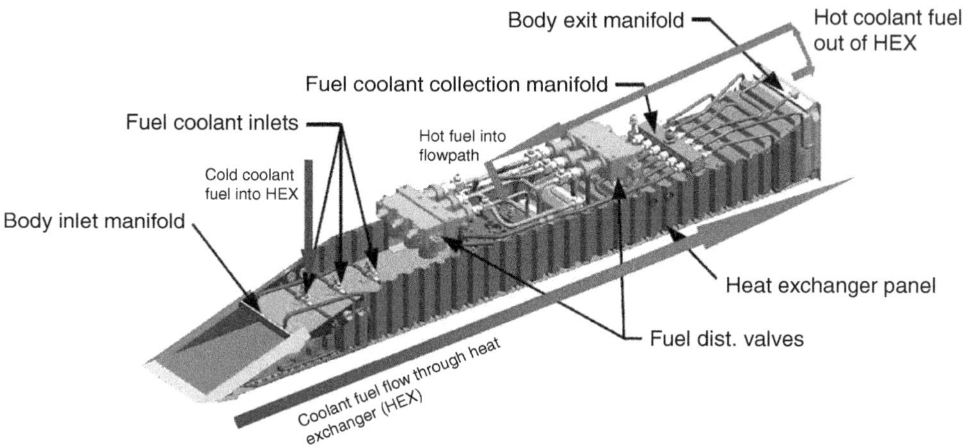

Figure 9.12 X-51A engine structure actively cooled with JP-7 fuel. Source: Mutzman et al. (2011)/ American Institute of Aeronautics and Astronautics.

Unlike the water/heat sink-cooled X-43A scramjet, the X-51A scramjet engine used JP-7 fuel in a regenerative or active cooling system. In this design approach, liquid JP-7 entered at the front of the scramjet and was distributed to each wall of the HEX. As the fuel circulated through the length of the HEX, it absorbed heat from the combustion chamber, keeping the thin walls below the melting point of the Inconel (1290–1350 °C or 2350–2460 °F) (Hank et al. 2008). The amount of heat carried off was to match the amount the burning JP-7 fuel added to the engine, achieving thermal balance, which is one of the keys characteristics in a practical scramjet propulsion system. The X-51A actively cooled design (Figure 9.12) included a control system to keep the pressure of the fuel in the HEX above a threshold to prevent boiling. The hot JP-7 fuel was then routed to the fuel distribution valves, expanding to a vapor across the valves, and the gaseous fuel was then delivered to various injector locations in the X-51A combustor (Hank et al. 2008).

9.6.3 LAPCAT 2: LH2 Mach 5 Civil Transport Concept

In Europe, the Long-Term Advanced Propulsion Concepts and Technologies (LAPCAT) and LAPCAT II programs are designing a Mach 5 aircraft and developing critical technologies to realize antipodal flights with a range of 18 700 km in less than four hours (Steelant 2007).

The 300 passenger A2 concept vehicle designed by British company Reaction Engines (REL) has a slender fuselage with a low aspect ratio delta wing positioned slightly aft of the mid fuselage section. The design was selected to obtain good supersonic and subsonic lift-to-drag ratios and acceptable low-speed handling qualities for takeoff and landing. The A2 will be propelled by four hydrogen-fueled precooled turboramjet engines named Scimitar, also designed by REL. The 139-m-long fuselage has a diameter of 7.5 m. The passenger cabin is 32 m long (arranged in two decks). The liquid hydrogen fuel tanks are split into two large pressurized tanks on either side of the passenger compartment. Storage of the fuel in the fuselage instead of in the wings allows tanks of circular cross-section, which minimizes insulation and pressure vessel mass (Verstraete et al. 2012).

Due to the high operating temperatures, aluminum is not an option for the structure of the A2 hypersonic transport. With the long exposure times, the complete structure would reach an unacceptable elevated temperature and heat would conduct throughout the entire vehicle. Hence, the LAPCAT team is considering titanium alloys as the primary candidate material because this material could operate effectively at temperatures up to 800 K and have, relative to most other metals, a significantly high specific strength. Beryllium alloys, carbon–carbon composites, and MMCs are also considered (Verstraete et al. 2012; Sharifzadeh et al. 2014).

9.6.4 SR-71: Mach 3.2 Military Aircraft

The structure of the SR-71 long-range, high-altitude, Mach 3+ strategic reconnaissance aircraft was constructed with titanium (85% of the structure) and polymer composite materials. Engineers had predicted temperatures on the aircraft's leading edges exceeding 538 °C (1000 °F), causing them to deal with seemingly insurmountable design and material challenges. Titanium alloy was the only option for the airframe, as it had the required strength, relatively lightweight, and promised durability at the excessive temperatures the supersonic aircraft would experience.

Titanium, however, was found to be a particularly sensitive material from which to build an airplane. The brittle alloy shattered if mishandled. Since conventional cadmium-plated steel tools embrittled the titanium on contact, this forced the design and fabrication of new tools made of titanium. It was reported that the SR-71 titanium fuselage increased in length by about 10 inches during the supersonic flight. For the cruise segment, the temperature gradient between fuselage windward and leeward surfaces caused that the vehicle's nose experienced a significant structural droop.

Interestingly, although aerodynamic heating was expected to generate high heat flux at the leading edges of the aircraft, the ambient temperature outside the cockpit window would be −51 °C (−60 °F). Engineer Ben Rich was responsible for tackling the problem of how heat could be dissipated across the entire airframe. He recalled a simple lesson from one of his university courses: black paint both emits and absorbs heat. Thus, the aircraft was painted black, and soon after the SR-71 earned its nickname: "Blackbird."

9.7 Materials and Structures Technical Challenges

Material capability and performance, used in concert with a viable thermal management approach, is a significant enabler for hypersonic flight. Research by Glass et al. (2014) predicted that, at Mach 6 cruise speeds, the temperature of scramjet combustor liners can exceed 2204 °C (4000 °F), with higher liner temperatures during the transient climb phase. They proposed to use coated C–C as the liner structural member, since C–C material has excellent high-temperature strength properties. However, the upper temperature limit on available oxidation-resistant coatings is about 1649 °C (3000 °F). Therefore, research and development must focus on extending the coating usable temperature range to

1927–2038 °C (3500–3700 °F). In addition, researchers consider cooling the liner to a temperature at which coated C—C can survive without significant ablation.

As we discussed throughout the book, the inlet of the scramjet engine must be designed with sharp leading edges in order to maximize capture and compression of the freestream air. Predictions indicate that aerodynamic heat fluxes at the leading edge stagnation line may be very high. Thus, the stagnation line temperature of uncooled leading edges could reach temperatures that would melt conventional materials. One alternative is to use metallic materials (titanium, titanium aluminides, and superalloys), which offer good elevated temperature capability, but they are heavier than composites. In fact, C—C composites and other CMCs offer best potential as lightweight, very high-temperature structural materials for hypersonic flight applications. However, some technical issues remain to be investigated, such as (i) oxidation resistance, (ii) coating compatibility, and (iii) material toughness, in order to advance the utilization of C—C composite materials in high-speed flight systems.

9.7.1 Thermostructural Analysis

The structural design of hypersonic vehicles requires a radical approach to address the challenges associated with the heating effects while flying at high Mach numbers. At hypersonic speeds, aerodynamic heating poses serious structural problems such as expansion and distortion of structures. At high temperatures, conventional structural materials, such as metals, suffer from reduced strength, reduced stiffness, increased creep, increased oxidation, increased thermal stresses, and other detrimental effects that impact the structural design.

High thermal gradients result in high thermal stresses. Such phenomenon was evident during testing of the Hyper X Mach 10 vehicle; the leading edges had a deformed geometry due to the thermal expansion resulting from the high-temperature environment. Thermal structural modeling predicted the leading edges would fail due to the high compressive thermal stresses resulting from the constraint of the thermal expansion (Glass 2008). Heat transfer generates large thermal stresses. Thermal strains can be destructive in materials that may be strong but brittle. If the material is not appropriate to work with those stresses, it may rupture the first time it is heated or may suffer thermal fatigue after a few thermal or hot–cold cycles.

Consider a simple convectively cooled flat wall at temperature T_w. A hot wall surface tends to expand relative to a cold wall, so the hot surface will develop a compressive stress and the cold wall will experience a tensile stress. The thermal stress on a surface is proportional to the material properties $\alpha E/k$, where E is the elastic modulus (Young's modulus), α is the CTE, and k is thermal conductivity. The thermal stress can be expressed as:

$$\sigma_t = \frac{E\alpha}{k}\dot{q}_w t \qquad (9.13)$$

Hence, knowing the heat transfer rate to a wall and the wall material properties, we can easily estimate the thermal stress. It is clear from Eq. (9.13) that to reduce the stress on a surface, we must strive to utilize materials that are strong and have low thermal expansion and high thermal conductivity.

A similar approximation can be obtained for the case of a curved surface, such as a leading edge or a vehicle nose. Consider the nose of a vehicle with inner and outer radii given by a and b, respectively, and made of Haynes 25. The thermal stress σ_t on the wall at the stagnation line can be calculated from the formula:

$$\sigma_t = \frac{E\alpha\Delta T_w}{2(1-\nu)}\left(1 + \frac{b-a}{3a}\right) \tag{9.14}$$

where E is the elastic modulus, α is the coefficient of expansion, and ν is Poisson's ratio. Using English units, Eq. (9.14) yields a thermal stress of 4510 psi. Compare this value with an estimated yield strength of 10 000 psi and an estimated ultimate strength of 30 000 psi, for Haynes 25 at 1800 °F. See thermophysical data for TPS materials in Williams and Curry (1992).

Hypersonic technology development must focus on cost-effective, environmentally durable, and manufacturable material systems capable of operating at temperatures from 1500 to 3000 °C, depending on the application (Rivers and Glass 2006).

Understanding and managing such enormous heating is a particular challenge on the flight development program of hypersonic air-breathing-propelled vehicles. Managing the thermal environment, e.g. whether the nose radius should be large or small and what trajectory should be flown, will depend on a variety of factors, including the propulsion system, the TPS, the time of flight, reusability requirements, and costs.

The structural design of future hypersonic air-breathing vehicles is a very demanding task. It requires an aggressive materials technology development for application to the primary structure. The technical challenges that will enable hypersonic flight include:

- Development of reusable, flight-weight cryogenic tankage,
- Demonstrate active cooling concepts for engine inlet cowl lip,
- Development of materials for hot structures,
- Establish factors of safety and probabilistic approach to structural design,
- Develop high-temperature, fast-response instrumentation,
- Improve hypersonic analytical tools and test methods,
- Establish structural certification requirements.

The need to protect a hypersonic vehicle against aerodynamic heating drives its design. Aerothermal modeling must provide an accurate prediction of the heating environment encountered by a vehicle. This modeling is coupled and entwined with TPS design. The TPS is designed to withstand the predicted environment with risk-appropriate margin. The value of heat flux (with pressure and shear) is used to select TPS material. The heat load determines the thickness of the TPS. The engineering design goal is to minimize conduction into the vehicle to minimize TPS mass/risk.

Approaches to determine the aerodynamic heating environment range from simple approximate analyses to detailed computational fluid dynamic (CFD) solutions, which are coupled with ground and flight testing. Approximate methods are widely used because they yield results with only a fraction of the computational time that is usually required to obtain computing-intensive numerical solutions. A number of approximate methods have been developed with algebraic equations that accurately predict stagnation point and laminar or turbulent heat transfer distributions for axisymmetric bodies. These equations assume that the flow field is a continuum and chemical nonequilibrium effects are insignificant. When the geometry and the flow field are more complicated, numerical solutions are required (Bouchez 2010).

There are powerful numerical tools used today for the design of the TPS of a hypersonic vehicle (see one example by Marley and Driscoll 2017). The current state of the art in numerical simulation of hypersonic flows involves the steady solution of the reacting Navier–Stokes equations via CFD or Direct Simulation Monte Carlo (DSMC) methods. DSMC simulations are performed on regions where CFD methods are not accurate. One can perform full 3-D simulations in hours to days, although longer time is required for the simulation of specific details.

Questions

1. Conventional TPS that includes ablative coatings (e.g. NASA's space rocket vehicles) and reusable reinforced carbon–carbon system (e.g. Space Shuttle Orbiter) keeps heat flux of susceptible aluminum airframe below a critical level. Is there an alternative TPS approach for hypersonic air-breathing propulsion-powered vehicles?

2. What is the most appropriate thermal management approach for a Mach 5 civil aircraft?

References

Amundsen, R.M., Leonard, C.P., Bruce, W.E., III (2004). Hyper-X hot structures comparison of thermal analysis and flight data. *Fifteenth Annual Thermal and Fluids Analysis Workshop (TFAWS)*, Pasadena, CA (30 August to 3 September 2004).

Anton, D. and Shah, D., "High temperature properties of refractory intermetallics." *MRS Proceedings*, (1990) 213, 733. doi:https://doi.org/10.1557/PROC-213-733

Bakos, R. (2008). Current hypersonic research in the USA. *Advances on Propulsion Technology for High-Speed Aircraft*, 10-1–10-26. Educational Notes RTO-EN-AVT-150, Paper 10. Neuilly-sur-Seine, France.

Bertin, J.J. (1994). *Hypersonic Aerothermodynamics (AIAA Education Series)*. American Institute of Aeronautics and Astronautics.

Bouchez, M. (2010). Scramjet Thermal Management. *Tenue thermique des superstatoréacteurs*. RTO-EN-AVT-185. Lecture of RTO-AVT-VKI course "High Speed Propulsion: Engine Design – Integration and Thermal Management" Von Karman Institute, Belgium (September 2010).

Das, D., Kestler, H., Clemens, H., and Bartolotta, P.A. (2004). Sheet gamma TiAl: status and opportunities. *JOM* 56 (11): 42–45.

Draper, S.L., Krause, D., Lerch, B. et al. (2007). Development and evaluation of TiAl. Sheet Structures for Hypersonic Applications. *Materials Science and Engineering: A*, 464, 1–2, 330–342.

Glass, D.E. (2008). Ceramic matrix composite (CMC) thermal protection systems (TPS) and hot structures for hypersonic vehicles. *Paper AIAA 2008-2682. 15th AIAA International Space Planes and Hypersonic Systems and Technologies Conference*, Dayton, Ohio (28 April to 1 May 2008).

Glass S.E., Capriotti, D.P., Reimer, T., Kutemeyer, M., and Smart, M. (2014). Testing of DLR C/C-SiC and C/C for HIFiRE 8 scramjet combustor. *19th AIAA International Space Planes and Hypersonic Systems and Technologies Conference*, Atlanta, GA (16–20 June 2014).

Grindle, L. (2011). X-43A final flight observations. NASA Presentation, *DFRC-E-DAA-TN3463. Flight Test Safety Workshop*, Pensacola, FL (3–6 May 2011).

Hallion, R.P. (2008). *The NACA, NASA, and the Supersonic–Hypersonic Frontier*, NASA's First 50 Years Program, vol. 273 (28–29 October 2008). NASA Book.

Hank, J.M., Murphy, J.S., and Mutzman, R.C. (2008). The X-51A scramjet engine flight demonstration program. *Paper AIAA 2008-2540. 15th AIAA International Space Planes and Hypersonic Systems and Technologies Conference*, Dayton, OH (28 April to 1 May 2008).

Heiser, W.H. and Pratt, D.T. (1994). *Hypersonic Airbreathing Propulsion*. AIAA Education Series.

Ho, S.Y., Kotousov, A., Nguyen, P., et al. (2007) FGM (Functionally Graded Material) Thermal Barrier Coatings for Hypersonic Structures: Design and Thermal Structural Analysis, 1–56. AOARD Report, The University of Adelaide (29 June 2007).

Hopkins, E.J. and Inouye, M. (1971). An evaluation of theories for predicting turbulent skin friction and heat transfer on flat plates at supersonic and hypersonic Mach numbers. *AIAA Journal* 9 (6): 993–1003.

Jacobson, N.S. (2014). High temperature chemistry at NASA: hot topics, NASA Glenn Research Center, Cleveland, OH Presentation at Missouri State University (23 October 2014).

Jozwick, P., Polkowski, W., and Bojar, Z., "Applications of Ni_3Al based intermetallic alloys – current stage and potential perceptivities." *Materials* 2015, 8, 2537–2568; doi:https://doi.org/10.3390/ma8052537.

Lane, J. (2007). Design processes and criteria for the X-51A flight vehicle airframe. *2007 UAV Design Processes/Design Criteria for Structures (pp. 1.7-1–1.7-14). Proceedings RTO-MP-AVT-145, Paper 1.7. Neuilly-sur-Seine*, France.

Marley, C.D. and Driscoll, J.F. (2017). Modeling an active and passive thermal protection system for a hypersonic vehicle. *Paper AIAA 2017-0118. 55th AIAA Aerospace Sciences Meeting*, Grapevine, Texas (9–13 January 2017).

Marshall, L., Bahm, C., Corpening, G., and Sherrill, R. (2005). Overview with results and lessons learned of the X-43A Mach 10 flight. *AIAA 2005-3336. AIAA/CIRA 13th International Space Planes and Hypersonics Systems and Technologies Conference,* Capua, Italy (16–20 May 2005).

Marshall, D., Cox, B., Kroll, P., et al. (2014). National Hypersonic Science Center for Materials and Structures Report No. ADA609952. Air Force Office of Scientific Research, Arlington Virginia.

McComb, H., Harold, N.M., and Card, M.F. (1990). Structures and materials technology for hypersonic aerospacecraft. NASA Technical Memorandum, NASA-TM-1-2583, 1990.

Messe, C. (2017). Thermostructural problem of hypersonic airbreathing flight systems, modeling and simulation. Ph.D. Thesis. Institute of Statics and Dynamics of Aerospace Structures. University of Stuttgart (4 August 2017).

Meysam, T., Nuruldiyanah, K., Zahra, D. et al. (2018). Conventional and advanced composites in aerospace industry: technologies revisited. *American Journal of Aerospace Engineering* 5 (1): 9–15.

Mital, S.K., Gyekenyesi, J.Z., Arnold, S.M. et al. (2006). Review of current state of the art and key design issues with potential solutions for liquid hydrogen cryogenic storage tank structures for aircraft applications. NASA/TM – 2006-214346, October 2006.

Mutzman, R., Murphy, S., and Hank, J. (2007). The X-51A scramjet engine flight demonstration program. *Presented at the National Institute of Aerospace, NIA Session on US Hypersonic Flight Test Programs*, Hampton, VI (14 September 2007).

Myers, D.E., Martin, C.J., and Blosser, M.L. (2000). Parametric Weight Comparison of Advanced Metallic, Ceramic Tile, and Ceramic Blanket Thermal Protection Systems. NASA Report TM-2000-210289, NASA LaRC.

Ohlhorst, C.W., Vaughn, W.L., Ransone, P.O., and Tsou, H.-T. (1997). Thermal conductivity database of various structural carbon–carbon composite materials. NASA-TM-4787, NAS 1.15:4787, L-17620 (November 1997).

Ohlhorst, C.W., Glass, D.E., Bruce, W.E., III et al. (2005). Development of X-43A Mach 10 Leading Edges. *Paper IAC-05-D2.5.06. 56th International Astronautical Congress*, Fukuoka, Japan, 2005.

Padture, N.P. (2016). Advanced structural ceramics in aerospace propulsion. *Nature Materials* 15: 804–809. www.nature.com/naturematerials.

Rasky, D.J. (1992). Advanced ceramic matrix composites for TPS. Current technology for thermal protection systems. *NASA Conference Publication CP 3157*, 1992.

Rivers, H.K. and Glass, D.E. (2006). Advances in hot structure development. NASA Technical Report, Langley Research Center. *5th European Workshop Thermal Management Systems and Hot Structures*, Noordwijk, The Netherlands (17–19 May 2006). (ESA SP-631, August 2006).

Rondeau, C.M. and Jorris, T.R. (2013). X-51A scramjet demonstrator program: waverider ground and flight test. *SFTE 44th International/SETP Southwest Flight Test Symposium,* Ft Worth, TX (28 October to 1 November 2013).

Sharifzadeh, S., Hendrick, P., D'Mello, S. et al. (2014). Structural design optimisation and aerothermoelastic analysis of LAPCAT A2 Mach 5 cruise vehicle. *AIAA AVIATION Forum. AIAA 2014-2294. 14th AIAA Aviation Technology, Integration, and Operations Conference*, Atlanta, GA (16–20 June 2014).

Steelant J. (2007). LAPCAT: High-Speed Propulsion Technology. *Advances on Propulsion Technology for High Speed Aircraft*, 12.1–12.38. Educational Notes RTO-EN-AVT-150. Paper 123. Presented at AVT/VKI Lecture Series held at the von Karman Institute, Rhode St. Genèse, Belgium (12–15 March 2007).

Thornton, E. (1996). *Thermal Structures for Aerospace Applications*. AIAA Press.

Throckmorton, D.A. and Hartung, L.C. (1993). Space shuttle orbiter entry lee-side heat-transfer data STS-28. NASA RP 1306 (September 1993).

Van Wie, D.M., Drewry, D.G., King, D.E., and Hudson, C.M. (2004). The hypersonic environment: required operating conditions and design challenges. *Journal of Materials Science* 39: 5915–5924.

Verstraete, D., Sharifzadeh, S., and Hendrick, P. (2012). Definition and aero-elastic optimization of the structure of the LAPCAT A2 Mach 5 airliner. *ICAS2012. 28th International Congress of the Aeronautical Sciences*, Brisbane, Australia (23–28 September 2012).

Vincent-Radonnier A., Moule Y., and Ferrier M. (2014). Combustion of hydrogen in hot air flows within LAPCAT-II dual mode ramjet combustor at Onera-LAERTE facility – experimental and numerical investigation. *AIAA 2014-2931. 19th AIAA International Space Planes and Hypersonic Systems and Technologies Conference*, Atlanta, GA (16–20 June 2014).

Williams, S. and Curry, D.L. (1992). Thermal Protection Materials: Thermophysical Property Data. NASA Report N93-18765, NASA Johnson Space Center.

10

Scramjets and Combined Cycle Propulsion

Imagine a reusable hypersonic vehicle that could takeoff from conventional runways, accelerate powered by air-breathing propulsion, and when the atmosphere oxygen level becomes too low, it would switch to rocket propulsion to reach low Earth orbit (LEO). Then, after completing the mission, the vehicle would return to Earth intact, landing horizontally. Let us refer to this craft as "air-breathing ascent and re-entry single stage to orbit vehicle" or simply a *spaceplane*.

From such spaceplane configuration, we can derive an aircraft for sustained hypersonic cruise. This hypersonic transport would takeoff, climb to a high altitude, and cruise through the atmosphere at hypersonic speeds to its destination, landing like an airplane. The vehicle could be automated or crewed and be capable of carrying passengers and/or cargo. Just like conventional aircraft, this hypersonic cruiser would be capable of repetitive missions with minimum turn-around time.

Spaceplanes and hypersonic cruise vehicles require propulsion systems that can accelerate them from subsonic runway takeoff velocity to their hypersonic top speed. Figure 10.1 gives a qualitative map of propulsion options in terms of specific impulse I_{sp} as a function of flight Mach number. Those trends represent the approximated performance range of each type of air-breathing engine as compared with the I_{sp} of rockets (~460 seconds). Air-breathing engines fueled by hydrogen have a higher I_{sp} mainly due to the heating value per unit mass of hydrogen (about 2.3 times that of jet fuel). As shown, no one propulsion system is optimum over the entire flight Mach number range $0 < M_0 < 25$. However, integrating air-breathing and rocket engines, the resulting combined cycle propulsion (CCP) systems can provide an alternative to conventional multistage rockets for access to space.

For acceleration and sustained cruise in the flight regime $3 < M_0 < 14$, dual-mode scramjet engines are most effective due to their high propulsive efficiency (higher I_{sp}). However, since the scramjet is incapable of self-acceleration, it must be integrated to an engine that can propel the vehicle from takeoff at ground level to scramjet takeover (TO) speed. For the reusable spaceplane, the final boost into orbit will require liquid propellant rocket engines. Moreover, because of the extremely high energies involved to access LEO, the fuel for rocket and scramjet engines must be hydrogen.

Thus, we seek a propulsion system that integrates both air-breathing engines and rockets, utilizing their distinct operating modes at high efficiency to match thrust demands and

Scramjet Propulsion: A Practical Introduction, First Edition. Dora Musielak.

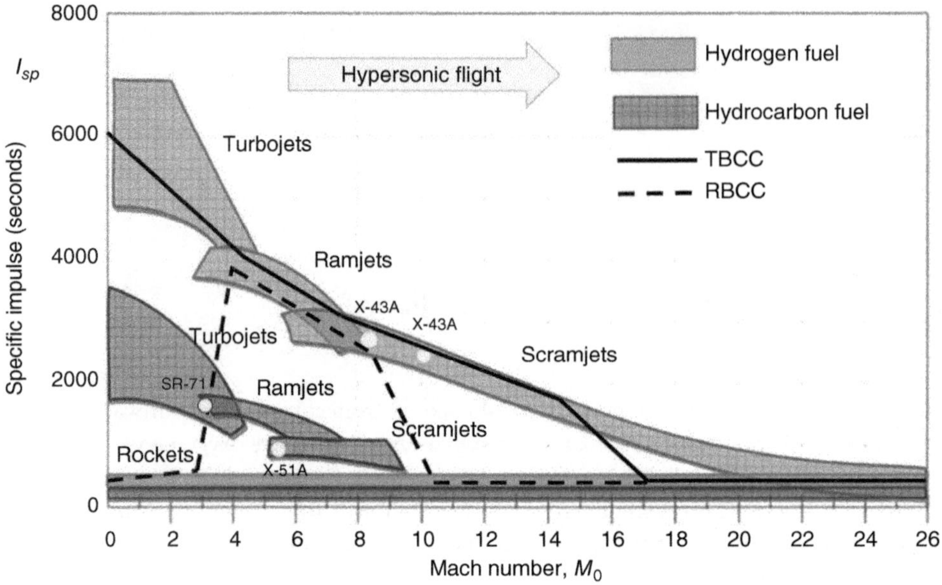

Figure 10.1 Propulsion options for hypersonic flight. Source: Adapted from a NASA image.

flight conditions. Having such integrated system would result in a single, lighter weight, more versatile engine that could compete with the traditional approach to reaching LEO.

The objective of this chapter is to present an overview of CCP concepts and to highlight basic technology challenges that must be overcome to develop them. The intent now is to review fundamental ideas to help us answer questions such as

- What are the technical limitations of modern turbojet engines?
- How can we "stretch" the operational envelope of the state-of-the-art low-speed air-breathing engines (e.g. turbojets) to provide initial acceleration to scramjets?
- How can we integrate rockets and air-breathing engines to obtain optimum hypersonic propulsion for a spaceplane?
- What technologies are crucial to build a spaceplane that would takeoff from a runway, accelerate powered by air-breathing propulsion, and achieve orbit?

10.1 Aerospace Propulsion

There are two classes of aerospace propulsion systems based on chemical energy: air-breathing and rocket engines. Air-breathing propulsion includes gas turbine jet engines (turbojet and turbofan) and no rotor jet engines (pulsejet and ram/scramjet). To date, air-breathing propulsion has exclusively been used for powering civil and military aircraft that fly at subsonic and supersonic speeds.

According to Figure 10.1, conventional turbojet/turbofan engines are limited to supersonic Mach number operation. The fastest military aircraft today can achieve

Mach 2.5+. The limitations on engine operating conditions, include air temperature and pressure, fuels, materials, and turbomachinery thresholds. For example, the McDonnell Douglas F15 Eagle tactical fighter aircraft is powered by two Pratt & Whitney F100 axial compressor afterburning turbofan engines, burning hydrocarbon fuel (JP-8). The engines can propel the 18 000 kg aircraft to more than 2.5 times the speed of sound.

The operational limit of turbojets/turbofans is set by the turbine temperature limits: $M_0 \approx 3.0$ with hydrocarbon jet fuel or $M_0 \approx 4.0$ with hydrogen. A hydrogen-fueled turbojet could probably be operated at a higher Mach number, owing to the greater cooling capability of hydrogen. However, the ramjet is a better choice above $M_0 \approx 3.5$. As indicated in Figure 10.1, ramjets yield a higher I_{sp} in the regime $2.0 < M_0 < 7.0$. Above Mach 7.0, the ramjet engine is limited by air dissociation and various aerothermodynamic losses. Scramjet engines have higher propulsion performance than rockets in the flight range $5 < M_0 < 12$. Since they are incapable of self-acceleration, scramjets must be integrated with another type of engine capable of operation from takeoff to the scramjet take-over speed.

It is conceivable to combine individual engine types that are best for each flight regime (turbojet/turbofan, ramjet, scramjet, and rocket), operating in parallel or separately. Such integrated propulsion can include supporting elements that may enhance system performance, such as ejectors, air liquefiers, precoolers, and thrust augmentators. Traditionally known as CCP, such combined cycle propulsion concepts can be as simple as those that "stretch" the capability of a given engine type (e.g. a turbojet that can operate as a ramjet), or as complicated arrangements of various types of engines into a single system (e.g. ram-rocket) invented specifically to meet the requirements of a given flight mission.

The first dual cycle propulsion system ever built and operated was the turboramjet designed by the French aircraft manufacturer Nord Aviation to propel the Nord 1500 Griffon II interceptor plane. Built with a shared inlet and nozzle flowpath, the Nord Stato-Réacteur ramjet (67.8 kN [15 200 lbf] thrust) was wrapped around a SNECMA Atar 101 E3 dry turbojet (34.3 kN [7700 lbf] thrust). This dual-cycle propulsor operated by controlling the fuel flow to the two jet engines. Hence, the fraction of the total thrust generated by the ramjet varied from 0 under static conditions to over 80% at the maximum flight Mach number. The Griffon II attained a maximum speed of Mach 2.19 and set a world record in 1959. It was last flown in 1961 and currently resides in the *Musée de l'air et de l'espace* outside Paris, France.

In the United States, the first combined cycle engine was designed by Pratt & Whitney (P&W) to meet the supersonic flight requirement of the Mach 3.2+ crewed SR-71 Blackbird aircraft (retired by USAF in 1990 and by NASA in 1999). The 151.24 kN (34 000 lbf) thrust-class, JP-7-fueled P&W J58 (JT11D-20) turbojet engine was fitted with an afterburner (AB), a uniquely designed duct downstream of the turbine used to augment the overall thrust. The J58 operated as a turbojet until the flight Mach number reached a certain value, at which point six large tubes bypassed the airflow flow from the engine inlet directly to the AB. Bypassing the compressor, burner, and turbine with its unique air bypass system allowed the J58 to operate as a ramjet with the AB acting as the ramjet's main burner. This dual-mode or variable cycle air-breathing engine allowed the SR-71 to operate in two flight regimes, subsonic and supersonic.

10.2 Combined Cycle Propulsion Concepts

To establish requirements for CCP, hypersonic air-breathing vehicles may be separated into two major groups:

- Accelerators ($0 < M_0 < 25$), exo-atmospheric, or space access vehicles that must continually produce excess thrust. These vehicles are characterized by low drag per unit engine inlet capture.
- Cruisers ($0 < M_0 < 12$), endo-atmospheric vehicles flying within Earth's atmosphere. These vehicles require high lift-to-drag ratio, thin flat fuselages with high fineness ratios. They must have high range at a determined cruise Mach number.

CCP is required for both types of vehicles. To develop propulsion capability for various flight regimes from takeoff to hypersonic speeds (using atmospheric air to supply oxidizer to the combustor) requires integration of several engine types. CCP integrates multiple engine cycles into a single propulsion system capable of operating in multiple modes (cycles) to extend the flight regime. These different modes allow a propulsion system to be extremely versatile and efficient while transitioning through varied flight conditions.

Combined cycle configurations can be classified as Turbine-Based Combined Cycle (TBCC), and Rocket-Based Combined Cycle (RBCC). At their core, TBCC integrates a turbojet or turbofan with a dual-mode scramjet, while the RBCC may combine an air-augmented rocket and a dual-mode scramjet. The TBCC and RBCC propulsion systems are conceived to transition from different cycle modes to achieve maximum performance. The TBCC propulsion uses high specific impulse turbine engines to boost the vehicle from Mach zero to TO speed, about Mach 3+, where the air-breathing propulsion transitions from gas turbine to ramjet/scramjet mode. For example, to propel a Mach 7 cruise aircraft requires an engine that combines turbojets (with or without fan, with or without thrust augmentation) and dual-mode scramjets. TBCC can be configured as a wraparound turboramjet engine (single flowpath), or as over–under (dual flowpath) turboramjet with an upper duct for mounting the turbine engine, and below this a duct for the ram/scramjet engine. As indicated in Figure 10.2, the over–under configuration requires a dual variable inlet and

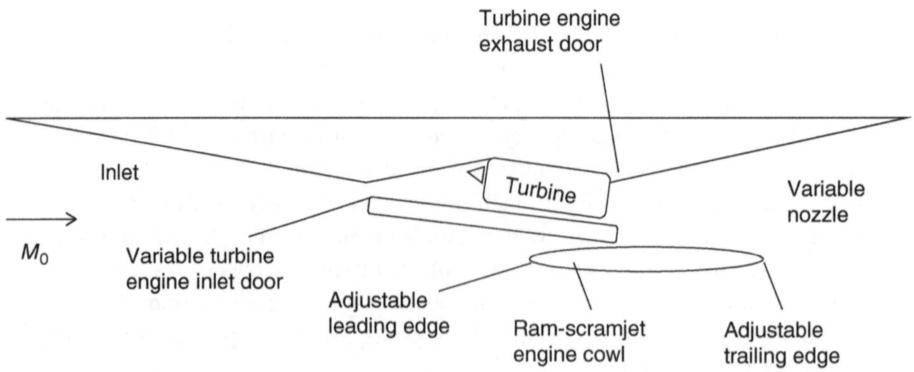

Figure 10.2 Turbine-Based Combined Cycle (TBCC) propulsion over–under concept.

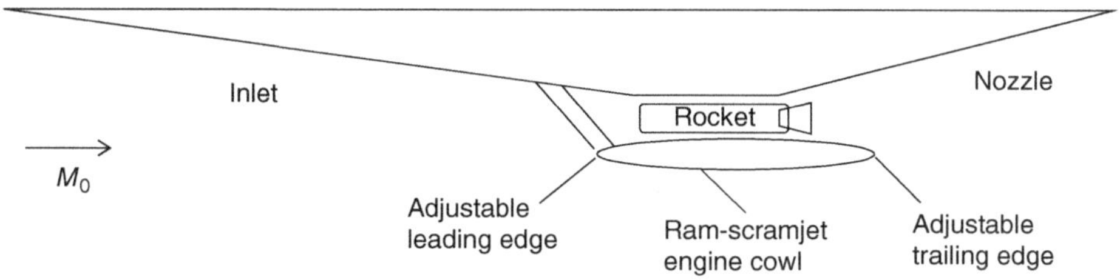

Figure 10.3 Rocket-Based Combined Cycle (RBCC) propulsion concept.

a dual variable nozzle: the turbo engine flowpath requires a low-speed inlet and low-speed nozzle, while the ram/scramjet flowpath integrates a high-speed inlet and a high-speed nozzle.

The Rocket-Based Combined Cycle (RBCC) utilizes a rocket engine as the primary propulsion element to boost the vehicle to takeoff speed and for the final propulsion mode. The scramjet operates in between these modes. The RBCC allows for a fully integrated engine with one single flowpath, as shown in Figure 10.3. Many possible modifications to the basic RBCC have been considered and studied by combining ejector rocket, ramjet, scramjet, and pure rocket cycle operation to cover all flight regimes.

Consider the air-augmented rocket concept sketched in Figure 10.4. At low speeds, air enters the engine and, after passing through the inlet, it mixes with the rocket flow stream. This mode of operation is referred to as ejector ram/scramjet or air-augmented rocket. Ejector refers to the operation used to multiply the original or primary air mass flow by drawing-in a supplemental or secondary mass flow from the surrounding atmosphere. The total pressure of the secondary flow is raised to a value between that of ambient and that of the primary flow. An ejector rocket behaves both as a rocket (for producing static thrust) and also as a ramjet because the inlet air is compressed and burned with the fuel to produce thrust.

Air-augmented rockets (also called rocket-ejector, ramrocket, ducted rocket, integral rocket/ramjets, or ejector ramjets (ERJs)) use the supersonic exhaust of a rocket engine to further compress air collected by ram effect during flight to use as additional working mass, leading to greater effective thrust for any given amount of fuel than either rocket

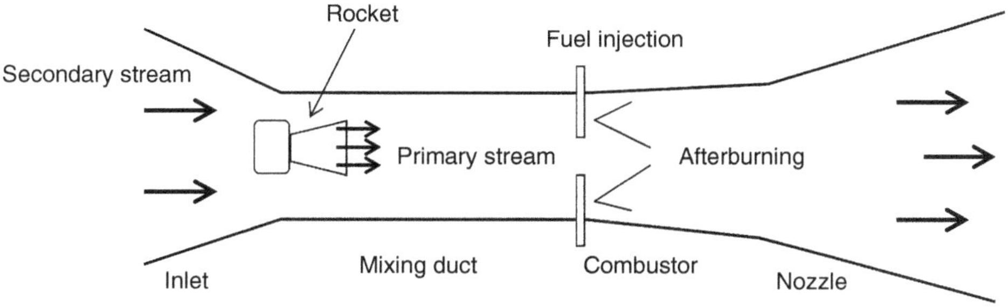

Figure 10.4 Air-augmented rocket, a basic RBCC propulsion concept.

or a ramjet alone. The ejector rocket represents a hybrid class of rocket/ramjet engines, similar to a ramjet, but able to give useful thrust from zero speed, and can also operate outside the atmosphere by using on-board oxidizer (see for example, Kanda and Tani 2007).

In an air-augmented rocket, fuel could be injected into the flowpath either upstream or downstream of the rocket engine. Or the incoming air could be mixed with the fuel-rich rocket exhaust. At speeds near Mach 3, there is sufficient ram compression that the engine could transition from ERJ to ramjet mode. Gradually, the rocket engine can be shut down and allow full ramjet operation. Ramjet mode can be continued until flight speeds reach Mach 5–6, where performance and materials limitations will require transition to supersonic combustion, or scramjet mode. Eventually, the performance benefits of air-breathing propulsion diminish and the propulsion system must transition to rocket only mode. The Mach number at which this mode transition occurs may be approximately 10 or 12. The transition Mach number for each of the four operating modes are design parameters for the propulsion system.

10.3 From Takeoff to Hypersonic Cruise

The idea of hypersonic cruise flight is not new, and many concepts for aircraft and their propulsion have been studied. In the 1970s, for example, McDonnell Aircraft Company (now the Boeing Company) carried out a study under a NASA contract to evaluate effects of fuselage cross section (circular and elliptical) and structural arrangement (integral and nonintegral fuel tanks) on the performance of actively cooled Mach 6 cruise aircraft fueled by liquid hydrogen (LH_2) (Nobe 1975). The propulsion system consisted of four variable cycle General Electric GE5/JZ6C turboramjets rated at 400 kN (90 000 lbf) thrust each. A two-dimensional, three ramp, external compression inlet was designed, with variable capture area and a translating cowl to enhance the airflow capture characteristics and minimize inlet drag over the entire mission. The inlet was located beneath the wing to benefit from the wing compression flowfield. The four engines, inlets, and exhaust ducts were integrated into the fuselage body to provide low nacelle drag on the total vehicle. The study considered three different aircraft designs carrying a 200-passenger payload and a constant fuel quantity of 108 862 kg (240 000 lbm).

With the primary objective of minimizing weight and maximizing aircraft range, the study conducted thermodynamic and structural trade-offs to refine the conceptual designs. It was determined that the major factors affecting range are vehicle weight, volumetric efficiency, and aerodynamic characteristics. These factors interact differently, depending on the fuselage shape and type of fuel tank structure.

The study indicated that integral fuel tanks combined with an elliptical-blended wing-body results in the lightest aircraft weight and longest-range configuration. It was found that integral tanks use the fuselage volume more efficiently than nonintegral tanks; the greater volumetric efficiency of the integral tank resulted in a smaller sized vehicle. The other concepts studied had similar time histories except for the cruise time, revealing the difference in range between all concepts. There were also small differences in acceleration time and descent time, but these were less than one minute (Nobe 1975).

More recently, the Boeing Company introduced their vision for a Mach 5+ commercial airplane to cruise at an altitude about 95 kft (~29 km). The propulsion system is an afterburning turbofan engine conceived to operate as a turboramjet. In such configuration, after reaching the turbofan performance limit, valves can bypass the incoming air and send it into the AB, without going through the compressor/turbine spool. The AB would function as a ramjet to propel the airplane to Mach 5. Since slowing down the high-speed airflow inside the engine produces excessive heat, the airframe skin may reach 593 °C (1100 °F) during flight due to air friction. For this reason, Boeing considered fueling the engine with liquid methane and use it also as coolant.

The American company Hermeus is also developing a Mach 5.0 aircraft to be powered by a TBCC ramjet designed around a flight-proven GE J85-21 single-shaft turbojet engine. One of GE's most successful military turbojets, the J85 can deliver up to 2950 lbf (13.1 kN) of thrust dry and with afterburning it can reach up to 5000 lbf (22 kN). It is unclear at this stage of development if the TBCC engine for the Hermeus aircraft will require the AB.

The European Space Agency is developing the LAPCAT-A2 Mach 5 airliner that will be powered by precooled air-turbo ramjets, and a Mach 8 airplane powered by a ramjet engine, both fueled by LH_2. In Japan, JAXA is also developing a precooled turbojet for a Mach 5 commercial plane with a variable inlet, AB, and a variable exhaust nozzle. The precooler is a 100 kW class heat exchanger comprised of 648 separate cooling tubes (2 mm diameter), using LH_2 fuel (−250 °C) as the coolant (Taguchi et al. 2022).

10.4 Ideal Cycle Analysis of Turbojet and Ramjet Engines

To enable air-breathing propulsion for a vehicle that will takeoff and reach hypersonic speeds, a high Mach turbine engine is required to accelerate the vehicle to scramjet TO speeds. A turboramjet, a hybrid cycle concept that combines a turbojet (based on gas turbines) and a ramjet (air-breathing engine that do not require turbomachinery) may enable a vehicle to fly up to Mach 5. The turboramjet can operate simultaneously or separately to accelerate and sustain high flight supersonic speeds, combining the advantages of both types of engines. An example of a turboramjet is the 151.24 kN (34 000 lbf) thrust-class Pratt & Whitney (P&W) J58 turboramjet engines that powered the Mach 3+ SR-71 Blackbird aircraft, as described in Section 10.1.

The amount of thrust generated by turbojets and ramjets depends on two parameters: the mass flow of atmospheric air ingested and processed by the engine, and the exit velocity of the hot gas leaving the engine. Consequently, these two engine cycles can be integrated with scramjets to extend the flight regime from subsonic to hypersonic.

A turbojet is an air-breathing engine consisting of a gas turbine (compressor + burner + turbine) with an air intake section (engine inlet) and an exhaust section (thrust nozzle). A mechanical compressor is required to increase the pressure of the incoming freestream air. The compressed air is mixed with the injected fuel in the combustor where the mixture undergoes chemical reactions. The hot gas (product of air/fuel combustion) expands through the turbine and then enters the nozzle where the exhaust gas is accelerated to high speed to provide the thrust force to move a vehicle. The turbine, which is attached to the compressor by a common driveshaft, provides power to run the compressor.

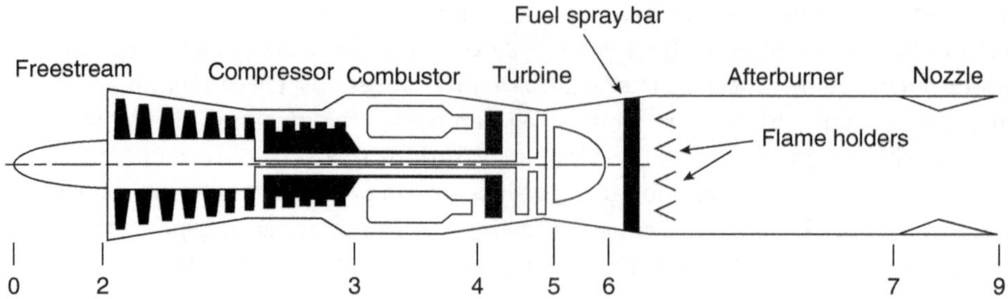

Figure 10.5 Turbojet engine with afterburner (AB).

In order to augment its thrust, the turbojet can include an AB. The AB is a duct section placed between the turbine and the nozzle where additional fuel can be injected and combusted with the oxygen-rich gas flow leaving the turbine. The AB combustion process increases the nozzle entry temperature significantly, resulting in an increase in temperature and velocity of the exhaust gas through the nozzle, thereby increasing the net thrust. Figure 10.5 depicts the station numbers for a turbojet engine with an AB. Not shown is station 1 at the engine inlet (or cowl) lip; station 2 corresponds to the exit of the air inlet system, which is the entrance to the axial compressor.

In addition to the increase in AB exit stagnation temperature, there is also an increase in nozzle mass flow and a decrease in stagnation pressure, which results from thermal, friction, and turbulence losses. The increase in flow must be accommodated by increasing the throat area of the engine nozzle. As the flight speed increases above a threshold supersonic Mach number, the ram pressure produced by the engine inlet decelerates the incoming air to subsonic speeds and increases the air pressure to values sufficient for combustion. At that point, it is no longer necessary to use a mechanical compressor; therefore, there is no need for the turbomachinery. In such case, the design yields a simpler jet engine we call *ramjet.*

10.4.1 Parametric Analysis of Turbojet and Ramjet Engines

In the study of air-breathing engines, a cycle analysis is conducted to obtain estimates of the performance parameters (primarily thrust and specific fuel consumption) in terms of design limitations, the flight conditions (the ambient pressure and temperature and the Mach number), and design choices. Initially, the parametric cycle analysis (thermodynamic design point) considers all engine components to be ideal, and thus the results of such analysis are rather optimistic. However, the general trends are valid and help us obtain values of the main performance parameters that are later refined when engine component losses are included.

Engine cycle analysis incorporates predefined parameter relationships such as ratios of total temperatures and pressures that serve as indicators of the flow conditions throughout the engine. The conventional notation denotes the ratio of total pressures across a component by π and the ratio of total temperatures across a component is denoted by τ, with a subscript indicating the component: d for diffuser (inlet), c for compressor, b for burner,

t for turbine, AB for AB, and n for nozzle. These ratios are given in terms of the standard engine station numbering (shown in Figure 10.5).

The total/static temperature and pressure ratios of the freestream air, denoted τ_r and π_r, respectively, are defined as

$$\tau_r = \frac{T_{t0}}{T_0} = 1 + \frac{\gamma_0 - 1}{2} M_0^2, \pi_r = \frac{p_{t0}}{p_0} = (\tau_r)^{\gamma_0/(\gamma_0 - 1)} \tag{10.1}$$

where the flight Mach number and the freestream sonic velocity are, respectively,

$$M_0 = \frac{V_0}{a_0}, a_0 = \sqrt{\gamma_0 R_0 T_0} \tag{10.2}$$

The freestream stagnation temperature and pressure conditions are written as $T_{t0} = T_0 \tau_r$, $p_{t0} = p_0 \pi_r$, respectively, to indicate the effect of the flight Mach number M_0.

Another figure of merit for cycle analysis is the ratio of the combustor exit enthalpy h_{t4} to the ambient static enthalpy h_0, denoted τ_λ. For the dry turbojet, this parameter is written

$$\tau_\lambda = \frac{h_{t4}}{h_0} = \frac{c_{p4} T_{t4}}{c_{p0} T_0} \tag{10.3a}$$

Hence, for turbojet engines designed with an AB, we define

$$\tau_{\lambda_{AB}} = \frac{h_{t7}}{h_0} = \frac{c_{p7} T_{t7}}{c_{p0} T_0} \tag{10.3b}$$

Similarly, for ramjet engines, we define the total to static enthalpy ratio as

$$\tau_{\lambda_{ramjet}} = \frac{h_{t7}}{h_0} = \frac{c_{p7} T_{t7}}{c_{p0} T_0} \tag{10.3c}$$

where the index 7 denotes the condition at the exit of the ramjet combustor (see Figure 10.7).

The τ_λ, $\tau_{\lambda_{AB}}$, and $\tau_{\lambda_{ramjet}}$ parameters provide the thermal design limits for the respective engines, and thus their values are specified and remain fixed by technology and operational limits (Section 10.4.6). These parameters establish the maximum allowable turbine inlet stagnation enthalpy for the turbojet, or the maximum cycle stagnation enthalpy to reduce dissociation losses in both turbojets with AB and ramjet engines.

The cycle analysis assumes:

1) Air, the engine working fluid, behaves as a calorically perfect gas with constant specific heats throughout, so $\gamma = 1.4$.
2) Compression and expansion processes are isentropic (reversible and adiabatic).
3) Combustion is idealized as a constant pressure heat input process, and we assume that the fuel to air ratio $f = \dot{m}_f/\dot{m}_0 \ll 1$. That is, the fuel flow rate is much less than the airflow rate through the engine such that $\dot{m}_9 \cong \dot{m}_0$.
4) The exhaust nozzle expands the working fluid to the ambient pressure ($p_9 = p_0$).

Under these idealized conditions, the following relationships are valid.

Inlet Diffuser: The process in the diffuser is considered to be adiabatic and isentropic. Hence,

$$\tau_d = \frac{T_{t2}}{T_{t0}} = 1, \pi_d = \frac{p_{t2}}{p_{t0}} = (\tau_d)^{\gamma/(\gamma-1)} = 1 \tag{10.4}$$

Compressor: The compression process is assumed isentropic:

$$\tau_c = \frac{T_{t3}}{T_{t2}} = (\pi_c)^{(\gamma-1)/\gamma} \tag{10.5}$$

For the turbojet, the compressor pressure ratio π_c is an engine design choice (prescribed).

Combustor: For an ideal combustor, the stagnation pressure is constant. Thus,

$$\pi_b = \frac{p_{t4}}{p_{t3}} = 1 \tag{10.6}$$

This assumption implies the combustion process occurs at very low Mach numbers. A typical combustor Mach number in a conventional ramjet may be ~0.2 to 0.3.

Turbine: The expansion process is assumed isentropic. Hence,

$$\tau_t = \frac{T_{t5}}{T_{t4}} = (\pi_t)^{(\gamma-1)/\gamma} \tag{10.7}$$

Afterburner: The stagnation pressure during combustion remains constant. Thus,

$$\pi_{AB} = \frac{p_{t7}}{p_{t5}} = 1 \tag{10.8}$$

Nozzle: The process in the ideal nozzle is assumed to be adiabatic and isentropic. Hence,

$$\tau_n = \frac{T_{t9}}{T_{t8}} = 1, \pi_n = \frac{p_{t9}}{p_{t8}} = (\tau_n)^{\gamma/(\gamma-1)} = 1 \tag{10.9}$$

Viewed as heat engines, the performance of ramjets and turbojets can be derived from the air-standard Brayton cycle. Figure 10.6 depicts a qualitative representation of the cycle's temperature–entropy (T–s) diagram, where, for a wet or afterburning turbojet, station 9′ represents the engine nozzle exit. In the case of a ramjet, we model the expansion process through the nozzle from the subsonic combustor to ambient conditions as $t_7 \rightarrow 9$. Moreover, we indicated in the T–s diagram the limit parameter τ_λ, representing the maximum main combustor temperature (maximum allowable turbine temperature) T_{t4}, the limit cycle parameters $\tau_{\lambda_{AB}}$ and $\tau_{\lambda_{ramjet}}$ representing the maximum allowable AB or ramjet combustor temperature T_{t7}.

The design inputs for the engine cycle analysis are grouped in four classes:

1) Flight conditions: p_0, T_0, M_0, c_p, τ_r, π_r.
2) Engine design limits: combustor exit enthalpy, $(c_pT_t)_4$ (turbojet); $(c_pT_t)_7$ (turbojet).
3) Engine component performance: π_d, π_c, π_n.
4) Design choices: π_c.

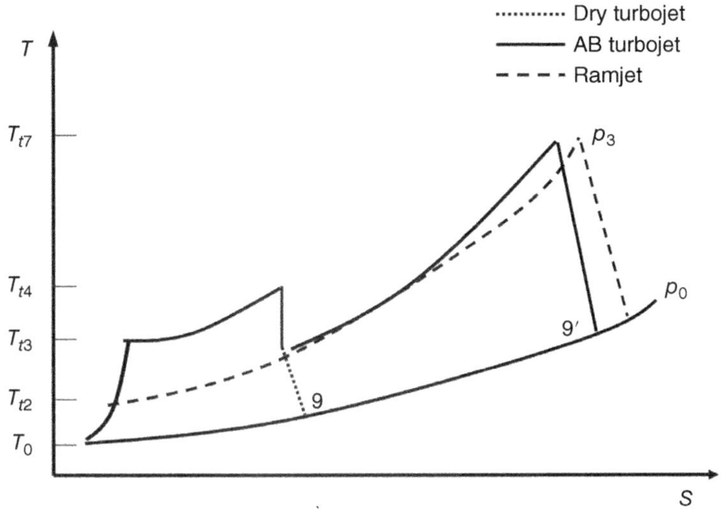

Figure 10.6 Temperature–entropy diagram for turbojet and ramjet engines.

For the ideal air-breathing engine, the uninstalled thrust equation is given by Eq. (2.1), that is $F = (\dot{m}_9 V_9 - \dot{m}_0 V_0) + A_9(P_9 - P_0)$. However, $p_9 = p_0$ and $\dot{m}_9 \cong \dot{m}_0$ for the ideal engine, thus, the specific thrust is written as

$$\frac{F}{\dot{m}_0} = (V_9 - V_0) = a_0\left(\frac{V_9}{a_0} - M_0\right) \tag{10.10}$$

where the velocity ratio V_9/a_0 is

$$\frac{V_9}{a_0} = M_9\frac{a_9}{a_0} = M_9\sqrt{\frac{\gamma RT_9}{\gamma RT_0}} = M_9\sqrt{\frac{T_9}{T_0}} \tag{10.11}$$

The parametric analysis proceeds to obtain the quantities in Eq. (10.11) in terms of the nondimensional parameters outlined above. With simple algebraic manipulation of those relations, an expression for the specific thrust for the ideal engine is obtained in terms of nondimensional parameters. The interested reader may consult the excellent book by Mattingly and Boyer (2016) who provides a step-by-step derivation of the required engine performance equations. Here, we simply recall the outcome of the analysis to illustrate engine performance trends.

10.4.2 Ideal Ramjet

A ramjet engine (Figure 10.7) operates on pressure produced by the dynamic action of the freestream airflow, the so-called "ram effect" of the vehicle's speed. The freestream airflow enters the ramjet at the speed of flight (V_0) and is decelerated so that the kinetic energy of the airflow is converted to provide the compression without the need of a mechanical compressor. The compressed air is then mixed with fuel in the combustion chamber, where the process of combustion raises the gas temperature. The hot gas is expanded through a

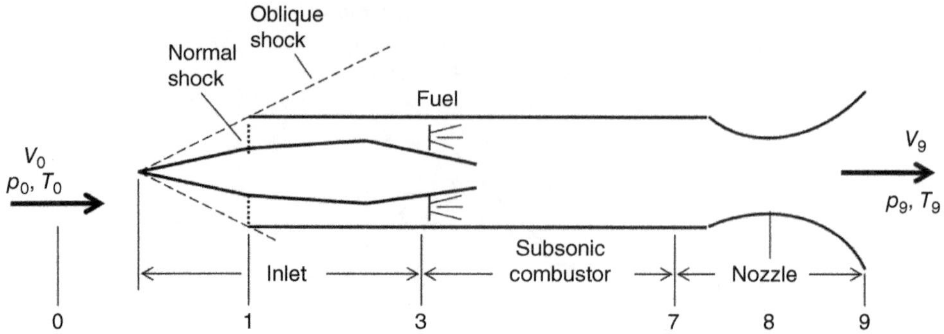

Figure 10.7 Ramjet engine.

nozzle, producing a high-velocity ($V_9 > V_0$) jet stream to produce thrust. By eliminating the turbomachinery, the ramjet can operate at a higher cycle temperature as compared with the turbojet. Higher temperatures result in an increase in thrust and efficiency.

The specific thrust for the ideal ramjet engine is given by the following expression:

$$\text{Ramjet } \frac{F}{\dot{m}_0} = a_0 M_0 (\sqrt{\tau_b} - 1) = a_0 M_0 \left(\sqrt{\frac{\tau_{\lambda_{ramjet}}}{\tau_r}} - 1 \right) \tag{10.12}$$

Not surprisingly, the specific thrust of a ramjet depends on both M_0 and the maximum cycle stagnation enthalpy via the parameter $\tau_{\lambda_{ramjet}}$, Eq. (10.3c).

In addition, by performing an energy balance in the combustor in the same manner we did in Chapter 2, we obtain an expression for the fuel/air ratio in the ramjet combustor:

$$f = \frac{c_p T_0}{h_{PR}} \left(\tau_{\lambda_{ramjet}} - \tau_r \right) \tag{10.13}$$

The performance of an ideal ramjet is illustrated in Figure 10.8 with a plot of the specific thrust $F/\dot{m}_0$ as a function of flight Mach number M_0 for three values of the total temperature leaving the combustor, T_{t7}, represented by the limiting parameter $\tau_{\lambda_{ramjet}}$, for which we chose values to range from 7.0 to 9.0. Figure 10.8 confirms that, at low speeds, a ramjet cannot produce thrust as it requires ram compression, which develops until the flight Mach number is $M_0 > 1$. At the same time, the ramjet ceases to produce enough thrust at high flight velocity, $M_0 > 5$, when it experiences very high ram compression.

By having complete dependency on the ram air to produce the necessary compression, the ramjet can operate only if the vehicle maintains a sufficient flight speed. This means that the ramjet cannot operate at takeoff nor land an aircraft on its own power. This explains why rocket boosters are typically used for a ramjet to takeoff; to circumvent the landing problem most ramjets have been used in one-way flight missions, such as powering guided missiles.

The plotted data in Figure 10.8 indicate that the ideal ramjet produces a maximum $F/\dot{m}_0$ at a certain M_0 for each value of $\tau_{\lambda_{ramjet}}$. The trend shows that the higher the $\tau_{\lambda_{ramjet}}$ (T_{t7}), the higher $F/\dot{m}_0$ produced, as Eq. (10.12) indicates. For an actual ramjet, the maximum temperature limit T_{t7} will be governed by the engine design and level of cooling technology, materials, and the protective thermal coatings used.

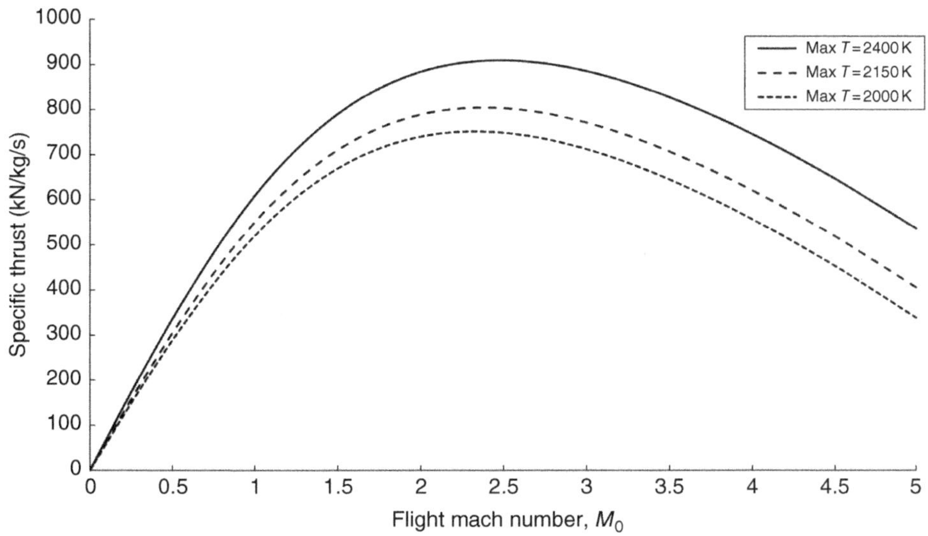

Figure 10.8 Effect of limiting total temperature T_{t7} on specific thrust of ideal ramjet.

For example, consider a ramjet engine fueled by JP-7 operating at an altitude where $p_0 = 7.565$ kPa, $T_0 = 216.62$ K. Using Eq. (10.12), we can assess the effect of combustor exit temperature T_{t7} on specific thrust for the flight Mach number range $0 < M_0 < 4$, subjected to the limiting values $T_{t7} = 2200$ K, and 2400 K. For JP-7 fuel, $h_{PR} = 42{,}800$ kJ/kg, and for air $\gamma = 1.4$, $c_p = 1.004$ kJ/(kg · K), and $R = 0.2869$ kJ/(kg · K). At the flight condition where the air temperature is 216.62 K, the local sonic velocity is $a_0 = \sqrt{\gamma R T_0} = 294.97\ \text{m/s}$.

For $M_0 = 3$, the freestream values τ_r and π_r are

$$\tau_r = \frac{T_{t0}}{T_0} = 1 + \frac{\gamma_0 - 1}{2} M_0^2 = 1 + 0.2(3)^2 = 2.8$$

$$\pi_r = \frac{p_{t0}}{p_0} = (\tau_r)^{\gamma_0/(\gamma_0 - 1)} = (2.8)^{3.5} = 36.73$$

The stagnation conditions of the freestream at $M_0 = 3$ are $T_{t0} = T_0 \tau_r = 216.62$ K (2.8) = 606.54 K, and $p_{t0} = p_0 \pi_r = 7.565$ kPa (36.73) = 277.86 kPa.

The ratio of combustor exit enthalpy to ambient enthalpy $\tau_{\lambda_{ramjet}}$ are

$$\tau_{\lambda_{ramjet}} = \frac{T_{t7}}{T_0} = \begin{cases} 10.15 \\ 11.08 \end{cases}$$

Hence, the specific thrust for the two limiting values of $\tau_{\lambda_{ramjet}}$ is

$$\frac{F}{\dot{m}_0} = a_0 \left(\frac{V_9}{a_0} - M_0 \right) = \begin{cases} 799.37\ \text{N/(kg/s)} \\ 876.06\ \text{N/(kg/s)} \end{cases}$$

By calculating the fuel–air ratio, we determine that for the given conditions, f is less than the stoichiometric fuel air ratio:

$$f = \frac{c_p T_0}{h_{PR}} \left(\tau_{\lambda_{ramjet}} - \tau_r \right) = \frac{1.004\ \text{kJ/(kg}\cdot\text{K)} \cdot 216.62\ \text{K}}{42\,800\ \text{kJ/kg}} (10.15 - 2.8) = 0.037$$

Now, by examining Figure 10.8, it is clear that the maximum specific thrust of the ramjet occurs at a certain flight M_0 for each value of T_{t7}, which is the maximum temperature for the propulsion system. We can easily find an analytic expression to find the optimum Mach number by taking the partial derivative of the specific thrust, Eq. (10.12), with respect to M_0, setting this equal to zero, and solving. Doing so, we obtain an expression that gives the **optimum Mach number** for maximum specific thrust:

$$(M_0)_{max\,F/\dot{m}_0} = \sqrt{\frac{2}{\gamma-1}\left(\sqrt[3]{\tau_\lambda}-1\right)} \tag{10.14}$$

10.4.3 Specific Impulse of Ramjet with Losses

We are interested in determining how the ramjet component losses affect the specific impulse. Let us begin with the engine inlet. The diffusion process in the ramjet inlet is not perfect. The losses in the supersonic inlet (or diffuser) arise because of the presence of wall friction, shock waves, and regimes of separated flow. It is customary to define $\pi_{d\max}$ as that portion of the ramjet inlet total pressure ratio π_d that is due to wall friction, and the ram recovery η_r as that portion of π_d due to shock waves. Hence, the overall pressure ratio of the inlet is the product of the ram pressure ratio and the diffuser pressure ratio:

$$\pi_d = \frac{p_{t2}}{p_{t0}} = \pi_{dmax}\,\eta_r \tag{10.15}$$

A typical value for supersonic aircraft with engine(s) in airframe is $\pi_{d_{max}} = 0.96$ (higher values are possible for subsonic aircraft). Therefore, for the performance analysis of a propulsion system operating at supersonic speeds, we must know the portion of the ram total pressure that can be recovered. The ram recovery η_r can be estimated with the military specification 5008B:

$$\eta_r = 1 - 0.075(M_0 - 1)^{1.35} \tag{10.16}$$

We use this relation as an ideal goal to achieve for inlet ram recovery for the flight regime $0 < M_0 < 5$. For modern supersonic aircraft with engines fit in the airframe, $\pi_{d\max} \sim 0.96$. There are other effects that must be considered in the ramjet inlet design analysis such as flow distortions, and these are included in a more advanced analysis.

The combustion process is far from ideal due mainly to incomplete combustion of fuel and a loss in total pressure across the combustor. The combustion efficiency η_b is defined by

$$\eta_b = \frac{(\dot{m}_0 + \dot{m}_f)h_{t7} - \dot{m}_0 h_{t3}}{\dot{m}_f h_{PR}} = \frac{(\dot{m}_0 + \dot{m}_f)c_{p7}T_{t7} - \dot{m}_0 c_{p3} T_{t3}}{\dot{m}_f h_{PR}} \tag{10.17}$$

Thus, we can rewrite the fuel/air ratio for the nonideal combustor as

$$f = \frac{c_p T_0}{\eta_b h_{PR}}\left(\tau_{\lambda_{ramjet}} - \tau_r\right) \tag{10.18}$$

The total pressure loss in the ramjet combustor is due to viscous losses, and the total pressure loss is due to the combustion process taking place at a finite Mach number. In performance analysis, these effects are combined in the parameter $\pi_b = p_{t7}/p_{t3} < 1$.

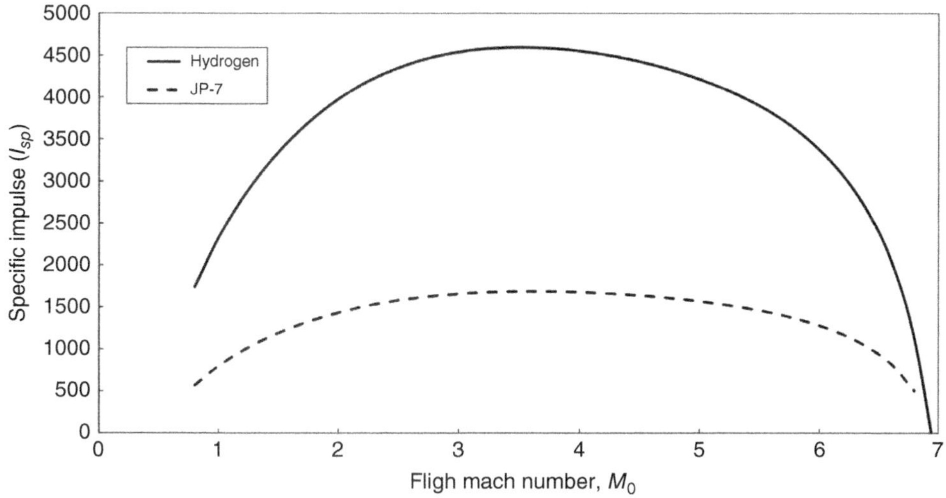

Figure 10.9 Specific impulse for a ramjet engine with component losses operating on hydrogen and JP-7.

Finally, in the nozzle, the main loss is due to over- or under-expansion so that $p_9 \neq p_0$ and a loss in total pressure, so that $\pi_n = p_{t7}/p_{t3} < 1$. Therefore, the derivation of the specific impulse I_{sp} for the ramjet with component losses begins with the general form of the specific thrust, Eq. (10.12). Figure 10.9 plots the specific impulse for the case when $T_{t7} = 2200$ K, $\pi_{dmax} = 0.9$ with η_r as a function of flight Mach number, according to Eq. (10.16), $\eta_b = 0.95$, and $\pi_n = 0.95$. As shown, the I_{sp} of the hydrogen-fueled ramjet is much higher than that burning JP-7 fuel. In both cases, the maximum value of I_{sp} occurs at a given flight Mach number and eventually drops to unacceptable levels as $M_0 > 6.0$. A small improvement is possible if the maximum cycle temperature is $T_{t7} > 2200$ K, and the losses in the ramjet are further reduced.

10.4.4 Ideal Turbojet

First, we consider the ideal turbojet operating without AB, a condition called "dry turbojet." The cycle analysis yields the specific thrust, which is governed by the total temperature ratios across the compressor and the turbine, the flight Mach number, and the limiting parameter τ_λ:

$$\text{Dry turbojet } \frac{F}{\dot{m}_0} = a_0\left[\sqrt{\frac{2}{\gamma-1}\frac{\tau_\lambda}{\tau_r\tau_c}(\tau_r\tau_c\tau_t - 1)} - M_0\right] \tag{10.19}$$

Figure 10.10 shows that for a given compressor pressure ratio and maximum cycle temperature $T_{t4} = 2200$ K, the turbojet's specific thrust decreases with increase in flight Mach number. Note that for $M_0 > 3$, the specific thrust has decreased to unacceptable level. Figure 10.11 shows the effect of compressor pressure ratio π_c on specific thrust $F/\dot{m}_0$. Observe that for a fixed flight Mach number M_0, there is a π_c that gives maximum

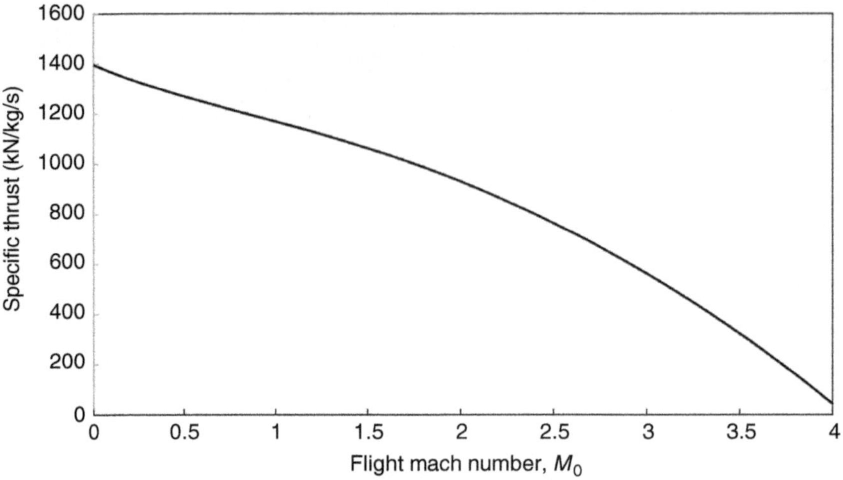

Figure 10.10 Ideal turbojet specific thrust with constant T_{t4} = 2200 K, π_c = 20.

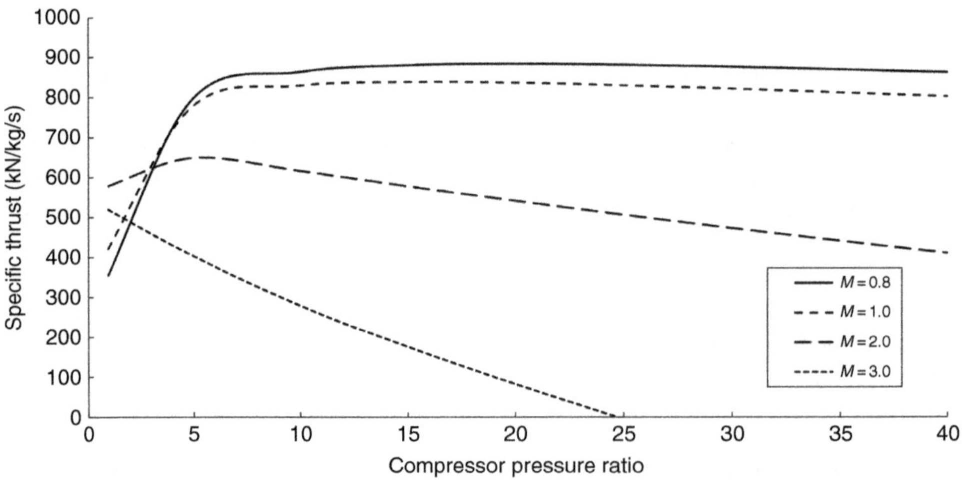

Figure 10.11 Effect of compressor pressure ratio π_c on specific thrust of ideal turbojet.

$F/\dot{m}_0$. Moreover, at higher M_0, a lower compressor pressure ratio is desired in order to obtain reasonable specific thrust. This trend suggests that the compressor of a turbojet designed to power a supersonic vehicle requires to produce a higher π_c for subsonic flight and a lower π_c for supersonic flight.

In addition, by performing an energy balance in the combustor in the same manner, we did in Chapter 2, we obtain an expression for the fuel/air ratio for the ideal turbojet:

$$f = \frac{c_p T_0}{h_{PR}}(\tau_\lambda - \tau_r \tau_c) \tag{10.20}$$

where we note the effect of the compressor total temperature ratio τ_c.

The compressor temperature ratio τ_c that gives maximum thrust for a turbojet engine is

$$\tau_{c_{optimum}} = \frac{\sqrt{\tau_\lambda}}{\tau_r} \tag{10.21}$$

This relation implies that for a given combustor temperature, the compressor temperature ratio requirement for the maximum cycle output (i.e. the optimum compressor pressure ratio) goes down with the flight Mach number. We can recast Eq. (10.21) in terms of the (optimum) compressor pressure ratio for maximum thrust as

$$\text{Dry turbojet } \pi_{c_{optimum}} = \tau_c^{\gamma/(\gamma-1)} = \left(\frac{\sqrt{\tau_\lambda}}{1+(\gamma-1/2)M_0^2}\right)^{\gamma/(\gamma-1)} \tag{10.22}$$

Figure 10.12 shows the optimum compressor pressure ratio variation with flight Mach number for three values of τ_λ. Note, $\pi_{c_{optimum}}$ drops nonlinearly with flight Mach number. At approximately $M_0 = 2.8$, the optimum compressor pressure ratio is $\pi_{c_{optimum}} \approx 1$, at which point there is no need for a mechanical compressor, e.g. in this limit the turbojet reaches the condition of the ramjet.

Now we wish to know the effect of the AB on specific thrust. For the afterburning (AB) turbojet, $F/\dot{m}_0$ is given by the following expression:

$$\text{AB turbojet } \frac{F}{\dot{m}_0} = a_0\left[\sqrt{\frac{2}{\gamma-1}\frac{\tau_{\lambda_{AB}}}{\tau_r\tau_c\tau_t}(\tau_r\tau_c\tau_t-1)} - M_0\right] \tag{10.23}$$

Observe the similarity between Eq. (10.23) and that obtained for the dry turbojet (without afterburning), Eq. (10.19). The minimum value for $\tau_{\lambda_{AB}}$ that occurs for no afterburning is $(\tau_{\lambda_{AB}})_{min} = \tau_\lambda\tau_t$. Substituting this minimum, then Eq. (10.23) reduces to Eq. (10.19).

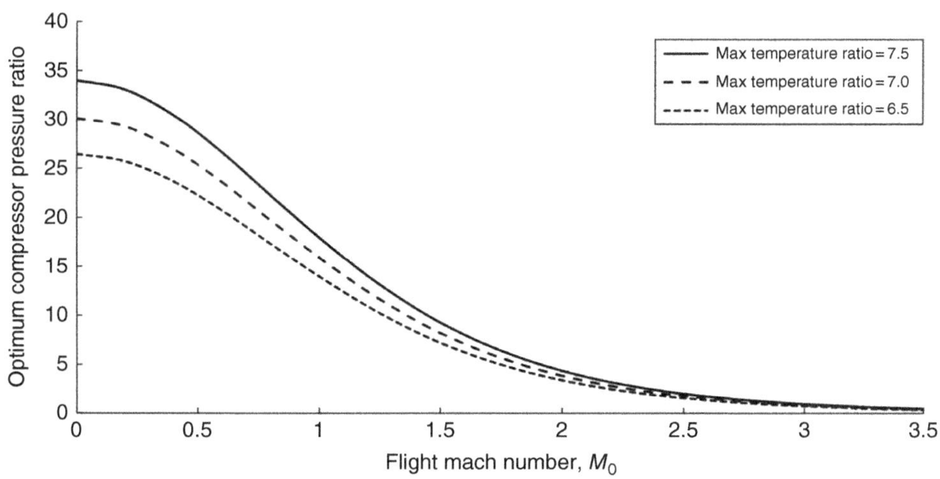

Figure 10.12 Optimum compressor pressure variation with the flight Mach number for ideal turbojet with three different values of τ_λ.

Now we examine the compressor pressure ratio that maximizes the specific thrust of the AB turbojet at a given flight Mach number (M_0), altitude (a_0), and both cycle thermal limits ($\tau_\lambda, \tau_{\lambda_{AB}}$). The (optimum) compressor pressure ratio for maximum thrust is given by

$$\text{AB turbojet } \pi_{c_{optimum}} = \tau_c^{\gamma/(\gamma-1)} = \left(\frac{\tau_r + \tau_\lambda}{2\tau_r}\right)^{\gamma/(\gamma-1)} \tag{10.24}$$

Note that the optimum compressor pressure ratio for the ideal afterburning turbojet engine is independent of the AB temperature limit $\tau_{\lambda_{AB}}$.

It is also interesting to compare the compressor pressure ratio for a maximum thrust of a dry turbojet with the afterburning engine. As shown in Figure 10.13, at low flight speeds, the AB turbojet requires unreasonable large compressor pressure ratios. This confirms what air-breathing propulsion professionals know that it is unpractical for an engine to use the AB at subsonic flight Mach numbers. Thus, a reasonable limit for the use of AB may be established at $M_0 > 1.8$.

Afterburning boosts the specific thrust of the turbojet engine. The magnitude of the increase depends on the compressor pressure ratio. As shown in Figure 10.14, for $\pi_c = 10$, afterburning increases the specific thrust by about 200 kN/kg/s at subsonic flight Mach numbers, and this increase is even greater at supersonic speeds.

It should be clear that the increase of thrust by afterburning is also accompanied by an increase in specific fuel consumption. Therefore, the AB is limited to periods of short duration. Afterburning can be used to improve thrust during climb, to accelerate beyond the sound barrier, and to improve combat performance (for military aircraft), and while transitioning to scramjet power for TBCC propulsion.

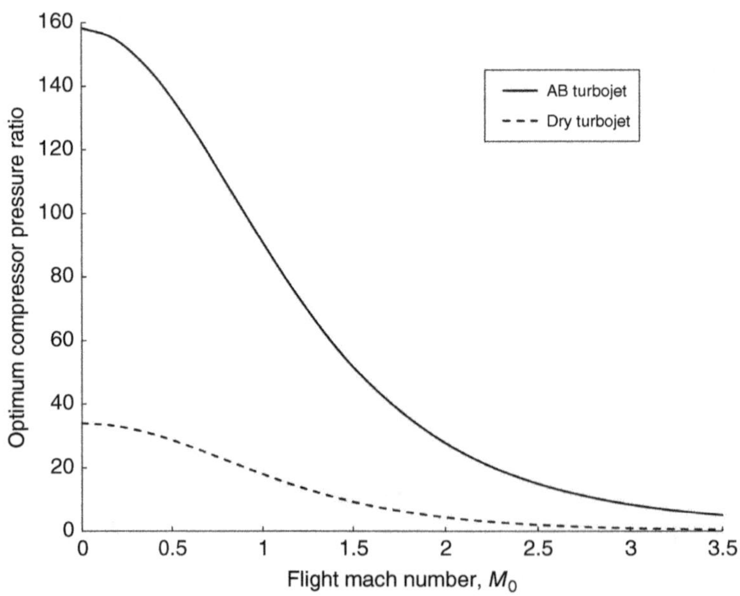

Figure 10.13 Compressor pressure ratio for maximum thrust of ideal turbojet with AB and without afterburner (dry) for τ_λ = 7.5.

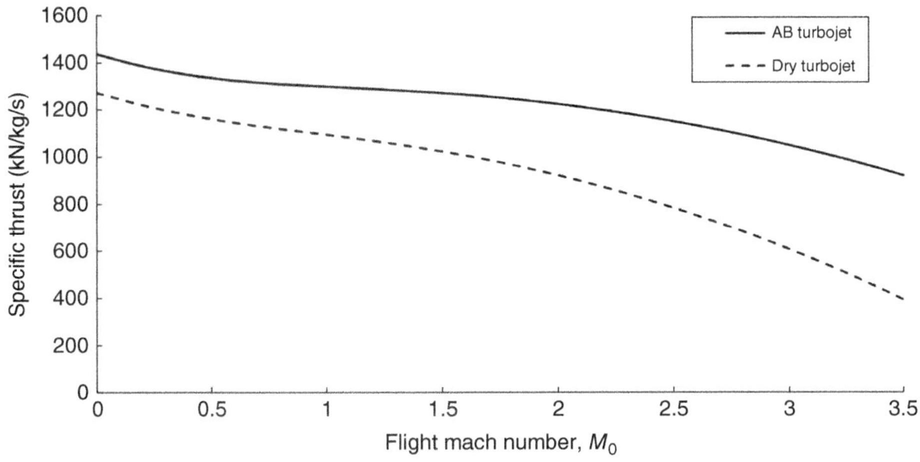

Figure 10.14 Specific thrust for turbojet operating at π_c = 10, T_{t4} = 2000 K and T_{t7} = 2220 K.

An afterburning turbojet requires a variable supersonic nozzle with a convergent–divergent (C–D) geometry. A C–D nozzle has a well-defined throat before a divergent cone or ramp, designed to accommodate the increase/decrease of mass flow rate resulting from having the AB on and off. Over a wide range of operating conditions, the nozzle has a choked throat. When the AB is on, the exhaust gas temperature rises, and the density decreases accordingly. Thus, the throat area needs to be opened to accommodate the lower density gas at the sonic condition at the nozzle throat and to satisfy the continuity equation.

Consider a turbojet engine fueled by JP-7 operating at an altitude where $p_0 = 7.565$ kPa, $T_0 = 216.62$ K, subjected to a limiting $T_{t4} = 2200$ K. Assume an ideal turbojet cycle and determine the specific thrust when the cruise Mach number is $M_0 = 3$. One can easily calculate the compressor discharge total temperature T_{t3} and study the effect of the compressor pressure ratio π_c on the turbojet specific thrust, considering a flight Mach number range $0 < M_0 < 4$, and assuming π_c is 10, 20, and 30.

For air $\gamma = 1.4$, $c_p = 1.004$ kJ/(kg · K), and $R = 0.2869$ kJ/(kg · K). At the flight condition, the local sonic velocity is $a_0 = \sqrt{\gamma R T_0} = 294.97\ \text{m/s}$.

The total/static temperature and pressure ratios of the freestream τ_r and π_r are

$$\tau_r = \frac{T_{t0}}{T_0} = 1 + \frac{\gamma_0 - 1}{2} M_0^2 = 1 + 0.2(3)^2 = 2.8 \rightarrow \pi_r = \frac{p_{t0}}{p_0} = (\tau_r)^{\gamma_0/(\gamma_0 - 1)} = (2.8)^{3.5} = 36.73$$

The freestream air stagnation conditions are $T_{t0} = T_0\tau_r = 216.62$ K (2.8) = 606.54 K, $p_{t0} = p_0\pi_r = 7.565$ kPa(36.73) = 277.86 kPa, respectively.

For isentropic compression, the compression temperature ratio values are

$$\tau_c = \frac{T_{t3}}{T_{t2}} = (\pi_c)^{(\gamma - 1)/\gamma} = \begin{cases} (10)^{0.2857} = 1.93 \\ (20)^{0.2857} = 2.35 \\ (30)^{0.2857} = 2.64 \end{cases}$$

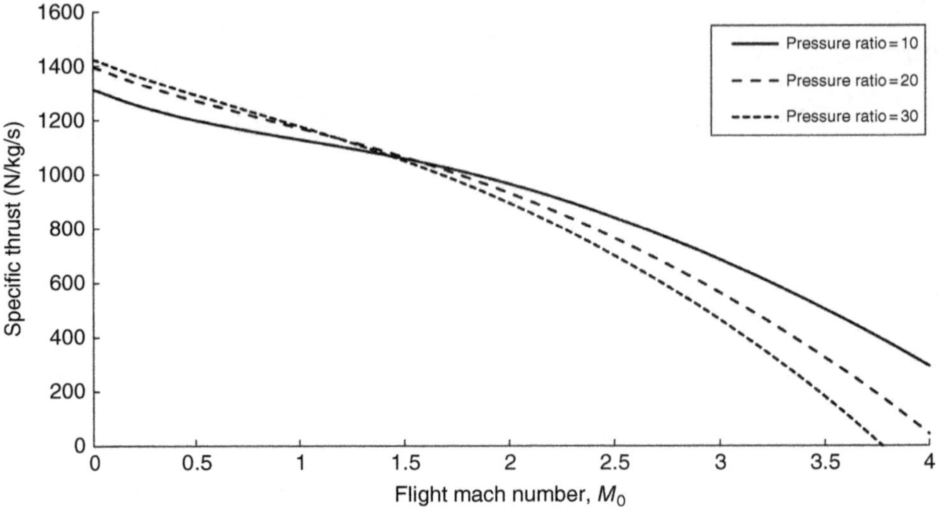

Figure 10.15 Effect of compression pressure ratio on specific thrust of ideal turbojet.

The compressor air discharge total temperature is 1170.6 K < T_{t3} < 1601.3 K. Verify that these values are not higher than the practical limit temperature of the compressor materials.

The ratio of the burner exit enthalpy to the ambient enthalpy τ_λ is

$$\tau_\lambda = \frac{T_{t4}}{T_0} = \frac{2200}{216.62} = 10.15$$

The specific thrust of the ideal turbojet at the given conditions is, from Eq. (10.19),

$$\frac{F}{\dot{m}_0} = a_0 \left[\sqrt{\frac{2}{\gamma - 1} \frac{\tau_\lambda}{\tau_r \tau_c} (\tau_r \tau_c \tau_t - 1)} - M_0 \right] = 294.97\,\text{m/s}\,(5.32 - 3.0) = 684.3\,\text{N/(kg/s)}$$

Figure 10.15 shows the effect of π_c on specific thrust for $0 < M_0 < 4$, with $\pi_c = 10, 20, 30$.

10.4.5 Performance Characteristics of Hydrogen-Fueled Turbo-Ramjet Engine

The selection of an appropriate CCP system for a hypersonic vehicle must be based on the requirement that fundamental technology for the accelerator engine is available in a predetermined schedule consistent with an appropriate development cycle. For a preliminary study, I selected a turbojet and a ramjet fueled by LH_2 to obtain baseline performance and assess their capability to accelerate a vehicle from takeoff at sea level to scramjet TO speed, $M_0 \approx 4.5$. The analysis includes engine component losses and processes inefficiencies (e.g. diffusion, combustion, and expansion) and assumes that the turbojet operates with a compressor pressure ratio $\pi_c = 20$.

As shown in Figure 10.16, consider that the turbojet and ramjet operate simultaneously above Mach 1.0, until transitioning to full ramjet power, which I estimate to occur at about Mach 3.0.

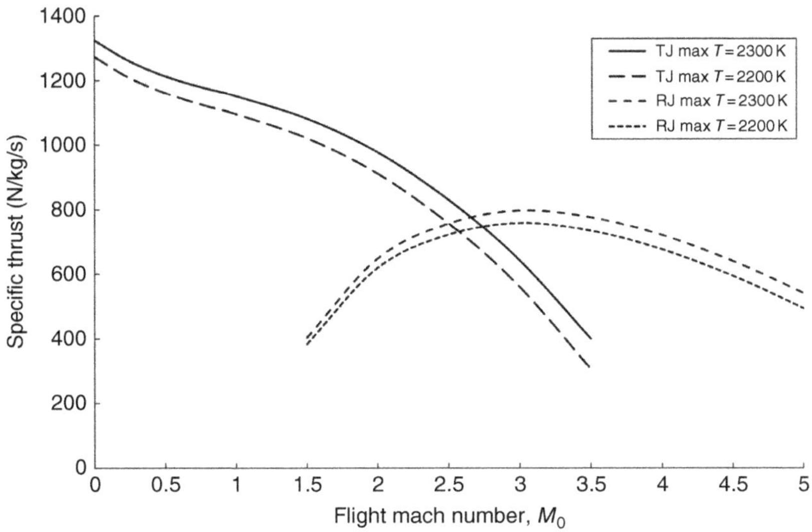

Figure 10.16 Uninstalled thrust: transition from turbojet to ramjet on hydrogen fuel.

A note on the maximum value of air-breathing engine cycle temperature. In a turbojet, the limitation imposed on the maximum temperature of the flow after combustion T_{t4} is due to the operational limits of the turbine, as the turbine blades are exposed to severe rotating tensile stresses that impose a limit on the gas temperature to avoid blade failure. Considerable research and development effort is ongoing, aimed to increase the maximum allowable turbine inlet stagnation temperature T_{t4} and the maximum cycle stagnation temperature T_{t7} in afterburning jet engines. For the state-of-the-art limits, Mattingly and Boyer (2016) give the following values: $T_{t4} = 2000$ K (3600 °R) and $T_{t7} = 2220$ K (4000 °R). Materials, coatings, and cooling technologies have improved in the past decades, and thus we expect that these limiting values will be surpassed. The ramjet does not have a turbine, and thus can operate at higher temperatures than are possible in a turbojet. However, too high a temperature would cause gas dissociation and impose unacceptable thermal loads in the ramjet walls. Therefore, for the review in this chapter, I restricted the ramjet cycle to the above given AB maximum temperature.

Before closing on this topic, let us remember that the actual air-breathing jet engine differs from the ideal cycle outlined primarily because of irreversibilities and pressure drops in the flow path and other parts of the engine. However, once we incorporate component losses, when comparing the turbojet with losses and the ideal turbojet model, we observe the variation of specific thrust with compressor pressure ratio or Mach number is not appreciably changed, and the magnitudes of $F/\dot{m}_0$ are nearly equal.

10.4.6 Turbojet for Supersonic Civil Transports

In 2018, Boom Technology showcased its future supersonic passenger jet, which is expected to achieve top cruise speeds of Mach 2.2. To address the sonic boom problem, the aircraft will only fly at top speeds over-water, unless flying in a supersonic corridor. The full-scale aircraft will be 170 ft (51.8 m) long and have a wingspan of 60 ft (18.3 m). It is expected to carry 45–55 passengers and have a maximum cruise range of 9000 nm (16 670 km). Interestingly, this is the same range of the Airbus A340-500; the Boeing 777-200LR is the world's longest range subsonic (Mach 0.84 cruise speed) airliner with a range of 9395 nm (17 395 km).

To demonstrate the concept, the company is building the XB-1 Supersonic Demonstrator, a 1/3-scale version of the Boom supersonic aircraft. Three General Electric J85-21 turbojet engines will power the 68-ft long XB-1 demonstrator. The J85 is a small single-shaft afterburning turbojet engine that has a military version delivering up to 2950 lbf (13.1 kN) of thrust. With the AB on, it can deliver up to 5000 lbf (22 kN). However, it is very likely that the three engines in the full-scale Boom airliner will not have ABs.

10.4.7 Thrust Augmentation Options

Air-breathing jet engines require excess thrust to accelerate through the transonic drag rise and carry other acceleration mission requirements. In addition to increasing the turbine inlet temperature (T_{t4}) and using afterburning, there are other operational or design approaches that can boost the thrust achieved by a turbojet engine. For the P&W J58 engine, for example, engineers considered these options: increase the maximum AB temperature; increase the compressor rotor speed; modify the compressor bleed and inlet guide vane schedules; increase the AB fuel flow, combined with nitrous oxidizer injection (Conners 1997).

Another thrust augmentation idea is to increase the mass flow rate through the engine. For example, the inlet area may be increased in order to ingest more air, but this would result in increased drag and weight. However, the vehicle airframe forebody can be designed to make the freestream capture area A_0 much larger than the physical opening of the engine inlet (see Chapter 4).

In addition, water injection or a mixture of water and alcohol could be added to increase engine thrust during takeoff. Water injection is an old technique by which a finely atomized spray of water or a mixture of water and methanol is injected upstream of the compressor, or at the entrance of the combustion chamber. In the first option, spraying the water in the compressor the mass flow rate increases and as a result the combustor entrance pressure increases. The higher p_3 and the increased $\dot{m}_0$ yields a higher thrust. In the second option, injecting the water mixture into the combustor produces the same effect, but to a lesser degree and with greater consumption of water. Water injection on a hot day can increase the takeoff thrust by as much as 50% because the original mass of air entering the jet engine is less for a hot day (Mattingly and Boyer 2016).

In the late 1990s, NASA Dryden Flight Research Center carried out performance analysis of the SR-71 aircraft for use as a launch platform for high-speed research vehicles, and for carrying captive experimental packages to high altitude and Mach number conditions. Using a payload drag profile example representative of a large hypersonic research vehicle designed for high-speed air-launch, the analysis predicted that, without the use of thrust enhancement, the vehicle could not accelerate through the transonic range. However, using fuel control modifications only, the vehicle was able to accelerate through the transonic range but was unable to reach the target test condition before reaching the return-to-base jet fuel limit. Using nitrous oxide injection from Mach 1 to Mach 1.23, which fully depleted the 2700 lbm (1225 kg) load, the SR-71 was able to reach the target test condition with approximately four minutes of cruise time available before having to return to base (Conners 1997).

10.4.8 Turbo Engine for Low-Speed Cycle of TBCC Propulsion Systems

The P&W J58 (JT11D-20) afterburning turbojet was capable of propelling the SR-71 aircraft to Mach 3+ flight speed operating in ramjet mode. By the late 1990s, the fastest supersonic military aircraft are propelled by low-bypass afterburning turbofans. The turbofan is a turbojet with a ducted fan at the front of the engine used to ingest additional air to augment the thrust. The turbofan was designed with an annular-shaped bypass duct that allows part of the fan's discharge air to flow around the high-pressure compressor, combustor, and both turbines. Some of the air leaving the fan enters the high-pressure compressor, while the remainder flows through the bypass duct. This bypass air is eventually accelerated through a nozzle to produce thrust. The ratio of the mass-flow of air bypassing the engine core compared to the mass-flow of air passing through the core is referred to as the bypass ratio (BPR).

The P&W F100-220 is an afterburning turbofan (0.36 BPR, $T/W = 7.6$). The F100 (JTF22) powers the McDonnell Douglas F-15 Eagle tactical fighter aircraft (maximum speed: Mach 2.5 [1650 mph, 2655 km/h] flying at high altitude) and the General Dynamics F-16 Fighting Falcon (maximum speed: Mach 2.05 [1318 mph; 2121 km/h] flying at 40 000 ft). The engine inlet is usually not shown in engine cutaways because each tactical aircraft requires a different inlet design to accommodate engine integration with the airframe design.

Although more complex than a turbojet, a turbofan engine is more efficient and has a higher thrust at zero speed (takeoff thrust), to accelerate quickly on the runway to allow high payloads with short runways. In addition, modern turbofan engines have a high thrust-to-weight ratio, which means high acceleration. For most flight conditions, an aircraft with a high thrust to weight ratio will also have a high value of excess thrust. High excess thrust results in a high rate of climb.

Turbofans are the only engines on military fighter aircraft designed with ABs for additional thrust boost capability. In afterburning turbofans, the portion of the fan's air that passes through the bypass duct is remixed with the core's combustion products in the AB, before the mixture is accelerated through the nozzle. When maximum thrust is needed, the AB injects additional fuel into these flows as they are mixing, and then this air–fuel mixture burns before it reaches the nozzle. Due to fuel efficiency (flight duration and range) considerations, the AB is used only for takeoff and when maximum acceleration is needed for a short period of time. The afterburning turbofan (P&W F119-100) is powerful enough to allow the Lockheed Martin F-22 Raptor to supercruise, that is, to fly supersonically without afterburning.

The GE YF120 (GE37) was a variable cycle augmented turbofan engine designed by General Electric Aircraft Engines (GEAE) in the early 1990s for the US Air Force's Advanced Tactical Fighter (ATF) project. Prototype engines were installed in the Lockheed YF-22 and Northrop YF-23 (the two competing technology demonstrator) aircraft. The P&W F119 was selected over the YF120 to power the ATF, which became the F-22 Raptor (Mach 2.25 maximum speed).

Turbofan engines for military aircraft are designed with low BPRs (typically 0.3–0.8) to balance engine efficiency, engine diameter, and weight (having acceptable thrust-to-weight ratios, T/W). The thrust-to-weight ratio indicates the maximum thrust performance available from an engine for each pound of engine weight. Increasing the T/W is desirable because it enhances overall aircraft performance and reduce lifecycle costs. Modern tactical aircraft engines, which normally place a great emphasis on thrust performance, have T/W of about 8 : 1.

In the past four decades, a number of air-breathing engine concepts have been considered to extend the operational envelope of turbojets/turbofans (see Section 10.7).

10.5 Single-Stage-To-Orbit and Two-State-To-Orbit Vehicles

The National AeroSpace Plane (NASP) was an ambitious program aiming to build the ultimate vehicle capable of flying at speeds greater than 17 000 miles per hour (27 359 km/h) or $M_0 = 25$. The main goal of the NASP program was to develop a reusable experimental vehicle (denoted the X-30) to explore the entire hypersonic velocity flight range. As a spaceplane, it would insert into LEO a crewed system with a combined cycle air-breathing propulsion system. This required maximum use of Earth's atmosphere for maneuvering and propulsion. To meet operational requirements, the early NASP program reviewed many vehicle configurations and studied extensively many CCP options, including:

1) **Single-Stage-To-Orbit (SSTO) vehicles**. These are reusable launcher concepts that can takeoff and land repeatedly and are able to boost payloads into orbit. Also called spaceplanes, SSTO vehicles could be powered by Rocket-Based Combined Cycle (RBCC) propulsion (integrating scramjets and rockets), or propelled by air-breathing rockets that utilize elements from air-breathing engine technology. An example is the Skylon launch vehicle powered by air-breathing rockets under development in the United Kingdom (see Section 10.6.3).
2) **Two-Stage-To-Orbit (TSTO) vehicles**. These are reusable launch system concepts composed of two matted vehicles (stages) powered independently by their own propulsion system and fuel: the first stage vehicle (booster) is conceived to provide horizontal takeoff capability, having a propulsion system that accelerates both stages to a predetermined altitude and speed (staging point). At staging, the second vehicle (orbiter) disengages from the first stage. After separation, the orbiter can use its own propulsion system for further acceleration and insertion into space, while the first stage vehicle returns to the launch site. NASA is currently developing a TSTO vehicle (see Section 10.6.1).

Viable and practical hypersonic air-breathing propulsion systems are required for such reusable spaceplanes. By incorporating air-breathing capability, SSTO or TSTO reusable vehicles have the potential to realize space launch operations with high payload fraction that may significantly reduce launch costs and improve safety, providing aircraft-like operations with expanded launch site capability and reduced system maintenance. Horizontal takeoff and landing would enhance launch, flight and ground operability.

After terminating the NASP program, NASA instituted the Integrated Space Transportation Plan, which provided a comprehensive, long-term strategy to meet future launch vehicle, and technology needs combined with reduced cost and improved safety and reliability. Considering the technology readiness in the areas of propulsion, structures, and materials, the technical risk for SSTO was determined to be high. Thus, the advanced reusable launch vehicle (RLV) design attention shifted back to TSTO concepts. The TSTO concepts ranged greatly with respect to orbiter/booster shape, propulsion system characteristics, and resulting nominal flight trajectories. In all cases, studies indicated a critical need to enhance the aerodynamic and aerothermodynamic knowledge base in two-body separation characteristics within the sensible atmosphere to satisfy both nominal and abort ascent separation scenarios.

To date, a fully reusable space transportation system has not been demonstrated in flight. Thus, we rely on system and conceptual designs to study them, including substantial data from component ground testing, analysis, and conceptual studies related to TBCC propulsion (see for example, Eklund and Boudreau 2005). For RBCC, many predictions are available based on various conceptual studies. As indicated qualitatively in Figure 10.1, TBCC propulsion systems exhibit significant performance improvements (higher I_{sp}) over RBCC propulsion in the subsonic takeoff and return mission segments. In the following sections, we provide highlights of some of the ideas that can motivate future endeavors.

10.6 Propulsion for Spaceplanes

To capitalize on the available oxygen in the atmosphere and fly a reusable hypersonic vehicle with the flexibility of an aircraft, it requires development of hypersonic air-breathing propulsion to power spaceplanes that can achieve orbital speeds. Compared with conventional rocket launchers, air-breathing spaceplanes offer unique benefits. Let us put this into perspective: 75% of the takeoff weight of the former Space Shuttle rocket launch system was propellant mass, and of this mass, 83% was oxidant (liquid oxygen LOX). Even with modern launch systems that utilize reusable first stages, the amount of LOX they must carry is in excess of 600 340 kg. An air-breathing vehicle would not carry all that oxidizer at launch, and this would give it a capability to reach orbit with a greater payload. This fact alone has inspired nations to conceive space transportation systems propelled by air-breathing propulsion that could offer horizontal takeoff and landing (HTOL) capability. Examples of such spaceplane ideas are the European SÄNGER, the American NASP, and Skylon, now under development by the British company REL.

Integration of elements from multiple engines into a single propulsion system using the same elements is preferable to avoid dead weight, with the purpose of optimizing flight performance over a wide range of speeds. Integration must benefit from each propulsion cycle in the flight regime where it has the best performance, eliminating undesirable characteristics (added weight or using multiple separate engines). This requires closely coupling elements of various cycles: turbo-machinery (from turbojets/turbofans), combustors (from turbojets, ramjets, and scramjets), gas generators (ejector rocket engines), heat exchangers, air-breathing compression/inlet systems, and shared propulsion nozzles. All these elements are subsystems of an overall power plant, and when used together in their respective modes constitute a combined-cycle engine.

However, disparities among engine types, performance, and flight requirements yield rather difficult propulsion integration solutions (Espinosa 2003). The fuel, for example, will impact the choice of propulsion arrangement for a given mission. Mach 8 is considered the limiting cruise speed to which a dual-mode ramjet (DMRJ)/scramjet can be fueled and cooled with endothermic fuels for optimum performance. On the other hand, hydrogen provides the ideal heat of reaction and heat sink for cooling engine components at all hypersonic velocities. Hydrogen fuel has also environmental benefits by elimination of CO, CO_2, SOx, unburnt hydrocarbons (UHC) and smoke. Overall, whether fueled by hydrocarbons or hydrogen, integration and transition through multiple propulsion cycles possess huge challenges for designers, requiring advanced optimization rules, and technologies still requiring maturation and demonstration in realistic conditions.

10.6.1 NASA Two-Stage Launch Vehicle

A fully reusable, TSTO vehicle concept was conceived under the NASA/U.S. Air Force Joint System Study, a program intended to evaluate the critical technologies for development. The first vehicle stage (the launcher or booster stage) is propelled by an air-breathing CCP system (turbo-ram–scramjet), and the second stage vehicle (the orbiter) is powered by conventional rocket engines. Depicted in Figure 10.17, this TSTO launch system is studied for a mission to reach a 100 × 100 nmi circular orbit at a 28.5° inclination, to deploy or retrieve a 20 000 lbm (9100 kg) payload (Snyder and Espinosa 2013).

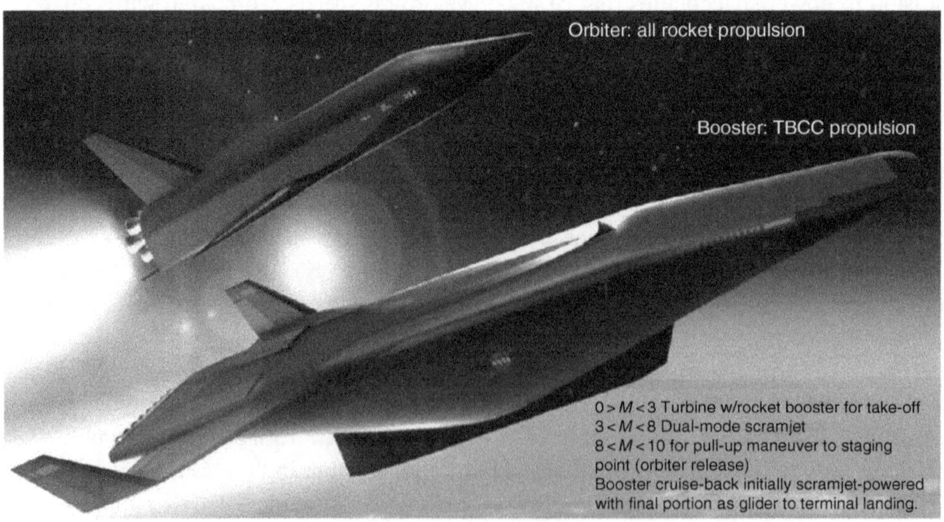

Figure 10.17 NASA's reusable TSTO vehicle concept. Source: From Snyder and Espinosa (2013).

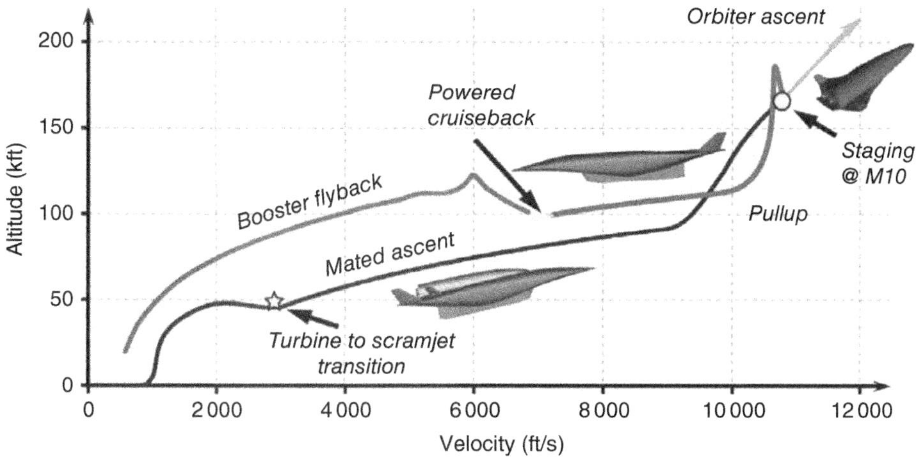

Figure 10.18 NASA TSTO vehicle mission profile. Source: From Snyder and Espinosa (2013)/NASA/ Public Domain.

The unnamed reusable TSTO launcher is conceived to perform horizontal takeoff in a mated configuration (orbiter sitting on top of the launch vehicle), using the booster stage TBCC propulsion and tail rockets. As depicted in the nominal flight profile of Figure 10.18, at approximately Mach 3, the turbine engine will transition to dual-mode scramjet, operating until just before vehicle stage separation. At Mach 10, 100 psf (4.788 kPa) dynamic pressure, and a 5° flight path angle, the two vehicle stages will separate. Igniting its rockets, the orbiter stage will continue its ascent to the designated orbit. Meanwhile, the booster stage vehicle will perform a complex maneuver with scramjet powered cruiseback to make its return to the launch site. At an altitude of approximately 125 kft (~38 km) and 6000 ft/s (1828.8 m/s) velocity, the booster stage will glide back to the ground, as depicted in Figure 10.18. Meanwhile, the orbiter stage will achieve orbit to deliver the payload to LEO, and it will re-enter on an unpowered glide back to the launch site.

10.6.2 Over–Under Dual Flowpath TBCC Concept

For propelling the first stage of the NASA TSTO launch system described in the previous section, the TBCC will be configured using an over–under dual flowpath concept, as depicted in Figure 10.19. In this configuration, two engines (dual-mode scramjet and turbojet) share a common variable inlet and variable area nozzle, therefore saving weight and reducing system complexity.

From takeoff to the first cycle transition, thrust is provided by the turbojet/turbofan. The low-speed inlet flowpath will be open, with the high-speed inlet flowpath closed. Near the transition flight condition (Mach 3–4), the high-speed inlet will open and begin stabilization of the isolator/dual-mode combustor flowfield, and the TBCC will transition from turbo to ram/scramjet propulsion with fuel injection starting the scramjet combustor. At the same time, the low-speed inlet will close to shut down the turboengine flowpath. The ram/scramjet will then be ready to provide adequate acceleration to cruise hypersonic speed.

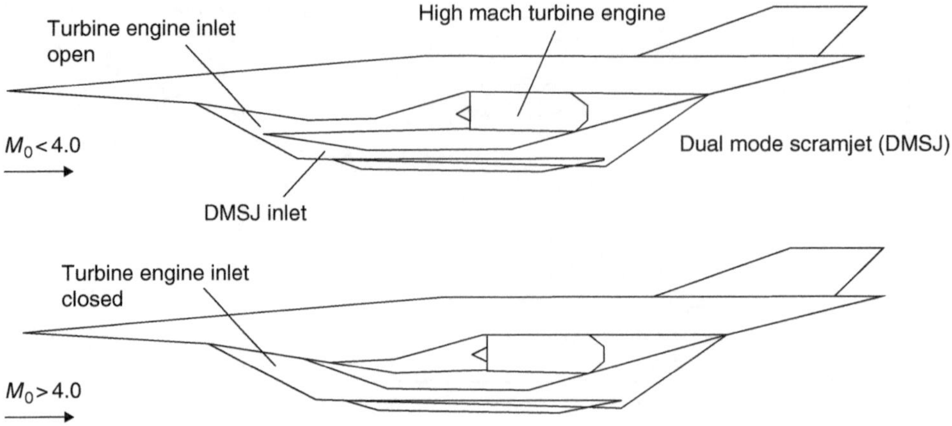

Figure 10.19 Over–under dual flowpath TBCC concept.

Although the dual flowpath requires a highly variable dual inlet and variable dual nozzle to accommodate the two separate and very distinct flowpaths, having the turbine engine in a separate duct will facilitate cocooning to isolate it from the high enthalpy flow at hypersonic speeds that the first stage of the TSTO vehicle will experience (up to Mach 10).

What type of turbine engine is capable of providing the low-speed performance for the initial supersonic climb of a TBCC system? In 2002, NASA established the Next Generation Launch Technology (NGLT) Program to pursue technologies for future space launch systems. NGLT funded research in key technology areas, including propulsion. A unique low bypass afterburning turbofan concept named the Revolutionary Turbine Accelerator (RTA) was part of that effort. The RTA was intended for ground demonstration of a Mach 4 accelerator in a TBCC engine configuration. The main objectives of the tests were to demonstrate mode transition from an augmented turbofan to a ramjet mode, and to demonstrate the Mach 4 thrust level required to accelerate a conceptual future X-vehicle to scramjet take over speed (Lee et al. 2005).

Designed by GE AE, the RTA variable cycle engine (VCE) depicted in Figure 10.20 has an internal flowpath that changes to allow for high specific impulse performance throughout the flight trajectory of an accelerator vehicle. The low-bypass RTA engine includes a new thrust augmentor called hyperburner. From takeoff to the point of transition to supersonic flight, the hyperburner serves as a conventional thrust augmentor. At high Mach numbers, the hyperburner is designed to operate as a ramjet to accelerate the vehicle to Mach 4.

Fueled by JP-8, the RTA engine has several strategically located variable area bypass injectors (VABIs) to achieve optimal balance between the core turbo machinery and the hyperburner. The VABIs are mechanical doors designed to divert airflow from various regions of the engine to another region. The combination of flow control and tailored fuel schedule is intended to enable the RTE to make a smooth and stable combustion transition from AB to ram combustor, producing the thrust required to accelerate the vehicle from sea level condition to Mach 4.

Using an existing YF120 (GE37) engine core, GE AE designed the ground demonstrator variable cycle-augmented turbofan engine for testing at NASA. Designated RTA-1, it can

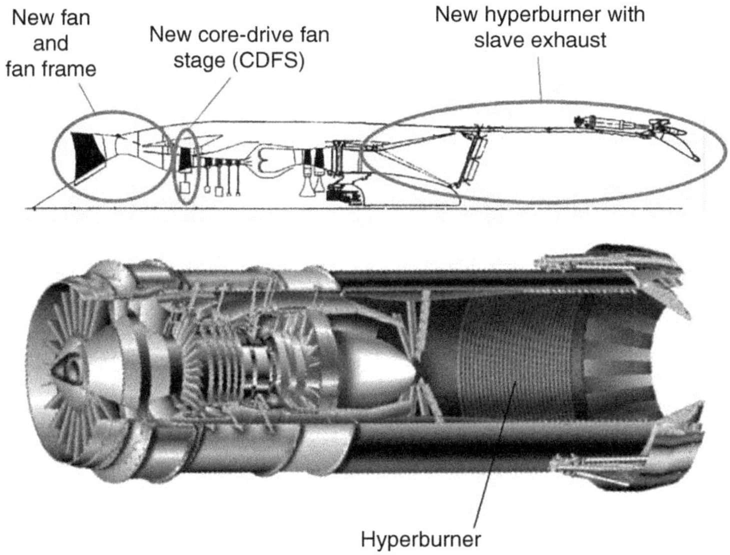

Figure 10.20 NASA RTA-GE 57 ground demonstration engine. Source: From Davoudzadeh et al. (2005)/NASA/Public Domain.

accommodate a large variation in BPRs and is designed with a new nickel alloy fan and new core drive fan stage, advanced trapped vortex combustor, radial flameholders with multiple fueling zones, and material upgrades to the YF120 stage 2 and 3 compressors to withstand the elevated temperatures at the higher Mach engine conditions. The RTA-1 has new hyperburner (augmentor/ramjet) that connects with an axisymmetric nozzle (Davoudzadeh et al. 2005).

Figure 10.21 shows the operating modes of the variable-cycle RTA-1, showing the test points that are critical to the flight demonstration of a TBCC engine for a TSTO vehicle. Operation allows for high fan pressure ratio mode for sea level static (SLS) to Mach 2 flight speeds and transitions to a lower fan pressure ratio mode for operation up to Mach 3. Above Mach 3, the hyperburner transitions from conventional augmentor to ramjet combustor, in order to accelerate a vehicle to Mach 4+. During acceleration from Mach 3 to 4+, the turbomachinery (compressor and turbine) is set to flight-idle speed (not completely shut down), so air restarts are not required after the orbiter vehicle separates from the first stage booster vehicle.

The notional engine temperatures shown on the right side of Figure 10.21 indicate the temperature limits on the compressor (T_3) and turbine (T_{41} and T_{40}). The highest limiting value corresponds to T_{41}, which denotes the inlet temperature condition of the high-pressure turbine (HPT), which is the first turbine stage downstream from the subsonic combustor (see the note on the maximum value of air-breathing engine cycle temperature in Section 6.4.5).

For the RTA-1, GE used components of its YF120 (GE57) turbofan engine integrated with new components in the definition of this low-cost, versatile, variable cycle Mach 4+ demonstrator. The GE37 is a 156 kN (35 000 lbf) class, variable cycle afterburning turbofan

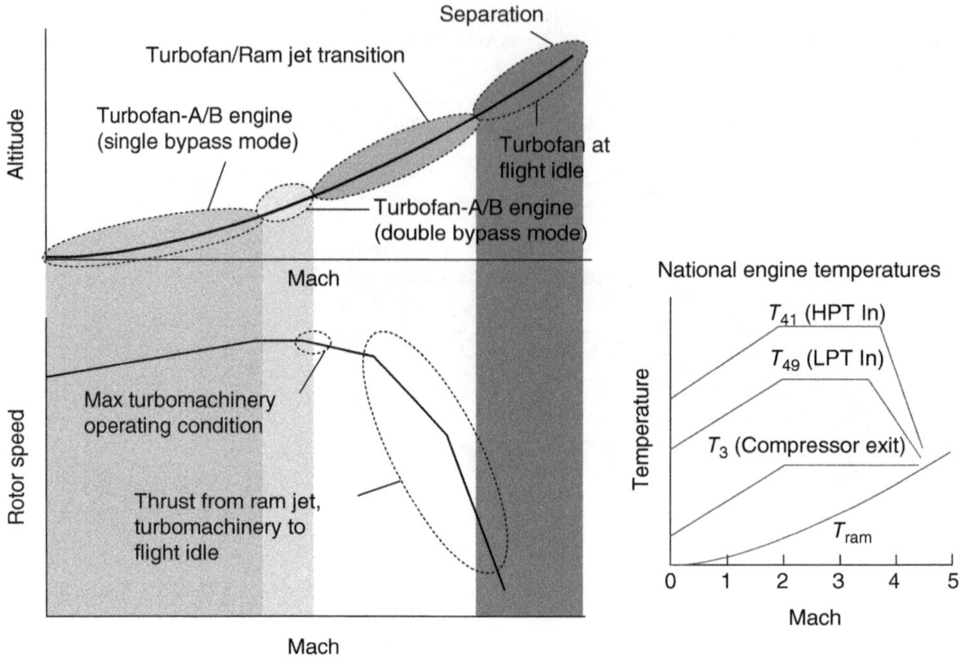

Figure 10.21 RTA-1 operation and basic engine temperatures. Source: From Davoudzadeh et al. (2005)/NASA/Public Domain.

engine designed by GE in the early 1990s for the U.S. Air Force's Advanced Tactical Fighter (ATF) project (Snyder and Espinosa 2013). Prototype engines were installed in the two competing technology demonstrator aircraft, the Lockheed YF-22 and Northrop YF-23. Operating on JP-4 or JP-8 fuel, the VCE would allow it to produce a large amount of dry thrust (without AB) with high off-design (cruise) efficiency. For the NASA RTA/TBCC program, studies were meant to compare the YF120 (GE57) engine performance capability for TSTO mission profiles and assess impacts to existing YF120 components.

10.6.3 Synergetic Air-Breathing Rocket Engine (SABRE)

The Synergetic Air-Breathing Rocket Engine (SABRE) is developed by Reaction Engines Ltd. (REL) to power the Skylon spaceplane. SABRE is a special type of air-augmented rocket engine or turbo-rocket, but many call it simply air-breathing rocket. It incorporates an air-inlet cone, turbo-compressor, preburner, and four rocket engines (Figure 10.22). The 44 000 lbf thrust engine is designed to operate in two modes: in air-breathing mode, SABRE will ingest atmospheric air as a source of oxygen to burn with its LH_2 fuel in the rocket combustion chamber.

In air-breathing mode, the ingested air must be compressed to about 140 atm before entering the rocket combustion chamber. Such pressure raises the air temperature so high that it would melt any known material. Hence, SABRE incorporates a precooler heat exchanger to first cool the ingested air from more than 1000 °C (1832 °F) down to −150 °C (−238 °F) in one one-hundredth of a second. The oxygen in the chilled air will become liquid in the process.

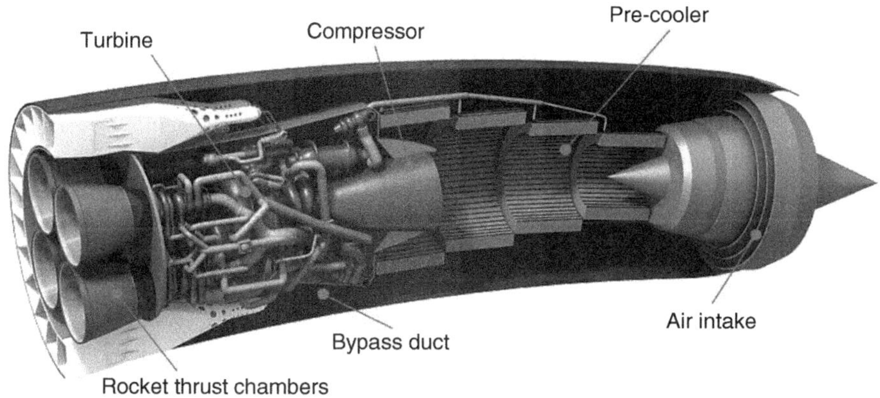

Figure 10.22 SABRE air-breathing rocket concept. Source: Reaction Engines Limited, REL.

Then a conventional turbo-compressor (using jet engine technology) will compress the cool air to the required rocket chamber pressure. After the air intake is closed down, SABRE will transition to pure rocket mode using conventional on-board liquid oxygen (Reaction Engines 2010).

Skylon is a reusable SSTO spaceplane concept capable of lifting 15 tons to LEO from a 5.5 km runway and returning to land at the same location, Skylon will be powered by two SABRE air-breathing rockets. The hybrid cycle propulsion has the potential to decrease the propellant fraction required for access to space by using oxygen in the atmosphere during part of the vehicle's ascent trajectory. According to REL estimates, the SABRE-powered Skylon spaceplane could save carrying over 250 tons of on-board oxidant on its way to orbit, in addition to removing the need for massive throw-away first rocket launch stages that are jettisoned once the oxidant has been used (Varvill and Bond 2008).

In its nominal ascent trajectory, Skylon will takeoff from a runway using the SABRE in air-breathing mode (burning hydrogen fuel with precooled compressed air) until it reaches an altitude of 28.5 km, at which point the vehicle will be traveling at about Mach 5.2. Then SABRE will close its air inlet and switch to rocket mode to accelerate for the remaining ascent trajectory. At about 80 km altitude, the rockets will shut off, achieving Main Engine Cut-Off (MECO), which will occur 980 seconds after launch. Figure 10.23 depicts the ascent trajectory from Reaction Engines (2010). As shown, after takeoff the vehicle will climb and accelerate propelled by air-breathing propulsion, following its predetermined trajectory for 694 seconds (approximately 11½ minutes), by which time Skylon would reach an altitude of 28.5 km (now 620 km downrange from the launch site).

Transitioning to pure rocket mode, burning LH_2 and liquid oxygen in the common combustion chambers, Skylon will climb rapidly and perform a gravity turn (gradual transition from vertical to horizontal flight caused by the force of gravity) in order to inject into an 80 by 300 km transfer orbit after a further 285 seconds (4¾ minutes), at which time main engine cut-off will occur; after jettisoning nonessential elements, the spaceplane will continue on a ballistic trajectory for a further 44 minutes until it can reach apogee at 300 km altitude. Immediately, its orbital maneuvering system (OMS) thrusters (small rockets producing a few kN thrust) will ignite in order to circularize the orbit at apogee.

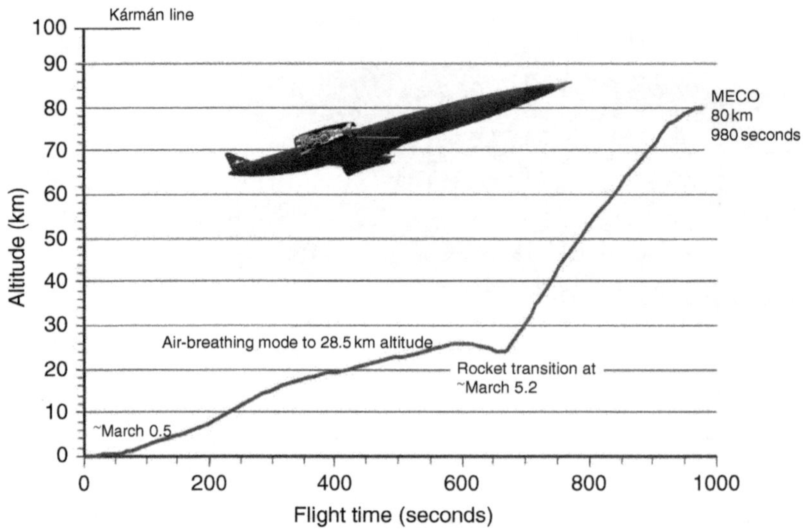

Figure 10.23 Artistic rendering of Skylon, a SSTO vehicle, and its ascent trajectory. Source: Adapted from image in Reaction Engines 2010 by Reaction Engines Ltd.

After routine system checkouts, Skylon will open its bay doors to deploy its payload (15 tonne or 150 00 kg). After completing its mission, the SSTO vehicle will maneuver to return at the typical re-entry interface altitude (120 km) and will glide on its descent to Earth, landing in a similar manner as the Space Shuttle Orbiter did (but with much increased maneuverability due to Skylon having a higher L/D ratio), and will land automatically (Reaction Engines 2010).

10.6.4 Australia Three Stage Space Launch System

In Australia, Hypersonix Launch Systems Pty Ltd. is developing a reusable three-stage launch system (Figure 10.24), which has the potential to reduce the costs of placing satellites into Earth orbit. On the launch pad, it will look and launch like a conventional rocket vehicle, but its mission profile will be radically different as it will incorporate an air-breathing stage.

The first stage consists of an Austral Launch Vehicle (ALV), a reusable rocket booster that lifts the upper stages to scramjet take-over speed (Mach 5). At this point in the mission, the first stage will separate and fly back to the launch pad. The second reusable stage, known as Scramjet Powered Accelerator for Reusable Technology AdvaNcement (SPARTAN), will unfurl its wings to fly like an airplane up to Mach 10–12, releasing the final rocket stage into the upper atmosphere. The Spartan air-breathing aircraft will return to the launch pad. The third stage, a rocket, will take the payload to its final location on LEO. The overall mission is depicted in Figure 10.25.

What makes the Hypersonix three-stage launch system concept more attractive is the reusability of the first two stages. Once completing their operation, both the scramjet stage and

Figure 10.24 The Delta-Velos Orbiter which contains the SPARTAN (ScramJet Powered Accelerator for Reusable Technology AdvaNcement) launching the third stage booster which will contain the satellite payload. Source: Hypersonix.

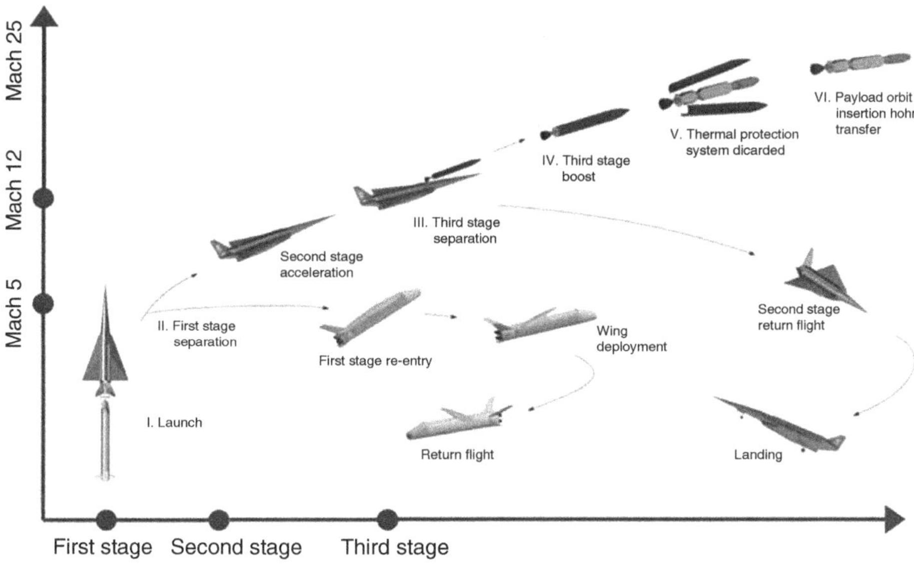

Figure 10.25 Hypersonix Delta-Velos, a three-stage space launch system. Source: Hypersonix.

the rocket booster will fly back to a runway, landing by deploying its aerodynamic wings and a small propeller. Only the final third rocket stage of the launch vehicle will burn up in the atmosphere after releasing its payload into orbit.

The R&D effort at Hypersonix is now concentrated in developing the SPARTAN second stage of the launch vehicle. Made of CMC composites, the reusable scramjet is designed to generate thrust to travel at speeds from Mach 5 to Mach 12.

10.7 Hydrogen for Hypersonic Air-Breathing Propulsion

In principle, a hydrogen-fueled engine has optimum propulsion performance because hydrogen has the highest heating value, and it has excellent combustion characteristics over a wide range of fuel–air mixtures. Hydrogen also provides an ideal heat sink for cooling propulsion-vehicle components. Equally important, hydrogen-air combustion does not produce carbon monoxide (CO), carbon dioxide (CO_2), sulfur oxides (SOx), UHC, or smoke.

10.7.1 Hydrogen for Fueling Entire Combined Cycle Propulsion Systems

In examining different reusable hypersonic space and global vehicle concepts that utilize air-breathing propulsion, we find that many of the CCP architectures consider multi-fuel arrangements. For example, hydrocarbon, methane, and hydrogen fuels are considered in the propulsion system for Boeing's two-stage to orbit (TSTO) Hypersonic Space and Global Transportation System (HSGTS): the first stage would employ eight hydrocarbon-fueled turbine engines to accelerate the mated second stage from horizontal takeoff to Mach 4. At this point, the HSGTS second stage from Mach 4 to orbit would be propelled by two top-mounted RBCC engines that each integrate a dual-mode methane (CH_4)-fueled scramjet and LOX/LO_2 rocket engines (Bowcutt et al. 2011).

The TBCC for the TSTO launch concept developed under the NASA/Air Force Joint System Study (JSS) considered JP-5 fuel for a Mach +4 turbofan. In the initial phase of the RTA program (see Section 10.6.2), General Electric (GE) Aircraft Engines designed a ground demonstrator engine for validation testing. The JP-7-fueled RTA demonstrator depicted in Figure 10.21 is a turbofan ramjet combined cycle engine, designed to transition from an augmented turbofan to a ramjet for peak thrust performance throughout the acceleration mission to Mach 4+. GE used an existing YF120 (GE57) engine core integrated with new components in the demonstration RTA (Lee et al. 2005). This suggests that the NASA TSTO vehicle will integrate a dual-fuel architecture, since the second stage is powered by conventional rocket engines.

Critical issues that must be considered is planning to use multiple fuels in a multistage hypersonic vehicle include:

- Various fuel tanks designs and manufacture;
- Complex vehicle configurations to accommodate distinct fuel tankage arrangements;
- Multiple type of fuel injectors, multiple fuel pumps, and fuel delivery and control complexities;
- Materials, structures, and thermal management.

In the face of these and other design and operational challenges, one must ask, why vehicle designers opted for multiple-fuel propulsion? Past research in the United States showed the feasibility of hydrogen turbojet, turbofan, and turboramjet engines, and some of those hydrogen-fueled engines were demonstrated in flight. Hence, we must establish whether hydrogen air-breathing propulsion systems are technologically mature enough to be considered as accelerators for contemporary hypersonic vehicles. It is then imperative to investigate the potential technology issues that must be resolved in order to establish how hydrogen-fueled gas turbine engines would benefit Turbine-Based Combined Cycle

(TBCC) propulsion considered for future spaceplanes and hypersonic aircraft. Ultimately, it is critically important to establish the kind of technical efforts required to develop hydrogen-fueled air-breathing accelerator engines for hypersonic flight.

10.7.2 Hydrogen Fuel for Air-Breathing Propulsion

Air-breathing hypersonic vehicles employing LH_2 fuel have the potential of satisfying a number of mission requirements for both global reach hypersonic transports and access to space with SSTO and TSTO configurations. Use of hydrogen fuel has benefits and drawbacks that we must acknowledge.

Because of its high specific impulse (I_{sp}), LH_2 is the best fuel for hypersonic flight when compared with conventional JP fuels. Additionally, hydrogen fuel offers significant heat capacity for engine and airframe cooling (see Chapter 6). Hydrogen fuel is more environmentally-benign by elimination of carbon monoxide (CO), carbon dioxide (CO_2), sulfur oxides (SOx), UHC, and smoke. The benefits of using hydrogen fuel for hypersonic air-breathing propulsion include

- Increased performance, e.g. higher I_{sp};
- Improved thermal management;
- Lower takeoff weight;
- Lesser adverse impact to the environment.

However, the low density of hydrogen (approximately 1/12 that of conventional JP fuel) and its required cryogenic storage temperature (20.3 K) represent a huge challenge for airframe design. The large volume required for fuel tankage imposes a requirement for the vehicle's shape, as it must provide high volume per total airframe surface area. The most important issues to consider are:

- Low volumetric energy density;
- Storage and handling;
- Production and operating costs.

In the past decades, engineers have identified optimal shapes which combine structural concepts with integral fuel tanks that yield vehicle structures with high volumetric utilization. Without considering production and cost of hydrogen fuel, we can concentrate on the lessons learned thus far regarding use of hydrogen for diverse air-breathing propulsion systems to power hypersonic aircraft.

10.7.3 Hydrogen for Orbital Flight Propulsion

The first efforts to design and develop a SSTO vehicle incorporating a CCP system (turbojet + ramjet + scramjet + rocket) was entirely fueled by hydrogen. This effort, inspired by the ideas of Antonio Ferri, was carried out in the 1960s as part of the U.S. Air Force's original AeroSpace Plane program. The conceptual vehicle was propelled by four hydrogen-fueled J-58 type turbojet (TJ) engines and four ramjet engines that transitioned to supersonic combustion for the Mach 7–25 range. The vehicle had a gross takeoff mass of 181 437 kg (400 000 lbm) and was designed to carry a payload of 9073 kg (20 000 lbm) (Erdos and Nucci 1992).

After the X-15 program ended in 1968, NASA Langley Research Centre launched two programs: the Hypersonic Research Facilities study (HYFAC) to establish a Mach 12 design point, and the High-Speed Research Aircraft (HSRA) focused on a Mach 8 concept. Under NASA contract, McDonnell Aircraft Company (now Boeing) defined several flight research vehicle concepts. The objective of that study was to assess the research and development requirements for hypersonic aircraft and (based on these requirements) to provide NASA with characteristics of a number of desirable hypersonic research facilities (McDonnel Aircraft Company 1970).

For the speed range up to Mach 4.5, a JP-fueled GE14/JZ8 was chosen. For the speed range up to Mach 5–8, the GE5/JZ6C fueled by LH_2 was selected. Both engines were wrap-around turboramjets (TRJs), with the annular ramjet concentric to the central turbojet. Both engines used separate but concentric nozzles. The engines operate the TJ to it maximum allowable speed per the engine specification: Mach 3.5 for the JZ8, and Mach 3.75 for the JZ6C. The ramjet (RJ) was to operate from Mach 1.0 to cruise speed. The JZ8 was air-cooled throughout, while the JZ6C was fuel-cooled (Hypersonic Study 1970).

In the 1990s, NASA shifted its attention to dual-fuel propulsion systems for several vehicle types: Mach 10 cruise airplane (first stage of TSTO launch system); and Mach 15 cruise airplane (first stage of TSTO launch system) using LH_2 or slush H_2, and a Mach 5 waverider fueled by liquid hydrocarbons (Hunt and Eiswirth 1996). The Dual-Fuel Study (phase I) considered three vehicle designs: (i) lifting body, (ii) waverider, and (iii) cone body. The lifting-body configuration received a substantial amount of development through several Air Force-funded studies. The waverider was developed under subcontract by the University of Maryland (UMD) and the Astrox Corporation (AC) using advanced waverider definition techniques.

The lifting-body vehicle was powered by air-core enhanced TurboRamjet (AceTR) engines fueled by hydrocarbon for low-speed flight (Mach 0–4.5), and slush hydrogen for dual-mode ram/scramjet engines for high-speed flight (Mach 4.5–10). The structural architecture was comprised of an integral graphite/epoxy cryogenic hydrogen tank with a mechanically bonded Thermal Protection System. Conclusions of the study were that, for missions which return to Continental United States (CONUS) bases, a dual-fueled vehicle is superior, due to its capability to in-flight refuel. However, for one-way mission, an all-hydrogen vehicle would be preferable because of its higher specific impulse (Hunt and Eiswirth 1996).

In the 1986–1995 time period, GE Aircraft Engines conducted several comprehensive Mach 4 to Mach 6 propulsion concept evaluations under the Air Force sponsored High-Speed Propulsion Assessment (HiSPA), and the NASA sponsored High Mach Turbine Engine (HiMaTE) programs. These studies considered a wide array of engine concepts, fuel types, and thermal-management issues. They also included detailed conceptual design studies on each engine, concluding that the best propulsion concept for the Mach 4-6 flight regime was a turbofan ramjet (Snyder and Espinosa 2013).

After that, in the 1999–2000 time period, GE completed extensive studies under another NASA contract that addressed the propulsion needs of several Mach 4 class vehicles defined by Boeing. GE proposed a midscale turbofan ramjet engine (35 to 38-in. fan diameter) as a suitable candidate size for demonstration. This would be the basis for a full-scale, 445 kN (100 000 lbf) class engine for the TBCC powering the first-stage vehicle of a TSTO vehicle.

The so called 100 000-lbf Vision Propulsion System (VPS) engine size would be developed to meet the required thrust while meeting the maintainability, cost, and safety requirements.

In the 2000s, NASA returned to use of hydrocarbon fuels for the TSTO launch vehicle. The Revolutionary Turbine Accelerator (RTA)/Turbine Based Combined Cycle (TBCC) project, under Next Generation Launch Technologies (NGLT), investigated a variable cycle Mach 4+ turbofan engine with a hyperburner to operate as a ramjet for the TBCC propulsion of a TSTO vehicle. The first phase focused on the technologies required for a turbofan ramjet engine to meet the access to space mission. NASA GRC along with GE, under the Revolutionary Aero-Space Engine Research (RASER) Task Order Contract designed a mid-scale engine (the RTA-1 in Section 10.6.2) that would provide a system-level validation of candidate advanced technologies.

10.7.4 Hypersonic Transport Aircraft for $0 < M_0 < 12$

Due to low volumetric energy density of hydrogen, aircraft must store the liquid fuel in the fuselage. This results in a larger fuselage length and diameter than a conventional hydrocarbon-fueled aircraft. A larger fuselage causes more skin friction drag and additional wave drag. On the other hand, hydrogen is about one-third of the weight of kerosene jet-fuel for the same amount of energy. This means that for the same range and performance (neglecting the effect of volume), the hydrogen-fueled aircraft would have about one-third of the fuel weight.

From March 1974 through June 1975, McDonnell Aircraft Company, (now Boeing) carried out a study under a NASA contract to evaluate the effects of fuselage cross section (circular and elliptical) and structural arrangement (integral and nonintegral tanks) on the performance of LH_2-fueled, actively cooled Mach 6 cruise vehicles (Nobe 1975). The propulsion system consisted of four variable cycle General Electric GE5/JZ6C turboramjets rated at 400 kN (90 000 lbf) thrust each. A two-dimensional, three ramp, external compression inlet was designed, with variable capture area, and a translating cowl to enhance the airflow capture characteristics and minimize inlet drag over the entire mission. The inlet was located beneath the wing to obtain the benefits of the wing compression flowfield. The four engines, inlets, and exhaust ducts were integrated into the fuselage body to provide low nacelle drag on the total vehicle. The study considered three different aircraft carrying a 200-passenger payload, and a constant fuel quantity of 108 862 kg (240 000 lbm).

With the primary objective of minimizing weight and maximizing aircraft range, the study conducted thermodynamic and structural trade-offs to refine the conceptual designs. The main factors affecting range were found to be weight, volumetric efficiency, and aerodynamic characteristics. These factors interact differently, depending on the fuselage shape and type of fuel tank structure.

Results of the study showed that integral fuel tanks, combined with an elliptical-blended wing-body, yields the lightest aircraft weight and longest-range configuration (Figure 10.26). Integral tanks use the fuselage volume more efficiently than nonintegral tanks; the greater volumetric efficiency of the integral tank results in the concept depicted in Figure 10.26, which had the smaller size. The other concepts studied had similar time histories except for the cruise time, revealing the difference in range between all concepts.

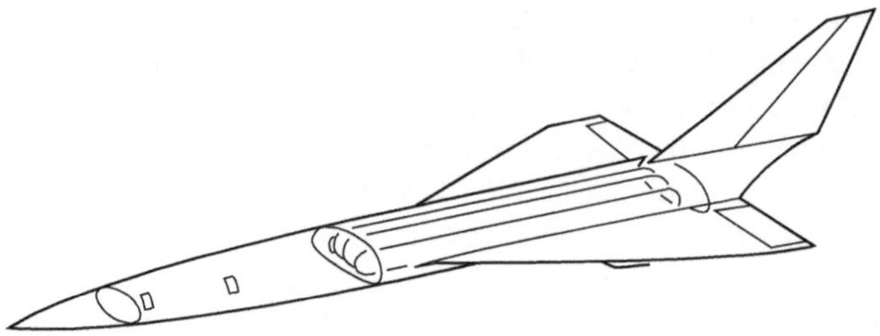

Figure 10.26 Fuselage/fuel tank optimal configuration for LH_2 Mach 6 aircraft. Source: Nobe (1975)/NASA/Public Domain.

There were also small differences in acceleration time and descent time, but these were less than one minute (Nobe 1975).

In the late 1980s, Boeing conducted a study of the potential for high-speed civil transports under a NASA contract. It reviewed three cryogenically fueled engine concepts proposed by three propulsion companies, General Electric, Pratt & Whitney, and Aerojet (Boeing 1989).

The General Electric (GE) Aircraft Engine Company proposed a Mach 4.5 liquid methane (CH_4) fueled, tandem turboramjet (Figure 10.27). At takeoff and subsonic climb, the core intake guide vanes would open for the engine to operate as an afterburning turbojet. As the vehicle would climb transonically and at low supersonic speeds, the open bypass would allow a fraction of the inlet air flow to divert around the turbomachinery and to mix with the core flow before entering the AB. At Mach 4.5 cruise, the intake guide vanes would close, the bypass would be fully open, and the engine could operate as a pure ramjet. This concept is similar to the P&W J58 turbojet that powered the SR-71.

Aerojet General proposed a LH_2-fueled Mach 6.0, air turboramjet (ATR) (Figure 10.28). The ATR is a continuous flow engine that involves ram plus mechanical air compression, constant pressure heat addition combustion, and gas expansion through a variable thrust nozzle. The unique feature of the engine is that the turbine of the turbo compressor is driven

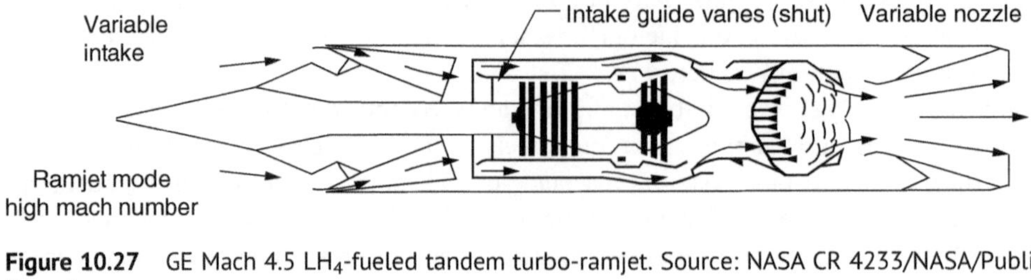

Figure 10.27 GE Mach 4.5 LH_4-fueled tandem turbo-ramjet. Source: NASA CR 4233/NASA/Public Domain.

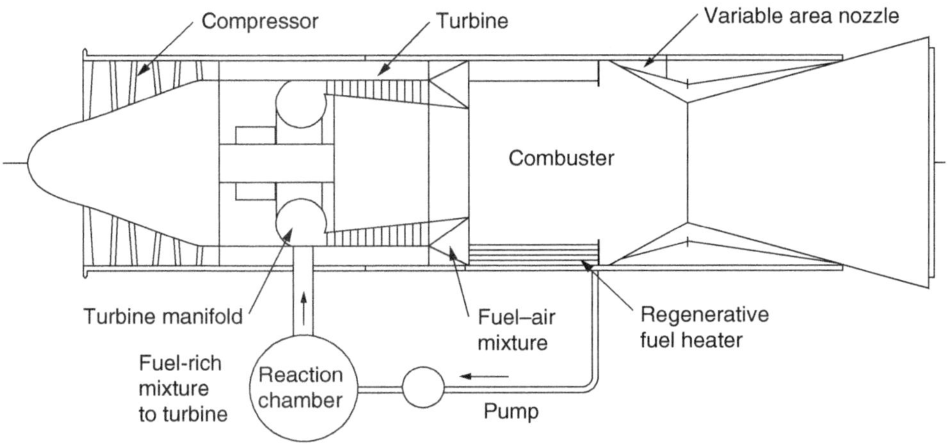

Figure 10.28 Aerojet Mach 6.0 LH_2-fueled air turbo-ramjet (ATR). Source: From NASA CR 4233/ NASA/Public Domain.

by high-temperature, fuel-rich gas from a separate gas generator rather than by the heated air from a fuel–air combustor as in a turbojet. After passing through the turbine, this fuel-rich gas is mixed with the airflow from the turbo compressor and burned in the combustor before expansion through the nozzle. This gas is formed by heating and vaporizing the LH_2 fuel in a dual-regenerator process. The first heat exchanger located at the turbine exit; the second heat exchanger would use waste heat from the combustion chamber.

For the high-speed civil transport study of 1989, Pratt & Whitney (P&W) proposed a LH_2-fueled Mach 10.0 turboramjet–scramjet (Figure 10.29). This design installed a turbojet above the ramjet–scramjet and incorporated a variable geometry inlet for combined turbo-jet/ram/scramjet operation. Initially, the ramjet–scramjet flowpath would be closed off, and the turbojet was to provide the thrust for takeoff, climb, and acceleration to transonic speeds. The inlet would vary to provide combined turbojet/ramjet operation for supersonic climb. Between Mach 3.5 and 4, the turbojet flowpath would be completely closed for the engine to operate as a pure ramjet for climbing to Mach 6. For the remaining climb to the Mach 10.0 cruise altitude, the engine would operate as a pure scramjet.

The results of the 1989 study indicated that cryogenically fueled aircraft were unacceptable. Hence, no further updating of the Mach 6.0 or 10.0 aircraft was pursued.

In the past two decades, innovative propulsion systems were sought for application to high-performance hypersonic aircraft. For example, the TriJet combined cycle engine concept was considered by the former Aerojet company (now Aerojet Rocketdyne) to propel aircraft flying in the range $0 < M_0 < 7.0$.

The TriJet engine integrates turbojet with an ERJ and a DMRJ. The turbine engine is sized to provide fuel-efficient takeoff, tanking, and powered landings at nonaugmented power levels. The ERJ with the turbine operate in full-augmented power mode to produce adequate transonic acceleration. The DMRJ operates over the hypersonic flight regime. The TriJet operating envelope of the three engine cycles overlaps to provide seamless propulsion. One inlet feeds all three engine flowpaths. While the DMRJ has an unobstructed flowpath, the turbojet and ERJ are concealed behind doors that open and close, depending on

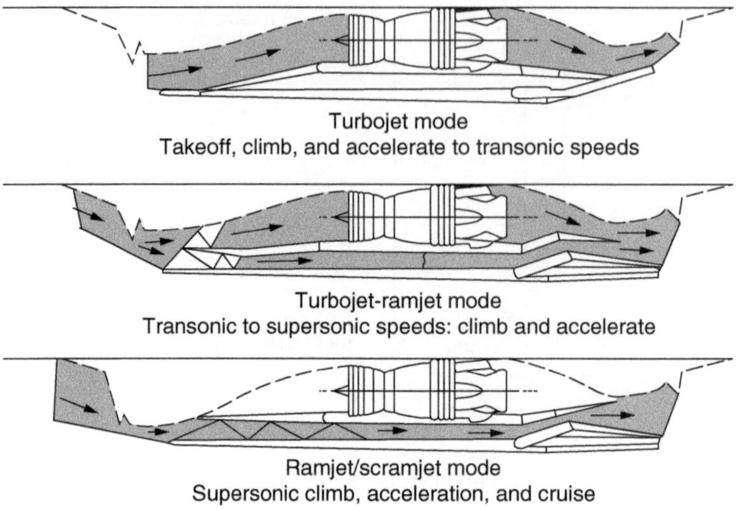

Figure 10.29 P&W Mach 10.0, LH_2-fueled turbo-Ram–scramjet. Source: From NASA CR 4233/NASA/Public Domain.

the phase of the flight. Powered by TriJet engines, an aircraft would takeoff on turbine power, then ignite the ERJ to push through the transonic drag rise and accelerate to the take-over speed to start the DMRJ. Studies by Bulman and Siebenhaar (2011) suggested that the TriJet engine could outperform other combined cycle engines.

10.7.5 Critical Areas Requiring Additional Research and Technology Development

Over the past several decades, hydrogen has been studied extensively as an alternative to kerosene. While investigating advanced propulsion concepts for high-speed transport aircraft in 1987, Strack concluded that conventional jet fuels could not withstand the high temperatures associated with flight speed above Mach 2. His conclusion was that, if subjected to temperatures above approximately 250 °C (time dependent also), these fuels thermally decompose and form coke deposits that clog fuel supply components (Strack 1987). Consequently, the challenge was to extend the thermal stability of conventional jet fuel (Jet A) to higher temperatures without incurring a significant fuel price increase, either in the fuel manufacture or associated with special fuel transportation and handling requirements (such as with JP-7 and cryogenics). Other studies concluded that

- The vehicle forebody geometry is tightly constrained by the large forward volume requirements in the payload bay to accommodate the LH_2 fuel tank.
- An all-body shape represents the limit in wing–fuselage blending to obtain the large volumes required for LH_2 fuel storage.
- Hydrogen-fueled hypersonic transports are capable of long-range flight, as can be predicted with the aid of the Breguet equation.
- A high degree of technological commonality is crucial for the development and operation of CCP for both cruise and acceleration applications.

- Compared with hydrogen, methane is less expensive and denser, but has a lower heating value (see Chapter 6). Therefore, the actual advantage due to using methane rather than hydrogen will depend on the trade-off between structural weight and fuel weight.

In general, LH_2 is the optimum fuel for hypersonic air-breathing propulsion. However, some R&D issues remain. For example, we need a clearer understanding of the effect of engine emissions on the atmosphere; this is essential to designing the optimum, environmentally acceptable propulsion system for any flight regime application.

In recent years, the U.S. Defense Advanced Research Projects Agency (DARPA) launched its Advanced Full Range Engine (AFRE) program. AFRE seeks to develop and demonstrate a new aircraft propulsion system that could operate over the full range of speeds required from low-speed takeoff through hypersonic flight. AFRE aims to explore a TBCC engine concept, which would use a turbine engine for low-speed operations and a DMRJ – which would work efficiently whether the air flowing through it is subsonic (as in a ramjet) or supersonic (as in a scramjet) combustion – for high-speed operations. I believe that LH_2-fueled air-breathing propulsion should be pursued vigorously, as the benefits seem to outweigh the conventional systems of today.

10.8 Technical Challenges of Combined Cycle Propulsion

The propulsion system for the ambitious, high-energy missions described in previous sections combine rockets with air-breathing propulsion, including turbojets and dual-mode scramjets. Turbojet engines would be used for takeoff and acceleration to speeds high enough for scramjet engines to start. Operation in several flight regimes poses many challenges for any propulsion system due to a wide range of aerothermodynamic conditions it will encounter. To date, flight tests for scramjet-powered vehicles has been limited to discrete Mach speeds. Although the hypersonic flights of the X43-A and X-51A aircraft were crucial for the demonstration of scramjet propulsion, much more flight testing is still needed, extending the duration of the flight, and especially to prove the viability of CCP. This requires maturing technologies for the high Mach turbine engines intended for TBCC propulsion.

There are a number of technical challenges that a high Mach turbine booster engine like the RTA must overcome. It requires superior performance over the flight regime $0 < M_0 < 4+$, integration with a dual-mode scramjet engine operating in the regime $4 < M_0 < 12$, and the dual-flowpath TBCC propulsion system must transition smoothly from one cycle to the other, controlling a sophisticated fueling schedule for both.

In this regard, efforts to develop turbine engine accelerator must focus on determining the upper speed limit capability, setting a goal of at least Mach 4. The NASA RTA provides the baseline to develop a flight-weight version engine. At the same time, and in a parallel effort, advancing TBCC propulsion requires to determine the low-speed limit of dual-mode scramjet having acceptable performance as an integrated cycle with the turbine engine.

The turbine engine must propel the vehicle from takeoff to up to Mach 4 flight in order to provide the ram conditions for the scramjet to operate. An issue that may be resolved is that of transonic thrust pinch. Concerns have been raised as to whether the turbine engines alone can provide sufficient thrust to accelerate through the transonic flight regime with a fully loaded hypersonic vehicle (Zheng et al. 2019).

10.8.1 Transonic Thrust Pinch

A hypersonic vehicle must be designed and configured properly to maintain an acceptable excess thrust capability through the transonic region. The curves in Figure 10.30 represent the thrust available T_A and thrust required T_R for a hypersonic vehicle. Transonic thrust pinch refers to the flight region near Mach 1, where the excess thrust T_{excess} is significantly reduced, that is where T_A and T_R curves are "pinched" closer together.

The excess thrust T_{excess} is defined as the difference between the thrust available (directly related to the engine) and the thrust required by the vehicle for flight, that is, $T_{excess} = T_A - T_R$. If $T_{excess} > 0$, the vehicle can accelerate and climb. Supersonic aircraft experience drag increases near Mach 1, which is known as transonic drag. The added drag is due to the shockwave(s) that forms at the location where airflow has accelerated to the speed of sound. Transonic drag increase results in a decrease of T_{excess}, which is detrimental for hypersonic vehicles.

The rapid increase of transonic drag results in a large increase of T_R and a significant decrease of T_{excess} near Mach 1 flight. In fact, T_{excess} can become very small, go to zero, or even become negative for some vehicles, depending on its aerodynamic configuration and on the atmospheric temperature conditions (the thrust available is reduced by higher air temperature). Transition thrust pinch is a rather difficult problem for TBCC powered vehicles operating on a predetermined flight path at constant dynamic pressure. Engineering solutions mainly focus on increasing thrust by adding auxiliary power when transition occurs. Zheng et al. (2019) have taken a different approach based on trajectory optimization, altering thrust and drag status, using a gravity-assist strategy, to gain additional thrust. No auxiliary power is required as the transition thrust pinch problem is resolved by optimizing the trajectory where the maximum terminal velocity is the cost function.

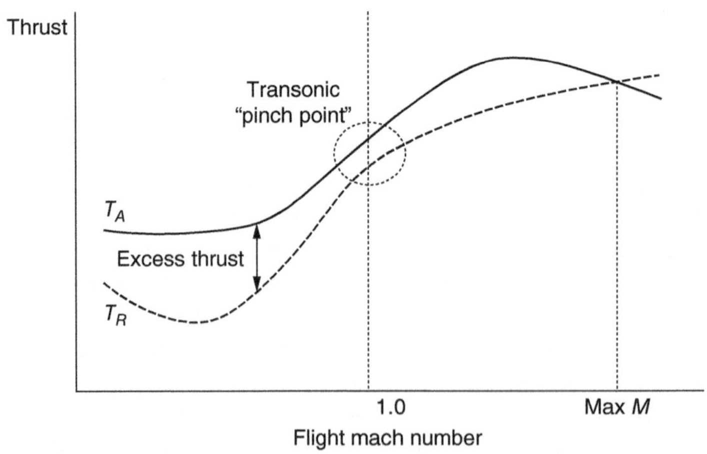

Figure 10.30 Transonic thrust "pinch" for high-speed vehicles.

10.8.2 TBCC Propulsion Mode Transition

A huge technical challenge for RBCC propulsion is its mode transition between the low-speed cycle (turbojet/turbofan) and the high-speed cycle (ram/scramjet), especially since the over–under dual flowpath RBCC configuration requires shared inlets and nozzles, which could stall compressors or unstart inlets (Li et al. 2018). These operational phenomena must be studied experimentally to demonstrate mode transition from a low-speed engine to a high-speed engine, and to assess the transition effect on overall vehicle performance and operability of the TBCC propulsion system (Mo et al. 2014; Huang et al. 2014; Lv et al. 2017). Ground-based altitude facilities are limited by size and operating range. Hence, TBCC mode transition presents a challenge that must be fully assessed to help design flight-ready engines.

In 2007, NASA began a research effort to demonstrate TBCC inlet mode transition. Known as the Large Scale Inlet Mode Transition (LIMX) experiment, this testing was conducted in the 10 × 10 foot supersonic wind tunnel at the NASA Glenn Research Center. The main goal was to evaluate inlet performance during inlet mode transition at Mach 4.0. (Foster et al. 2012).

The inlet for the LIMX experiments has two flowpaths: one to provide flow to a turbine engine and one to provide flow to a DMRJ/scramjet (see Figure 10.19). The design incorporates a rotating splitter cowl to close the turbine flowpath needed when the engine transitions from turbine power to ramjet/scramjet power at Mach 4. The goal of LIMX inlet testing was to characterize the performance of the turbine flowpath (total pressure recovery and distortion at the engine face location) as factors such as bleed rates and configuration and vortex generators are varied during the inlet mode transition. The performance of the inlet had to be examined at off-design Mach numbers (2.5–3.0) and at a determined angle-of-attack. See also the unique dual-flow Mach 7 inlet system designed by Sanders and Wier (2008) for discussion on constraints to divide the low-speed turbofan flow from the high-speed scramjet inlet.

One question related to integrated inlet technology that needs to be addressed is whether the Mil-Spec performance measured by the inlet total pressure recovery is achievable over the wide operating range that is required for TBCC space access. As we saw in Chapter 4, Mil-Spec total pressure recovery can be achieved for an inlet carefully designed for single-point operation. However, for a TBCC system, the inlet must operate over a wide range of flight conditions, and distribute the flow with acceptable distortion and unsteadiness to both the turbine engine as well as the dual mode ram/scramjet engine. Common practice is to use variable geometry and designed bleed configurations to enable a wide range of inlet operability and meet performance requirements, including maximizing pressure recovery and minimizing flow spillage. However, this added design and operational complexity results in additional weight, complex vehicle integration, and an increase in drag. In addition, the high-enthalpy airflow at high speeds may make conventional bleed controls difficult to achieve.

10.8.3 Materials for Combined Cycle Propulsion

One of the greatest technology challenges for development of hypersonic vehicles and their CCP systems is related to high temperature materials and thermal management (see

Chapter 9). The selection of materials for turbojet/turbofan engines and heat exchangers plays a primary role in determining the performance, weight, life, and costs of TBCC propulsion systems.

Modern jet engines are made of a variety of advanced materials, including high-temperature nickel-based superalloys, titanium, aluminum, steel, composites, and ceramics for hot engine parts. For future engines, there will be greater use of ceramics that can withstand temperatures over 1093 °C (2000 °F) without cooling and could provide substantial weight savings compared with conventional metallic alloys. However, ceramics are more brittle than metals, precluding their use to date as structural elements in turbines (where the combination of high temperatures and material stresses are greatest) and in most other safety-related engine components. In addition, joining ceramic and metal parts without damaging the ceramics is difficult, and the difference in rates of thermal expansion further complicates their integration.

Intermetallic alloys – including titanium aluminides, nickel aluminides, and niobium intermetallic composites – offer strength, temperature endurance, and weight advantages over current materials. Titanium aluminides are useful at temperatures higher than temperatures at which aluminum can be used and offer substantial weight savings over pure titanium or nickel-based alloys, making them good candidates for combustor cases, compressor blades in the last (highest pressure and temperature) stages, and other such applications.

In addition, high-temperature materials and TPS will pose additional technical challenges for integrating such advanced TBCC propulsion with the airframe. The maximum operating temperature is especially critical for the HPT, whose turbine entry temperature (TET) represents the maximum cycle temperature. The blades of the HPT must be made of high-temperature materials and be effectively cooled. The EJ200, which powers the Mach 2 "Eurofighter" Typhoon aircraft, is a low bypass (0.4) augmented turbofan with a TET of 1800 K (2780 °F). The evolution of TET and future trends indicate that, with advanced materials and improved manufacturing, current engines may have a nominal TET of about 2000 K (cooled) or higher.

10.9 Closing Remarks

Civil applications of advanced high-speed air-breathing propulsion are seen as an answer for the aviation industry seeking to develop high-speed transport aircraft capable of flying passengers or cargo across continents in a fraction of the time that is possible today. And of course, the need to increase access to space with reusable launch vehicles that can takeoff and land horizontally and facilitate repeated transport of payloads to orbit (for space tourism and commercialization of space) motivates the advances in the field to compete with conventional vertical rocket-propelled launch vehicles.

Innovative CCP systems are the foundation for future high-speed vehicles to achieve their mission over a wide range of flight speeds, from subsonic to hypersonic Mach numbers. This requires a synergetic combination of different types of engine cycles, integrating diverse operation requirements while exploiting the performance advantages of each cycle into a viable propulsion system that can adjust to the varying flight conditions. Comparing the two classes of CCP concepts, by offering a potentially higher specific impulse at takeoff and subsonic flight velocities, TBCC propulsion configurations are most favored.

Through the years, a broad range of concepts and technologies have been conceived for meeting access to space goals, and studies concluded that a TSTO concept is one of the most promising concepts to develop today. Those efforts have addressed different options for integration of multiple engine systems, including turbojet cocooning, air augmentation, inlet systems, and nozzle systems (Whitlow et al. 2001). Ultimately, new advances in scramjet engines, propulsion-airframe integration, structures, materials, thermal management concepts, and advanced design methods will contribute to the readiness for design of such vehicles. The development of the scramjet is aimed at providing propulsion technology for hypersonic vehicle concepts that include cruise aircraft, and both TSTO and SSTO launch vehicles.

It is urgent to investigate pollutant generation, atmospheric impact, and combustion processes for high-speed flight. Hydrogen is the ideal fuel for all types of air-breathing propulsion. Emissions from combustion of hydrogen and air are relatively benign, comprising of water and nitrogen oxides (NOx), each of which has significantly less impact on the environment as compared with emissions produced by conventional hydrocarbon/air burning. Advancing hydrogen-fueled turbo-engines will contribute to our efforts to reduce CO_2 emissions and other negative effects on global warming.

Various technology advancements are crucial to develop future hydrogen-fueled air-breathing accelerator propulsion for either sustained hypersonic cruise aircraft or for ascent to orbit vehicles. Developing adequate performance trends for such accelerator engines will require extensive analyses of design tradeoffs, inlet-engine matching, cooling requirements, material considerations, and operation optimization (Marshall et al. 2004). It is clear that R&D will help us to determine with certainty whether a turbo accelerator engine fueled with LH_2 will indeed be superior to a conventional hydrocarbon-fueled engine.

Questions

1. How can we "stretch" the operational envelope of state-of-the-art turbojets or turbofans to integrate them with scramjets? Give at least two feasible options that would be technologically practical to implement.

2. How can we combine rockets and air-breathing engines to obtain an effective hypersonic propulsion system for a single-stage space launch vehicle?

3. Range is a figure of merit for endo-atmospheric vehicles (for given payload at a given cruise Mach number). How is Range impacted by fuel choice?

4. CCP aims to integrate several operation modes into a single engine system with higher efficiency and lower weight. Is this approach doable in a practical design for a SSTO spaceplane? How would you do it?

5. In a TBCC configuration for TSTO vehicle, what kind of turbine engine can we use? Turbojet? Turbofan? Which jet engine has a better performance for the TBCC system?

6. In developing the inlet for a TBCC, is Mil-Spec performance, measured by total pressure recovery, achievable over the wide operating range required for TBCC propulsion? What is the trade-off between bleed and inlet performance/operability? Can available design and analysis tools adequately predict the TBCC inlet operability limits?

7. For a multistage launch vehicle, how do you select the fuel for the first stage powered by a TBCC engine? What are the advantages and disadvantages of operating the overall launch system with multiple fuels?

References

Boeing (1989). High-Speed Civil Transport Study. NASA Contractor Report 4233. Boeing Commercial Airplanes.

Bowcutt, K.G., Smith, T.R., Kothari, A. et al. (2011). The hypersonic space and global transportation system: a concept for routine and affordable access to space. *AIAA 2011-2295. 17th AIAA International Space Planes and Hypersonic Systems and Technologies Conference*, San Francisco, CA (11–14 April 2011).

Bulman, M.J. and Siebenhaar, A. (2011). Combined cycle propulsion: aerojet innovations for practical hypersonic vehicles. *AIAA Paper 2011-2397. 17th AIAA International Space Planes and Hypersonic Systems and Technologies Conference*, San Francisco, CA 11–14 April 2011).

Conners, T.R. (1997). Predicted Performance of a Thrust Enhanced SR-71 Aircraft with an External Payload. NASA Technical Memorandum 104330, June 1997. NASA Dryden Flight Research Center, Edwards, CA.

Davoudzadeh, F., Buehrle, R., Liu, N.-S., and Winslow, R. (2005). Numerical Simulation of the RTA Combustion Rig. *NASA/TM – 2005-213899. 40th Combustion, 28th Airbreathing Propulsion, 22nd Propulsion Systems Hazards, 4th Modeling and Simulations Joint Subcommittees Meetings sponsored by the Joint Army, Navy, NASA, and Air Force Interagency Propulsion Committee*, Charleston, South Carolina (13–17 June 2005).

Eklund, D.R. and Boudreau, A.H. (2005). A turbine-based combined cycle solution for responsive space access. *AIAA Paper 2005-4186. 41st AIAA/ASME/SAE/ASEE Joint Propulsion Conference & Exhibit*, Tucson, Arizona (10–13 July 2005).

Erdos, J.I. and Nucci, L.M. (1992). Pioneering scramjet developments by Antonio Ferri. In: *Rocket-Based Combined-Cycle (RBCC) Propulsion Technology Workshop. Tutorial Session, January 1992, Cleveland, OH.* NASA, Lewis Research Center.

Espinosa, A.M. (2003). Implications of turbine integration for hypersonic airbreathing vehicles. *AIAA Paper 2003-4408. 39th AIAA/ASME/SAE/ASEE Joint Propulsion Conference and Exhibit*, Huntsville, Alabama (20–23 July 2003).

Foster, L.E., Saunders, J.D. Jr., Sanders, B.W., and Weir, L.J. (2012). Highlights from a Mach 4 experimental demonstration of inlet mode transition for turbine-based combined cycle hypersonic propulsion. NASA/TM – 2012-217724, NASA Glenn Research Center, Cleveland, OH (December 2012).

High-Speed Civil Transport Study (1989a). NASA Contractor Report 4233. Boeing Commercial Airplanes.

High-Speed Civil Transport Study (1989b). NASA Contractor Report CR 4233. NASA CR 4233. Boeing Commercial Airplanes.

Huang, W., Yan, L., and Tan, J.G. (2014). Survey on the mode transition technique in combined cycle propulsion systems. *Aerospace Science and Technology* 39: 685–691.

Hunt J.L. and Eiswirth, E.A. (1996). NASA's dual-fuel airbreathing hypersonic vehicle study. *AIAA 96-4591. Space Plane and Hypersonic Systems and Technology Conference*, Norfolk, VA (18–22 November 1996).

Kanda, T. and Tani, K. (2007). Conceptual study of a rocket-ramjet combined-cycle engine for an aerospace plane. *Journal of Propulsion and Power* 23: 301–309.

Lee, J-H., Winslow, R., and Buehrle, R.J. (2005). The GE-NASA RTA hyperburner design and development. *NASA/TM – 2005-213803. 40th Combustion, 28th Airbreathing Propulsion, 22nd Propulsion Systems Hazards and 4th Modeling and Simulation Joint Subcommittee Meetings*, Charleston, SC, USA (13–17 June 2005).

Li, N., Chang, J.T., Jiang, C.Z. et al. (2018). Unstart/restart hysteresis characteristic analysis of an over-under TBCC inlet caused by backpressure and splitter. *Aerospace Science and Technology* 72: 418–425.

Lv, Z., Xu, J.L., and Mo, J.W. (2017). Study of the unsteady mode transition process for an over-under TBCC exhaust system. *Acta Astronautica* 136: 259–272.

Marshall, A.W., Gupta, A.K., and Lewis, M.J. (2004). Critical Issues in TBCC Modeling. *AIAA Paper 2004-3827. 40th AIAA/ASME/SAE/ASEE Joint Propulsion Conference and Exhibit*, Fort Lauderdale, Florida (11–14 July 2004).

Mattingly, J.D. and Boyer, K.M. (2016). *Elements of Propulsion: Gas Turbines and Rockets*, 2e. AIAA Education Series.

McDonnell Aircraft Company (1970). Hypersonic Research Facilities Study. Volume II, Part 2, Phase I: Preliminary Studies – Flight Vehicle Synthesis. NASA Contractor Report CR 114324 (October 1970).

Mo, J.W., Xu, J.L., and Zhang, L.H. (2014). Design and experimental study of an over-under TBCC exhaust system. *Journal of Engineering for Gas Turbines and Power* 36: 014501-1-8.

Nobe, T. (1975). A fuselage/tank structure study for actively cooled hypersonic cruise vehicles. *NASA CR-132668*. NASA Contractor Report (January 1975).

Reaction Engines (2010). SKYLON User's Manual. Doc. Number - SKY-REL-MA-0001, Revision 1.1, January 2010, Abingdon, Oxon, OX14 3DB UK.

Sanders, B.W., and Weir, L.J. (2008). Aerodynamic Design of a Dual-Flow Mach 7 Hypersonic Inlet System for a Turbine-Based Combined-Cycle Hypersonic Propulsion System. *NASA Contractor Report NASA/CR-2008-215214*, (June 2008).

Snyder, C.A. and Espinosa, A.M. (2013). Lessons learned during TBCC design for the NASA-AFRL joint system study. *NASA/TM-2013-218100. JANNAF Propulsion Meeting*, Arlington, VA (December 2013).

Strack, W.C. (1987). Propulsion challenges and opportunities for high-speed aircraft. Chapter in Aeropropulsion 87. *NASA CP-3049. Conference at NASA Lewis Research Center*, Cleveland, OH (17–19 November 1987).

Taguchi, H., Harada, K., Kobayashi, H. et al. (2022). Mach 4 simulating experiment of pre-cooled turbojet engine using liquid hydrogen. *Aerospace* 9: 39.

Varvill, R. and Bond, A. (2008). The skylon spaceplane: progress to realization. *JBIS* 61: 412–418.

Whitlow, Jr., W., Blech, R.A., and Blankson, I.M. (2001). Innovative airbreathing propulsion concepts for access to space. *NASA TM-2001-210564. Novel Aero Propulsion Systems International Symposium sponsored by the Institution of Mechanical Engineers*, London, England (3–4 September 2000).

Zheng, J.L., Chang, J.T., Yang, S.B. et al. (2019). Trajectory optimization for a TBCC-power supersonic vehicle with transition thrust pinch. *Aerospace Science and Technology* 84: 214–222.

11

Ground Testing and Evaluation

11.1 Introduction

In order to advance and certify the design tools to develop future hypersonic air-breathing propulsion systems, we need a carefully designed program that incorporates extensive ground test and evaluation techniques combined with computational modeling and simulation analysis, a monumental undertaking that is required to reduce technology risks associated with air-breathing hypersonic flight demonstration testing.

The development of the NASA X-43A scramjet engine and its integration with the airframe required several engine models that were tested in the most advanced wind tunnel facilities within and outside NASA. The ground testing program was designed to allow an integrated test program to isolate and measure the effects on engine operability and performance caused by geometric scale, dynamic pressure, and test differences encountered in each tunnel, as those differences are due to limitations of the test technique and facilities. Figure 11.1 shows one of the X-43A scramjet engines tested in the NASA 8-Foot high temperature tunnel prior to the X-43 demonstration flights. The full-length/width propulsion flowpath test model, known as the Hyper-X Flight Engine (HXFE) with Vehicle Flowpath Simulator (HXFE/VFS), is shown mounted inverted on its pedestal during a test simulating Mach 7 flight conditions.

11.2 Airframe/Propulsion-Integrated Vehicle Design Requirements

System studies are conducted to identify potential vehicle configurations, components, and approaches and focus on technology development. For rocket- and gas-turbine-propelled vehicles, where propulsion performance is independent of the vehicle aerodynamics, such system studies are straightforward and well understood. The vehicle aerothermal environment is simply correlated to its flight trajectory. For such systems, the dominant design drivers are well understood and therefore properly optimized.

System studies for hypersonic air-breathing propulsion are much more complex than those for gas turbines and rockets. This complexity is due to the tightly integrated nature

Scramjet Propulsion: A Practical Introduction, First Edition. Dora Musielak.

Figure 11.1 Mach 7 wind tunnel test of the full-scale X-43A model with spare flight engine in the NASA Langley Research Center's 8-Foot High Temperature Tunnel. Source: Jeff Caplan/NASA/Public Domain.

of the engine to the airframe: optimum flight operation requires the engine to be fully aero-thermodynamically integrated with the fuselage. In other words, the vehicle and the propulsion system must be developed together, as one cannot decouple propulsion performance from vehicle performance, stability, and control due to the shared surfaces and the inherent flow field interactions.

It is challenging to evaluate propulsion–airframe flow field interactions and the integrated aero-propulsive performance of the given system. That is why integrated vehicle aero-propulsive performance requires tip-to-tail design and analysis, an approach pioneered by NASA to develop the X-43A vehicles. The tip-to-tail methodology includes an integration of aerodynamic and propulsion flow fields in and around a hypersonic air-breathing vehicle, including propulsion flowpath, and all external flow physics. Nose-to-tail propulsion analysis together with integrated thermal management and structural analysis of the entire hypersonic air-breathing vehicle require an innovative set of analytic tools and ground testing.

Systems analysis methods for air-breathing vehicles can be executed at several levels. For the Hyper-X program, the NASA team conceived design methods to assess the level of fidelity for the different tools. For example, ideal or approximate cycle analysis is considered the lowest level scramjet design tool. The next level is cycle analysis using plug-in efficiencies, which are externally estimated or determined for a particular configuration as a function of flight Mach number only. The next level is computational fluid dynamics (CFD) analysis, which, if validated by comparison with representative unit and engine component test data, it represents a step up in fidelity. The highest fidelity in system analysis is achieved by propulsion tests in a wind tunnel facility and culminating with powered flight. The wind tunnel tests are a lower fidelity than flight test because the results must be scaled by analysis to the actual flight vehicle and flight conditions.

Obviously, the uncertainty in predicted performance and operability decreases with higher level analysis methods. At the lowest system analysis level, the predictions could easily be off by 50% or more, and the conceived propulsion system would not be operable. In contrast, at the highest system analysis level, the performance is expected to be within 1–2% of predictions (Hunt and McClinton 1997).

The design process requires synergistic utilization of experimental, analytical, and computational analysis. We will address the CFD tools in Chapter 12. Here we just want to emphasize that CFD analysis requires dedicated validation with ground test experiments. This chapter is intended to provide a brief overview of key ground testing, presented with reference to airframe-integrated scramjet programs developed to date.

11.3 Ground Testing Overview

Extensive ground testing is required to advance the development of hypersonic air-breathing propulsion systems. Wind tunnels and other test facilities on the ground are used to duplicate and simulate the aerodynamic and aerospace flight environments for a fully integrated vehicle or a propulsion component, allowing for the testing of aerospace systems, subsystems, and components, including hypersonic materials.

Ground testing is useful to answer questions related to basic thermophysical processes, those developing in the external vehicle flowfield, and those occurring within the engine flowpath, including aerothermodynamics, turbulence, and combustion. Ground facilities are expected to duplicate physics and chemistry with a high degree of fidelity in order to investigate engine design features, or the degree of system operation or mechanization desired, and to test system component fabrication and mechanization. Previous chapters presented some of the most important phenomena associated with a fully integrated hypersonic air-breathing vehicle in flight, including aerothermal and propulsion challenges. *How can ground facilities duplicate those phenomena to develop a powered flight vehicle?*

The NASA Hyper-X (X-43A), the USAF Waverider (X-51A), and the United States/Australia HIFiRE flight programs included extensive utilization of wind tunnels to support the development of the research vehicles and scramjet engine designs. The ground experiments provided critical input to the predicted engine flight performance and development of the control laws for the vehicles in flight.

11.3.1 Flow Physics Fidelity, Scale, and Chemistry

The three most important aspects of a ground testing program are (a) scale, (b) fidelity, and (c) chemistry. The greatest technical issue for a subscale scramjet experiment is reproducing the proper flow physics and chemistry. A subscale component can mask or totally miss simulating key flow parameters that affect the full-scale system operation and performance fidelity. For instance, for studies of propulsion and thermal management, the ground experiment must replicate the viscous interaction vehicle forebody compression process ahead of the scramjet inlet where Reynolds-number-regulated viscous effects begin, to develop the inlet boundary layer in order to simulate the engine operation in flight.

A wind tunnel must duplicate the stagnation conditions of hypersonic flight along a constant dynamic pressure trajectory. Consider first the enthalpy requirement. For hypersonic flows, the total enthalpy is approximately equal to the kinetic energy of the flow stream. In a per mass basis, the specific total enthalpy is given by

$$h_t = e_t + p_t/\rho_t = e + p/\rho + V^2/2 \approx V^2/2$$

Figure 11.2 plots the specific total enthalpy as a function of flight velocity along a constant dynamic pressure path equal to 47.88 kPa (0.47 atm or 1000 psf). For example, for a velocity of 2210 m/s (~Mach 7), the total enthalpy of the airflow is 2.44 MJ/kg. In general, for the flight velocities of interest ($M_0 < 15$), the required total enthalpy is less than 10 MJ/kg (4299 BTU/lbm).

High-stagnation enthalpy in a ground facility can be achieved by raising the stagnation temperature of the test gas. This principle is applied to arc-heated and combustion-heated tunnels. Another way of obtaining high-stagnation enthalpy flow is to elevate the stagnation pressure. Heiser and Pratt (1994) estimated the power requirements for a continuous-flow facility providing 45.36 kg/s (100 lbm/s) of test flow to duplicate Mach 10 flight at 47.88 kPa dynamic pressure. Their result gives the compressor power as 210 MW (281 kHP). This estimate does not include the power required to return the exhaust flow to the atmosphere, which, as they stated, is an amount that equals or exceeds the upstream value. By comparison, a wind turbine can generate 2 MW of power and provide electricity for about 400 homes.

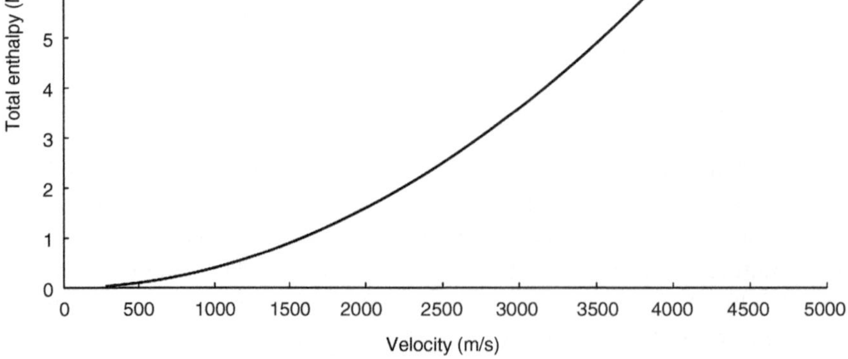

Figure 11.2 Total enthalpy along a constant dynamic pressure path q_0 = 47.88 kPa (0.47 atm).

It is important to distinguish the difference between the aerodynamic Mach number and the enthalpy associated with an actual vehicle flying at a certain Mach number. In flight, these states are equivalent; but in the wind tunnel tests, these states may not be the same. For example, specifying Mach 6 test requirement may indicate either the aerodynamic Mach number being 6 or the enthalpy associated with flight at that Mach number at a particular altitude. Hence, we should specify the required facility's operating condition. Aerodynamics and aeroheating facilities often just simulate the aerodynamic Mach number (usually defined by the area ratio of the facility nozzle), while combustion facilities and materials testing facilities usually simulate the correct enthalpy associated with flight at a certain Mach number. This flow enthalpy is determined by the flowrate and the amount of heat addition. For example, a Mach 7 enthalpy flow can pass through a Mach 2 nozzle to simulate some part of a hypersonic vehicle such as the combustion chamber.

11.3.2 Types of Wind Tunnels

Ground test facilities can be classified depending on their operating characteristics. They can be supersonic or hypersonic wind tunnels (depending on the Mach number and temperature required in the test section); they can be low- or high-enthalpy wind tunnels (depending on the level of test gas stagnation enthalpy); they can be blowdown or continuous operating wind tunnels (depending on their run time); and they can be freejet type or direct-connect wind tunnels. Standard direct-connect propulsion tests are those where the airflow is fed directly through and into the engine (internal flowpath). A tunnel can be characterized by a combination of operating features. For the purpose of this overview, we will focus on continuous and blowdown hypersonic tunnels.

Continuous-Flow Ground Testing. This tunnel produces a long duration, continuous flow of gas that is thermally and calorically (virtually) perfect, as well as both viscous and compressible. The flow is primarily in the continuum regime with the correct Mach number, and a Reynolds number that is sufficiently high for the general class of problems for which the tunnel is applicable. The test air is conditioned to duplicate the actual freestream stagnation conditions, and then it is introduced to the test section from an upstream plenum by a facility nozzle that brings the flow to the freestream Mach number, static pressure, and direction required for the given test. The gas is then removed along with any combustion products by an exhauster system containing vacuum pumps and water spray cooling that maintains freestream static pressure downstream of the test article. In order to duplicate the conditions of the flight, the total airflow required for freejet testing must exceed that being swallowed by the engine.

Continuous-flow, full-scale testing provides the most precise control over the flight environment simulation and propulsion system variables, and it permits to test the actual hardware. A long duration test also allows the propulsion system to come to mechanical and thermal equilibrium, which is one way of precisely defining the configuration angle of attack in order to simulate the effect of transient maneuvers on the propulsion system, of the hardware, and is the condition most likely to be found in flight. An example of continuous-flow facility is the AEDC von Karman Gas Dynamics Facility (VKF). Tunnel B is a 50-inch, closed-circuit hypersonic tunnel with continuous-flow capability with a Mach number capability of 6 and 8. Provided with air heated to a maximum of 900 °F (482 °C) with natural gas-fired heaters, Tunnel B is primarily utilized to support aerodynamic design efforts.

Blowdown Tunnel. In this type of tunnel, the test air is compressed and stored in high-pressure vessels until ready, and then it is heated as it flows to the test section. There, the flow Mach number is determined by pressure and temperature in the plenum and the area ratio between the test section on the nozzle throat. Test times are limited in blowdown wind tunnels compared with those in continuous-flow tunnels. However, the run times are of the order of a few seconds to a few minutes, depending on the test conditions, and thus long enough to approach aerothermodynamic equilibrium. The test air can be heated by passing it through a pebble bed heater, or it can burn in a vitiated heater, a combustor that replenishes the consumed oxygen in order to supply the engine with suitable chemical constituents. The NASA Langley 8-Foot High-Temperature Tunnel (HTT) is the nation's largest hypersonic blow down test facility.

11.3.3 Duplicating Hypersonic Flight Environment in Ground Facilities

The flight state of a vehicle is defined by its velocity or Mach number and the dynamic pressure conditions that define its trajectory. Ground tests should simulate hypersonic flight at the correct altitude conditions: p_0, T_0, M_0, Re. However, due to many operational and test technique limitations, ground facilities cannot simulate all the proper operating conditions of hypersonic flight with sufficient fidelity.

For test and evaluation facilities, limitations involve the requisite combination of size, run time, and dynamic pressure at the enthalpy level representative of flight. In addition, facility limitations include lack of real gas quiet tunnels for boundary-layer transition investigation, the presence of flow contaminants (heater induced), and dynamic disturbance fields that affect the air-breathing propulsion performance simulation. Thus, a test plan must consider what must be replicated and what can be relaxed.

In the hypersonic flight regime, test facilities and measurement techniques are subjected to a number of practical limitations such as test duration, size of the test article they can accommodate, and ability to provide the correct gas flow. For example, real air cannot be simulated correctly because of the need to test with other gases, or to heat the air in vitiated heaters in order to obtain the appropriate level of total enthalpy. Vitiated air contains combustion products or contaminating particulates, and thus the chemical composition of the working fluid is not the same as that found in flight, and this can affect the correct determination of scramjet performance.

Table 11.1 summarizes the type of hypersonic flow tests and the requirements in terms of flow parameter duplication and test time. Specific needs within each type of test have demands that determine the types of facilities in which the tests can be performed. These characteristics include size requirements for model accommodation, needs for technical support, and test type. Facility limitations must be accounted for in the design and analysis methodologies when using wind tunnel test results as an integral part of hypersonic air-breathing vehicle/engine design.

In general, the main goal of ground testing is to simultaneously provide:

- **Correct Test Medium**. This means the facility must provide the correct levels of pressure, temperature, velocity, flow composition, etc.

Table 11.1 Generic hypersonic ground test requirements.

Type of test	Test requirements	Simulation: Duplicate	Simulation: Relax	Test time required
Aerodynamic	Reproduce force coefficients, boundary-layer transition, turbulence	Mach number, Reynolds number	Temperature	Milliseconds, seconds
Aerothermal	Duplicate heating rates and skin shear	Total temperature, surface pressure	Mach number	Minutes
Aeropropulsion	Provide conditions for fuel injection, chemical reactions, mixing, boundary layers, and shocks	Gas composition, pressure, temperature, Mach number, velocity, scale		Minutes
Structures and materials	Duplicate loads (mechanical, thermal, acoustic), temperature gradients	Loads, temperature, heating rates	Flow velocity	Minutes

- **Adequate Scale**. A component reduced in scale to fit the tunnel test section could overlook the simulation of key flow parameters that are crucial to establishing the operation and performance of the full-scale system.
- **Adequate Test Duration**. Ideally, a wind tunnel should provide long run times. Testing time affects thermal effects on performance/operability and durability. Test duration also affects the choice of instrumentation and diagnostics tools.

No one ground facility does all this 100%. Among other operational challenges, there are fundamental physics related to the simulation of hypersonic flight in a wind tunnel that we must consider such as those involving total temperature, boundary layers, chemical kinetics, and real-gas effects.

To simulate Mach 7 flight conditions at an altitude of 30 km, the total temperature of the test air flow should be about 2300 K; the stagnation temperature increases to 4500 K for Mach 10. As we discussed in Chapter 2, air at 2000 K has its translational, rotational, and vibrational energy states fully excited, and at 2300 K we expect that oxygen dissociation ($O_2 \rightarrow 2O$) has already begun. Molecular oxygen is essentially totally dissociated at 4000 K. At this temperature, nitrogen dissociation ($N_2 \rightarrow 2N$) begins.

Boundary-layer transition from laminar to turbulent flow conditions must be correctly characterized in order to simulate the levels of heat flux and friction drag on the vehicle and on the forebody surfaces leading to the engine, establishing the correct boundary-layer displacement thickness and turbulent velocity profile at the inlet entrance. The size of the test model must be chosen carefully to prevent model-scaled effect associated with transition phenomena.

For the engine flowpath simulation, the ground test must duplicate forebody shock patterns and air compression process, positioning the inlet shocks to avoid boundary-layer separation and shock interaction effects on flow distortion. The system of generated and reflected shocks must also be accurately simulated through the inlet and isolator to the combustor entrance, as these shocks are propagated downstream and affect the overall engine performance.

Ground facilities are also limited by time (shock tunnels can only test in milliseconds) and limited by operational Mach number (a maximum of Mach 8 for long-duration wind tunnels). Other test problems include accurate Reynolds number simulation; pressure and temperature, or both, (enthalpy) simulation at high Mach number and dynamic pressure conditions; tunnel interference problems; and engine exhaust effects simulation.

A wind tunnel must simulate what happens during flight using reduced-scale models. The quality of such simulations is governed by similarity laws (Mach number, Reynolds number, and Stanton number). Scaling issues were noted by flight programs that tested vehicle models in wind tunnels at less than full-scale Reynolds number, and engine components at less than full scale at actual flight conditions. Some of the scaling issues occurred due to inability to accurately model flight conditions or chemically reacting gas effects (Bertin and Cummings 2006).

Since complete duplication of all flight parameters in ground test facilities may not be possible, we must identify parameters that impact physical processes of supersonic combustion: flight Mach number (M_0), total enthalpy (h_t), Reynolds number (Re), Stanton number (St), Damköhler first and second numbers (Da_1, Da_2), and wall enthalpy ratio. Mach number is used to simulate compressibility effects. Reynolds number is used to simulate proper boundary-ayer characteristics; Knudsen number is needed in flows involving low-density conditions.

The Damköhler number (Da) is a useful ratio for determining whether diffusion rates or reaction rates are more important for defining a steady-state chemical distribution over the length and time scales of interest. The first Damköhler number (Da_1) is the ratio of flow residence time to chemical reaction time. The second Damköhler number (Da_2) is the ratio of chemical reaction rate to mass transfer rate, or simply the kinetic-to-thermal energy ratio.

The process of matching flight total enthalpy and Mach number allows proper simulation of Da_2. If flight dynamic pressure is not matched due to power requirements or facility constraints and mass flow limitations forces, testing a smaller scale engine becomes necessary. Then Da_1 is not simulated properly. If combustion is kinetically limited, then ignition delay characteristics of the fuel and the reaction times become a critical issue, and proper simulation of Da_1 becomes critical. However, if combustion is mixing limited, proper simulation of Da_1 is not an issue (Voland and Rock 1995).

Simulating the conditions in the dual-mode scramjet (DMSJ) combustor is challenging. A characteristic of dual-mode DMSJ operation is the time required for a precombustion shock train to establish and stabilize, a process that cannot be properly simulated. There are issues related to scaling that affect the kinetics and combustor burning length. This length involves residence time for proper chemical kinetics simulation and adequate fuel–air mixing length. Fuel injection, cooling, turbulence, jet–shock interactions, and other combustion processes have to be properly characterized as they determine the assessment of the propulsion capability of the engine. This requires pretest analysis to determine how the flow parameters will be affected by the limitations of the facility and its instrumentation.

Hypersonic boundary-layer transition studies should be conducted in quiet wind tunnels that provide nearly clean air, quiet flow at high Reynolds number, and with low noise levels comparable to flight. Quiet tunnels with low-freestream disturbances are necessary to provide accurate laminar to turbulent flow transition measurements. The facility should simulate the true enthalpy of flight and have test sections large enough to characterize the full vehicle and its subsystems.

A hypersonic wind tunnel facility must provide at least a reasonable fraction of mission time to prove system durability and concept feasibility. Appropriate run time is needed to provide an opportunity to demonstrate critical technologies, to provide aerothermodynamic test data, and to validate computer models. The time scale (t) for fluid dynamic establishment of attached flows can be estimated. For example, a 1 m test model at Mach number 10, the time scale is less than 1 ms. The test time requirement in the one to two second range is driven by structure thermal response, not by fluid dynamic or air-breathing propulsion performance.

Properly designed ground tests contribute to successfully designing and developing a flight vehicle. Often, these tests isolate and address technical problems or design issues using component tests or by controlled experimental conditions. The results yield databases and design tools, measurement and analysis techniques, and system design validations, and together these tools and data lead to developing the integrated flight vehicle.

For test and evaluation, there are hypersonic facilities in the United States, France, China, Japan, and Russia. In 2008, the Library of Congress published "Wind Tunnels of the Western Hemisphere," a report compiling data on subsonic, supersonic, and hypersonic wind tunnels. This report was prepared for NASA's Aeronautics Research Mission Directorate. More recently, Chazot and Panerai (2015) reviewed high-enthalpy facilities and plasma wind tunnels. These publications and references therein provide much more detailed information about hypersonic flow facilities.

In the United States, the following are true free-jet facilities that are capable of long-duration ground testing of small-scale scramjet engines: The Aerodynamic Propulsion Test Unit (APTU) at Arnold Engineering Development Center (AEDC) in Tennessee, the 8-ft HTT at NASA Langley, and the Leg VI large hypersonic tunnel at Northrop Grumman Propulsion Test Complex (formerly GASL) in Ronkonkoma, New York. The following sections will focus on the facilities at NASA LaRC that provided experimental data for parametric design and optimization of vehicles, CFD validation and uncertainty assessment, flight database construction and technology development required for the NASA Hyper-X (X-43A), and the USAF HyTech X-51A scramjet demonstration vehicles.

> Ground testing in the hypersonic regime is limited by facility size, energy, and test duration as well as vitiation of the test media. Testing can only be done at a fixed Mach number ... it cannot simulate flight transients. Active structural cooling exacerbates the test challenges.

11.4 Ground Testing for the NASA Hyper-X Program

The Hyper-X program demonstrated and validated the technology, the experimental techniques, and computational methods and tools for design and performance predictions of hypersonic aircraft with an airframe-integrated, hydrogen-fueled DMSJ propulsion system. In order to demonstrate the powered hypersonic aircraft in flight, the technology portion of the program focused on risk reduction (preflight analytical and experimental verification of the predicted performance and operability of the Hyper-X flight vehicle), and flight validation of design predictions, which also required continued development of the tools used to advance scramjet powered vehicle designs.

To meet budget and schedule, the Hyper-X flight research vehicle was based on existing design databases and off-the-shelf materials and components. For example, the copper alloy engine materials and fabrication methods were similar to those used for many other experimental wind-tunnel programs, rather than the flight weight actively cooled material propulsion system that is required for an operational hypersonic aircraft (McClinton et al. 1998). The Hyper-X efforts included extensive wind tunnel verification of the engine, vehicle aerodynamic and thermal data base development, thermal-structural design, boundary-layer analysis and control, flight control law development, and flight simulation models. The following sections highlight some of the tests carried out.

11.4.1 Aerodynamic Testing

For the NASA Hyper-X program, aerodynamic wind tunnels and wind tunnel models were utilized to develop the aerodynamic database, to provide limited aero-heating data, develop the boundary-layer trip design, and to calibrate the flush air-data system (FADS). The FADS requires surface pressure measurements from the nose cap of the vehicle for deriving the air data parameters of the vehicle such as angle of attack, angle of sideslip, Mach number, and other mission specific parameters.

The primary wind tunnels for aerodynamic testing the Hyper-X research vehicle (HXRV) and launch vehicle stack included 3 hypersonic tunnels: NASA Langley's 20-Inch Mach 6, 31-Inch Mach 10 facilities, and the AEDC VKF Tunnel B. Several transonic-supersonic tunnels were also used: the Boeing Polysonic Wind Tunnel (PSWT) facility, the Lockheed Martin High Speed Wind Tunnel, and the National Transonic Facility (NTS) Trisonic Wind Tunnel. In addition, two subsonic tunnels were used: the Vigyan low-speed tunnel in Hampton, Virginia, and the Boeing North American subsonic tunnel. Table 11.2 summarizes the test capabilities for these facilities.

The NASA Langley Aerothermodynamic Laboratory (LAL) consists of four hypersonic blow-down-to-vacuum tunnels that represent 100% of NASA's and over half of the Nation's conventional aerothermodynamic test capability. LAL includes the 20 Inch Mach 6 Air Tunnel and the 31 Inch Mach 10 Air Tunnel. Both are blowdown facilities that utilize dried, heated, and filtered air as the test gas. Table 11.3 gives their operating conditions.

The 20-Inch Mach 6 Air Tunnel is used for both parametric and benchmark aerodynamic, aerothermodynamic, and fluid dynamic research. It allows assessment of compressibility effects at constant freestream Reynolds number (Re) and specific heat ratio (γ). The 20-inch Mach 6 Tunnel allows determination of real-gas aerodynamic effects at constant freestream

Table 11.2 Aerodynamic test facilities for NASA Hyper-X program.

Facility	Mach number	Test section	Unit Re	Type
NASA LaRC Mach 10	10	31-inch × 31-inch	0.5–2.2 × 10^6/ft	Blowdown
NASA LaRC Mach 6	6	20-inch × 20-inch	0.5–7.1 × 10^6/ft	Blowdown
Boeing PSWT	0.4–4.6	48-inch × 48-inch × 9-ft	2.0–50 × 10^6/ft	Blowdown
Lockheed Martin HS	0.4–4.6	48-inch × 48-inch × 5-ft	2.0–38 × 10^6/ft	Blowdown
NTS Trisonic WT	0.4–4.6	48-inch × 48-inch	1.0–43 × 10^6 /ft	Blowdown
AEDC VKF B	6 and 8	50-inch diameter	0.3–4.7 × 10^6/ft	Continuous
Vigyan low speed	0.16	3 ft × 4 ft × 5 ft	1.145 × 10^6/ft	

Table 11.3 NASA Langley hypersonic aerodynamic test facilities.

Test condition	20-inch tunnel	31-inch tunnel
Mach number	6	10
Unit Reynolds number	0.5 – 8.0 × 10^6/ft	0.25 – 2.2 × 10^6/ft
Total pressure (psia)	30 – 475	125 – 1450
Total temperature (°F)	410 – 475	1350
Test gas	Air, $\gamma = 1.4$	Air, $\gamma = 1.4$
Run time	Up to 15 minutes	120 seconds
Test section dimensions	20.5 × 20 in.	31 × 31 in.

Mach number and Reynolds number. The test section is 20.5 by 20 inches. A bottom-mounted model injection system can insert models from a sheltered position to the tunnel centerline in less than 0.5-seconds. Run times up to 15 minutes are possible with this facility, although for most heat transfer and flow visualization tests, run times are only a few seconds. Optical access to the model for both thermographic phosphors and oil flow is viewed through a high-quality window on the top of the tunnel (Berry et al. 2000b).

The 31-inch Mach 10 Air Tunnel provides uniform, clean flow, having a five-micron particle filter to remove flow contaminates. It has a three-dimensional, contoured, water-cooled nozzle. This facility is suited for both parametric and benchmark aerodynamic, aerothermodynamic, and fluid dynamic studies. It allows assessment of compressibility effects at constant freestream Re and γ. The tunnel has a closed 31- by 31-inch test section with a contoured three-dimensional water-cooled nozzle to provide a Mach number of 10. A hydraulically operated side-mounted model injection mechanism can inject the model into the flow in 0.6 seconds. The maximum run time for this facility is approximately two minutes; however, only approximately five seconds of this time is required for transient heat transfer tests. Optical access to the model is viewed through a high-quality window on the side opposite the injection mechanism (Berry et al. 2000a).

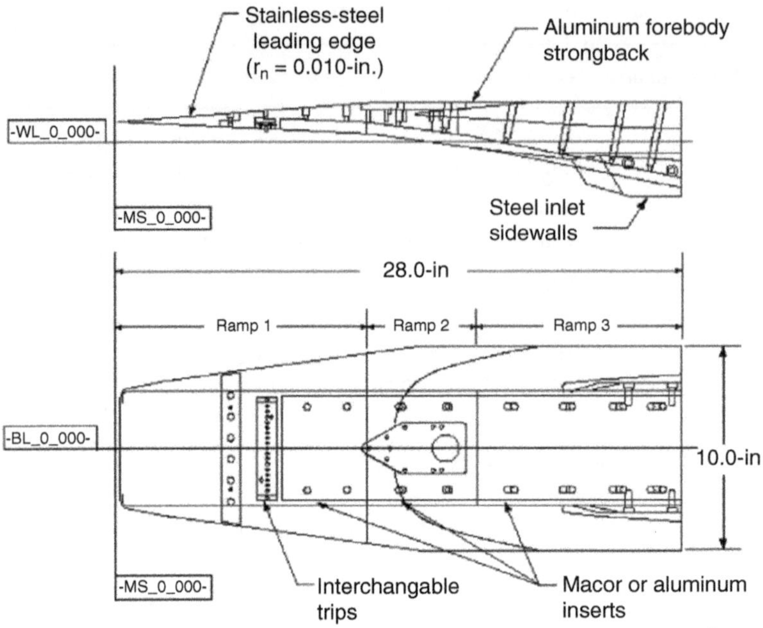

Figure 11.3 Schematic of Hyper-X forebody model tested in both the 20-Inch Mach 6 air and the 31-Inch Mach l0 air tunnels. Source: Berry et al. (2000a)/Public Domain.

Crucial tests utilizing the 20-inch and 31-inch tunnels were conducted to investigate the aeroheating characteristics of the Hyper-X forebody and to examine the effect of discrete roughness elements on the windward surface boundary layer. As we found in Chapter 3, in order to achieve optimum scramjet performance, a turbulent boundary layer at the inlet interface is required. Estimates for natural transition on the Hyper-X flight vehicle forebody suggested that boundary-layer trip devices were necessary to ensure a turbulent boundary layer at the inlet for both Mach 7 and 10 flight conditions (Berry et al. 2000a,b).

For a forebody length of 6 ft (1.8288 m), $Re_L \sim 5.5 \times 10^6$ at a freestream Mach number $M_0 = 7$, and $Re_L \sim 4.0 \times 10^6$ at $M_0 = 10$. These conditions can be approximated in the 20-Inch Mach 6 Air Tunnel, which provides $1.2 \times 10^6 < Re_L < 18.4 \times 10^6$, and in 31-Inch Mach 10 Air Tunnel with a range $1.2 \times 10^6 < Re_L < 5.1 \times 10^6$, for a model length of 28 inches (2.3 ft or 0.701 m).

A sketch of the 33% scale Hyper-X forebody model is shown in Figure 11.3. The forebody strongback was constructed from aluminum to save weight, and the leading edge (detachable to allow replacement if damaged) was machined from stainless steel with a nose radius of 0.010-inch; the length of the detachable leading edge section was 5-inch. The location of the trips was placed another 2.418-inch aft of the attachment point (for a total length from the model leading edge of 7.418-inch). In order to obtain accurate heat transfer data using the one-dimensional heat conduction equation, models need to be made of a material with low thermal diffusivity and uniform, isotropic thermal properties. The models must be durable for repeated use in the wind tunnel and not deform when thermally cycled (Berry et al. 2000a).

Thermographic phosphors were used to obtain global surface heating images, Schlieren imagining provided detailed shock shapes, and surface oil flow provided surface streamline information. The tests were done with inlet cowl door open and closed to study the effect of angle of attack variation ($\alpha = 0$-deg, 2-deg, and 4-deg), unit Reynolds number variation, and both discrete and distributed trip roughness (including configuration and height). The discrete roughness effects were included in these tests to provide information to develop an efficient trip device for the Hyper-X flight vehicle. Results from the 31-Inch Mach10 Tunnel were reported by Berry et al. (2000a), and results from 20-Inch Mach 6 Air Tunnel tests are found in Berry et al. (2000b).

The airframe-integrated scramjet propulsion system effects on vehicle aerodynamic performance, stability, and control were evaluated in the two tunnels. These aerodynamic tests were conducted using 12.5% scale (18 inch), high-fidelity models of the HXRV. To study wind tunnel mount effects interference on the HXRV Mach 6 basic longitudinal characteristics, the model was mounted in the 20-inch Mach 6 wind tunnel with a blade strut balance and tested in the presence of a removable, nonmetric, false sting, such that sting interference effects could be directly computed (Figure 11.4). The basic longitudinal aerodynamic characteristics for the HXRV airframe (inlet door closed configuration) at Mach 6 conditions provided relatively linear lift characteristics over the anticipated flight angle-of-attack and elevator deflection angle range. The effect of the blade mount was found to be rather dramatic in pitching moment and to a lesser degree in the lift and drag coefficient data. From this series of tests, a set of increments were derived to account for the sting mount interference by taking the differences of the blade mount + false sting and the blade mount alone data (Engelund et al. 1999).

A 3%-scale model of the booster rocket mated to the X-43A research vehicle (the Hyper-X launch vehicle) was tested in the Vigyan Low-Speed Wind Tunnel. Intended to obtain force and moment data, the low subsonic test was conducted both in free-stream air and in the presence of a partial plastic model of the B-52B aircraft. The test data were used to generate structural loads affecting the pylon of the B-52B and to determine the aerodynamic influence it had on the Hyper-X launch vehicle to evaluate launch separation characteristics (Davis et al. 2007).

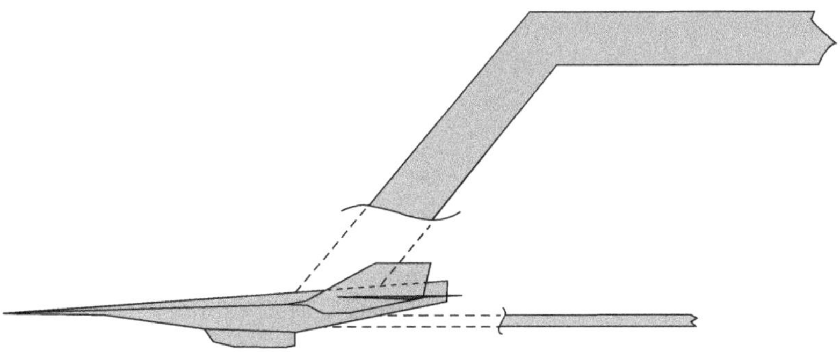

Figure 11.4 Sting and blade mount adapter hardware for the NASA Hyper-X Research Vehicle wind tunnel tests. Source: Engelund et al. (1999)/Public Domain.

The Vigyan is an open-circuit facility with a rectangular open-jet test section 3 ft tall and 4 ft wide in cross-section and 5 ft long. The tunnel can provide a freestream air velocity up to 54.864 m/s (180 ft/s), corresponding to a Mach number of 0.16; a dynamic pressure of 38.5 psf; and a unit Reynolds number of 1.145×10^6 per foot on a standard day. The model attitude in the test section can be varied from −10 deg to +90 deg in pitch and −20 deg to +20 deg in yaw. The models for the Hyper-X launch vehicle were an existing 3%-scale HXLV model, and a partial model of the B52-B carrier airplane consisting of approximately one-half of the fuselage length and one-third of the starboard wing. These ground tests are described by Davis et al. 2007.

The low-speed test showed that the B-52B airplane imparts a strong downwash onto the Hyper-X launch vehicle, reducing the net lift of the Hyper-X launch vehicle. Also, pitching and rolling moments are imparted onto the booster and are a strong function of the launch–drop angle of attack. The test data were used to formulate a simple launch separation model, which was then used to examine the possibility of re-contact (Davis et al. 2007). The researchers concluded that recontact was not likely because of the decreased lift capability demonstrated, while the model was under the influence of the B-52B airplane. In addition to providing test data to formulate a launch separation model, the results of the low-speed aerodynamic tests were used to aid in clearing the X-43-A vehicle for flight.

11.4.2 AeroPropulsion Testing

To satisfy real-gas and propulsion test requirements, ground testing aims for full simulation of flight velocity, altitude, atmospheric environment, and vehicle size. In addition, the operability of an air-breathing propulsion system requires measuring/assessing off-design operation, inlet starts, unstarts, and instabilities, light-off and blowout, combustion stability, and fuel control.

For the NASA Hyper-X program, wind tunnel testing was used to verify the scramjet design (performance and operability), evaluate the engine operating characteristics required to develop autonomous propulsion controls for flight test, and to generate the hypersonic air-breathing propulsion aerodynamic database. Although a substantial database existed, the proposed flight test scramjet engine required specific tailoring for the small scale to provide the desired thrust efficiency to assure vehicle acceleration. Hence, to develop the Hyper-X scramjet engine flowpath (vehicle external lower surfaces and body and cowl internal surfaces), six hypersonic facilities were utilized (Table 11.4). All of these facilities have vacuum capability for simulation of engine exhaust to flight static pressure.

The Direct-Connect Supersonic Combustion Test Facility (DCSCTF) at Langley is used to test ramjet and scramjet combustor models in flows with stagnation enthalpies duplicating that of flight at Mach numbers between 4 and 7.5. Direct connect means that the combustor model is connected to the entire facility test gas, which flows directly into the combustor. Thus, testing is done to assess mixing, ignition, flameholding, and combustion characteristics of the combustor models. The combustor model may exhaust freely (into the test cell), or directly (connected) to an air-ejector, or to a 70-ft diameter vacuum sphere.

The Combustion-Heated Scramjet Facility (CHSTF) at Langley is used to test complete (inlet, combustor, and partial nozzle) subscale scramjet component-integrated models in

Table 11.4 Propulsion test facilities used by NASA Hyper-X Program.

Facility	Primary use	Flow energizing method Max T_{t0} (°R)	Simulated flight Mach no.	Nozzle exit Mach no.	Nozzle exit size (in.)	Test section dimensions (ft)	Dynamic pressure PSF
Direct-connect module GASL	Combustor tests	H_2/O_2/air combustion (3800)	4.0–7.5	2.2	4.71 × 6.69		>4000 PSF @ M5 & 7
Combustion-Heated Scramjet Test Facility (CHSTF)	Engine tests	H_2/O_2/air combustion (3800)	3.5–5.0 4.7–6.0	3.5 4.7	13.26 × 13.26	2.5 W × 3.5 H × 8 L	1500 PSF @ M5
Arc-Heated Scramjet Test Facility (AHSTF)	Engine tests	Linde ($N = 3$) arc heater (5200)	4.7–5.5 6.0–8.0	4.7 6.0	11.17 × 11.17 10.89 × 10.89	4 dia. × 11 L	800 600
8-ft high temperature tunnel (8′ HTT)	Engine tests	CH_4/O_2/air combustion (3560)	4.0 5.0 6.8	4.0 5.0 6.8	96 diameter	8 dia. × 12 L (26 dia. Chamber)	1500 PSF @ M7
Hypersonic pulse facility (HYPULSE)	Engine and combustor tests	RST (15 550)	7 10	6.5	24 dia.	7 dia.	>2000 PSF @ M7 ~800 @ M10
GASL Leg IV	Engine tests	Pebble-Bed + H_2/O_2/ air combustion (5200)	5 7	4.7 6	13.26 × 13.26	25 W × 3.5 H × 8 L	1200 1700

Source: McClinton et al. (1998)/NASA/Public Domain.

flows with stagnation enthalpies duplicating that of flight at Mach numbers from 3.5 to 6. The (hydrogen, air, and oxygen) heater allows the CHSTF to obtain the flight stagnation enthalpy required for engine testing. Oxygen is replenished in the heater to obtain a test gas with the oxygen mole fraction of air (0.2095). The facility may be operated with either a Mach 3.5 or 4.7 nozzle. The CHSTF is equipped to test the scramjet models with either gaseous hydrogen or gaseous hydrocarbon (both at ambient temperature). A 20% silane, 80% hydrogen mixture (by volume) is also available for use as an igniter/pilot gas to aid in the combustion of the primary fuel.

The Arc-Heated Scramjet Test Facility (AHSTF) at Langley is used to tests component models of airframe-integrated scramjets at conditions experienced at flight Mach numbers of 4.7 to 8. Test models may include the inlet, isolator, combustor, and a significant portion of the nozzle. The flow at the exit of the facility nozzle simulates the flow that would enter a scramjet engine in flight, air flow which has been processed by the vehicle's forebody shock. Hence, tests results are used to assess the performance of the scramjet, to optimize the design of the components, and to optimize fueling configurations. The total enthalpy of the flight condition is achieved by electrically heating the air with a Linde arc heater. Run times in the AHSTF range from 30 seconds at flight Mach number of 8 simulated conditions, to 60 seconds at flight Mach number of 4.7 simulated conditions. Gaseous hydrogen fuel, with a 20/80% silane/hydrogen mixture, is available to promote ignition.

During 2008, the AHSTF was reconfigured to run not only on free jet flow mode but also in a direct-connect mode, which simulates inlet conditions at the scramjet engine's combustor for tests involving fuel ignition and flame stability. To support the ground tests of the HIFiRE-2 hydrocarbon-fueled flowpath, the AHSTF operated in direct-connect mode with a gaseous fuel mixture as a surrogate for a cracked liquid JP-7 fuel. Test conditions covering the range of interest to the HIFiRE-2 experiment (Mach 6 to 8) are achievable in AHSTF using three facility nozzles (nozzle exit Mach 2.51, 3.00, and 3.46).

For air-breathing propulsion and aerothermal/thermostructural testing, the 8-foot high temperature tunnel (8-ft HTT) at Langley produces true enthalpy environments simulating flight from Mach 4 to Mach 7. To achieve those conditions, the facility uses a methane–air heater and nozzles that can produce aerodynamic Mach numbers of 4, 5, or 7 and have exit diameters of 8 feet or 4.5 feet. Housed inside a 26-ft vacuum sphere, the 12-ft long free-jet test section can accommodate large test articles. Figure 11.5 provides a schematic representation of this facility.

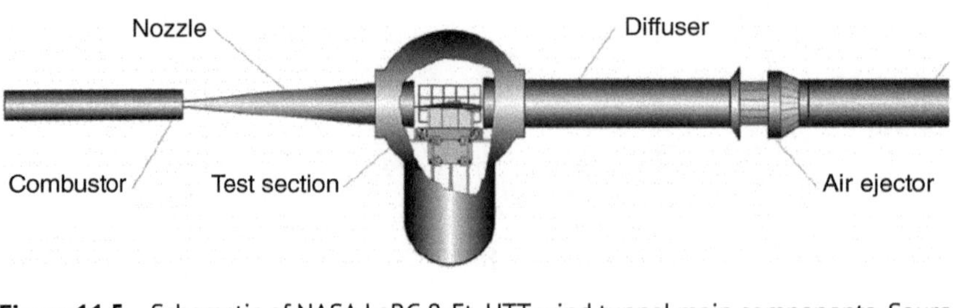

Figure 11.5 Schematic of NASA LaRC 8-Ft. HTT wind tunnel main components. Source: Harvin et al. (2006)/Public Domain.

> The NASA LaRC 8-Foot High Temperature Tunnel (HTT) provides combustion-heated hypersonic blowdown-to-atmosphere simulation for Mach numbers of 3, 4, 5, and 6.5 through a range of altitude from 50 000 to 120 000 feet (15–37 km). The facility has tested the National Aerospace Plane Concept Demonstration Engine, the X-43 Hyper-X Scramjet, the Office of Naval Research HyFly Dual Combustor Ramjet Engine, the X43C program Ground Demonstrator Engine No. 2, and the Air Force Research Laboratory SJX61-1 and SJX61-2 scramjets (X-51A waverider).

The 8-ft HTT can provide stable flow test conditions up to about 60 seconds. The facility can provide several different fuel systems for the test article. Systems are currently configured for gaseous hydrogen, a mixture of gaseous hydrogen and silane, gaseous ethylene, and liquid JP-7, but could be used to deliver other gaseous and liquid fuels. Each of these systems provides ambient temperature fuel and has automated control systems to provide preprogrammed fuel and ignition sequences. The HTT meets engine system development and flight clearance verification requirements, which were defined by the NASA-USAF X-43C Hypersonic Flight Demonstrator Project, and later by the Air Force X-51A Program (Harvin et al. 2006).

The NASA Hypersonic Pulse Facility (HYPULSE) is a dual-mode shock tunnel facility. It can operate in both as a reflected-shock tunnel (RST) or as a shock-expansion tunnel (SET). In the RST mode, stagnation enthalpies in the test gas corresponding to flight speeds from Mach 5 to 10+ are achievable. In the SET mode, it delivers stagnation enthalpies in the test gas corresponding to flight speeds from Mach 12 to near orbital. Figure 11.6 shows the operational envelopes of the HYPULSE shock tunnel simulation for both RST and SET modes. The range of dynamic pressure between 600 and 2000 psf (~30–100 kPa) covers the generally accepted air-breathing flight corridor. Note the test points for the scramjet tests

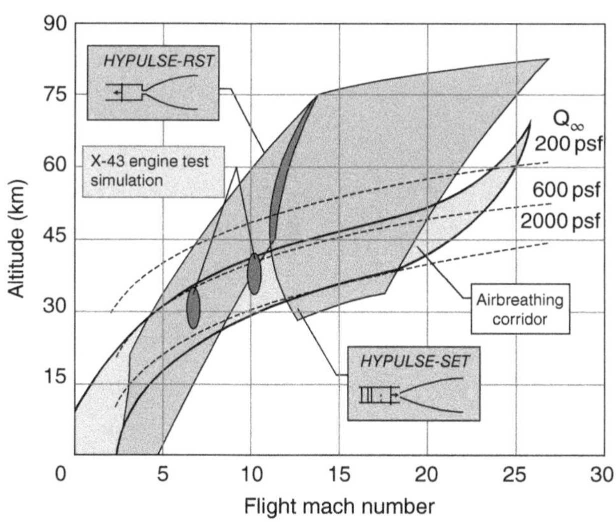

Figure 11.6 NASA HyPulse shock tunnel test envelope. Source: Rogers et al. (2005)/Public Domain.

supporting the Hyper-X scramjet tests at Mach 7 and 10. The 7 ft diameter test section can accommodate models up to 15 ft long for aerothermodynamic and propulsion testing. When operated in the tunnel mode, HYPULSE expands the test gas to Mach 6.5 using a 26-inch diameter axisymmetric nozzle.

Optical access is provided for Schlieren images and for laser diagnostics, and instrumentation is available for collecting measurements from pressure, heat transfer, and temperature transducers. Since the HYPULSE run times are on the order of a few milliseconds of established flow, the test article remains at nearly room temperature, allowing for visualizations of the combustor flow through uncooled windows.

The Leg IV at Northrop Grumman Propulsion Test Complex (formerly GASL) is a storage heated facility for clean-air combustion and aerodynamic heating (Roffe et al. 1999). Tests in this facility provide results for full pressure and enthalpy simulation, comparisons of performance for high-to-low pressure tests, and a direct, low-pressure comparison of H_2-Air-O_2 combustion-heated facility results with arc-heated facility data.

11.4.3 Hyper-X AeroPropulsion Test Simulation Method

The NASA Hyper-X ground test approach to replicate the scramjet engine inflow conditions expected in flight is described by Rogers et al. (2005). This approach builds a test article that models a portion of the engine flowpath. The partial width, truncated length segment of the actual flowpath model is tested at conditions that duplicate the stagnation enthalpy, and match flow profiles of static pressure and Mach number at the cowl entrance plane expected in flight.

Figure 11.7 depicts this simulation method, where the upper sketch shows the underside of the Hyper-X Research Vehicle (HXRV), with the shaded rectangle indicating the partial

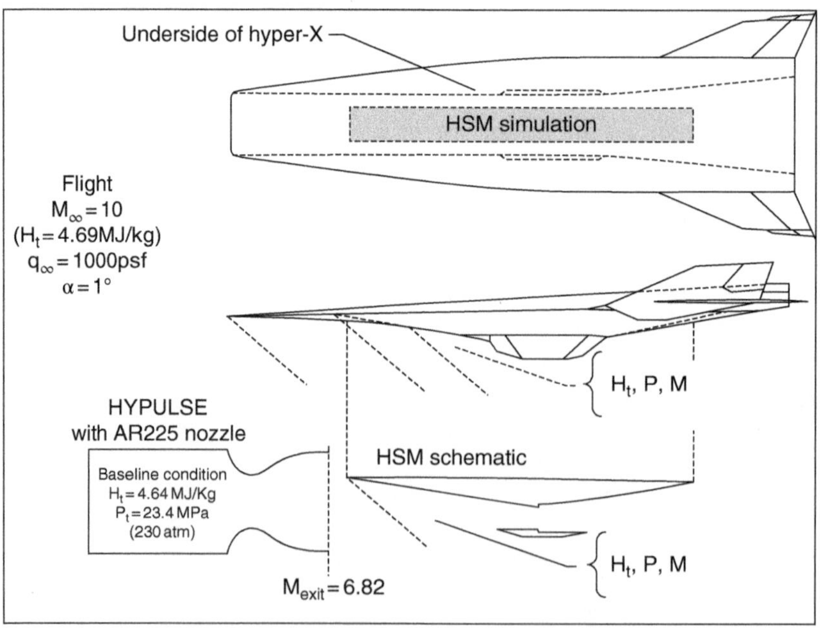

Figure 11.7 Aero-propulsion simulation method in ground test facilities. Source: Rogers et al. (2005)/Public Domain.

width and truncated length segment of the HYPULSE Scramjet Model (HSM) hardware. Below this sketch there is the side view schematically illustrating the HXRV forebody shock system, and it indicates the location of the HSM forebody leading edge right at the beginning of the first forebody ramp on the HXRV. The rationale in ground tests is to match the flight flowpath values of Mach number, pressure, and total enthalpy at the entrance to the test hardware.

For the Mach 10 flight condition, the required total enthalpy is 4.69 MJ/kg along a trajectory where the dynamic pressure is 1000 psf (47.88 kPa), and the vehicle having a 1°angle of attack. On the ground test, the stagnation enthalpy is duplicated, and the local Mach number is obtained by facility nozzle design and by the model mounting angle. However, facility pressure limitations constrain the simulated flight dynamic pressure capability. Therefore, to compensate for the lower pressure, tests are conducted with higher model mounting angles that yield higher inflow pressures but lower Mach numbers. The generated conditions at the HSM cowl-leading edge were varied by changes in facility conditions and model orientation (Rogers et al. 2005).

Figure 11.8 shows the HSM in the HYPULSE test section (cowl side up). The 40% partial width scramjet section model was designed with a fenced forebody that is 2 inches wider (1 inch each side) than the engine inlet to allow for bleed slots to remove the fenced forebody boundary layer. A removable boundary-layer trip strip is placed in the forebody to transition the boundary layer before entering the inlet. The model flowpath surfaces are stainless steel while the sidewalls are aluminum. Two window sets show views of the isolator/combustor and the internal nozzle (Bakos et al. 1999).

The HSM was identical to the Hyper-X Experimental Model (HXEM) which was tested in the Langley 8-Foot HTT, and thus they share instrumentation locations to facilitate data comparisons.

Figure 11.8 The scramjet model in the HyPulse test section. Source: From Northam et al. (2007), NASA.

11.4.4 Hyper-X Risk Reduction Testing for the Mach 7 Flights

The first fully integrated X-43A vehicle underwent a number of tests in a flight-like environment in order to reduce risks. This required the design of the HXFE, the only full-width Mach 7 scramjet (Figure 11.1) tested prior to the X-43 flights. Mounted on the VFS, the HXFE depicted in Figure 11.1 was tested in the 8-Foot HTT. VFS represented a geometrically accurate tip-to-tail X-43 propulsion flowpath, including forebody compression surfaces, and aftbody expansion that replicates the entire undersurface of the X-43 (Huebner et al. 2001).

Figure 11.9 shows the 8-Ft. HTT configured for air-breathing propulsion tests, with the propulsion test model mounted on the pedestal. The pedestal houses fuel-control system and model instrumentation and provides model access for internal cavity purging/cooling of airframe structures and water cooling for engine leading edges. The pedestal is attached to the facility force measurement system with angle of attack spacers. The tests articles are stored beneath the hypersonic flow stream in order to facilitate tunnel starting and to protect the models from startup and shutdown dynamic loads. Once steady-state hypersonic flow is established in the test section, a hydraulic elevator system inserts the model mounted on a three-component force measurement system into the test stream in approximately 1.5 seconds (Voland et al. 1998).

Table 11.5 gives the nominal tunnel combustor conditions and resulting simulated air freestream flow parameters for the HXFE/VFS tests.

The scramjet engine restart capability of an airframe-integrated scramjet engine was successfully demonstrated by the HXFE. To show this operation, the HXFE inlet was purposely caused to unstart (via excess fueling), followed by rapidly throttling down the fuel, restarting the inlet flow by actuating cowl door, and re-igniting the engine. The restart process was achieved in 1.76 seconds, and researchers reported that the data suggested that restarting could be achieved significantly faster (Rausch et al. 2000). These ground tests provided

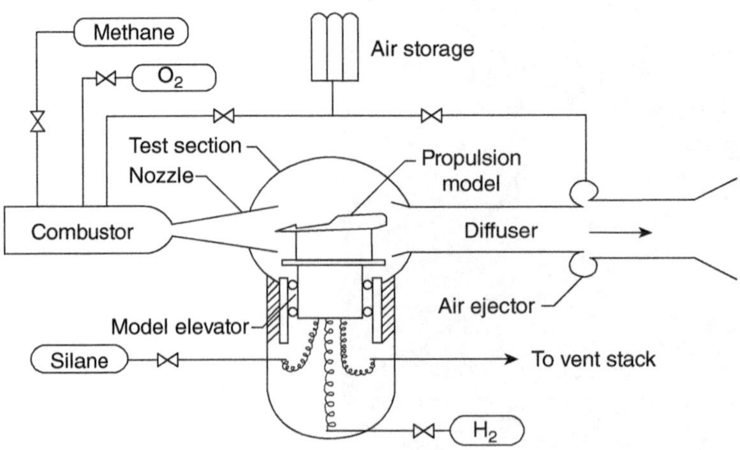

Figure 11.9 Schematic of 8-Ft. HTT for Hyper-X air-breathing propulsion testing. Source: Huebner et al. (2000)/NASA/Public Domain.

Table 11.5 Simulated freestream conditions for 8-Ft. HTT testing of HXFE/VFS and comparison to flight.

Simulation case	Tunnel low q_0	Tunnel flight q_0	Target flight point	Tunnel high q_0
p_{comb}, psig (MPa)	1000 (6.895)	1585 (10.93)	Air	2000 (13.79)
T_{comb}, °R (K)	3250 (1806)	3350 (1861)		3350 (1861)
M_0	6.84	6.92	7.00	6.87
p_0, psia (Pa)	0.140 (965)	0.211 (1455)	0.204 (1407)	0.263 (1813)
q_0, psf (kPa)	647 (30.98)	1000 (47.88)	1000 (47.88)	1230 (58.89)
T_0, °R (K)	434 (241)	423 (235)	408 (227)	434 (241)
h_t, BTU/lbm (MJ/kg)	1064 (2.47)	1052 (2.44)	1052 (2.44)	1064 (2.47)

Source: Adapted from Huebner et al. (2001).

engine performance and operability data, and design and database verification, giving the first-ever airframe-integrated scramjet data in flight.

11.4.5 Hyper-X Mach 10 Flowpath Testing in HYPULSE Shock Tunnel

The final design of the X-43 Mach 10 scramjet was tested in the NASA HYPULSE Shock Tunnel. Operating in the RST configuration at conditions that duplicated nominal Mach 10 flight speed, the engine flowpath development and evaluation testing was carried out with a test article referred to as the HSM. This was a partial-width replica of the HXRV that extended from about mid-length of the research vehicle forebody and included a portion of the nozzle expansion ramp. The shock-induced detonation driver delivers a test gas of shock-heated air at stagnation conditions of 3740 K and up to 30 MPa, with a useful test period of up to 3 ms. The axisymmetric contoured facility nozzle expands the test gas flow through an area ratio of 225 to an exit diameter of 67 cm, of which about 40 cm is a uniform core flow at the nominal Mach 10 test condition (Rogers et al. 2005).

Tested in a semi-direct-connect arrangement, the HSM was a full-scale height, partial width engine model with truncated forebody and aftbody sections. The test conditions included Mach 9 total enthalpy conditions, and Mach 10 total enthalpy conditions at 6 different model mounting angles and cowl leading edge positions to vary the combustor entrance properties, and tests with slightly greater than Mach 10 total enthalpy. Fifty-five test runs were completed: 9 were without hydrogen fuel injection, 20 with hydrogen only, 24 with 2% silane by volume, 1 with 1% silane, and 1 with 20% silane at low fuel equivalence ratio to model the flight fuel sequence prior to starting the hydrogen fuel (Ferlemann 2005).

The X-43 Mach 10 scramjet flowpath was tested in the Large Energy National Shock Tunnel (LENS) reflected shock tunnels located at Calspan University of Buffalo Research Center (CUBRC). Tested in a free-jet configuration with the facility nozzle producing Mach 10 freestream conditions, the scramjet model was a full-scale height, width, and length replica of the X-43 Mach 10 flight engine. The tests at LENS included a range of dynamic pressure at 1° angle of attack, and an angle of attack of 2° at the nominal dynamic pressure. Twenty test runs were completed: 3 were without fuel injection, 2 mixing (hydrogen into nitrogen test gas), 5 with cold hydrogen, 3 with heated hydrogen, 5 with 2% silane, and 2 with 5% silane (Ferlemann 2005).

To obtain surface heating and pressure data for the X-43A cowl-closed configuration at nominal Mach 10 flight conditions, tests were conducted in CUBRC's LENS I hypervelocity tunnel. This facility is a reflected shock tunnel with an electrically heated driver section capable of duplicating flight total enthalpy conditions up to a Mach number of 12. The test article consisted of a two-dimensional forebody plate, used to duplicate the forebody compression conditions of the Mach 10 X-43 vehicle. The cowl-closed configuration was designed to model the Mach 10 vehicle outer mold line in the vicinity of the cowl flap and sidewalls. The model was tested with and without the boundary-layer trip. The experimental results were compared with CFD analysis to predict the aeroheating for the X-43A vehicle in flight (Cockrell et al. 2002). The capabilities of LENS I are described by Holden 1990.

11.4.6 X-43A Cowl Actuation Simulated at Flight Condition

The X43A vehicle was designed with a cowl door that closed the inlet flowpath to protect the scramjet engine during the boost and descent phases of the flight. To determine if heat loads experienced during the flight would affect the cowl door actuation, ground testing of the HXFE was required. The wind tunnel test had to simulate the aerodynamic heating that the X-43 scramjet engine would encounter during the ascent part of the flight trajectory as the Pegasus rocket boosted the X-43A to its flight test point. This test was important to reduce the potential for inlet starting during the cowl opening sequence due to boundary-layer effects, internal contraction ratio effects, and cowl opening speed.

The actuation design had a linear electromechanical actuator motor and controller that rotated a torque tube connected with cams to connecting links that were attached through the engine sidewalls to the cowl door. Two slider blocks (flush with the internal engine sidewalls) moved with the connecting links, allowing for proper freedom of movement. HXFE/VFS testing verified the successful design of the sliders, establishing the specifications for the actuator settings (speed and torque levels) to ensure the cowl door would work as required during flight operation. The cowl door was actuated 355 times under no aerodynamic load (primarily during engine internal inspection and prerun preparation) and 52 times under Mach 7 aerothermodynamic loads (during testing). For all cases, the cowl was successfully actuated as required (Huebner et al. 2001).

According to analysis performed by the scramjet manufacturer, subjecting the engine to 25 seconds of Mach 7 enthalpy tunnel flow at a dynamic pressure of 1280 psf (61.29 kPa) equates to about the same amount of aerodynamic heating that the X-43A vehicle would encounter during boost. During ground tests, the engine model was in flow 26 seconds before the cowl door was commanded open. It opened in less than 0.5 seconds to full-open position with no problems (Rausch et al. 2000).

11.4.7 Hyper-X Stage Separation

Flight controls is a key enabling technology for flight testing. The X-43A vehicle had to separate from its booster at high altitude and at hypersonic speeds; the X-43A separated at Mach 6.946 at an altitude of 94 069 ft (28.67 km) during the Mach 7 mission (Marshall et al. 2005a), and at Mach 9.736 at an altitude of 109 440 ft (33.36 km) during the Mach 10 mission (Marshall et al. 2005b). Smooth separation from the booster stage in only 500 milliseconds was critical to the success of the X-43A hypersonic flight.

The stage separation system was based on cruise missile ejectors designed for the B-1B supersonic aircraft program, but never used. The ejectors were mounted so that the pistons bear on the titanium aft bulkhead of the X-43A vehicle, providing approximately 9 g acceleration when they are actuated. In approximately 350 milliseconds, this separation mechanism propelled the X-43A vehicle proper clear of the adapter–booster combination (Harsha et al. 2005).

To support multidegree-of-freedom simulation of the booster/scramjet powered vehicle separation process and to define the stage separation aerodynamic environment, the following wind tunnel tests were required for complete aerodynamic characterization: preseparation Hyper-X Launch Vehicle (HXLV) aerodynamics, mutual interference aerodynamics of the HXRV in close proximity with the booster, post-separation (interference-free) booster aerodynamics, and post-separation (interference-free) HXRV aerodynamics. Woods et al. (2001) reported on the chronology of the wind tunnel test program for risk reduction of the stage separation event, providing representative results from each series of tests, and they discussed issues and concerns during the process.

Tests were conducted at both Math 6 and 10 to evaluate Mach number effects on the hypersonic interference. The NASA team described the stage separation interference test setup in the Langley 31-1nch Mach 10 Tunnel. They reported that although the balance/sting assembly interfered with the flow between the HXRV and the adapter and allowed only axial separations, the model provided an important preliminary assessment of the order of magnitude of the interference aerodynamics and helped to develop the required parametric analysis for the interference aerodynamics database (Woods et al. 2001).

Numerous parametric and Monte Carlo analyses were conducted in order to select a separation scenario. A fixed adapter scheme and a more elaborate mechanized two-piece adapter approach were considered. The later approach had the potential to eliminate the possibility of re-contact, but it was too risky and uncertain. Therefore, the fixed adapter approach was adopted, and the separation analysis effort then focused on defining a viable fixed adapter separation strategy with sufficient robustness to handle uncertainties (Davidson et al. 1999). Analysis resulted in a separation controls/sequencing strategy that satisfied the desired objectives. Preliminary separation concepts were tested in the NASA Langley 20-inch Mach 6 tunnel.

The majority of the force and moment data used in the separation aerodynamic database were obtained from tests in the AEDC von Karman Facility Tunnel B, using the dual support/force balance Captive Trajectory System (CTS) rig. This system allowed to test the relative positioning of the research vehicle and the booster, yielding force data from each. The Mach 6 tunnel flow conditions were 104°R static temperature and $Re = 2.2 \times 10^6$/ft.

A 8.33% scale model of the 12-inch research vehicle (HXRV), booster-to-research vehicle adapter, and booster were fabricated for the AEDC stage separation test. This model could be combined into the full, 40-inch launch vehicle (HXLV) or stack, or tested in various combinations (Davidson et al. 1999).

Having as main objective to quantify aerodynamic loads during the HXRV-HXLV stage separation, the tests at AEDC included variations in interference loads due to the effects of adapter position, effects of relative position (translation and rotation) of the HXRV and booster-adapter combination, and variations in initial flight conditions (AOA and sideslip), which might result from booster dispersion from the nominal trajectory. These tests provided data to compare with CFD predicted loads and to compare with the limited stage separation results from the LaRC tunnels (Davidson et al. 1999).

After separation from the HXLV, closed-loop flight control was also required to stabilize the HXRV and achieve the engine test condition, and then to maintain this condition during the engine test sequence. These control laws are required to maintain the desired angle of attack and sideslip to within ±0.5 degrees during the entire engine test sequence. After the engine test sequence, the control laws follow steering commands from the guidance system to maintain a desired descent trajectory (Davidson et al. 1999).

11.5 Ground Testing for the USAF X-51A Waverider

The USAF X-51A WaveRider research vehicle evolved from an extensive development program that included testing in the best ground facilities in the United States, including at the NASA Langley 8-ft HTT, the Leg VI freejet tunnel at Northrop Grumman Propulsion Test Complex (formerly GASL). Prior to the flight demonstration, the integrated vehicle was further tested at other facilities. An overview of the ground testing carried out was published by Hank et al. (2008), from where we extracted the following highlights.

The first prototype of the hydrocarbon scramjet intended for ground testing was the Ground Demonstration Engine 1 (GDE-1). In 2003, the GDE-1 completed testing at Mach 4.5 and 6.5 simulated flight conditions in the Leg VI freejet tunnel (Boudreau 2005). The GDE-1 survived over 50 ground test attempts, proving the manufacturing processes used to construct the flight-weight design and showed the required durability to survive thermal and pressure loads for the X-51A flight (Hank et al. 2008).

Between 2004 and 2006, an extensive test program was conducted at the United Technologies Research Center (UTRC) Cell V direct-connect combustor rig in East Hartford, Connecticut. These tests were crucial because the direct-connect rig facilitated the simulation of combined isolator/combustor operation, having the ability to adjust fueling distributions with the X-51A fuel distribution valves. Over 30 different flight conditions were simulated at UTRC, investigating issues related to engine ignition on ethylene and transitioning to JP-7, JP-7-only performance sensitivity to total combustor equivalence ratio, fuel distribution, flight enthalpy and dynamic pressure, and scramjet mode transition at high Mach flight conditions.

In order to prove the viability of the fuel-cooled scramjet, the next model built was the Ground Demonstration Engine 2 (GDE-2), a flight-weight, fuel-cooled, closed-loop fuel system. In 2006, the GDE-2 was tested at Mach 5.0-simulated flight conditions in the Langley 8-ft HTT (Figure 11.10). Undergoing over 300 seconds of combustion time, this test validated the integrated closed-loop fuel system with the combustor working as required, providing valuable lessons that were fed into the X-51A engine development program, yielding the JP-7 hydrocarbon fuel-cooled SJX61-1 (X-1) development engine designed and built by Pratt & Whitney Rocketdyne.

Figure 11.10 The SJX61-2 scramjet built by Pratt and Whitney Rocketdyne successfully completed ground tests in simulated Mach 5 flight conditions at NASA LaRC. Source: Kathy Barnstorff/NASA/Public Domain.

To demonstrate ignition and transition to JP-7 fuel, the SJX61-1 (X-1), which included the vehicle forebody, combustor, and integrated nozzle, was tested at the Langley 8-ft HTT. The X-1 was run at simulated flight conditions of Mach 4.6, 5.0, and 6.5 at dynamic pressures close to the planned X- 51A mission. These tests checked the closed-loop fuel system, assessing the thermal and structural integrity of the engine and controls hardware, and verified the flowpath performance and operability required for the X-51A mission. Testing was completed in 2007; the SJX61-1 (X-1) had survived over 40 ground test cycles and had a total combustion time of over 1000 seconds.

The last ground test model was the SJX61-2 (X-2), the X-51A flight clearance engine. Its JP-7 fuel system was the same flight configuration and included a flight-like simulation of the vehicle ethylene ignition system. In its closed-loop thermal management system, the JP-7 both cools engine hardware and then is injected in the supersonic combustor. The scramjet engine (Figure 11.10) was evaluated in a simulated full flight propulsion configuration at Mach 5, which includes a Boeing-designed, full-vehicle forebody/inlet and nozzle, at dynamic pressures consistent with the X-51A trajectory, and at off-nominal dynamic pressures to give additional confidence for success of the X-51A flight.

Other ground testing investigated particular areas of analysis of X-51A, such as testing a 14% scale model of the vehicle stack at the Boeing's North American Aviation Research Tunnel (NAART) in California, at NASA-LaRC Tunnel 16T, and at Boeing's PSWT, to determine stack aerodynamics. Testing also included a 20% model of the cruiser at PSWT, the VKF at AEDC, and the NASA LaRC Unitary Plan Wind Tunnel (UPWT) for cruiser aerodynamics and fin deflections. The wind tunnel test program accrued over 1700 hours over the course of 3200 runs of both the vehicle stack and cruiser alone (Hank et al. 2008).

The AFRL Air Vehicles Directorate also tested a full-scale X-51 waverider model in CUBRC's Large Energy National Shock Tunnel 2 (LENS-2) to determine temperatures and heat loads. A test of boundary-layer transition study was also conducted at Purdue University's Mach-6 Quiet Tunnel (Borg 2009).

In addition to ground wind tunnel and CFD modeling and simulation, hypersonic air-breathing propulsion technology development requires substantial flight testing. Such flights must begin with the development of small-scale experimental vehicles since a demonstrator of an operational vehicle would be very expensive and the associated technical risk are very high. In Chapter 13, we will review important aspects of the flight testing associated with the NASA X-43A aircraft and the Air Force X-51A waverider.

11.6 ONERA Ground Testing for the European LAPCAT2 Combustor

The ONERA-LAERTE facility is equipped with the LAPCAT-II dual-mode combustor to carry out experiments that are representative of scramjet flight conditions (higher air temperature and static pressure, longer test duration). The combustor, built within LAPCAT2 European program, is fed with hot vitiated air (heating by H_2/air combustion, O_2 mol fraction kept at 21% by upstream extra injection of oxygen). Total temperatures up to 1800–1900 K can be obtained, and the total pressure can reach 1.0–1.2 MPa (Vincent-Randonnier et al. 2014). This facility (Figure 11.11) is operated in the blow-down mode, the test section working as a heat sink. Overall duration of a test run is around one minute, but the useful duration (required temperature) lasts between 5 and 15 seconds, depending on test combustion requirements (no water-cooling of the combustor). The constant width (40 mm) combustor is shown in Figure 11.11, showing the optical accesses windows.

Leonov et al. (2014) present the profiles of Mach number, static temperature, and static pressure along the combustor for nonreacting flow. For a reservoir pressure $P_0 \geq 0.9$ MPa, airflow is supersonic up to the exit of the combustor. For $P_0 < 0.9$ MPa, the combustor flow is overexpanded: the transition from supersonic to subsonic flow occurs within the combustor and pressure increases up to 0.1 MPa (pressure level in the exhaust pipe).

Within the LAPCAT-II project, studies focus on supersonic combustion in the dual-mode ramjet (DMR) engine for M = 5–6 flight conditions (reservoir conditions:

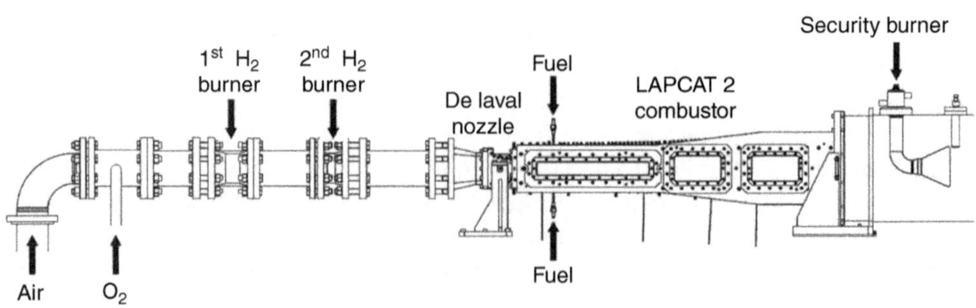

Figure 11.11 Schematic of ONERA's LAERTE-LAPCAT2 facility. Source: Adapted from Vincent-Randonnier et al. (2014).

$p_0 = 0.3 - 1.0$ MPa; $T_0 = 1300 - 1800$ K), both experimentally at LAERTE facility and numerically. The LAPCAT-II combustor is made of a copper alloy, and the inner walls include a 0.3 mm thick YSZ (Yttria Stabilized Zirconia) thermal barrier coating (Vincent-Randonnier et al. 2014). The combustor surface finish is very rough. As it is well known, airflow temperature and wall roughness both promote boundary-layer thickening. Moreover, wall temperature also modifies the ignition delay, affecting the location of pressure increase due to heat release. Consequently, both parameters have to be adjusted in the tunnel to fit pressure profile baseline and heat release position.

As described by Leonov et al. 2014, the LAPCAT combustor can be fueled with pure hydrogen, or with a mixture of hydrogen and gaseous hydrocarbons (methane, ethylene, and propane). Reported research related to LAPCAT-II program was conducted with hydrogen fuel injected through two 2 mm diameter holes, one on the top wall, the other on the bottom one. More details on this research are included in Chapter 12.

11.7 Vitiated versus Clean Air Hypersonic Wind Tunnel

To achieve the high enthalpy conditions associated with hypersonic flight, many ground test facilities burn fuel in air upstream of test chamber. For example, the Langley 8-Ft HTT is a methane vitiated facility with oxygen replenishment. The compositions of clean and vitiated air caused by combustion heating are listed in Table 11.6. The products of combustion contaminate the test gas and alter the gas properties, impacting the prediction of heat fluxes associated with aerodynamic heating. The effect of vitiates on the combustion process is of great interest to the engine designer, as the ground test results must be extrapolated to flight conditions.

The effect of vitiation on the net thrust on a JP-7-fueled scramjet was studied by Norris (2017) who found that thrust was lower by about 20%. This is the same effect seen in a study of hydrogen-fueled scramjets (Pellet et al. 2002).

Research has been carried out to assess the difference in heating rates between clean air and a vitiated test medium, as this affects fuel ignition, and the design of the TPS system. In 2008, Cuda and Gaffney performed an analytical investigation comparing clean air to a gas vitiated with methane/oxygen combustion products to determine if variations in gas properties contributed to changes in predicted heat flux. Their results confirmed that the

Table 11.6 Composition of vitiated versus nonvitiated air used in computer simulations.

Species	Clean air (mass fraction)	Vitiated air (mass fraction)
CO_2	0.0	0.0552
H_2O	0.0	0.0455
O_2	0.2330	0.2330
N_2	0.7660	0.6663

Source: Norris (2017)/Public Domain.

methane-vitiated test medium could alter the heat transfer rate by as much as −10% (Cuda and Gaffney 2008).

The approach to produce hot air flows in the ONERA's LAERTE facility (hydrogen burning) induces a vitiation of air by water vapor. Researchers (Vincent-Randonnier et al. 2014) have evaluated numerically the effect of this vitiation on ignition delay using experimental data for validation. Air composition and airflow conditions for pure and vitiated air were calculated from isentropic flow equations and NIST-Refprop V9.0 for this study. The Refrop program is NIST Standard Reference Database 23, which calculates thermochemical properties, including the NIST-equations of state for many of the pure fluids and mixtures.

Reported results by Vincent-Randonnier et al. (2014) showed the effects of vitiation and wall conditions on pressure and temperature profiles. The data showed that with pure air, the ignition distance is significantly increased. Moreover, at high temperature, the effect of air temperature on supersonic combustion is less pronounced, whereas wall temperature role remains crucial. This study showed that hydrogen ignition distance was significantly shorter in vitiated air than in pure air.

For clean, nonvitiated hypersonic testing, the NASA Glenn Hypersonic Tunnel Facility (HTF) can simulate true enthalpy conditions for flight up to Mach 7. The HTF is a blowdown, free jet wind tunnel that is capable of simulating Mach 5, Mach 6, and Mach 7 true enthalpy conditions. The facility generates these conditions by flowing nitrogen gas through a 3 MW graphite core storage heater. This heated nitrogen is then mixed with ambient temperature oxygen and ambient temperature nitrogen to yield a flow of synthetic (true composition) air at the requisite stagnation temperature. Three interchangeable nozzles are used to establish the facility Mach number condition of 5.0, 6.0, and 7.0.

11.8 Diagnostics and Measurements for Scramjet Combustion

An air-breathing scramjet poses a formidable challenge for ground testing due to the engine operational principle. If we consider the scramjet as based on thermal compression, we determine that this process is inherently intertwined with other complex processes that operate under a delicate balance of aero-thermodynamics and chemical kinetics interactions. Moreover, the high Reynolds number turbulent combustions fields are characterized by unsteadiness and spatial inhomogeneity that require transient and three-dimensional data to understand such complex physics. Hence, experimental measurements of scramjet combustion are rather difficult to obtain. Hypersonic flow instrumentation requires fast-response, minimally invasive microsensors able to survive the harsh hypersonic flow environment. Thin-film thermocouples (TFTCs), for example, provide a minimally intrusive means of measuring surface temperature in hostile, high temperature environments.

How can we measure mixing of fuel and air in a DMSJ combustor? In such device, the high-speed combustion flame produces a thermal blockage due to the heat release of chemical kinetics, which is manifested in a pressure rise that propagates into the isolator. That is why a shock train is generated upstream of the combustor, which in turn affects the quality of fuel–air mixing. The pressure rise can be measured with high-responding sensors and from this data the combustion characteristics are inferred.

Combustor measurement data are scarce and selective, as to obtain those measurements requires nonintrusive optical techniques, such as CARS (coherent anti-Stokes Raman scattering), PLIF (planar-induced laser fluorescence), and Schlieren systems.

CARS are laser techniques used to measure temperature and the mole fractions in a combustor. To obtain more information about multispecies composition of the combusting flow, the three most used methods are dual-Stokes CARS, pure-rotational CARS, and dual-pump CARS. In dual-Stokes CARS, two broadband beams interact with two pump beams to produce two spatially and spectrally separated CARS signals simultaneously. CARS can be used in the harsh supersonic combustion environment because its focused laser-based system usually uses a 10-nano-second pulsed laser system.

Planar laser-induced fluorescence (PLIF) is an instantaneous, spatially resolved optical measurement technique that involves molecular excitation for both quantitative measurements and flow visualization. A laser, first expanded into a planar sheet and then focused in the measurement region, is accurately tuned to a specific absorption transition of the target molecule; for a certain finite time (fluorescence life time), a spontaneous emission of radiation from an upper energy level (fluorescence) is emitted from the illuminated region at a longer wavelength. When the molecule emits energy at the same wavelength, the process is called resonance fluorescence (Cantu et al. 2015).

PLIF images are obtained by using a laser to excite a particular molecular species present in a flow. This species either occurs naturally or can be seeded. Commonly used species include nitric oxide (NO), hydroxil (OH), iodine, and acetone. The resulting fluorescence is imaged with a scientific grade camera. The resulting flow visualizations and quantitative measurements are commonly used to compare with computation fluid dynamics (CFD) codes. Research at NASA Langley has developed a NO-PLIF technique specifically applied to combustion systems. Investigating scramjet fuel injection and mixing physics, researchers take advantage of the fact that NO naturally exists in low concentrations in the AHSTF air as a result of the electric-arc heating process. Thus, the NO acts as an in situ flow tracer that can be imaged using PLIF. The NO PLIF technique uses an ultraviolet (UV) laser sheet to illuminate a planar slice in the flow. The UV light excites fluorescence from the NO molecules, which is detected by a digital camera (Drozda et al. 2017).

Schlieren is a flow visualization technique similar to the shadowgraph technique. It is based on the principle that light rays are bent whenever they encounter changes in density of a fluid. Schlieren systems are used to visualize the flow away from the surface of an object. The earliest Schlieren images of shock waves were black and white, but by putting a prism near the slit and breaking the white light into different colors, Schlieren images are therefore colored. The resulting image is two dimensional, while, in reality, shock waves are three dimensional. Hence, the Schlieren provides some valuable information about the location and strength of the shock waves, but it requires some experience to properly interpret the results of the physical process.

The filtered Rayleigh scattering (FRS) is a laser diagnostic technique where the intensity of predominantly elastically scattered light is measured after it passes through a molecular absorption filter. FRS is now used in the Isolator Dynamics Research Laboratory (IDRL) at the NASA Langley Research Center. FRS is considered because it can provide spatially- and temporally resolved quantitative data on an imaging plane without intrusively seeding the flow. This allows detailed studies into the complex three-dimensional shock train structure.

The IDRL is configured as a Mach 2.5 cold flow, direct-connect isolator model designed for fundamental shock train research studies. FRS can provide accurate and repeatable data of the isolator flow field, as it is a relatively simple yet powerful planar measurement technique that can provide spatially and temporally resolved quantitative information. Thus, it is an ideal tool to study the detailed instantaneous flow structure of the isolator shock train, extending our knowledge of this complex flow field past the wall-pressure measurements and path-integrated visualization techniques (e.g., Schlieren and shadowgraphy). Of particular interest is FRS measurements on cross-sectional planes in the isolator to study any asymmetry in the isolator shock front. The quantitative information gathered using FRS is an ideal way to validate CFD methods (Hunt et al. 2019).

In their recent review, Seleznev et al. (2019) summarize the experimental database developed in the United States, Russia, China, Germany, Australia, France, and Japan that will be useful for studying and simulating the flowfields in the scramjet. They also included the theoretical and the computational simulations directly related to the experiments.

Additional diagnostics for scramjet combustors include particle imaging velocimetry (PIV), line of sight (LOS), and tomographic tunable diode laser spectroscopy (TDLAS), techniques that provide unique databases highly suited for model development and CFD validation. The nonintrusive optical techniques can only provide planar or zonal flowfield information, but the three-dimensional flow structures as well as their evolution physics cannot be revealed solely by these measurements. Hence, high-fidelity numerical modeling is necessary to gain deep and comprehensive insights into the internal turbulent high-speed flow, fuel–air mixing, and combustion processes in scramjet combustors. The following Chapter 12 will address CFD and related topics.

Questions

1. Is it possible to duplicate all hypersonic air-breathing propulsion phenomena in a wind tunnel? If not, how would you compensate to obtain necessary operational data?

2. Which type of ground test facility would you select to test the TPS of a new scramjet design?

3. What are the most important instrumentation and diagnostics tools to incorporate in a scramjet propulsion test plan?

4. How do test engineers determine the reaction efficiency in the scramjet combustor?

5. Explain the ground testing required for a flight vehicle prototype that will take off and reach hypersonic cruise conditions.

6. What scaled engine components and initial conditions must be tested to provide adequate scaling parameters with respect to performance parameters of interest in order to obtain a desired level of performance at a particular scale of the scramjet?

7. If ground testing data are not available, can one rely exclusively on CFD analysis to make performance predictions for a prototype DMSJ engine?

References

Bakos, R.J., Tsai, C.-Y. Rogers, R.C., and Shih, A.T. (1999). The Mach 10 component of NASA's Hyper-X ground test program. *ISABE 99-7216, International Symposium on Air Breathing Engines*, Florence, Italy (5–10 September 1999).

Berry, S.A., DiFlubio, M., and Kowalkowski, M. (2000a). Forced boundary-layer transition on X-43 (Hyper-X) in NASA LaRC 31-inch Mach 10 air tunnel. *NASA/TM-2000-210315. August 2000. National Aeronautics and Space Administration, Langley Research Center*, Hampton, Virginia (August 2000).

Berry, S.A., DiFlubio, M., and Kowalkowski, M. (2000b). Forced boundary-layer transition on X-43 (Hyper-X) in NASA LaRC 20-inch Mach 6 air tunnel. *NASA/TM-2000-210316. August 2000. National Aeronautics and Space Administration, Langley Research Center*, Hampton, Virginia (August 2000).

Bertin, J.J. and Cummings, R.M. (2006). Critical hypersonic aerothermodynamic phenomena. *Annual Review of Fluid Mechanics* 38: 129–157.

Borg, M.P. (2009). Laminar instability and transition on the X-51A. Ph.D. Thesis. Purdue University, School of Aeronautics and Astronautics, 21 July 2009.

Boudreau, A.H. (2005). Hypersonic air-breathing propulsion efforts in the Air Force Research Laboratory. *AIAA 2005-3255. AIAA/CIRA 13th International Space Planes and Hypersonic Systems and Technologies Conference*, Capua, Italy (16–20 May 2005).

Cantu, L., Gallo, E., Cutler, A.D. et al. (2015). Nitric oxide PLIF visualization of simulated fuel-air mixing in a dual-mode scramjet. *AIAA 2015-0354. 53rd AIAA Aerospace Sciences Meeting*, Kissimmee, FL (5–9 January 2015).

Chazot, O. and Panerai, F. (2015). High-enthalpy facilities and plasma wind tunnels for aerothermodynamics ground testing. In: *Hypersonic Nonequilibrium Flows: Fundamentals and Recent Advances* (ed. E. Josyula), 329–342. American Institute of Aeronautics and Astronautics, Inc., AIAA.

Cockrell, C.E., Jr., Auslender, A.H., and White, J.A. (2002). Aeroheating predictions for the X-43 cowl-closed configuration at Mach 7 and 10. *Paper AIAA 2002-0218. 40th AIAA Aerospace Sciences Conference & Exhibit*, Reno, Nevada (14–17 January 2002).

Cuda, V. Jr. and Gaffney, R.L. Jr. (2008). Analysis of the effects of vitiates on surface heat flux in ground tests of hypersonic vehicles. *55th JANNAF Propulsion Meeting/42nd Combustion/30th Airbreathing Propulsion/30th Exhaust Plume Technology/24th Propulsion Systems Hazards/ 12th SPIRITS User Group Joint Subcommittee Meeting*, Boston, MA (12–16 May 2008).

Davidson, J., Lallman, F., McMinn, J.D., et al. (1999). Flight control laws for NASA's Hyper-X research vehicle. *AIAA-99-4124. Guidance, Navigation, and Control Conference and Exhibit*, Portland, OR (9–11 August 1999).

Davis, M.C., Sim, A.G., Rhode, M., and Johnson, K.D. Sr. (2007). Wind tunnel results of the B-52B with the X-43A stack. *Journal of Spacecraft and Rockets* 44 (4): 871–877.

Drozda, T.G., Cabell, K.F., Ziltz, A.R. et al. (2017). Comparisons between NO PLIF imaging and CFD simulations of mixing flowfields for high-speed fuel injectors. *AIAA 2017-4647. 53rd AIAA/SAE/ASEE Joint Propulsion Conference*, Atlanta, GA (10–12 July 2017).

Engelund, W.C., Holland, S.D., Cockrell, C.E., and Bittner, R.D. (1999). Propulsion system airframe integration issues and aerodynamic database development for the Hyper-X flight research vehicle. *ISOABE 99-7215. XIV ISOABE*, Florence, Italy (5–10 September 1999).

Ferlemann, P.G. (2005). Comparison of Hyper-X Mach 10 scramjet preflight predictions and flight data. *AIAA Paper 2005-3352. AIAA/CIRA 13th Int. Space Planes and Hypersonics Systems and Technologies,* Capua, Italy (16–20 May 2005).

Hank, J.M., Murphy, J.S., and Mutzman, R.C. (2008). The X-51A scramjet engine flight demonstration program. *AIAA Paper 2008-2540. 15th AIAA Int. Space Planes and Hypersonics Systems and Technologies Conference*, Dayton, OH (28 April to 1 May 2008).

Harsha, P.T., Keel, L.C., Castrogiovanni, A., and Sherill, R.T. (2005). X-43A vehicle design and manufacture. *AIAA 2005-3334. AIAA/CIRA 13th International Space Planes and Hypersonics Systems and Technologies Conference*, Capua, Italy (16–20 May 2005).

Harvin, S.F., Cabell, K.F., Gallimore, S.D., and Mekkes, G.L. (2006). Test capability enhancements to the NASA Langley 8-foot high temperature tunnel. *JANNAF 41st Combustion/29th Airbreathing Propulsion/23rd Propulsion Systems Hazards Joint Subcommittee Meeting*, San Diego, CA (2006).

Heiser, W.H. and Pratt, D.T. (1994). Hypersonic Airbreathing Propulsion, AIAA Education Series. Washington, DC: American Institute of Aeronautics and Astronautics (AIAA).

Holden, M.S. (1990). Large Energy National Shock Tunnel (LENS), Description and Capabilities. NASA Technical Report ADA338839.

Huebner, L., Rock, K., Witte, D. et al. (2000). Hyper-X engine testing in the NASA Langley 8-foot high temperature tunnel. *Paper AIAA 2000-3605. 36th AIAA/ASME/SAE/ASEE Joint Propulsion Conference and Exhibit*, Las Vegas, NV (24–28 July 2000).

Huebner, L.D., Rock, K.E., Ruf, E.G. et al. (2001). Hyper-X flight engine ground testing for X-43 flight risk reduction. *AIAA Paper 2001-1809. 10th AIAA/NAL-NASDA-ISAS International Space Planes and Hypersonic Systems and Technologies Conference*, Kyoto, Japan (24–27 April 2001).

Hunt, J.L. and McClinton, C.R. (1997). Scramjet engine/airframe integration methodology. *AGARD Conference*, Palaiseau, France (14–16 April 1997).

Hunt, R., Ground, C.R., Baurle R.A., and Danehy, P. (2019). Using computational flow imaging to optimize filtered Rayleigh scattering measurements of an isolator shock train. *AIAA 2019-4016. AIAA Propulsion and Energy 2019 Forum*, Indianapolis, IN (19–22 August 2019).

Leonov, S., Sabelnikov, V., Vincent-Randonnier, A., Firsov, A., and Yarantsev, D. (2014). Mecanism of flameholding in plasma-assisted supersonic combustor. *29th Congress of the International Council of the Aeronautical Sciences (ICAS 2014)*, SAINT PETERSBOURG, Russia (September 2014). hal-01111451.

Marshall, L.A., Corpening, G.P., and Sherrill, R. (2005a). A chief engineer's view of the NASA X-43A scramjet flight test. *AIAA/CIRA 13th International Space Planes and Hypersonics Systems and Technologies Conference* (January 2005).

Marshall, L.A., Bahm, C., Corpening, G.P. and Sherrill, R. (2005b). Overview with results and lessons learned of the NASA X-43A Mach 10 flight. *AIAA/CIRA 13th International Space Planes and Hypersonics Systems and Technologies Conference* (January 2005).

McClinton, C.R., Holland, S.D., Rock, K.E. et al. (1998). Hyper-X wind tunnel program. *NASA/TM-1998-207317, AIAA Paper 98-0553. 36th AIAA Aerospace Sciences Meeting and Exhibit*, Reno, NV (12–15 January 1998).

Norris, A.T. (2017). Evaluation of the Uncertainty in JP-7 Kinetics Models Applied to Scramjets. *2017 JANNAF - Interagency Propulsion Committee meeting*, Newport News, VA (4 December 2017).

Northam, G.B., Andrews, E., Guy, W. et al. (2007). An overview of hypersonic propulsion research at NASA Langley Research Center. *20th Anniversary Meeting and 8th Symposium on Propulsion System for Reusable Launch Vehicles, Japan Society for Aeronautical and Space Sciences*, Sendai, Japan (7 March 2007).

Pellet, G.L., Bruno, C., and Chinitz, W. (2002). Review of air vitiation effects on scramjet ignition and flameholding combustion processes. *38th AIAA/ASME/SAE/ASEE Joint Propulsion Conference and Exhibit*, Indianapolis, Indiana (7–10 July 2002).

Rausch, V.L., McClinton, C.R., Sitz, J., and Reukauf, P. (2000). NASA's Hyper-X program. *Paper No. IAF-00-V.4.0. 51st International Astronautical Congress*, Riode Janeiro, Brazil (2–6 October 2000).

Roffe, G., Bakos, R.J., Erdos, J.I., and Swartwout, W. (1999). The propulsion test complex at GASL. *ISABE 97-7096. International Symposium on Air Breathing Engines*, Florence, Italy (5–10 September 1999).

Rogers, R.C., Shih, H.T., and Hass, N.E. (2005). Scramjet development tests supporting the Mach 10 flight of the X-43. *AIAA Paper 2005-3351. AIAA/CIRA 13th International Space Planes and Hypersonics Systems and Technologies Conference*, Capua, Italy (16–20 May 2005).

Seleznev, R.K., Surzhikov, S.T., and Shang, J.S. (2019). A review of the scramjet experimental data base. *Progress in Aerospace Sciences* 106: 43–70.

Vincent-Randonnier, A., Moule, Y., and Ferrier, M. (2014). Combustion of hydrogen in hot air flows within LAPCAT-II Dual Mode Ramjet combustor at Onera-LAERTE facility-experimental and numerical investigation. *AIAA 2014-2932. 19th AIAA International Space Planes and Hypersonic Systems and Technologies Conference*, Atlanta, Georgia (16–20 June 2014). https://doi.org/10.2514/6.2018-5208. hal-02420953.

Voland, R. and Rock, K. (1995). NASP Concept demonstration engine and subscale parametric engine tests. *AIAA Paper 95-6055. International Aerospace Planes and Hypersonics Technologies*, Chattanooga, TN (3–7 April 1995).

Voland, R., Rock, K., L. Huebner, et al. (1998). Hyper-X engine design and ground test program. *Paper AIAA 1998-1532. 8th AIAA International Space Planes and Hypersonic Systems and Technologies Conference*, Norfolk, VA (27–30 April 1998).

Woods, W.C., Holland, S.D., and DiFlubio, M. (2001). Hyper-X stage separation wind-tunnel test program. *Journal of Spacecraft and Rockets* 38 (6): 811–819.

12

Analysis, Computational Modeling, and Simulation

Hypersonic flows are characterized by strong shocks, high enthalpy, thick boundary layers, chemical reactions, turbulence, transitional flows, shock–boundary interactions, and a myriad of other thermophysical interactions. Such complex flows therefore present a significant challenge for analysis and modeling. These challenges are magnified in a hypersonic air-breathing vehicle, as its design imposes a different set of demands on computational tools than does a rocket-powered vehicle. The dominant requirements for the hypersonic air-breathing vehicles result from the close integration of the propulsion with the airframe. An integration driven by very small performance margins requires ac highly accurate prediction of vehicle aerodynamics and engine performance. Such prediction effort is compounded by the requirement to optimize such performance over a Mach number regime, and especially when we impose vehicle reusability as an operational requirement. Figure 12.1 illustrates the complexity of the analysis effort that was carried out to develop the first hypersonic vehicle propelled by a scramjet engine, autonomously flying to Mach 10. This iconic figure represents the advances in analytical and simulation tools that were required for this monumental achievement in 2004.

In hypersonic air-breathing propulsion, we deal with significant temperature and flow nonuniformities arising from combustion, turbulence, and shock waves. The close integration of airframe and propulsion produces highly interacting, complex three-dimensional (3-D) viscous external and internal flows, thus further complicating the prediction efforts. Accurate computational prediction of hypersonic air-breathing system performance and operability limits must account for airframe and engine–component coupling. Inability to fully explore scramjet performance over its entire operational envelope before flight introduces high risk into flight test programs. As a result, the design and development of scramjet engines relies on computational fluid dynamics (CFD) to predict engine performance more than ever before.

Figure 12.2 summarizes phenomena of great importance to resolve numerically. Physical models of these phenomena must yield the wall properties (pressure, skin friction, and heat transfer) for accurate prediction of forebody aero and heating loads. At the same time, the air mass and momentum flux entering the engine must be correctly calculated to predict propulsion performance. The physics of high-speed combustion span a wide range of phenomena that are challenging to model accurately. These include the interaction between

Scramjet Propulsion: A Practical Introduction, First Edition. Dora Musielak.

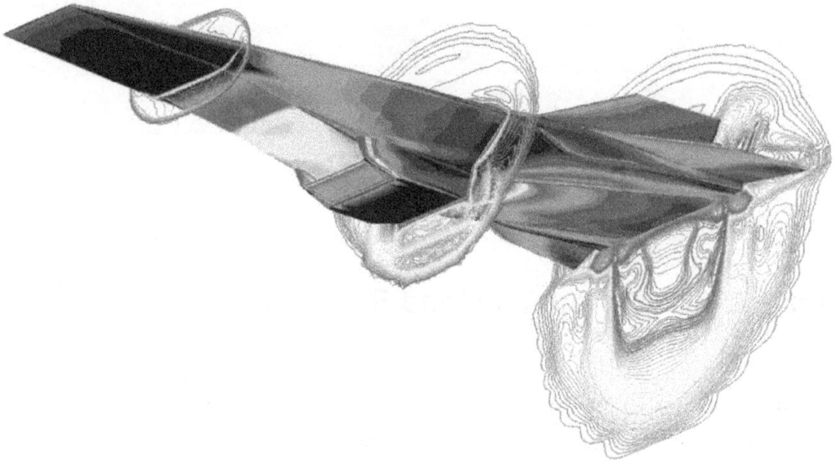

Figure 12.1 Tip-to-tail flowfield pre-flight simulation performed for X-43A vehicle with scramjet operating at Mach 7. Source: NASA image ED97 4396801; from Jentink and Bittner (1997)/Public Domain.

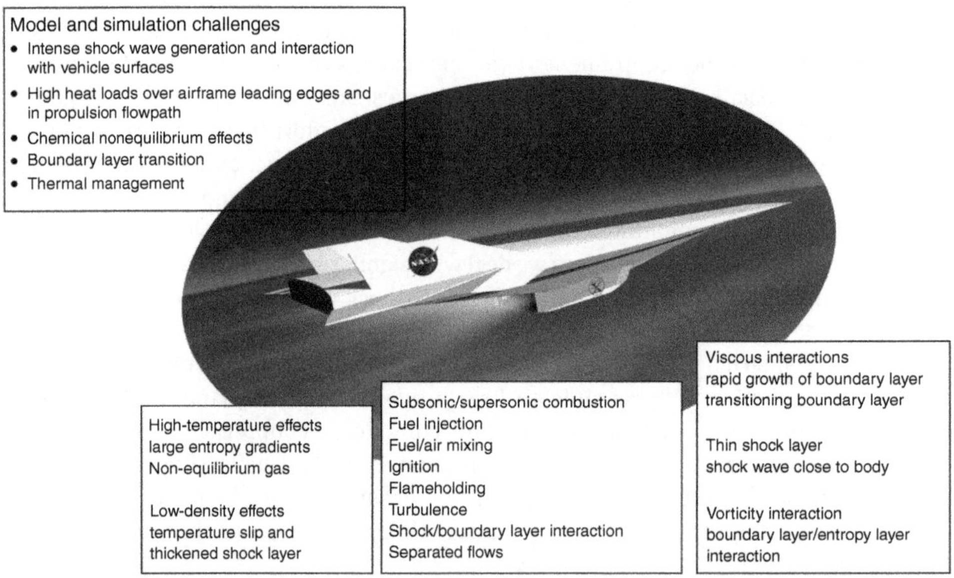

Figure 12.2 Important physical phenomena for fully-integrated hypersonic air-breathing vehicle. Source: NASA Dryden Flight Research Center, Image ED97 43968-03.

chemical kinetics, fuel injection, fuel–air mixing, thick boundary layers, high-intensity turbulence, shock–boundary layer interactions, and strong flow compressions and expansions. Modeling the scramjet engine exhaust is also challenging. For example, analyzing the X-43A aftbody flowfield contours (Figure 12.1) suggest to us that the exhaust plume

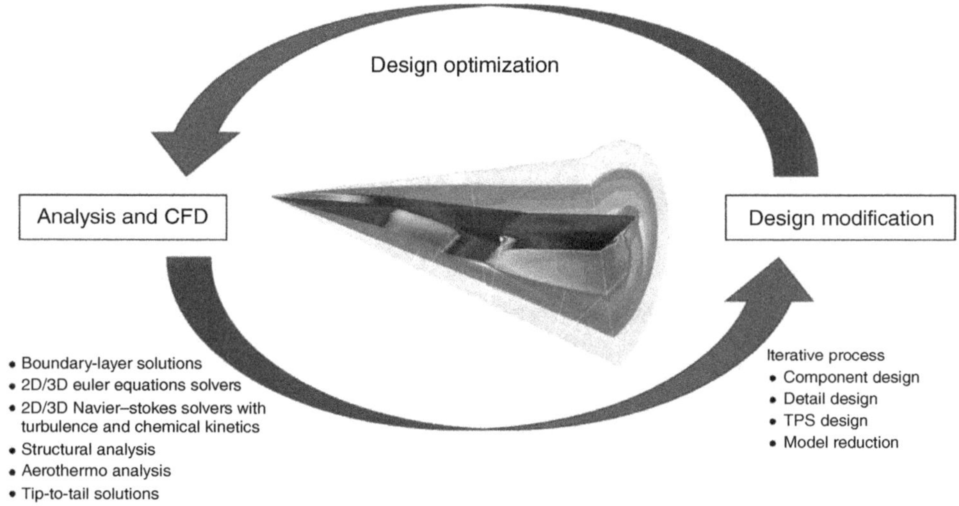

Figure 12.3 CFD is a tool for design analysis and optimization.

(modeled as a single species thermally perfect gas) expands beyond the boundaries of the external nozzle surface. This creates uncertainty in the force accounting assumptions that must be used to couple propulsion flowpath and airframe CFD predictions.

For air-breathing hypersonics, predictive methodologies – CFD and other analysis tools – must encompass a wide range of modeling capabilities to capture all of the relevant flow physics of the complete scramjet flowpath as well as the external airframe. This analysis is normally accomplished using a multilevel approach, increasing in complexity and fidelity as the design is matured. The preliminary analysis phase may employ different tools for the various flowpath components, which necessitates the development of force accounting systems appropriate for specific configurations. Figure 12.3 illustrates the interactive nature of vehicle design optimization and the range of computational solvers that must be used.

It may be helpful to begin with a quick overview of CFD modeling, just enough to put in perspective the tools that are more appropriate for simulating the complex flow phenomena and physical process inherent in hypersonic air-breathing propulsion-integrated vehicles.

12.1 Overview of Computational Fluid Dynamics and Turbulence

CFD is the analytical tool of choice for simulation of continuum flowfields from subsonic to hypersonic conditions, using a wide range of mathematical models and numerical methods (discretization methods, solvers, numerical parameters, and grid generations, etc.) to characterize complex flow phenomena. CFD codes solve the appropriate governing equations, including allowances for both chemical and thermal nonequilibrium processes, coupled with a wide variety of turbulence models. Robust CFD codes developed for the design and analysis of high-speed propulsion systems, together with massively parallel computing capabilities, make it possible to perform simulations of complete engine flowpaths. Results from CFD analysis allow us to investigate trends observed during ground testing, and we use those results to extract various measures of engine efficiency.

For many viscous flow problems where boundary-layer equations are not applicable, it is possible to solve a reduced set of Navier-Stokes (NS) equations that, in terms of complexity, fall between complete multidimensional NS equations and boundary-layer equations. These reduced equations belong to a class of equations referred to as "thin-layer" (TLNS) or "parabolized" Navier–Stokes (PNS). Sets of equations are applicable to both inviscid and viscous flow regions. Having fewer terms to compute leads to reduction in required computation time. For steady flows, all PNS equations (except TLNS) are a mixed set of hyperbolic–parabolic equations in the streamwise direction. The NS equations are "parabolized" in the streamwise direction. A substantial reduction in computation time and storage is achieved when using PNS models.

12.1.1 Turbulence and Computational Approaches

In simple terms, flow turbulence refers to the unsteady, irregular (aperiodic) motion in which transported quantities (mass, momentum, and scalar species) fluctuate in time and space. Turbulent flow is identifiable by the swirling patterns that characterize turbulent eddies. Turbulence leads to enhanced mixing (matter, momentum, energy, etc.). In turbulent flow, fluid properties and velocity exhibit random variations, and the statistical averaging results in accountable, turbulence-related transport mechanisms. It is this characteristic that allows for turbulence modeling.

Turbulent flow contains a wide range of turbulent eddy sizes (scales spectrum). The size/velocity of large eddies is on the order of mean flow. Engineering analysis of turbulence may use three distinct computational methods for solving the NS equations: Reynolds-Averaged Navier-Stokes Simulation (RANS), Direct Numerical Simulation (DNS), and Large Eddy Simulation (LES).

RANS is a method that solves ensemble-averaged (or time-averaged) NS equations. RANS include a form of NS equations in which additional terms (Reynolds stresses) are included to account for time-averaged effects of turbulence. With this approach, all turbulent length scales are modeled. RANS is the most widely used approach for computing hypersonic flows.

DNS is a method in which turbulent flow is directly simulated by numerically solving the full N-S equations, without any form of time or length averaging, i.e. both mean flow and all turbulent fluctuations (eddies) are simulated. Since turbulent eddies are both 3-D and unsteady (time-variant), simulations using DNS must also be both 3-D and unsteady. Moreover, since length and time scales of turbulent eddies cover a large range, both grid size and time-step size must be very small to account for the smallest fluctuations (resolves the whole spectrum of scales). Hence, DNS is computationally expensive and is only practical for simple flows at low Reynolds numbers. DNS is not practical for hypersonic flows

LES solves the spatially averaged N-S equations. LES reduces the computational cost by ignoring the smallest length scales, which are the most computationally expensive to resolve, via low-pass filtering of the Navier–Stokes equations. Filtering the equations is done to dissipate the minor scale vortices via a subgrid model (Sagaut 2006). A low-pass filtering, which can be viewed as a time- and spatial averaging, effectively removes small-scale information from the numerical solution. Such information is not irrelevant, and its effect on the flowfield must be modeled. This feature of LES is an active area of research for problems in which small scales play an important role, such as near-wall flows, reacting flows, and multiphase flows.

LES turbulence modeling gives a compromise between DNS and RANS. Unsteady flow equations are solved for mean flow and larger eddies, and a subgrid scale model is used to simulate effects of smaller eddies. Since the largest eddies contain most energy and interact most strongly with the mean flow, LES approach results in a good model of main effects of turbulence. Grid size does not need to be small enough to allow for smallest turbulent eddies, so LES is much less computationally expensive than DNS and may be applied to a wider range of flows. However, time-dependent simulations using relatively fine meshes are still necessary, so computational requirement remains high. LES modeling shows that its results are in better agreement with some experiments compared to RANS if the grid resolution is sufficiently fine. However, one has to assess with a priori knowledge of the flow characteristics what grid resolution is actually optimum.

Detached Eddy Simulation (DES) is a hybrid simulation strategy, a modification of a RANS in which the model switches to a subgrid scale formulation in regions fine enough for LES calculations. Regions near solid boundaries and where the turbulent length scale is less than the maximum grid dimension are assigned the RANS mode of solution. As the turbulent length scale exceeds the grid dimension, the regions are solved using the LES mode. Therefore, the grid resolution is not as demanding as pure LES, thereby considerably reducing the cost of the computation.

Hybrid RANS/LES models were developed to permit much more freedom in the placement of RANS and LES subdomains in a computation, so that the more costly LES need only be performed where the RANS model is inadequate to capture the necessary physics, such as in turbulent boundary layers that exhibit inherent unsteadiness, turbulent shock–boundary layer interactions, separation, and other complex flow phenomena. In general, hybrid RANS/LES methods involve modifications to existing Reynolds-averaged closures to provide a scale-resolving functionality in regions of the computational domain with sufficient grid fineness to support the resolution of turbulent structures. This method attempts to resolve a fraction of the turbulence in free shear flows and possibly the outer portion of turbulent boundary layers. Many hybrid RANS/LES models are represented by some blend between the equations governing the Reynolds-averaged and spatially filtered NS equations (or a subset of these equations). For example, Choi et al. (2008) and Gieseking et al. (2011) obtain their hybrid formulation by blending of the turbulent viscosity. Georgiadis et al. (2011) developed a hybrid method for compressible mixing layer simulations. Using RANS, they calculated wall-bounded regions entering a mixing section, and LES to calculate the mixing dominated regions.

Improved Delayed Detached Eddy Simulation (IDDES). Finding that the interface between the RANS and LES models lead to incorrect predictions of turbulent wall parameters, Shur et al. (2008) introduced a hybrid model called Improved Delayed Detached

Eddy Simulation (IDDES), intended to correct this problem. In this strategy, they combined the one-equation Spalart–Allmaras (S-A) model (Spalart and Allmaras 1992) to model only the wall boundary layers. A wall destruction term was used to reduce the turbulent viscosity in the laminar sublayer and logarithmic sublayer, providing a smooth transition from laminar to turbulent statuses. IDDES redefines the subgrid length scale to depend on local cell sizes and the wall distance, and then it uses a shielding function to avoid deteriorating the detection of the boundary-layer edge. By doing that IDDES ensures that the RANS can fully model the boundary layer.

The wall-modeled LED part of the IDDES method is intended to resolve issues related to the log-layer mismatch and underprediction of skin friction (Shur et al. 2008). However, some researchers have found difficulties applying this model. Work reported by Han et al. (2020) revealed a flaw in IDDES resulting from the LES model, which cannot fully resolve the energetic, large-scale, and stress-carrying turbulent structures in the vicinity of the RANS/LES interface. Thus, Han et al. (2020) proposed a modification to IDDES, essentially by moving the RANS/LES interface further away from the wall so that, at the RANS/LES boundary, the wall-parallel grid size is sufficient to resolve the turbulent structures. They found that the modified IDDES model did solve the log-layer mismatch and underprediction of skin friction problem, concluding that the new strategy is applicable to flat plate boundary layer flows at low-, moderate-, and high-Reynolds numbers. We should note that IDDES has been successfully applied for modeling the wall boundary layers of scramjet combustors (Yao et al. 2021).

Although hybrid RANS-LES and IDDES methods have made analysis of scramjet flowpath more affordable, especially to simulate hydrocarbon dual-mode combustors, there is not yet a single, practical turbulence model that can reliably predict *all* turbulent flows with sufficient accuracy.

12.1.2 RANS Modeling – Time Averaging

The Navier–Stokes are nonlinear equations governing the velocity and pressure of a fluid flow. For turbulent flow, each of these quantities is decomposed into a mean part and a fluctuating part. Averaging the equations gives the RANS equations, which govern the mean flow. However, the nonlinearity of the Navier–Stokes equations means that the velocity fluctuations still appear in the nonlinear term arising from the convective acceleration. This term is known as the Reynolds stress. Its effect on the mean flow is like that of a stress term, such as from pressure or viscosity.

Therefore, to obtain equations containing only the mean velocity and pressure, we need to close the RANS equations by modeling the Reynolds stress term as a function of the mean flow, removing any reference to the fluctuating part of the velocity. This is referred to as the closure problem. In 1877, French mathematician J.V. Boussinesq proposed relating the turbulence stresses to the mean flow to close the system of equations, introducing the concept of eddy viscosity. Eddy viscosity is the proportionality factor describing the turbulent transfer of energy as a result of moving eddies, giving rise to tangential stresses. Applying the Boussinesq hypothesis to model the Reynolds stress term in the NS equations, it introduces a new proportionality constant, the turbulence eddy viscosity. Such turbulence models are known as eddy viscosity models (EVM).

The Reynolds-averaged momentum equations can be written as

$$\rho\left(\frac{\partial \overline{u}_i}{\partial t} + \overline{u}_k \frac{\partial \overline{u}_i}{\partial u_k}\right) = -\frac{\partial \overline{p}}{\partial x_i} + \frac{\partial}{\partial x_j}\left(\mu \frac{\partial \overline{u}_i}{\partial x_j}\right) + \frac{\partial R_{ij}}{\partial x_j}$$

where the Reynolds stress tensor is $R_{ij} = -\rho\overline{u_i' u_j'}$

The Reynolds stresses are additional unknowns introduced by the averaging procedure; hence, they must be modeled (related to the averaged flow quantities) in order to close the system of governing equations.

The RANS equations can be closed via the Boussinesq hypothesis, which lead to EVMs. In these models, the Reynolds stresses are modeled using an eddy (or turbulent) viscosity

$$R_{ij} = -\rho\overline{u_i' u_j'} = \mu_T\left(\frac{\partial \overline{u}_i}{\partial u_j} + \frac{\partial \overline{u}_j}{\partial u_i}\right) - \frac{2}{3}\mu_T \frac{\partial \overline{u}_k}{\partial u_k}\delta_{ij} - \frac{2}{3}\rho k \delta_{ij}$$

where μ_T is the eddy or turbulent viscosity.

The Boussinesq hypothesis is reasonable when applied to simple turbulent shear flows such as boundary layers, round jets, mixing layers, and channel flows. Closure can also be done via transport equations for Reynolds stresses, yielding the so-called Reynolds-stress models (RSM). Modeling is still required for many terms in the transport equations, and thus models are more complex and computationally intensive. However, in 3-D turbulent flows with large streamline curvature and swirl, RSM are more advantageous although more difficult to converge than EVM.

12.1.3 Selection of Turbulent Model

There are many closure models for the Reynolds-averaged equations, varying in complexity from simple algebraic (zero-equation) models, which require specification of a turbulent velocity and length scale, to full second-order closures that involve transport equations for the Reynolds stress tensor and flux vectors. The turbulence closure models typically used in hypersonics air-breathing simulations include two-equation models and "complete" one-equation models. The most common models that use Boussinesq hypothesis and offer a relatively low-cost computation for the turbulence viscosity are the Spalart-Allmaras one-equation model, the Menter $k - \omega$ (k–omega) two equation models, where k is the turbulence kinetic energy and ω is the turbulence frequency, the $k - \varepsilon$, (k-epsilon) where ε is the turbulence dissipation rate (one of the most common two-equation models), and the Wilcox k-omega models. There are three variants of Menter k-omega: the baseline (BSL), shear stress transport (SST) models, and a modified SST model, where the shear stress limiter (a parameter to calculate the turbulence shear stress) was altered. These and other models and their common implementations are described on the NASA Langley Research Center Turbulence Modeling Resource website (Rumsey 2017).

Each turbulence model calculates the eddy or turbulent viscosity μ_T differently, as it can be determined from a turbulence time scale (or velocity scale) and a length scale.

The Spallart–Allmaras model solves a transport equation for a modified turbulent viscosity denoted by $\mu_T = f(\tilde{\nu})$. The standard $k - \varepsilon$, and RNG $k - \varepsilon$ solve transport equations for k and ε with $\mu_T = f(\rho k^2/\varepsilon)$, while the standard $k - \omega$ and SST $k - \omega$ define $\mu_T = f(\rho k/\omega)$.

Renormalization Group (RNG) is a mathematical technique that can be used to derive a turbulence model, similar to the k-epsilon model. The RNG methods developed by Yakhot et al. (1992) renormalize the NS equations to account for the effects of smaller scales of motion. In the standard k-epsilon model, the eddy viscosity is determined from a single turbulence length scale, so the calculated turbulent diffusion is that which occurs only at the specified scale, whereas in reality all scales of motion will contribute to the turbulent diffusion. The RNG approach results in a modified form of the epsilon equation which attempts to account for the different scales of motion through changes to the production term.

Which turbulence model is more appropriate to simulate supersonic combustion?

A number of scramjet analyses have found that the underlying turbulence model and modeling of scalar transport (i.e. settings for the turbulent Prandtl number and Schmidt number) have substantial effects on flow path predictions (Georgiadis et al. 2011). The turbulent Schmidt number is a dimensionless number defined as the ratio of momentum diffusivity and mass diffusivity, $Sc_t = \mu_t/D_t$. This number is used to characterize fluid flows in which there are simultaneous momentum and mass diffusion convection processes.

The application of CFD using the RANS model to turbulent flows with mass transfer requires estimating the turbulent scalar flux, assuming the gradient diffusion hypothesis, which requires definition of the turbulent Schmidt number, Sc_t. Many CFD studies assume $Sc_t = 0.7$. However, no universal value of Sc_t has been established, and empirical values are typically used.

While applying the Wind-US RANS solver to the HIFiRE-2 scramjet ground test configuration, Georgiadis et al. (2014) found that the Menter-BSL model produced the highest pressures, and the Menter-SST model produced the lowest pressures. Moreover, using small values for the Sc_t enabled more rapid mass transfer, faster combustion, and higher flowpath pressures. Their conclusion was that optimal settings for turbulence model and Sc_t are rather case dependent. The same conclusion has been drawn in other scramjet flow simulations.

Turbulence–chemistry interactions are also of great importance and must be investigated when simulating the reacting flows in ram/scramjet combustors. Steady-State Reynolds-Averaged Simulations (RAS) for scramjet engine development are typically used to determine fuel–air mixing, combustion flowfield, and to determine optimal fuel injector configurations. However, RANS present some limitations. There are cases where RANS models have failed to even qualitatively mimic the fundamental flow physics present in scramjet engines. For combustion simulation problems in particular, Direct Numerical Simulation (DNS) and LES are higher order numerical approaches which offer significant advantages that overcome many of the shortcomings associated with RANS.

Many original ideas regarding simulation of turbulent reacting flows were developed in the CFD Laboratory at the University at Buffalo led by Prof. Peyman Givi. In 2001, Tom Drozda first conceived Figure 12.4 to compare his LES and RANS models of a turbulent

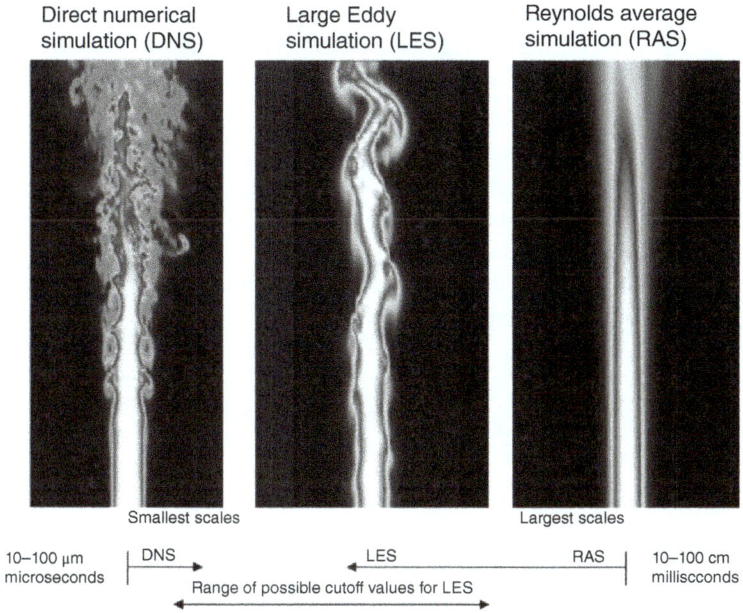

Figure 12.4 Turbulent jet flame simulated with DNS, LES, and RANS. Source: Used with permission of Prof. Peyman Givi.

jet flame with the DNS of the same flame model created by Laurent Ciquel, also a member of the CFD Lab research at that time.

The effect of the three numerical approaches on the final solution are made clear in the comparison of the flame structure obtained with the three simulation methods. As shown in Figure 12.4, with DNS all scales of fluid's motion are fully resolved; with LES, the large scales of fluid's motion are resolved, while small scales are averaged; with RAS, all flow scales of fluid's motion are averaged in time. This illustration also emphasizes that LES requires less computational effort than DNS, while delivering more detail than the less expensive RANS.

It is important to note that, in the study of turbulent combustion in supersonic flow, the most standard closures for combustion modeling, which are based on the fast chemistry approximation, are not appropriate since supersonic combustion is largely governed by finite-rate chemistry and self-ignition phenomena. In a scramjet combustor, chemical reaction timescales tend to have the same order of magnitude as turbulent timescales, which means that the Damköhler number (Da) values approach unity, that is, the characteristic turbulent time scale τ_t is about the same order as the chemistry time scale τ_c. Thus, researchers seek finite-rate chemistry-based closures to describe turbulent supersonic combustion.

12.1.4 Representation of the Flame Structure in Turbulent High-Speed Flow

Visualization of unsteadiness in a flowfield is done by identifying coherent vortical structures. The vorticity magnitude might be used to attempt to visualize these structures either through a maximum vorticity threshold or isosurfaces of vorticity. Vortices can be identified using the Q-criterium to discern vortices based on the distinction it makes between shear and swirling flow. The first 3-D vortex criterion is known as the Q-criterion of Hunt et al. (1988), which defines a vortex as a spatial region where the Euclidean norm of the vorticity tensor dominates that of the rate of strain, that is,

$$Q = 0.5\left(\omega^2 - S^2\right) > 0$$

an expression to represent the relative dominance of the rotational component over the stretching component in deformation of a fluid element.

Formally, the velocity gradient tensor ∇u is decomposed into the symmetric rate of strain tensor S and antisymmetric rate of rotation tensor, Ω, as

$$\nabla u = S + \Omega$$

where

$$S = \frac{1}{2}\left[\nabla u + (\nabla u)^{\mathrm{T}}\right], \quad \Omega = \frac{1}{2}\left[\nabla u - (\nabla u)^{\mathrm{T}}\right]$$

The Q-criterion is then defined as

$$Q = \frac{1}{2}\left[\|\Omega\|^2 - \|S\|^2\right]$$

In simple flows, especially simulations in two-dimensions, contours of $Q > 0$ are often used to define vortices, interpreted as where local rotation dominates over local strain. In more complex flows, notably 3-D or turbulent experimental flows, contours of a certain percentage of Q_{max} are often used, for example giving $0.1Q_{max}$.

Scherrer et al. (2016) published an instantaneous representation of the flame structure obtained from their hydrogen-fueled supersonic combustion experiments at ONERA. We reproduce in Figure 12.5 their snapshot of the instantaneous temperature field superimposed with a H_2 mass fraction iso-surface (white). In the bottom image, the Q-criterion and OH mass fraction iso-surfaces are represented, both colored by temperature. The researchers noted four regions in the flame structure: the induction zone ($0 < X/D < 10$), the auto-ignition zone $10 < X/D < 18$, the stabilization region ($18 < X/D < 26$), where the flame anchors at the beginning of a shock diamond, and the end of the combustion zone ($30 < X/D < 34$). For this study, all the computations were performed with the ONERA in-house code CEDRE.

12.1.5 Flamelet Models for Turbulent Combustion

Flamelet models are increasingly popular for analysis of turbulent combustion. The flamelet method models turbulent reactions, assuming that chemical time scales are shorter than turbulent time scales, such that flame in a combustor can be approximated as an ensemble of laminar flamelets. Flamelet models present an alternative to traditional finite rate chemistry approaches for CFD (where reaction rates are determined by Arrhenius kinetic expressions, as shown in Chapter 5). The flamelet approach allows chemistry to be computed independently of the flow simulation and stored in tables as functions of a small number of scalars. During CFD simulation, thermochemical quantities are interpolated from a

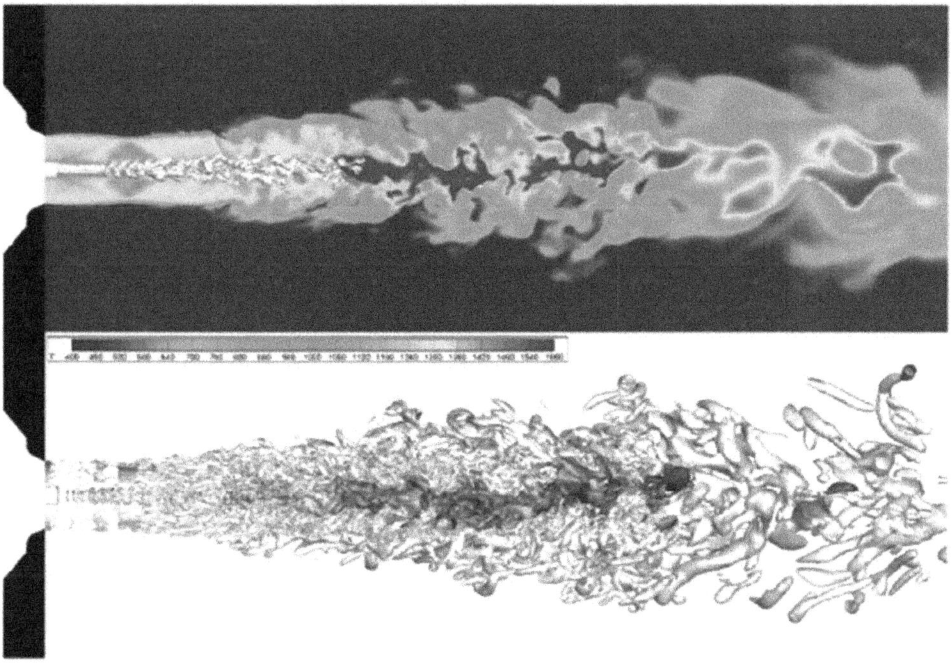

Figure 12.5 Flame structure – instantaneous field of temperature and H_2 mass fraction iso-surface [0.05] (top) - Iso-surfaces of Q-criterion [1×10^9 (s^2)] and OH mass fraction [0.01] colored by temperature (bottom). Source: From Scherrer et al. (2016)/ONERA.

chemistry table, dramatically decreasing overall computational cost, and allowing use of complex chemical mechanisms.

The flamelet model conceives the turbulent flame as an aggregate of thin, laminar, locally one-dimensional (1-D) flamelet structures present in the turbulent flow field (Peters 1986). When the chemistry of the flame is assumed to occur much faster than other flow transport processes, the scalar dissipation rate provides the only means of influence for the fluid mechanics on the flamelet structure.

If those conditions are met, flamelet equations are derived and used to construct a multidimensional manifold prior to incorporating it into a CFD simulation. This manifold is usually parameterized by a number of scalar variables and tabulated, so that these scalars may be accessed at runtime to retrieve relevant thermochemical data. Such an approach requires solving transport equations only for the parameterizing variables instead of solving the much larger system of chemical species transport equations. This results in a significantly less-stiff and less computationally expensive system of governing equations to solve.

To determine the applicability of a flamelet model for a turbulent combusting flow requires one to consider the extent to which the flowfield of interest meets the fundamental assumptions of flamelet modeling. For example, for nonpremixed combustion, in which the flamelet is attached to the surface of a stoichiometric mixture fraction, and for which the scalar dissipation rate couples the flame dynamics to that of the fluid dynamics, the characteristic chemical time scale must be considerably smaller than that of the representative diffusive and turbulent transport processes. This means that the characteristic turbulent time scale τ_t must be greater than the chemistry time scale τ_c. Hence, the Da must be

$$\mathrm{Da} = \frac{\tau_t}{\tau_c} > 1$$

The Da is used to relate chemical reaction timescale to other phenomena occurring in a system. In analysis and simulation of scramjet combustion flowfields, the Da number is used for determining whether diffusion rates or reaction rates are more important for defining a steady-state chemical distribution over the length and time scales of interest.

In premixed combustion, the chemical time scale and thermal diffusivity effectively govern the flame thickness. In the flamelet regime, the flame thickness must be considerably smaller than the representative turbulent length scales. Thus, for premixed turbulent flames, the Karlovitz (Ka) number should be much less than unity. The Ka describes the physical interaction of flow and combustion on the smallest turbulent scales, and thus is used as a measure of the curvature or shape of the flame.

The Ka (named after Hungarian engineer Béla Karlovitz) is defined as the ratio of chemical time scale τ_c to the smallest turbulent time scale τ_k (Kolmogorov scale):

$$\mathrm{Ka} \equiv \frac{\tau_c}{\tau_k}$$

Kolmogorov microscales are the smallest scales found in turbulent flow. At the Kolmogorov scale, viscosity dominates and the turbulent kinetic energy is dissipated into heat. Hence, when $\mathrm{Ka} \ll 1$ the chemical reactions occur much faster than all turbulent scales. The Karlovitz and Damköler numbers are related as

$$\mathrm{Ka} = \frac{1}{\mathrm{Da}}$$

The first step in characterizing the combustion fields is to devise a metric indicative of flame activity, which can be used to identify regions of chemical reactivity. Once the data are filtered using the flame index, a flame-weighted Takeno index is calculated to distinguish regions of premixed and nonpremixed fuel/air conditions. The Takeno index is a product of the gradient of the fuel mass fraction and the gradient of the oxygen mass fraction, giving information on whether a flame regime is premixed or of the diffusion-type.

To apply the flamelet model for simulating turbulent combustion requires to estimate local Da and Ka numbers, using, for example, the approach outlined by Peters (2000). With these values, the next step is to construct Borghi diagrams for the premixed combustion, and proxy combustion diagrams are devised for the nonpremixed combustion using the flame-weighted Takeno index and Da number. Borghi diagrams define regimes of premixed turbulent combustion in terms of velocity and length scale ratios.

Combustion regimes are typically represented on a diagram that qualitatively indicates the turbulence intensity versus turbulence length scale of a combustion system. Such diagram is called Borghi diagram, after the author who first introduced it (Borghi 1984). As shown in Figure 12.6, the upper left corner represents the combustion conditions of the well-stirred reactor, while the lower left corner represents the laminar flame regime.

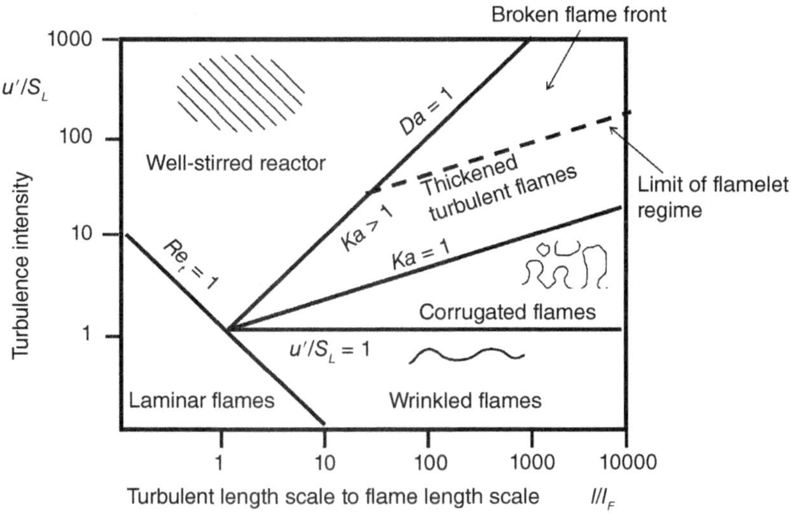

Figure 12.6 Borghi diagram.

The well-stirred reactor model is that in which the residence time is coupled with the local strain rate.

The boundary between laminar and turbulent flames is set at a turbulent Reynolds number $Re_t = 1$, although the precise value depends on the geometry of the flow domain. The remaining four combustion regimes of the Borghi diagram are used to identify the flamelet regime. The wrinkled and corrugated flames belong to the flamelet regime, characterized by $Re_t > 1$ (i.e. turbulence), Da > 1 (i.e. fast chemistry), and Ka < 1 (i.e. weak flame stretch). For premixed combustion, if Ka < 1, the turbulent flame is known as corrugated flamelets and wrinkled flamelets.

When Da > 1, the time scale of the macro structure is large in comparison with the chemical time scale. This means that (i) the macro structure is not rapid enough to destroy the laminar flame structure to such a degree that the laminar burning velocity becomes an irrelevant parameter, and (ii) the chemistry is so fast that every change in the flame shape due to the large eddies is being reflected in the turbulent burning rate, as the flame propagates normal to itself. Conversely, if the chemical time scale is smaller than the turbulence time scale, the flame may have changed its shape many times before any significant amount of reactants has been consumed.

In the flamelet regime, if $u' < S_L$, where u' can be considered the rotation speed of the largest eddies, the turbulence wrinkles the flame front, and the turbulent burning velocity is largely determined by laminar flame propagation. In such condition with a weak influence of the flowfield, the average flame thickness will be greater than in the laminar case. The turbulent burning velocity associated with this flame regime can be assessed by the Karlovitz number for large scale, low-intensity turbulence.

Since flamelet models rely on precomputing the entire thermochemical statespace, the total simulation cost is largely insensitive to complexity of the chemical reaction mechanism employed, it makes LES affordable for high-speed, compressible, reacting flows applications. RAS utilize a finite-rate chemical kinetics model from which typical quantities defining a flamelet, such as mixture fraction and progress variable, can be calculated. For example, Quinlan et al. (2014) simulated dual-mode combustion using the thermally perfect Viscous Upwind aLgorithm for Complex flow ANalysis (VULCAN-CFD) code and further characterized the flowfield using flame and combustion mode indices (see Sect. 12.5.1).

12.2 Surrogate-Based Analysis and Optimization (SBAO)

The successful full-scale development of hypersonic systems depends on the accuracy of the physical models employed and on the ability of the computational tools to simulate complex phenomena. Accurate, high-fidelity models are typically time consuming and computationally expensive. Hence, the computational analysis of the components of such systems (air-breathing engines and rocket propulsion) is expensive, making it challenging to search for optimal designs. One approach to reduce the computational burden is to approximate and replace expensive simulation models with a cheaper-to-run surrogate model. In the past decades, surrogate-based analysis optimization or SBAO tools have demonstrated to be effective for the design of computationally expensive models. SBAO can reduce the number of expensive, high-fidelity simulations by using a succession of approximate (surrogate) models, which are created from data or the results of submodels.

Surrogate models are increasingly required for applications in which first-principles simulation models are prohibitively expensive to employ for uncertainty analysis, design, or control. They can also be used to approximate models whose discontinuous derivatives preclude the use of gradient-based optimization or data assimilation algorithms.

12.2.1 Surrogate Modeling

Surrogate modeling uses inexpensive "surrogates" to represent the response surface of simulation models. The surrogates are constructed using data drawn from high-fidelity models and provide fast approximations of the objectives and constraints at new design points, thereby making sensitivity and optimization studies feasible. There are two broad families of surrogates: (i) statistical or empirical data-driven models emulating the responses of a high-fidelity simulation model, and (ii) lower fidelity physical-based surrogates, which are simplified models of the original system. Queipo et al. (2005) provide a comprehensive discussion of the fundamental issues that arise in surrogate-based analysis and optimization (SBAO), highlighting concepts, methods, techniques, as well as practical implications. They summarize the main steps in surrogate modeling:

1. Design of experiments (DOE). The design of experiment is the sampling plan in design variable space. Latin hypercube sampling (LHS), augmented by the corner points of the design space (full factorial design for four two-level factors), are used to efficiently fill the multidimensional design space. LHS a statistical method for generating a near-random sample of parameter values from a multidimensional distribution. For four design variables, the minimum number of data samples to form a fully determined linear system and to solve for polynomial coefficients. The key question in this step is how to assess the goodness of such designs, considering the number of samples is severely limited by the computational expense of each sample.

2. Compute numerical simulations at selected locations. In this step, the computationally expensive model is executed for all the values of the input variables in the DOE specified in the previous step. For example, in a scramjet flowpath design, RANS simulations can be run to obtain solutions to determine fuel injection thrust potential and combustion efficiency (see Sect. 12.5.3).

3. Construction of surrogate model. In this step, two questions are important: (a) what surrogate model(s) should be used (model selection), and (b) how to find the corresponding parameters (model identification). For example, using the DAKOTA code, four different surrogate models can be applied to fit the data: the quadratic and cubic polynomial models, the Kriging model (Gaussian process model with a reduced quadratic trend option), and the artificial neural network (ANN) model. The surrogate models are fitted to approximate both objective functions (such as thrust potential and combustion efficiency in a scramjet flowpath design) by minimizing the root-mean-squared errors. The overall quality of the fit can be evaluated by computing the fit residuals and 10-fold cross-validation (CV). CV is a technique used to estimate how well a given surrogate model performs when tested against data unused in the surrogate model construction.

4. Model validation. This step has the purpose of establishing the predictive capabilities of the surrogate model away from the available data (generalization error). The models are fitted to approximate both objective functions by minimizing the root-mean-squared errors. The overall quality of the fit can be evaluated by computing the fit residuals and 10-fold CV. In the 10-fold CV technique, the simulation data are divided into 10 partitions, and the surrogate model is computed using 9 of the partitions, with the 10th partition set aside for model CV.

An example of software utilized to carry out surrogate modeling is the Design Analysis Kit for Optimization and Terascale Applications (DAKOTA) code, developed at Sandia National Laboratory (DAKOTA 2022). The DAKOTA manages and analyzes ensembles of simulations to provide broader and deeper perspective for analysts and decision makers. According to the user manual, the DAKOTA toolkit provides a flexible, extensible interface between analysis codes and iteration methods. Surrogate models are categorized as data fitting or hierarchical. Each of these surrogate types provides an approximate representation of a truth model, which is used to perform the parameter to response mappings. This approximation is built and updated using results from the truth model, called the training data.

To date, SBAO methods have been applied successfully to supersonic turbines (Papila et al. 2002) and rocket propulsion components (Shyy et al. 2001) including propellant injectors (Vaidyanathan et al. 2004). For air-breathing hypersonics, NASA researchers are exploring the application of SBAO methods in the design of fuel injectors for high Mach number scramjet applications (see Section 12.5.3). SBAO is an active area of research, and its methods have made progress in addressing the analysis and optimization of a variety of aerospace systems. Although the surrogate-based approach for analysis and optimization can play a very valuable role in advancing the design of hypersonic systems, approximations generally have a limited range of validity. Hence, SBAO methods must be judged on a case-by-case basis.

12.3 Flowfield in Highly Integrated Hypersonic Air-breathing Vehicle

Figure 12.2 highlights some of the most important aerothermodynamics flow phenomena that required computational modeling and simulation to predict the performance of closely integrated hypersonic air-breathing vehicles. To get a sense of the analytical and computational modeling issues present in the flowfield of such vehicles, we consider the physics of the four major components: vehicle forebody, inlet/isolator, combustor, and nozzle/vehicle afterbody.

12.3.1 Vehicle Forebody

The hypersonic flowfield around the forebody includes the bow shock, the compression ramp shocks, laminar-transitional-turbulent boundary layers, and shock–boundary layer interaction, and associated flow separation.

Transition from laminar to turbulent flow at hypersonic speeds is not well understood. Accurate prediction for transition onset is needed to design the forebody and its thermal protection system (TPS). Some analytical methods require a transition onset database from quiet wind tunnels and flight tests. However, there is very limited flight test data available, and tunnel data may not be representative of flight conditions. Empirical transition prediction methods are typically used in the preliminary design of hypersonic vehicles. Real gas effects on the forebody flowfield become important as the flight Mach number increases. For aeroheating analysis, high-temperature effects should be handled with chemical equilibrium (CE) or chemical nonequilibrium (CNE) models.

For hypersonic cruise aircraft, SSTO or TSTO vehicle applications, the flowfield over the forebody will vary from subsonic to hypersonic, presenting a wide range of thermophysical phenomena, including the conditions at the walls, which must be modeled accurately to help design the TPS and determine the vehicle's performance to carry out its challenging mission. For reusable vehicles with nonablative surfaces, the wall boundary conditions are simplified as there is no need to consider surface ablation and associated shape change. However, wall boundary conditions are rather complicated when considering surface catalysis and thermal deformation of the forebody surface, a problem which becomes important at high hypersonic flight Mach numbers.

12.3.2 Inlet/Isolator

The inlet of a pure scramjet engine collects and process large amounts of atmospheric air. It is required to provide the precompressed air to the supersonic combustor, and it does it by utilizing the vehicle forebody surface. The inlet flowfield is characterized by compression shocks produced by the forebody ramps, which intersect the engine cowl leading edge to achieve the shock-on-lip condition (optimal inlet configuration). The compression shocks may produce extremely high pressure and heating on the cowl leading edge. Peak pressure and heating depend on the flow Mach number, Reynolds number, gas composition, and the impinging shock strength.

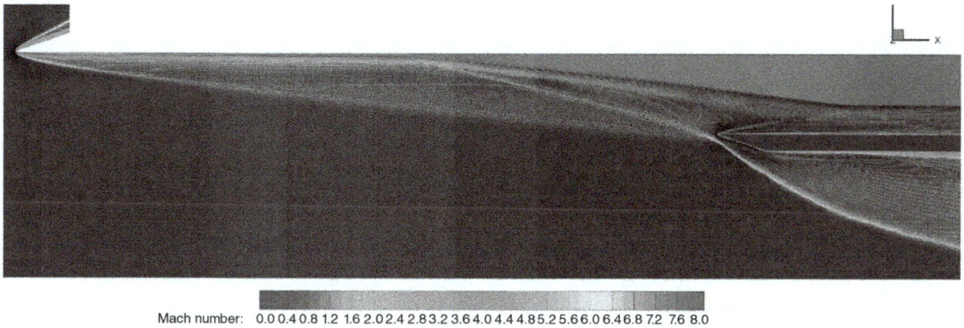

Figure 12.7 HIFiRE 7 REST scramjet engine forebody/inlet symmetry plane computational Mach contours with unstructured grid superimposed computed using the fn2 stencil. Source: From White et al. 2020. NASA.

Let us consider a dual-mode ram/scramjet intended to operate in the lower hypersonic regime between flight Mach numbers 3 and 7. Transitioning from subsonic to supersonic combustion modes requires an isolator between the inlet and combustor sections. Operating in ramjet mode, combustion heat release can drive the supersonic inflow to sonic conditions, using thermal choking, whereby a precombustion shock train forms in the isolator. The shock train in the isolator consists of a series of shocks, which terminate with a normal shock that drives the flow to subsonic conditions. This precombustion shock train and associated heat release are strongly coupled, and this coupling helps flame stabilization by increasing the static pressure and temperature while decelerating the flow. Placed downstream of the inlet, the isolator reduces the interaction between the incoming airflow and the combustor flowfield and prevents inlet unstart (see Chapter 4).

Figure 12.7 illustrates the flowfield for the HIFiRE 7 REST inlet, showing contours of the flow Mach number using VULCAN-CFD with unstructured grid superimposed computed using the *fn*2 stencil, e.g. a least-squares cell-average gradient stencil method for the computational grid (White et al. 2020). The simulation depicted by Figure 12.7 illustrates the small size of the forebody and cowl leading edges relative to the forebody boundary-layer thickness. The contours also show that the shocks are captured without significant oscillations and that, using a small number of cells, the model captured the forebody leading edge and forebody compression corner shocks. Chan et al. (2014) provide an excellent description of the HIFiRE 7 REST inlet.

Now, if we focus on the isolator section, we find that the flowfield is dominated by large separation regions and multiple shock–boundary layer interactions. The boundary-layer separation and the shocks will cause a high static enthalpy flow upstream from the combustor. These processes will increase the wall heat flux, up to 2 times. Thus, it is important to precisely predict the isolator heat flux that occurs with mode transition in order to develop strategies for thermal management.

Hunt et al. (2019) carried out CFD simulations of an isolator shock train flow field to input into a physics-based model to predict the values of filtered Rayleigh scattering (FRS) intensity measurements in a proposed experiment. FRS is a laser diagnostic where the intensity of elastically scattered light is measured after it passes through a molecular absorption filter. Giving a quantitative measurement capability, FRS can provide spatially- and temporally resolved quantitative data on an imaging plane without intrusively seeding the flow, allowing greater insight into the complex 3-D shock train structure in the scramjet isolator section and help researchers validate their CFD simulations.

The flow physics in the inlet/isolator include 3-D compressible turbulent boundary layers, transitioning boundary layer on some inlet surfaces, shock and expansion waves, shock–boundary layer interactions, and associated flow separation. The leading edge and engine cowl are subjected to high thermal loads due to shock–shock interactions. Inlet unstart will cause unsteady flows, and when operating over a Mach number range, air spillage will result in internal–external flow interactions. Therefore, computational studies of inlet unstart due to the unsteady nature of the flow must be part of the design process. CFD codes require to model strong shock–boundary layer interactions with large regions of separated flow, and accurate models to capture incipient transition (see Chang 2004; Fiévet et al. 2015). The complexity of the inlet/isolator flow physics makes high-resolution, time-dependent CFD simulations not only costly to acquire but also difficult. To model the physics of the inlet/isolator flow field requires the use of robust, full NS codes with compressible turbulence models.

12.3.3 Combustor

Turbulent high-speed reacting flows remain the focus of intensive research, driving the development of new models aiming to resolve, among other issues, whether combustion chemistry and turbulence models used in state-of-the-art CFD codes are realistic enough to use in the design of scramjet combustors (see Chapters 5–7).

Scramjet combustor flowfield studies require modeling and simulation of high Reynolds number supersonic wall-bounded flows, which exhibit inherent unsteadiness, turbulent shock–boundary layer interactions, chemical kinetics, and other complex flow phenomena. Today, researchers frequently use a blended LES/RAS approach, in order to address some of the fundamental limitations of RAS while taking advantage of the strengths offered by large-eddy simulations (LES).

Considering the relatively short residence time, and the complex dynamics of the flowfield in a scramjet combustor, much effort is devoted to study the issue of fuel/air mixing. This problem is important to understand and characterize because enhancing the rate at which the fuel and air mix will reduce the combustor length and achieve significant gains in scramjet performance. To this end, a variety of mixing performance metrics are calculated from the results of CFD simulations of different fuel injector configurations, including hypermixers. As we saw in Chapter 5, a hypermixer is a device that generates streamwise vortices for enhancing supersonic mixing and combustion.

To simulate the physics of fuel injection to assess mixing efficiency, should we use chemical reacting models to resolve the interaction of the fuel jet with both the boundary layer and the supersonic airstream? Consider a simple crossflow injection scheme such as the one depicted in Figure 5.5. The high momentum of the supersonic airstream causes the injected fuel jet to be turned downstream, limiting mixing and combustion efficiency. The CFD simulation must clearly show that, as the fuel jet penetrates the thick boundary layer, the obstruction by the fuel jet on the primary airflow causes a bow interaction shock in the main flow. The flowfield that develops also includes a jet expansion fan and a barrel shock, accompanied by upstream and downstream separated boundary-layer regions with recirculation, and bow shocks in the separation and reattachment regions. The Mach disk height is indicative of the extent of mean fuel penetration (see Chapter 5).

CFD studies have shown that, if the pressure rise on the surface caused by the shock is greater than about 3 to 1, the boundary layer will separate, and an oblique shock off the separated region will intersect the main stream. In the upstream and downstream regions of the fuel jet, the boundary layer may separate. With a separated boundary layer, the fuel jet will be partially blocked from the main flow momentum, depending on the size of the separation region. The size of the separation region also affects the penetration height. This is because, for a constant dynamic pressure and mass flow rate, few parameters change the height of the Mach disk, and the momentum losses across the disk limit the penetration of gaseous jets into the supersonic stream. Thus, the fuel must be injected with a supersonic flow injector having a large initial jet momentum.

Clearly, the flow field of multiport transverse injection into a supersonic crossflow is more complex than the single injection flow field (Lee 2006). This is due to the strong interactions among the injected fuel flows, the various shock wave structures and vertical flows that form around the injection flows.

Staggered fuel injectors may improve mixing and combustion efficiency. However, performance may be reduced if opposite wall injectors fail to fuel the space between them. A CFD simulation of staggered injection revealed that the near-field mixing of transverse jets is dominated by large-scale jet-shear layer vortices. In the cases investigated, due to the close proximity between injectors the interaction of the fuel jets at each interface between pairs generates vorticity. This causes some regions of the flow to roll up into counter-rotating vortex pairs, which stir and mix the fuel with the high-speed air. A large region of separated flow between the forward and aft injectors appeared to provide improved flameholding. Moreover, the bow shock caused by the aft jet had a higher angle with the horizontal, causing it to interact with the shocks originated by the forward jets.

An efficient injection system setup for single- or dual-mode combustor must include the selection of many parameters, i.e. the position of the injection port, the distribution of the mass flow rate and momentum flux, the injection angle, and the combination of the injection angles. CFD analysis must consider the effects of those variables in the selected models. For example, decreasing fuel injection angle aligns the injected fuel stream with the air stream and thus increases its streamwise momentum. On the other hand, increasing the injection angle directs the fuel stream normal to the air stream and, although this promotes more mixing (and thereby combustion) by increasing fuel penetration and blockage, it also increases total pressure losses due to stronger shocks and decreases the axial momentum augmentation.

Moreover, accurate simulation of chemical processes in the combustor is critical because the fuel–air chemistry determines the rate of heat release, flame stability, ignition, extinction, and pollutant emissions. The finite-rate chemical reaction of fuel and air is of concern for CFD analysis, as it requires the choice of adequate chemistry models that are representative of the combustion process. However, fully detailed chemical kinetic descriptions of fuel oxidation may require the tracking of hundreds of chemical species and thousands of reaction steps. This is especially true for hydrocarbon-air combustion. CPU and memory limitations make it unfeasible to implement the full detailed chemistry of scramjet fuels into 3-D CFD simulations, even using massively parallel computers.

Therefore, chemical reactions are normally resolved with reduced reaction, finite rate models. For the reactions occurring in a hydrogen-fueled scramjet, a typical hydrogen–air reaction mechanism includes nine chemical species and eighteen chemical reactions, although other less or more elaborate mechanisms may be used depending upon the research goals. For early scramjet development, gaseous hydrogen and air reaction was modeled by a 9-species, 18-reaction model (a shortened version of the Jachimowski's original 13-species and 33 reactions hydrogen-air model). In this model, eight of the chemical species (H_2, 0_2, H_20, OH, H, O, HO_2, and H_2O_2) are active, and the ninth (N_2) is assumed inert. Another reduced model that ignores the kinetics of HO_2 and H_2O_2 is also used due to its low computational power requirements relative to detailed kinetic models.

Important advances in CFD techniques have been done to simulate turbulent combustion in scramjets. For example, researchers at North Carolina State University developed a LES strategy with their reactive flow code REACTMB that include a low-dissipation discretization schemes to resolve small turbulent eddies. Figure 12.8 illustrates some of their results. The top image shows the fuel plume, where the silver isosurfaces indicate gaseous hydrogen mass fraction, and the colored isosurfaces indicate water vapor mass fraction. The bottom image gives centerplane slices of the combustor flow temperature. Researchers noted that, with high-performance computing and parallelization, their solution required several tens of thousands of iterations to achieve a steady-state solution, completing it in a few days.

In a hydrocarbon-fueled scramjet, combustion is much more complex and requires carefully reduced kinetics mechanisms to obtain practical computations. For example, the USC ethylene–air detailed reaction model from the University of Southern California consists of 75 species and 529 elementary reactions. Such large reaction model is only practical for 0-D and 1-D simulations. Hence, 3-D computations require a substantial reduction of the number of species and reactions. Hence, for the JP-7-air reactions, reduced kinetics models have been developed based on the ethylene–air USC reaction mechanism. Endothermic JP-7 is a

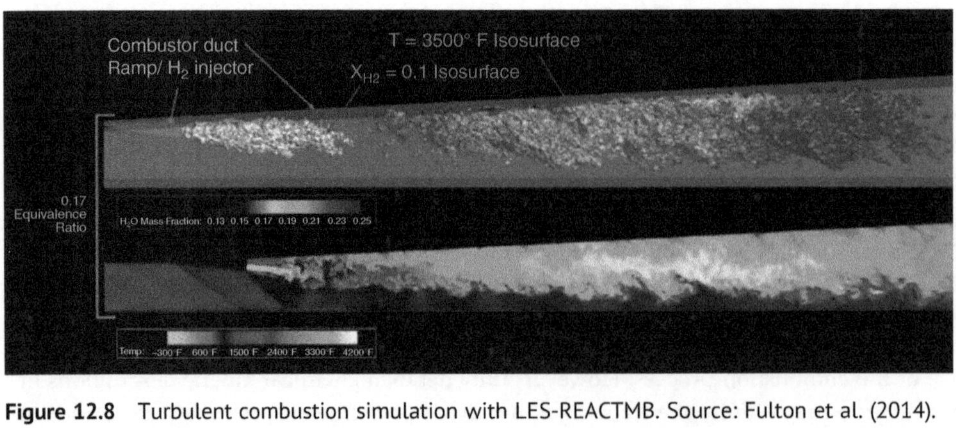

Figure 12.8 Turbulent combustion simulation with LES-REACTMB. Source: Fulton et al. (2014).

highly refined hydrocarbon fuel originally developed for the SR-71 and later chosen for the HIFiRE program (Gruber et al. 2008), and used as a fuel and coolant for the X-51A scramjet. For ground testing, Pellett et al. (2008) developed a gas mixture consisting of 36% methane and 64% ethylene by volume to serve as a surrogate for the endothermically cracked JP-7 fuel. A surrogate fuel is one that mimics the properties of the cracked JP-7 (see Chapter 6).

For the kinetics of the cracked JP-7 surrogate, Norris (2017) used the 22 species, 206 reduced reaction ethylene mechanism of Luo et al. (2012). With this model, he investigated the uncertainty of the cracked JP-7 chemical kinetics used in the modeling of a hydrocarbon fueled scramjet. To simulate the HIFiRE Direct Connect Rig (HDCR) flowpath test, Quinlan et al. (2014) modeled the chemistry of the JP-7 fuel surrogate, using a reduced 22 species, 18-step chemical reaction mechanism.

In addition to simulating the overall combustor flowfield, highly detailed analysis of very localized processes is required such as in the neighborhood of fuel injectors and flame holders. Moreover, for applications where the vehicle must fly within a range of flight Mach number requiring propulsion transition from ram/scram to scramjet operation, the dual-mode combustor (subsonic to supersonic combustion) is designed with multistaged fuel injectors and flameholders. The design becomes more complicated for hydrocarbon-fueled combustors. Thus, the numerical simulations must capture the thermochemical physics of combustion, including the transition from dual-mode to scram-mode operation, and yield accurate data to predict combustion efficiencies.

Previous CFD modeling of scramjet combustion was very limited in terms of including adequate chemical representations of the combusting flow. Modeling turbulence–chemistry interactions for high-speed reacting flows remains challenging due to both the computational cost of the models and the uncertainties of the models themselves.

In performing numerical simulation to support an experimental effort, other factors must be considered that take into account the actual test article design and manufacture. For example, wall friction is reflected in a shift of the near-wall velocity profiles. Hence, RANS simulations must take wall friction into consideration. Research related to the LAPCAT-II dual-mode scramjet combustor has studied the roughness effect caused by the coated metallic surfaces of the combustor. Using Menter's $k - \omega$ model, Pelletier et al. (2021) modified the boundary values of k and ω at the wall and found an increased level of turbulent viscosity and a velocity shift that are consistent with the LAPCAT-II experimental results.

Nowadays, design of scramjet combustors is aided by numerical simulations performed by means of codes that solve the RANS equations, or using LES, which is now the tool of choice to help in understanding how to improve mixing, flame anchoring, and combustion efficiency in supersonic reacting flows. RANS calculations require a small fraction of the computational resources required for LES and thus remain the choice for most R&D work. However, high-fidelity full-scale combustor modeling is required for the understanding of internal flow characteristics and the scramjet performance-based design. Therefore, LES is preferred. To save computational cost, the hybrid RANS-LES approach (Shur et al. 2008) has become the standard, especially for simulating hydrocarbon-fueled combustion (see Chapter 6).

LES has been successfully applied to study different dual-mode combustors, using flamelets and different finite rate chemistry approaches (Bermejo-Moreno et al. 2013; Fulton et al.

2014; Vincent-Randonnier et al. 2018; Yao and Fan 2017). LES is more accurate than RAS for dual-mode combustion because LES captures well the unsteady nature of the shock train in the isolator.

12.3.4 Nozzle/Afterbody

The overall propulsive efficiency of the vehicle is determined, to a large extent, by the exhaust plume flow over this vehicle afterbody section (see Chapter 8). The physics of the flow field in the nozzle/afterbody of a hypersonic vehicle retains the characteristics acquired in the processes taking place upstream in the combustor, and also includes additional requirements of modeling highly expanding nonuniform chemically reacting flow. Other processes may be present such as the possibility of relaminarization and then retransition back to turbulent flow. In addition, nozzle/aftbody models must consider flow divergence, skin friction losses, energy-bound chemical radicals that will not relax in the nozzle, and excited vibrational states and their relaxation.

Nozzle/afterbody flow field simulations must be carried out with the same NS with chemical kinetics code used in the analysis of combustor flow field. This is because in the nozzle finite-rate reaction chemistry is still required to assess the degree of reaction that continues to take place as the flow expands, and to determine the extent of recombination reactions that add to the available thrust. Some reduced kinetics models applied to nozzle/afterbody flow have shown to be reasonably accurate. However, the codes must be improved to better model and assess recombination reactions.

A high-fidelity CFD analysis must be used to determine the expanding flow through the engine nozzle. However, such solution is computationally expensive. Dalle et al. (2010) proposed a reduced-order model of the scramjet nozzle that is accurate to within 10% and requires less than a few seconds of computational time. This reduced-order model is used for the solution of a steady two-dimensional (2-D) supersonic flow through a nozzle. In the nozzle geometry they considered, the difference in pressure at the cowl trailing edge would allow the exhaust plume to adapt to different conditions, and to operate in a similar manner as an aerospike nozzle.

The reduced-order 2-D model can analyze shock waves, expansion fans, and finite rate chemistry. The authors (Dalle et al. 2010) describe their model as follows: Instead of solving directly for the flow conditions at each point in the flow (as is done in CFD), the reduced-order method solves for the positions of relevant waves directly, and the waves separate regions in which the flow properties are uniform. Using 2-D supersonic theory, this model finds solutions for the locations of the shock and expansion waves. It approximates expansion fans as a number of discrete waves, and curved surfaces are modeled as a number of straight sections. After determining the locations at which two or more waves intersect, the program solves the interactions among the waves as 2-D Riemann problems. As the authors noted, this approach can be considered as a generalized and automated version of the method of characteristics. For the chemistry of the flowfield, the model assumes that chemistry occurs along each streamline independently of neighboring streamlines, using a 1-D chemistry model for each streamline.

The 2-D model of Dalle et al. was accurate in determining the boundary of the exhaust plume, which is essential to thrust calculations. However, it does not have the capability to analyze boundary layers and other flow phenomena, and thus it provides a rater limited analysis of the expansion process in a scramjet nozzle.

12.4 NASA Hyper-X Program Computational Modeling Requirements

The origin of today's state-of-the-art computational tools for scramjet research and development began with the Hyper-X program. The NASA Hyper-X Program promoted high-speed computational research and development, placing a greater effort in developing new combustion codes. These analytical tools were advanced to study the fuel injection process, and the mixing and combustion of fuel and air downstream of the injectors. Detailed fuel injector design was also considered in order to enhance fuel–air mixing and enable the highest level of mixing and combustion efficiency (Drummond 2014).

Supersonic combustion models were developed to establish the requirements for and performance of fuel injection, mixing and reaction (ignition and flame holding) in the scramjet combustor environment. Those efforts were based on a dual approach of experimental and computational technology development since the overall propulsion flowpath had to be designed, and that design depended on both experimental research and computational analyses. However, because ground-based facilities operated in the lower Mach number range of the X-43A vehicles and testing were expensive, computational tools were needed to establish initial designs for testing, and to fill in the regions between test points in the various facilities (Drummond 2014).

At NASA Langley, one of the first efforts to develop computational tools for scramjet applications yielded the General Aerodynamic Simulation Program (GASP), a code that solved the steady and unsteady Euler, parabolized Navier–Stokes, thin-layer Navier–Stokes, and Navier–Stokes equations. GASP included both algebraic and two-equation turbulence models with wall function options and had sufficient capability to help design the flowpath of the X-43A vehicles, including a set of thermochemical kinetic models for air chemistry, hydrogen–air combustion, and various hydrocarbon reactions in a database containing 455 reactions and 34 species.

GASP was validated for a number of external and internal flowfields. Huebner and Tatum (1993) tested GASP computational capabilities, predicting complex 3-D hypersonic configuration flowfields under simulated scramjet-powered conditions. Srinivasan et al. (1993) studied fuel injection schemes for scramjets and compared results with measured wall pressure and heat flux data and with SPARK CFD solutions, finding that GASP solutions compared favorably with both the experimental data and the SPARK solutions. GASP was also used to analyze the external flowfields of a number of other hypersonic vehicles; Cockrell et al. (1996) modeled the aerodynamics of two waverider hypersonic cruisers; Loomis et al. (2004) used the code to predict both aerothermal heating, and aerodynamic characteristics in the hypersonic regime for vehicles with geometries under consideration for the X-38 program.

Researchers at NASA Langley undertook another effort to develop a CFD code for scramjet flowpath design. The Langley Algorithm for Research in Chemical Kinetics (LARCK) code development project began in the early 1990s as a replacement for the SPARK combustion code. LARCK was a cell-centered, finite volume, multiblock, multigrid code to solve the full RANS equations for turbulent nonequilibrium chemically reacting flows. It was validated against a number of 2- and 3-D flow problems and then used to model individual scramjet component flows, as well as the entire scramjet flowpaths. In 1996, LARCK served as the basis for VULCAN-CFD, a more advanced analysis code for high-speed flows that has become the gold standard for simulating external and internal flows to the present day (VULCAN-CFD 2022).

The VULCAN-CFD code was part of a ramjet–scramjet CFD code development effort at the Wright Patterson Air Force Base (Drummond 2014). In 1997, the code development effort moved to the NASA Langley Research Center. One of the first applications of the VULCAN-CFD code was the investigation of advanced fuel injection schemes for scramjet propulsion. The SCHOLAR Model was built with the sole purpose of validating the code and its turbulence and chemistry models. Initial validation concluded that different turbulence models gave qualitatively dissimilar solutions. Furthermore, a relatively high value of 1.0 had to be used for the turbulent Schmidt number Sc_t, if the ignition delay was to be captured at all. Since then, VULCAN-CFD has evolved and matured with both improved mathematical and physical models, effective algorithms, and advances in computer architectures playing a major role.

VULCAN-CFD solves the equations governing 2- and 3-D calorically perfect or thermally perfect nonequilibrium chemically reacting flows. The code uses a structured grid, cell-centered, finite volume, and density-based method. VULCAN-CFD employs a probability density function approach for species modeling. Its specialized focus is on turbulent mixing for accurate hypersonic combustion modeling. The code solves full spatially elliptic Euler or full NS equations by integrating conservative form of unsteady equations in real or pseudo-time. Pseudo-transient is a time-stepping approach to accelerate the solution. Real-time step size resolves flow features satisfactorily, and pseudo-time step size is dependent on the general stability condition. At every real-time step code performs pseudo-time iterations to find out converged solutions. VULCAN-CFD solves spatially hyperbolic Euler or PNS equations by integrating conservative form of unsteady equations in pseudo-time (VULCAN-CFD 2022).

12.4.1 Nose Tip-to-Tail Analysis Methodology

Development analysis for hypersonic air-breathing engines is much more complex than analysis methods for gas turbines and rockets. Hypersonic air-breathing propulsion operation requires engine to be fully aero-thermodynamically integrated with the airframe, that is, the vehicle and the engine must be developed together. Accurate computational prediction of hypersonic air-breathing propulsion system performance and operability limits must account for the coupling of all engine components. NASA Hyper-X program researchers conceived a viscous nose tip-to-tail calculation methodology to predict the performance of the fully-integrated X-43A vehicles, including simulation of propulsion effects. The tip-to-tail analysis yields the integration of aerodynamic and propulsion flow fields, using diverse analytical tools that examine flow-field characteristics and scramjet flowpath component performance and the external flowfields, combined to obtain preflight predictions of the X-43A longitudinal performance increments (Cockrell et al. 2000).

A variety of CFD codes and engineering analytical tools were utilized for the tip-to-tail analysis of the X-43A vehicles. The primary CFD tool used for the preflight performance analysis was GASP, while VULCAN-CFD was used for scramjet flowpath design and analysis. In addition, two other tools provided supporting analysis for the internal flowpath. To assess fuel injection in the combustor, the supersonic hydrogen injection program (SHIP), which uses the SIMPLE (semi-implicit method for pressure-linked equations) method was used to solve the parabolized, mass-averaged equations for conservation of mass, momentum, total energy, total fuel, and turbulence fields in a variable area domain of rectangular cross section.

The second tool used for flowpath analysis was the SRGULL code, a scramjet-cycle-analysis engineering tool developed at NASA Langley capable of nose tip-to-tail simulation of hydrogen-fueled, scramjet engine integrated with its airframe. SRGULL utilizes a set of codes to predict most flow phenomena, including inlet mass capture, boundary layers (shear, heat flux, and transition), inlet kinetic energy efficiency, isolator performance, combustor flow distortion, nozzle expansion, and divergence losses. Ferlemann (2005) provides a full description of the design tools, results of the ground tests, and comparison of the predictions with the flight results from the successful X-43A vehicle Mach 7 flight test.

The propulsion predictions for the X-43A Mach 7 flight test were obtained with SRGULL with input from CFD and wind tunnel tests. Figure 12.1 shows the predicted flowfield for the X-43A vehicle on hypersonic flight along a constant dynamic-pressure trajectory of 1000 psf (47.88 kPa). The nose tip-to-tail CFD solution includes both internal (scramjet engine) and external flow fields, showing the interaction between the engine exhaust and vehicle aerodynamics. This contour plot also shows surface heat transfer on the vehicle skin (red is highest heating) and flowfield contours at local Mach number. The last contour illustrates the engine exhaust plume shape. This solution approach is one method of predicting the overall vehicle performance, and the best method for the determination of vehicle structural, pressure, and thermal design loads. The flowfield in Figure 12.1 shows planar Mach contours ranging from 0.4 to 7.1 and surface heat flux contours ranging from 1.0 to 12.0 BTU/(ft^2-sec) (11.35 to 136.27 kW/m^2) (Anderson et al. 2000).

Many design and performance characteristics of the X-43A flight vehicle could not be tested in the wind tunnels. For example, due to the relatively small scale of the wind tunnel models, it was not possible to test the inlet-open position (Engelund et al. 1999). Thus, a comprehensive CFD study was carried out to provide estimates of the inlet open unpowered and powered flight aerodynamic characteristics, including the effects of Mach number, angle-of-attack, and sideslip on the Mach 7 vehicle in flight. CFD analysis investigated issues associated with Reynolds number scaling, scramjet propulsion flowfield-induced effects, force and moment increments associated with opening the inlet door, tunnel hardware interference, and unsteady flow effects. Monte Carlo simulations using test database along with a model of the mechanical process used to develop control systems to produce clear separations at minimum risk (Woods et al. 2001)

CFD predictions were also used to address other aspects of vehicle performance and flight test development, including boundary-layer trip design and assessment, thermal and structural loads, and scramjet flowpath component performance. Nose tip-to-tail NS calculations provided details on scramjet exhaust plume expansion in the aftbody region. Such CFD analysis provided a qualitative assessment of lateral-directional stability characteristics, including a simulation of the powered scramjet flight test condition. Cockrell et al. (2000) reported on the overall CFD methodology for the Hyper-X flight experiment.

To obtain a full understanding of the dominant phenomena of scramjet combustor flows, we must examine the interaction of turbulent flow with the combustion process. Interactions between the turbulence field and chemistry must be modeled to account for the effects of temperature fluctuations on chemical reaction rates and for the effect of species fluctuations on species production.

One of the key challenges in scramjet flowpath design is centered in fuel injectors, as the design of these components must lead to efficient fuel–air mixing, combustion, and flameholding. Attempts to improve the fuel–air mixing, while simultaneously reducing total pressure losses, have received a great deal of attention over the years (Lee et al. 2015). Computational methodologies to support these efforts continue to evolve. It is clear that a proper reduced kinetic scheme for H_2/air and for hydrocarbon fuel/air is necessary to correctly reproduce combustion temperature and species using acceptable CPU times.

The following section provides highlights from three research programs, intended to provide additional details on the different CFD tools that have been and are now used to study various aspects of scramjet combustion phenomena.

12.5 Overview of Selected CFD Analysis Cases

12.5.1 Flamelet Model for HIFiRE-2 Direct Connect Rig (HDCR) Flowpath

Quinlan et al. (2014) performed an a priori investigation of the applicability of flamelet-based combustion models to dual-mode scramjet, utilizing RAS. They considered the HIFiRE-2 Direct Connect Rig (HDCR) flowpath fueled with a JP-7 fuel surrogate and operating in dual- and scramjet mode (see Figure 7.6). The HDCR is a full-scale, heat sink, direct-connect ground test article that duplicates both the flowpath lines and a majority of the instrumentation layout of the isolator and combustor portion of the flight test hardware. The HDCR was tested at the NASA Langley Arc-Heated Scramjet Test Facility (AHSTF) at Mach 6 to 8 flight enthalpies (Cabell et al. 2011). The flight hardware and other details of the HIFiRE-2 program are given in Chapter 1, and results of the CFD modeling of the flowfields are given in Storch et al. (2011).

The analysis based on RAS utilized a finite-rate chemical kinetics model from which typical quantities defining a flamelet were calculated (Quinlan et al. 2014). In an effort to make LES of hydrocarbon-fueled scramjet combustors more computationally accessible (using realistic chemical reaction mechanisms), researchers proposed a compressible flamelet/progress variable (FPV) model that extends current FPV model formulations to high-speed, compressible flows. For the computational study, the RAS data were obtained using a reduced 22-species/18-step reaction mechanism, which was validated using available experimental data.

All simulations were performed using the thermally perfect VULCAN-CFD code, using RAS over a 6.6 million cell, quarter-geometry, structured grid. The governing RANS and scalar transport equations were closed using the Menter blended $\kappa - \omega/\kappa - \epsilon$ turbulence

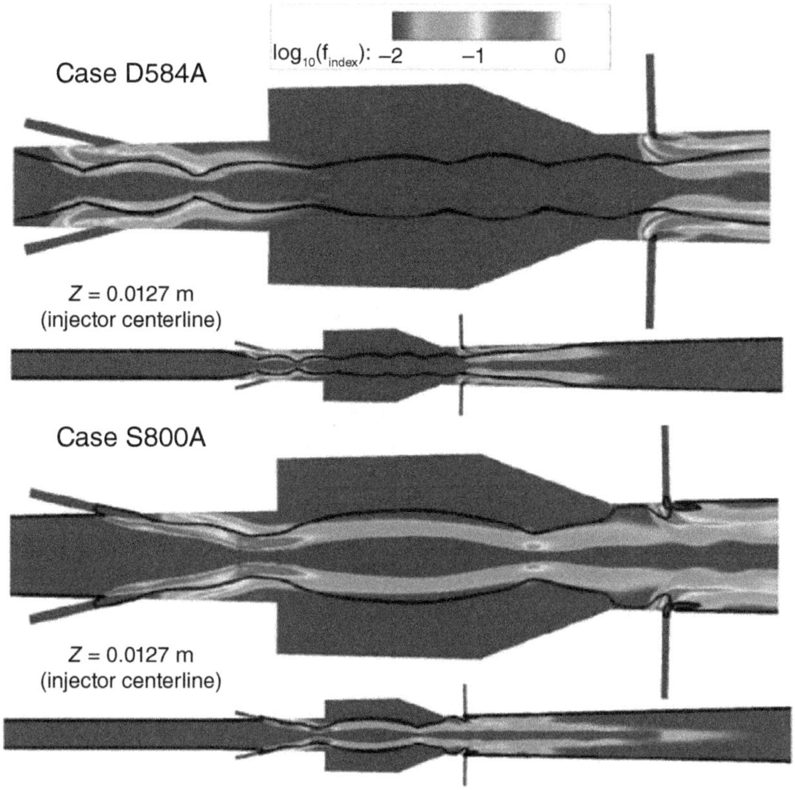

Figure 12.9 Contours of the flame index for HIFiRE direct connect rig flowpath with JP surrogate fuel: (top) dual-mode combustion RAS simulation at M_0 = 5.84;primary injection ϕ = 0.15; secondary injection ϕ = 0.50; (bottom) scramjet-mode combustion RAS simulation at M_0 = 8.00; primary injection ϕ = 0.40; secondary injection ϕ = 0.60. Black lines correspond to the sonic line. Source: Adapted from Quinlan et al. (2014).

model. Some of their results are illustrated in Figure 12.9, showing contours of the logarithm of chemical heat release normalized by its global maximum.

In the HIFiRE HDCR flowpath experimental and analytical study, results suggested that for scramjet-mode flowpath operation, the primary injector flames exhibit mixed combustion modes, in which significant heat release was found in regions of both nonpremixed and premixed conditions and at both moderate (Da = 1) and high (Da ≫ 1) Damköhler numbers.

For the flamelet computation, the flames from both primary and secondary injectors were characterized with a flame-weighted Takeno index and combustion regime diagrams. For dual-mode case, the combustion was found to occur primarily at high Da numbers and in a nonpremixed mode. Researchers concluded that the assumptions of nonpremixed flamelet models are likely satisfied and such models may sufficiently predict the combustion physics governing the dual-mode operation of the HDCR flowpath (Quinlan et al. 2014).

Not surprisingly, combustion in scramjet mode was found to be more complex. For the primary injectors (ϕ = 0.40), the combustion occurs over a range of Da numbers and includes both nonpremixed and premixed modes. A significant portion of the heat release was determined to be due to the primary injectors, which corresponds to premixed regions of combustion occurring near Da = 1, meaning that the characteristic flame time scale is on the same order of magnitude as that of the integral turbulence. However, a significant portion of the combustion occurs at high Da numbers in a nonpremixed mode, as well. For the secondary injectors (ϕ = 0.60), the combustion occurs at a range of Da numbers and primarily in a nonpremixed mode. The conclusion by the researchers is that simulation of the HDCR flowpath for scramjet-mode operation would require both premixed and nonpremixed flamelet models (Quinlan et al. 2014). Thus, the assumptions made for these models may only be valid for limited regions of supersonic combustion.

In addition, the nature of the premixed combustion data was derived from modified Borghi combustion regime diagrams, for dual-mode and scramjet-mode simulations, respectively. As shown in the diagrams, for both cases dual mode and scramjet mode, the premixed data lie within the broken reactions and thin reaction regimes, where $Re_t > 1$ and $Ka > 1$. In these regimes, the flame thickness is larger than the Kolmogorov scale, which allows Kolmogorov eddies to penetrate the flame partially for the thin reaction regime or completely for the broken reaction regime. As noted by the researchers, in such state the smallest turbulent eddies may alter the internal flame structure, causing localized extinction, resulting in segmented, broken flame regions.

In dual-mode combustion, the heat release corresponding to premixed data is relatively small compared to that of the nonpremixed data. Hence, the flamelet models may provide reasonable approximation. However, this is not the case for scramjet-mode combustion (Figure 7.8d). With supersonic combustion, much of the premixed combustion occurs within the thin reaction regime, and the heat release is significant. These regions of highly turbulent premixed thin and broken reaction zones challenge the flamelet assumptions.

These findings (Quinlan et al. 2014) suggest that a hybrid nonpremixed/premixed flamelet model may be necessary to model the flames for scramjet-mode flowpath operation. The effects of compressibility and heat losses were found to have a significant effect on combustion, suggesting that a suitable flamelet manifold should be parameterized by pressure and enthalpy.

12.5.2 LES for LAPCAT-II Dual-Mode Combustor

Experiments of the LAPCAT II dual-mode combustor at the ONERA-LAERTE facility are carried out to study transverse mixing, self-ignition, flame stabilization and compressible turbulent combustion, and the dependence of these phenomena on total temperature at fixed values of fuel/air ratio at the different operating conditions. Vincent-Randonnier et al. (2018, 2019) compared experimental measurements of high-speed combustion at different operating conditions with the LES method, using finite rate chemistry models and new skeletal H_2–air combustion chemistry for wall injection of hydrogen in a Mach 2.0 vitiated airflow. The operating conditions considered were on the range of observed subsonic and supersonic combustion modes at $\phi \approx 0.15$ and $p_0 \approx 0.40$ MPa with T_0 between 1414 and 1720 K.

The computational domain begins at the combustor entrance (exhaust of the facility nozzle) and ends at the combustor exit plane (facility dump tank). The hydrogen fuel injectors are located at $x = 0.20$ m from the origin (inlet). Computation is done on Hex-dominant grids refined along the walls and in the vicinity of the injectors. Researchers estimated the boundary conditions for the LES from the experiments, which they acknowledge may be detrimental to the modeling approach since the wall temperatures are not accurate; the combustor walls have a roughness due to the use of a thermal barrier coating (not considered in these computations).

Hydrogen is injected at sonic velocity both from the upper and lower combustor walls through two flush-mounted 2.0 mm porthole injectors. The flow in the combustor is over-expanded, and the transition from supersonic flow to subsonic flow occurs in the downstream part of the combustor (in the third combustor section) at $x \approx 0.75$–0.80 m. The filtered reaction rates are modeled using the Partially Stirred Reactor (PaSR) model. This is a multiscale model that assumes combustion takes place in dispersed or even intermittent fine-structure regions surrounded by low reaction rates. In PaSR, the interaction between turbulence and chemistry is represented with a factor, which is defined as the ratio between the chemical time scale and the sum of mixing and chemical scales. The PaSR model assumes that reactions are confined in a specific region of the computational cell, whose mass fraction depends both on the mixing and the chemical time scales. Hence, it is crucially important to select the appropriate mixing and chemical time scales to ensure the accuracy of the numerical simulation predictions.

In 1990, Chomiak proposed the PaSR model as an alternative to the Eddy Dissipation Concept (EDC) model, a formulation that attempts to incorporate the significance of fine structures in a turbulent reacting flow in which the focus is the combustion process (Chomiak 1990). Using different formulations of the PaSR model has yielded better prediction of temperatures and main species concentrations using RANS and LES, which were validated against experimental data collected in flames and reactors operated in moderate and intense low-oxygen dilution (MILD) combustion conditions.

Figure 12.10 illustrates the experimental Schlieren and OH^* chemiluminescence (top image) and the numerical simulations (combustor side view) of a case with a high total temperature ($T_0 = 1697$ K) with shock-induced combustion in the LAPCAT II combustor.

Vincent-Randonnier et al. (2018) solved the LES-PaSR model equations, using a fully-explicit finite-volume code based on the Open-FOAM C++ library. They modeled the H_2–air combustion process using a new H_2–air reaction mechanism called Z22, composed of 9 species and 22 irreversible reactions. The boundary conditions for the LES simulation were estimated to represent the LAPCAT II/LAERTE experiments. However, the study considered this a weakness of the simulation because the wall temperatures are poorly known. Moreover, the combustor walls have an unknown roughness due to the use of a thermal barrier coating.

The results of the numerical simulation in Figure 12.10 first superimpose the axial velocity (v_x) and the vorticity, represented by iso-surfaces of the second invariant of the velocity gradient tensor (λ_2), colored by v_x. The next image shows contours of temperature (T), followed by contours of heat release (Q). The bottom image superimposes contours of pressure (p) with the OH^* photon emission rate (k_{OH^*}). The OH^* and Schlieren images indicate that combustion starts at $x \approx 0.26$ m (the injector is at $x = 0.20$ m) due to shock-induced ignition; this is confirmed with the numerical simulations; but occurs much earlier, at around $x = 0.22$ m, and show intermittent combustion occurring in front of the

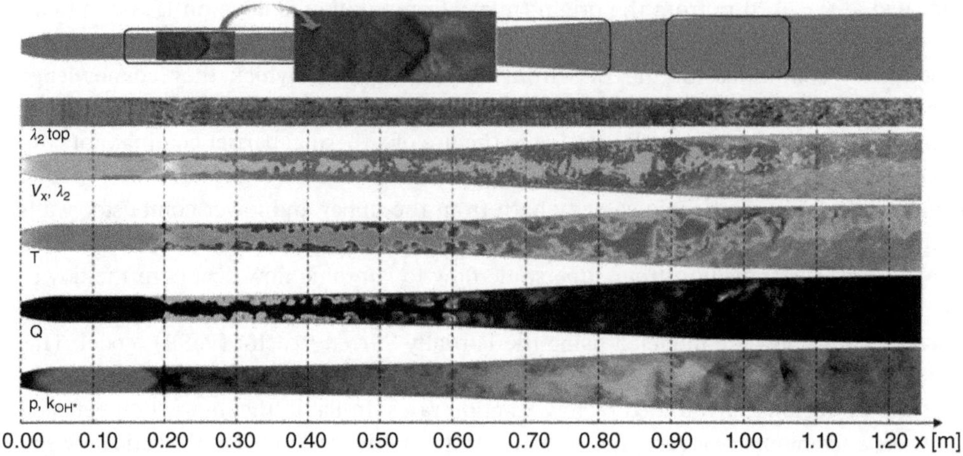

Figure 12.10 LAPCAT II combustor with hydrogen injection at $x = 0.20$ m from the origin. Experimental and computational results, from top to bottom: Schlieren and OH* Chemiluminescence (λ_2), predictions of axial velocity (v_x) and vorticity, represented by iso-surfaces of second invariant of gradient tensor, (λ_2), temperature (T), heat release (Q), and pressure (p) with superimposed OH* photon emission rate (k_{OH^*}). Source: From Vincent-Randonnier et al. (2018)/American Institute of Aeronautics and Astronautics.

injector, beneath the bow-shocks and partially in the horseshoe vortices surrounding the hydrogen rich jet.

Researchers concluded that the difference in results may be related to inconsistencies in the reaction mechanism, inconsistencies in the estimated wall temperatures used in the numerical simulations, or to some yet unknown factor (Vincent-Randonnier et al. 2018). For all cases studied, the authors found excellent agreement between predicted and measured reference values of total pressure and temperature. The results from LES and experiments both reveal a strong sensitivity of the combustor to variations in total temperature reminiscent to turbulent supersonic hydrogen–air combustion.

12.5.3 NASA LaRC Enhanced Injection and Mixing Project (EIMP)

The NASA Langley Enhanced Injection and Mixing Project (EIMP) is a comprehensive effort aimed to develop techniques to achieve more rapid mixing at high speeds (Cabell et al. 2014). In particular, the EIMP includes computational and experimental techniques to investigate scramjet fuel injection and mixing physics, with the goal of improving the understanding of underlying physical processes, and to develop enhancement strategies relevant to flight Mach numbers greater than eight. A shorter combustor results in a lighter vehicle. Hence, the ultimate goal of a propulsion system is to minimize the overall combustor length, while at the same time producing sufficient thrust with minimal losses. It is also important to obtain functional relationships between the relevant combustor performance metrics, such as combustion efficiency and thrust potential, and the flowpath geometrical parameters, such as spanwise injector spacing and duct height, in order to guide designs.

In the experiments conducted at Langley's AHSTF, various fuel injection devices are tested on an open flat plate located downstream of a Mach 6 facility nozzle. An open flat plate geometry was chosen, as opposed to a duct, in order to facilitate optical access for non-intrusive diagnostics and to simplify the experiment.

A CFD study was conducted to investigate the sensitivity of the mixing characteristics and calculate the performance of three types of fuel injectors at hypervelocity flow conditions to turbulence modeling choices (Drozda et al. 2019). The hypervelocity flow conditions match the high Mach number flow of the experiments conducted as a part of the EIMP. The injectors consist of a strut, ramp, and flushwall injector. As described in Chapter 5, these injectors represent three main categories of injectors typically considered individually or in combination for fueling the propulsive devices used for high-speed flight combustors. The CFD analysis study was performed with the VULCAN-CFD solver using RAS. Five turbulence models typically used in practical applications were used along with a prescribed turbulent Schmidt number Sc_t, which is **a key modeling parameter in simulations of turbulent mixing and reacting flows**.

Because the RAS solutions are computationally expensive, surrogate models were recently incorporated to the EIMP effort. In order to demonstrate the use of design and analysis of computer experiments (DACE) methods in the DAKOTA code for surrogate modeling and optimization, researchers applied them to a scramjet flowpath fueled with an interdigitated flushwall injector. To study combustion efficiency and optimizing thrust potential, the design variables included flight Mach number, duct height, spanwise spacing, and injection angle. A RAS database was created to build a surrogate model. The sequence of the design variables comprising the database were generated using LHS augmented with the corners of the design space. Researchers also created efficient computational grids using a newly developed automated geometry and structured grid generation methodology (Drozda et al. 2019).

Fuel injection performance was evaluated using two metrics: the thrust potential and the combustion efficiency. Although a certain amount of total pressure loss is unavoidable due to the desired effect of molecular mixing of the fuel and air, losses reduce the thrust potential of the engine, and thus total pressure loss must be minimized. The thrust potential for the flowpath is computed from the simple formula $TP = \dot{m}_e u_e + p_e A_e - \dot{m}_i u_i - p_i A_i$, where $\dot{m}$, u, p, and A are the mass flow rate, velocity, static pressure, and the area, respectively, with subscripts e and i denoting conditions at the thrust nozzle exit plane and the flowpath entrance (inflow).

Combustion efficiency η_c quantifies how completely a given flowpath is able to process a mixture of fuel and air into combustion products, thereby enabling heat release into the flow. In this work, where the equivalence ratio is less than one, η_c was computed based on the depleted mass flow rate. Four parameters were used as design variables: the flight Mach number ($8.0 \leq M \leq 15.0$), duct height ($1.0 \leq h(in) \leq 3.0$), spanwise spacing ($0.8 \leq w$ $(in) \leq 2.0$), and the injection angle ($30° \leq \alpha \leq 90°$).

For the design variables of the RAS database, the objective functions (TP, η_c) were determined by performing one-dimensionalization of the 3-D simulation data. For the surrogate modeling, the quadratic and cubic polynomials, Kriging and ANN were evaluated with respect to their fit and error characteristics. It was found that the Kriging model exhibited the least error, while the ANN and the cubic models had the greatest errors. The duct height was found to be the primary driver of both the thrust potential and combustion efficiency.

To assess the accuracy of the Pareto fronts, researchers selected three design points that were predicted by the cubic and Kriging models in order to characterize the flowfield with RAS. These design points are called challenge points as they are compared against the predictions given by the models. In this study, only the cubic and Kriging models were challenged because they predicted the greatest improvement with respect to the training data. For example, the point ($M_0 = 14.94$, $h = 2.969$ in, $w = 1.2771$ in, $\alpha = 30.312°$) is the challenge point for the cubic polynomial model where thrust potential is high and combustion efficiency is moderate. The researchers plotted the flowfield of this challenge point in the form of Mach contours.

For the case with the highest flight Mach number ($M_0 = 14.94$) investigated, a relatively high thrust potential was obtained by opening up the duct, as this reduces the total pressure losses that result from shock reflections, and also the low injection angle that augments the fuel momentum in the streamwise direction. The low mixing efficiency resulted because the fuel did not penetrate into the air stream and the fuel plume extended up to the combustor exit. The other two challenge points at lower flight Mach numbers were subjected to similar analysis. As reported (Drozda et al. 2019), the objective functions were not expected to exactly coincide with the values obtained from the surrogate models, and thus they had to quantify the extent of the mismatch in order to construct confidence intervals for the surrogate models.

Moreover, since the design variables have interactional effects that contribute to the thrust potential response, this research required to perform multiobjective design optimization using the surrogate models via a multiobjective genetic algorithm. It was determined that optimal solutions exist at both the upper and lower flight Mach number limits investigated. The Kriging model resulted in a Pareto front that yielded high values for both TP and η_c that was not captured by the other models.

In conclusion, the surrogate modeling approach carried out for the injection and mixing project (EIMP) gave relatively large uncertainty. This was attributed to the simplicity of the models used, the limited RAS database size, and the large extent of the design space. Hence, researchers believe that further sampling among the designs predicted by the Pareto fronts of the surrogate models are required in order to improve modeling and perform more accurate surrogate model-based optimization of this nature.

12.6 Closing Remarks

The design of the scramjet-powered vehicles flown to date (e.g. X-43A, X-51A) relied on varied computational tools, which had to be validated and verified with experimental data obtained from wind tunnel testing. In the case of the X-51A Scramjet Engine Demonstrator-WaveRider (SED), for example, such effort was possible because the design was based on the Affordable Rapid Response Missile Demonstrator (ARRMD) heritage, for which some CFD and wind tunnel aerodynamics data already existed (Hank et al. 2008). However, since the program lacked a detailed design, many configuration changes to the ARRMD conceptual design required further testing. To finalize the design and develop a comprehensive aerodynamics database for the X-51A vehicle, both Euler solutions (using the CART3D code) and full NS solutions (using the OVERFLOW code) were carried out. Nearly 2000 runs with over 80 different grids were used to determine vehicle safe separation, aeroheating, vehicle performance, boundary-layer transition, and fin deflection studies.

OVERFLOW is a NASA CFD flow solver, a 3-D time-marching implicit NS code that can also operate in 2-D or axisymmetric mode. The code uses structured overset grid systems. Several different inviscid flux algorithms and implicit solution algorithms are included in OVERFLOW 2.3. The code has options for thin layer or full viscous terms. A wide variety of boundary conditions are also provided in the code. The code may also be used for multi-species and variable specific heat applications. Algebraic, one-equation, and two-equation turbulence models are available. Low-speed preconditioning is also available for several of the inviscid flux algorithms and solution algorithms in the code. The code also supports bodies in relative motion and includes both a six-degree-of-freedom (6-DOF) model and a grid assembly code. Collision detection and modeling are also included in OVERFLOW 2.3. The code is written to allow the use of both MPI (distributed memory) and OpenMP (shared memory) for parallel computing applications.

The NASA CART3D is a high-fidelity inviscid analysis code used for conceptual and preliminary aerodynamic design. It allows users to perform automated CFD analysis on complex geometry and supports steady and time-dependent simulations.

VULCAN-CFD will continue to lead in modeling and simulation of high-speed reacting flows. The code includes a variety of PDE-based turbulence models, including explicit algebraic Reynolds stress models (ERASM) for RAS. It also includes one-equation and algebraic subgrid closures (including dynamic variants) for LES and hybrid RANS/LES capabilities. For a detailed description of the current capabilities of VULCAN-CFD, see the VULCAN-CFD Theory Manual (Baurle et al. 2020)

RAS methods, which attempt to model all of the scales present in turbulent flows, have become the standard in simulation of hypersonic air-breathing propulsion phenomena. RAS, however, have shown to be inadequate in resolving many challenging problems such as supersonic turbulent reacting flows, shock/boundary layer and shock/jet interactions, and unsteady turbulent flow simulations, including transition and separation.

Since LES aims to resolve only the large-scale turbulent structures while modeling only the smallest scales, these methods are used in order reduce the modeling uncertainty inherent to RAS. However, since wall-resolved LES is computationally expensive, hybrid RAS/LES methods were developed to alleviate the problem. Hybrid RAS/LES methodologies allow LES content to be resolved in areas that require a rigorous modeling approach, while maintaining a more cost-effective RAS approach for benign regions of the flow (Baurle 2017). Nevertheless, hybrid RAS/LES is still computationally expensive, and its framework requires further development

Today, the most critical area in CFD simulation capability is the ability to adequately predict viscous turbulent flows with boundary-layer transition and flow separation present (Slotnick et al. 2014). Moreover, the limitations associated with the turbulence closure models prevent the use of RAS as a true predictive tool in the scramjet complex flowfields (Baurle 2017). Hence, experts recommend to extend scale-resolving simulation approaches such as hybrid RAS/LES to problems of interest, something that will only become practical when substantial advancements to both the numerical and physical models are realized. To support this effort, the VULCAN-CFD flow solver continues to advance in its numerical framework and inclusion of hybrid RAS/LES, as established for both low- and high-speed benchmark flows of interest to hypersonic air-breathing propulsion (Baurle et al. 2020).

Today, hypersonic air-breathing propulsion flowfields continue to be analyzed with Reynolds-Averaged Navier–Stokes (RANS) codes, hybrid RAS/LES codes, and LES codes. Hybrid RANS/LES codes evolved into the computational pillars for scramjet flowpath development. Turbulent combustion is also treated with LES and turbulence–chemistry interactions in terms of the filtered density function (FDF). Emerging studies have shown that LES/FDF methodology offers the most promise.

Questions

1. How can you model (mathematically) and simulate all hypersonic flow phenomena? Include external and internal flowfields, reacting and nonreacting multiphase flows with turbulent and transitioning boundary layers. Explore the methodologies and the capabilities of available CFD codes to help you predict the performance of scramjet-powered vehicles.

2. What CFD code features can capture the flow/shock phenomena in the scramjet isolator?

3. What CFD capability is required to predict effectively the scramjet dual-mode combustion flowfield?

4. How do you select the correct value of the turbulent Schmidt number to model a dual-mode scramjet combustor? This is a key modeling parameter in simulations of turbulent mixing and reacting flows.

5. Explain the LES/FDF approach for turbulent combustion and its importance for scramjet propulsion analysis.

References

Anderson, G.Y., McClinton, C.R., and Weidner, J.P. (2000). Scramjet performance Chapter 6 in. In: *Scramjet Propulsion* (ed. E.T. Curran and S.N.B. Murthy). AIAA Progress in Astronautics and Aeronautics.

Baurle, R.A. (2017). Hybrid reynolds-averaged/large-eddy simulation of a cavity flameholder: modeling sensitivities. *AIAA Journal* 55 (2): 524–543.

Baurle, R. A., White, J.A., Drozda, T.G., and Norris, A.T. (2020). VULCAN-CFD Theory Manual: Ver. 7.1.0, NASA/TM-2020-5000766; User Manual, https://vulcan-cfd.larc.nasa.gov.

Bermejo-Moreno, I., Larsson, J., Bodart, J., and Vicquelin, R. (2013). Wall-Modeled Large-Eddy Simulations of the HIFiRE-2 Scramjet. CTR Annual Research Briefs.

Borghi, R. (1984). On the structure of turbulent premixed flames. In: *Recent Advances in Aeronautical Science* (ed. C. Casci), 117–138. New York: Plenum.

Cabell, K., Hass, N., Storch, A., and Gruber, M. (2011). HIFiRE Direct-Connect Rig (HDCR) phase I scramjet test results from the NASA langley arc-heated scramjet test facility. *17th AIAA International Space Planes and Hypersonic Systems and Technologies Conference*, San Francisco, CA (11–14 April 2011).

Cabell, K., Drozda, T.G., Axdahl, E.L., and Danehy, P.M. (2014). The Enhanced Injection and Mixing Project at NASA Langley. JANNAF 46th CS/34th APS/34th EPSS/28th PSHS Joint Subcommittee Meeting, NTRS Report Number NF1676L-25526, Albuquerque, NM.

Chan, Y.K., Razzaqi, S., Smart, M.K. and Wise, D. (2014). Freejet testing of the 75%-scale HIFiRE 7 REST scramjet engine. *AIAA-2014-2931. 19th AIAA International Space Planes and Hypersonic Systems and Technologies Conference* (June 2014).

Chang, C.-L. (2004). Langley Stability and Transition Analysis Code (LASTRAC) Version 1.2 user manual. NASA/TM-2004-213233, L-19022.

Choi, J., Edwards, J., and Baurle, R. (2008). Compressible boundary layer predictions at high reynolds number using hybrid LES/RANS methods. *Paper AIAA 2008-4175. 38th Fluid Dynamics Conference*, Seattle, WA (23–26 June 2008).

Chomiak, J. (1990). *Combustion: A Study in Theory, Fact and Application*. New York: Abacus Press/Gordon and Breach Science Publishers.

Cockrell, C.E., Jr., Huebner, L.D., and Finley, D.B. (1996). Aerodynamic characteristics of two waverider-derived hypersonic cruise configurations. NASA Technical Paper 3559.

Cockrell, C.E., Jr., Engelund, W.C., Bittner, R.D. et al. (2000). Integrated aero-propulsion CFD methodology for the hyper-X flight experiment. *AIAA Paper 2000-4010. AIAA 18th Applied Aerodynamics Conference*, Denver, Colorado (14–17 August 2000).

DAKOTA (2022). DAKOTA software for SBAO. https://dakota.sandia.gov/ (accessed 16 May 2022).

Dalle, D., Fotia, M., and Driscoll, J. (2010). Reduced order modeling of reacting supersonic flows in scramjet nozzles. *Paper AIAA 2010-6958. 46th AIAA/ASME/SAE/ASEE Joint Propulsion Conference & Exhibit*, Nashville, TN (25–28 July 2010).

Drozda, T.G., Shenoy, R.R., Axdahl, E.L., and Baurle, R.A. (2019). Numerical investigation and optimization of a flushwall injector for scramjet applications at hypervelocity flow conditions. *Paper AIAA 2019-4196. AIAA Propulsion and Energy 2019* Forum, Indianapolis, IN (19–22 August 2019).

Drummond, J.P. (2014). Methods for prediction of high-speed reacting flows in aerospace propulsion. *AIAA Journal* 52 (3): 465–485.

Engelund, W.C., Holland, S.D., Cockrell, C.E., and Bittner, R.D. (1999). Propulsion system airframe integration issues and aerodynamic database development for the hyper-X flight research vehicle. *ISOABE 99-7215 presented at XIV ISOABE*, Florence, Italy (5–10 September 1999).

Ferlemann, P.G. (2005). Comparison of hyper-X Mach 10 scramjet predictions and flight. *28th JANNAF Airbreathing Propulsion Subcommittee Meeting*, Charleston, SC (13–17 June 2005).

Fiévet, R., Koo, H., and Raman, V. (2015). Numerical simulation of a scramjet isolator with thermodynamic nonequilibrium. *AIAA 2015-3418. 22nd AIAA Computational Fluid Dynamics Conference*, Dallas, TX (22–26 June 2015).

Fulton, J.A., Edwards, J.R., Hassan, H.A. et al. (2014). Large-eddy/reynolds-averaged navier-stokes simulations of reactive flow in dual-mode scramjet combustor. *Journal of Propulsion and Power* 30: 558–575.

Georgiadis, N.J., Yoder, D.A., Vyas, M.A., and Engblom, W.A. (2011). Status of turbulence modeling for hypersonic propulsion flowpaths. *AIAA Paper 2011-5917. 52nd Aerospace Sciences Meeting*, National Harbor, Maryland (13–17 January 2014).

Georgiadis, N.J., Mankbadi, Mina R., Vyas, Manan A. (2014). Turbulence model effects on RANS simulations of the HIFiRE flight 2 ground test configurations. *AIAA Paper 2014-0624. 52nd Aerospace Sciences Meeting*, National Harbor, Maryland (13–17 January 2014).

Gieseking, D.A., Choi, J., Edwards, J.R., and Hassan, H.A. (2011). Compressible-flow simulations using a new large-eddy simulation/reynolds-averaged navier-stokes model. *AIAA Journal* 49 (10): 2194–2209.

Gruber, M., Jackson, K., Jackson, T., and Liu, J. (2008). Hydrocarbon fueled scramjet flowpath development for Mach 6-8 HIFiRE flight experiments. AFRL-RZ-WP-TP-2010-2243.

Han, Y., He, Y., and Le, J. (2020). Modification to improved delayed detached-eddy simulation regarding the log-layer mismatch. *AIAA Journal* 58 (2): 1–10.

Hank, J.M., Murphy, J.S., and Mutzman, R.C. (2008). The X-51A scramjet engine flight demonstration program. *AIAA Paper 2008-2540. 15th AIAA Int. Space Planes and Hypersonics Systems and Technologies Conference,* Dayton, OH (28 April–1 May 2008).

Huebner, L.D. and Tatum, K.E. (1993). CFD code calibration and inlet-fairing effects on a 3D hypersonic powered-simulation model. *AIAA Paper 1993-3041. 23rd Fluid Dynamics, Plasmadynamics, and Lasers Conference*, Orlando, FL (6–9 July 1993).

Hunt, J.C.R., Wray, A., and Moin, P. (1988). Eddies, Stream, and Convergence Zones in Turbulent Flows. Center for Turbulence Research Report CTR-S88.

Hunt, R., Ground, C.R., Baurle R.A., and Danehy, P. (2019). Using computational flow imaging to optimize filtered rayleigh scattering measurements of an isolator shock train. *AIAA 2019-4016. AIAA Propulsion and Energy 2019 Forum,* Indianapolis, IN (19–22 August 2019).

Jentink, T. and Bittner, R. (1997). Hyper-X Mach 7 powered tip-to-tail 3D CFD analysis. *34th JANNAF Combustion Subcommittee Meeting* (27–31 October 1997).

Lee, S.-H. (2006). Characteristics of dual transverse injection in scramjet combustor, part 1: mixing. *Journal of Propulsion and Power* 22 (5): 1012–1019.

Lee, J., Lin, K.C., and Eklund, D. (2015). Challenges in fuel injection for high-speed propulsion systems. *AIAA Journal* 53 (6): 1405–1423.

Loomis, M., Venkatapathy, E., Papadopolous, P., et al. (2004). Aeroheating and aerodynamic CFD validation and prediction for the X-38 program. *Presented at 32nd Thermophysics Conference,* Atlanta, GA (23–25 June 1997).

Luo, Z., Yoo, C.S., Richardson, E.S. et al. (2012). Chemical explosive mode analysis for a turbulent lifted ethylene jet flame in highly-heated coflow. *Combustion and Flame* 159 (1): 265–274.

Norris, A.T. (2017). Evaluation of the uncertainty in JP-7 kinetics models applied to scramjets. *2017 JANNAF - Interagency Propulsion Committee meeting*, Newport News, VA (4 December 2017).

Papila, N., Shyy, W., Griffin, L., and Dorney, D.J. (2002). Shape optimization of supersonic turbines using global approximation methods. *Journal of Propulsion and Power* 18: 509–518.

Pelletier, G., Ferrier, M., Vincent-Randonnier, A. et al. (2021). Wall roughness effects on combustion development in confined supersonic flow. *Journal of Propulsion and Power* 37 (1): 151–166.

Pellett, G.L., Vaden, S.N., and Wilson, L.G. (2008). Gaseous surrogate hydrocarbons for a hiFire scramjet that mimic opposed jet extinction limits for cracked JP fuels. *55th JANNAF Propulsion Meeting*, Boston, MA (12–16 May 2008).

Peters, N. (1986). Laminar flamelet concepts in turbulent combustion. *Proceedings of the Combustion Institute* 21: 1231–1250.

Peters, N. (2000). *Turbulent Combustion*. Cambridge University Press.

Queipo, N.V., Haftka, R.T., Shyy, W. et al. (2005). Surrogate-based analysis and optimization. *Progress in Aerospace Sciences* 41: 1–28.

Quinlan, J.R., McDaniel, J.C., Drozda, T.G. et al. (2014). A priori analysis of flamelet-based modeling for a dual-mode scramjet combustor. *AIAA 2014-3743. 50th AIAA/ASME/SAE/ ASEE Joint Propulsion Conference*, Cleveland, OH (28–30 July 2014).

Rumsey, C. (2017). NASA langley research center turbulence modeling resource. https:// turbmodels.larc.nasa.gov (accessed 1 March 2022).

Sagaut, P. (2006). *Large Eddy Simulations for Incompressible Flows*, 3e. Springer ISBN 978-3-540-26344-9.

Scherrer, D., Dessornes, O., Ferrier, M., Vincent-Randonnier, A., Sabel'nikov, V., and Moule, Y., Research on supersonic combustion and scramjet combustors at ONERA, Journal Aerospace Lab, Issue 11, 2016. DOI: https://doi.org/10.12762/2016.AL11-04.

Shur, M.L., Spalart, P.R., Strelets, M.K., and Travin, A.K. (2008). A hybrid RANS-LES approach with delayed-DES and wall-modelled LES capabilities. *International Journal of Heat and Fluid Flow* 29 (6): 1638–1649.

Shyy, W., Papila, N., Vaidyanathan, R., and Tucker, P.K. (2001). Global design optimization for aerodynamics and rocket propulsion components. *Progress in Aerospace Science* 37: 59–118.

Slotnick, J. Khodadoust, A., Alonso, J. et al. (2014). CFD vision 2030 study: A path to revolutionary computational aerosciences. NASA/CR-2014-218178.

Spalart, P. R. and Allmaras, S.R. (1992). A one-equation turbulence model for aerodynamic flows. *AIAA Paper 1992-0439. 30th Aerospace Sciences Meeting and Exhibit*, Reno, NV (6–9 January 1992).

Srinivasan, S., Bittner, R., and Bobskill, G. (1993). Summary of GASP code application and evaluation effort for scramjet combustor flowfields. *AIAA Paper 1993-1973. 29th Joint Propulsion Conference and Exhibit*, Monterey, CA (28–30 June 1993).

Storch, A., Bynum, M., Liu, J. et al. (2011). Combustor operability and performance verification for HIFiRE flight 2. *Presented at 17th AIAA International Space Planes and Hypersonic Systems and Technologies Conference*, San Francisco, CA (11–14 April 2011).

Vaidyanathan, R., Tucker, P.K., Papila, N., and Shyy, W. (2004). CFD based optimization for a single element rocket injector. *Journal of Propulsion and Power* 20 (4): 705–717.

Vincent-Randonnier, A., Ristori, A., Sabelnikov, V. et al. (2018). Combined experimental and computational study of the LAPCAT II supersonic combustor. *Presented at 22nd AIAA*

International Space Planes and Hypersonics Systems and Technologies Conference (September 2018). Orlando, US. Published as HAL Id: hal-02420953

Vincent-Randonnier, A., Sabelnikov, V., Ristori, A. et al. (2019). An experimental and computational study of hydrogen–air combustion in the LAPCAT II supersonic combustor. *Proceedings of the Combustion Institute* 37: 3703–3711.

VULCAN-CFD (2022). http://vulcan-cfd.larc.nasa.gov/ (1 February 2020).

White, J., Baurle, R., and Nishikawa, H. (2020). A 3-D nodal-averaged gradient approach for unstructured-grid cell-centered finite-volume methods for application to turbulent hypersonic flow. *AIAA Paper No. 2020-0652. AIAA Scitech 2020 Forum*, Orlando, FL (6–10 January 2020).

Woods, W.C., Holland, S.D., and DiFlubio, M. (2001). Hyper-X stage separation wind-tunnel test program. *Journal of Spacecraft and Rockets* 38 (6).

Yakhot, V., Orszag, S.A., Thangam, S. et al. (1992). Development of turbulence models for shear flows by a double expansion technique. *Physics of Fluids A* 4 (7): 1510–1520.

Yao, W. and Fan, X. (2017). Development of zone flamelet model for scramjet combustor modeling. *Paper AIAA 2017-2277. Presented at the International Space Planes and Hypersonic Systems and Technologies Conferences,* Xiamen, China (6–9 March 2017).

Yao, W., Liu, H., Xue, L., and Xiao, Y. (2021). Performance analysis of a strut-aided hypersonic scramjet by full-scale IDDES modeling. *Aerospace Science and Technology*, Vol. 117, 106941.

13

Hypersonic Air-Breathing Flight Testing

Autonomous hypersonic air-breathing flight was demonstrated in 2004. In that year, NASA's two X-43A hypersonic research aircraft flew powered by hydrogen-fueled scramjet engines. The first 11-second hypersonic flight was the culmination of years of planning, ground testing, and analytical work. On 16 November 2004, the second X-43A vehicle cruised at Mach 9.6 for 20 seconds, the highest hypersonic speed attained with hydrogen fueled air-breathing propulsion. In addition to demonstrating controlled accelerating flight at Mach 7 and Mach 10, the two nonaxisymmetric vehicles achieved a number of other crucial goals, including the first successful stage separation at high altitude and at high velocity. Figure 13.1 shows one of the most iconic representations of a hypersonic aircraft. It is an artist's rendition of the NASA X-43A research vehicle during its flight test.

13.1 Introduction

The ultimate performance and operational validation of any vehicle can only be achieved through flight testing. This is especially crucial for developing vehicles powered by hypersonic air-breathing propulsion. For such highly integrated vehicles, flight provides the conditions to conduct research and to advance the needed technologies in real atmospheric air under actual flight operating conditions. In this chapter, we wish to answer the question, is flight test the perfect simulation that duplicates true system performance?

13.2 Flight Operational Envelope

The flight envelope for the test vehicle considers both its design and operational characteristics subjected to the actual flight conditions. These issues must be fully explored, especially when ground testing cannot simulate all the proper flight and operating conditions with sufficient fidelity. The flight envelope considers the flight state, vehicle state, atmospheric effects, local flow conditions, and vehicle systems. The flight state is defined by vehicle velocity or Mach number and dynamic pressure conditions. The vehicle state along its flightpath includes such items as angles of attack and sideslip, acceleration load factors,

Scramjet Propulsion: A Practical Introduction, First Edition. Dora Musielak.

Figure 13.1 Artist's depiction of a Hyper-X research vehicle (X-43A) under scramjet power in free flight. Source: NASA Dryden Flight Research Center, Image ED97 43968-03.

and flight altitude and rates. As we have learned, the hypersonic air-breathing flight regime is the most severe for any flight vehicle. To produce sufficient levels of thrust, the hypersonic air breather must fly deep in the atmosphere at high dynamic pressures to capture and process enough airflow (see Figure 2.3). This high dynamic pressure results in high aerothermodynamic heating and large structural airloads. Moreover, hypersonic speeds cause changes in the chemical and physical conditions on the airflow around the vehicle and through the propulsion flowpath. In addition, localized atmospheric effects such as wind shears, turbulence, and density pockets can significantly affect flight vehicle performance and stability.

13.3 Flight Test Technique Concepts

In the following sections, we review the different approaches that have been conceived to flight test scramjet-powered vehicles and hypersonic flight technologies. These approaches include subscale captive carry, using both rockets and aircraft, and air-launched free flights. We will discuss the use of subscale hardware, its issues, and limitations and address the value of full-scale integrated flight vehicles.

13.3.1 Subscale Captive Carry

Captive carry refers to a flight testbed platform in which one or several test articles can be mounted. The carriage launch vehicle propels the test article to the experimental test conditions for a finite time. These flight experiments can complement and correlate with ground tests and can achieve such conditions as real gas effects, or higher test Mach numbers that are not attainable on the ground. Many useful flight experiments could be conducted with small subscale devices in a captive-carry mode aboard flight vehicles. Carriage vehicles can be expendable rockets or aircraft, depending on the experimental article, test conditions, risk assessment, cost, integration, infrastructure, and degree of mission control needed.

Aircraft captive-carry flight experiments are limited to approximately Mach 3 and below. During the 1990s, NASA used two SR-71 Blackbird aircraft as testbeds for high-speed and high-altitude aeronautical research at Dryden. The piloted aircraft included an SR-71A and SR-71B (the trainer version), loaned to NASA by the US Air Force. The Blackbirds were designed to cruise at Mach 3.2 (more than 2200 mph) and at altitudes up to 85 000 ft (25.91 km). The extreme operating environment in which they flew made these supersonic aircraft excellent platforms for conducting research and experiments in a variety of disciplines: aerodynamics, propulsion, structures, thermal protection materials, high-speed and high-temperature instrumentation, atmospheric studies, and sonic boom characterization. The NASA SR-71 aircraft are no longer in operation.

Using rocket launch vehicles for captive-carry flight experiments expands the test regime to hypersonic Mach numbers. Rocket launch vehicles usually cost less and provide a less risky approach to carrying experiment packages aloft than aircraft. These vehicles achieve significant lift weight and accelerate to very high hypersonic Mach numbers. NASA conceived a flight test configuration using the Boeing B-52B aircraft to air drop a rocket booster (to which a test article is mounted) from 43 kft at Mach 0.8. An example of such configuration is the Pegasus Piggyback Experiment. The Pegasus Hypersonic Experiment (PHYSX) consisted of a smooth glove installed on the first-stage delta wing of the Pegasus space rocket booster, which in turn was attached to the wing pylon of NASA's B-52 launch aircraft at Dryden Flight Research Center (DFRC). The glove was used to gather data at speeds of up to Mach 8 and at altitudes approaching 200 000 ft (60.96 km). The flight took place on 22 October 1998. The PHYSX experiment aimed to determine where boundary-layer transition occurs on the glove and to identify the flow mechanism causing transition over the glove. Data from this flight-research effort included temperature, heat transfer, pressure measurements, airflow, and trajectory reconstruction. Pegasus is an air-launched space booster designed by Orbital Sciences Corporation and Hercules Aerospace Company (initially; later, Alliant Tech Systems) and used to provide small satellite users with a cost-effective, flexible, and reliable method for placing payloads into low Earth orbit.

There are advantages and disadvantages for each flight test approach, and those must be evaluated considering cost, integration, flight time, available infrastructure, and test objectives.

13.3.2 Air-Launched Free Flight

Air-launched free flight is a test approach in which the experimental aircraft is taken by a launcher/carriage vehicle to its flight initial condition and allowed to separate to fly and operate independently at the goal-test conditions. Air-launched free-flight experiments are necessary for testing large subscale, highly integrated hypersonic systems or complete flight vehicle concepts. A free-flight vehicle allows direct assessment of the net thrust minus drag production without being hindered by effects from the launch vehicle such as carriage vehicle drag, control requirements, and aerodynamic or structural interference, or both. The ultimate hypersonic system or vehicle performance and proof-of-concept demonstration can be achieved by free flight.

Flight is the only way to demonstrate that an air-breathing propelled hypersonic vehicle is capable of flying on its own scramjet power. We can distinguish three approaches to hypersonic air-breathing propulsion free flight testing: (a) air-lifted, rocket-boosted; (b) sounding rocket launch; and (c) surface-to-air missile launch.

The air-lifted, rocket-boosted approach to flight testing requires a subsonic aircraft carrier and a rocket launcher. The aircraft is required to lift a stack (the scramjet vehicle attached to the booster solid rocket) to a predetermined altitude and take-off speed appropriate for safe flight testing. After the stack is dropped, the booster rocket ignites and thrusts the scramjet-powered vehicle to its altitude and hypersonic takeoff speed. Utilizing this approach to achieve hypersonic insertion conditions, the NASA X-43A and the Air Force X-51A programs had a goal of validating design methods in flight, using small-scale hypersonic air-breathing experimental vehicles fueled for powered flight over a narrow speed range and rather short time. Constrained by budget, those programs did not consider recovery of the spent vehicles despite having to build a new one for each flight test. It should be clear that because the X-43A and X-51A programs aimed to demonstrate technically sophisticated scramjet-integrated vehicles in flight, they required flight testing infrastructure of extraordinary scale and complexity.

The air-launched free flight successfully demonstrated by the NASA Hyper-X research program provided the first opportunity to obtain data on an airframe integrated scramjet propulsion system at true flight conditions. The program was conceived with a hypersonic vehicle that could be launched from a B-52B airplane and accelerated to hypersonic speeds by a small solid rocket. It was the first program for flight validation of experimental ground testing coupled with numerical and analytical methods to design a hydrogen-fueled hypersonic aircraft. Figure 13.2 illustrates the air-lifted, rocket booster approach, with the Pegasus solid propellant rocket booster positioned under the wing of the B-52B aircraft and with the

Figure 13.2 Air-lifted, rocket boosted approach to hypersonic air-breathing propulsion flight testing. NASA's B-52B launch aircraft cruises to a test range over the Pacific Ocean carrying the X-43A vehicle attached to a Pegasus rocket on 27 September 2004. Source: Gray Creech/NASA/Public Domain.

Hyper-X vehicle mounted to the booster through a load carrying adaptor. The complete launch system, referred to as the Hyper-X Launch Vehicle (HXLV) consisted of the Research Vehicle (best known as the X-43A) with its airframe integrated scramjet engine, a modified, stage one Pegasus booster rocket, and an adapter connecting the two.

A substantial portion of the Hyper-X program was based on wind tunnel tests, both for database development and performance validation. Wind tunnel verification of the propulsion-airframe integration included performance and operability, vehicle aerodynamic and thermal database development, thermal-structural design, boundary layer transition analysis and many other phenomena and operational characteristics. The databases so developed were used to baseline the scope of the flight testing that was necessary to validate the wind tunnel results and CFD predictions.

Flight is the ultimate application of hypersonic air-breathing propulsion; thus, flight testing is the only way to verify and validate the technologies integrated into a flying vehicle. However, flight testing of scramjet engines is extremely complicated, e.g. the integrated scramjet must be boosted to operating conditions; the test system must be built to survive the boost environment, achieve takeover conditions, have autonomous flight controls, transmit recorded data to ground stations, and use flight proven subsystems (Joyce et al. 2005). Moreover, extensive and expensive infrastructure and range assets must be available to support the flight. Nevertheless, flight data is required to compare with analytical predictions and ground and CFD data, and used to examine operability of key subsystems in flight, and feed back to the design of the protype flight system. Such a synergetic program reduces risk and ensures the delivery of an optimized, reliable flight ready vehicle.

The second most common method to test in flight utilizes a sounding rocket launch to accelerate a captive-carry (attached) payload through a hypersonic air-breathing propulsion flight corridor. The Free Flight Atmosphere Scramjet Test Technique (FASTT) approach was developed by GASL, now part of Northrup Grumman. It used an unguided, sounding rocket boost system to carry a scramjet-powered vehicle to hypersonic test conditions along a ballistic trajectory. FASTT was conceived to demonstrate the use of ground-launched ballistic rockets as a low-cost solution for maturing hypersonic air-breathing components and systems in flight, raising the engine Technology Readiness Level or TRL before transitioning to full-scale flight-testing (Foelsche et al. 2006).

In Australia, scramjets are flight tested with sounding rockets (Paull et al. 2002), an approach first pioneered through the HyShot (Hass et al. 2005) and HYCAUSE (Walker et al. 2008) programs utilizing the Woomera Test Facility in South Australia. These efforts were followed by the very successful Hypersonic International Flight Research Experimentation (HIFiRE) program. Envisioned and established during 2005–2006, this program integrated the efforts of the US Air Force Research Laboratory (AFRL), NASA, and the Australian Defence Science & Technology Organisation (DSTO) in the execution of a flight research activity intended to support foundational research in the hypersonic sciences. HIFiRE planned to carry out up to ten flights using various launch vehicles (two stage and three stage sounding rockets) designed to satisfy different flight experiment requirements. HIFiRE was jointly established by DSTO and AFRL, opening the door for new flight configurations to support future hypersonic flight research.

In the Russian CIAM program, scramjet engines were flown captive-carry atop the SA-5 surface-to-air missile that included an experiment flight support unit known as the "Kholod" or Hypersonic Flying Laboratory (HFL), (Roudakov et al. 1993).

Berry et al. (2011) provided a review of existing rocket launch capability for conducting future hypersonic boundary layer transition flight experiments, including small sounding rockets and larger launch vehicles. They identified 11 operational launch vehicles in the NASA Sounding Rocket Program, unguided rockets which are flown out of the Goddard's Wallops Flight Facility (WFF), located on Virginia's Eastern Shore.

A variation of the aircraft-rocket configuration approach was proposed in 2007. Named the Phoenix Air-Launched Small Missile (ALSM) flight testbed, it would utilize the NASA DFRC F-15B Boeing airplane, and the US Navy Phoenix AIM-54 long range, guided air-to-air missile to boost the testing article to its hypersonic condition. This aircraft-missile arrangement was proposed to help address the lack of quick-turn around and cost-effective hypersonic flight research capabilities (Bui et al. 2007).

In the 1960s, the gun-launch was proposed to accelerate a projectile to scramjet take-over speeds and for flight in an atmospherically controlled ballistic range. In 2003, DARPA's ScramFire, a gun-launched, scramjet-powered projectile, flew at Mach 6 and Mach 8 using gaseous ethylene fuel.

Combinations of these flight test configurations are possible, depending on available infrastructure, test objectives, size of the experimental payload, funding, and many other practical considerations related to a given development program.

> Only flight demonstration can fully validate new or immature technologies and operational techniques developed for a scramjet-powered hypersonic vehicle.

The following section provides some highlights of the Hyper-X (X-43A) program, in order to emphasize the tremendous effort that took place to achieve Mach 9.8, the fastest air-breathing hypersonic flight ever. The NASA Hyper-X program provided the know-how for flight testing the hydrocarbon-fueled scramjet that powered the Air Force X-51A waverider. X-51A achieved the highest hypersonic speed with a JP-7-fueled scramjet. On its final flight test (1 May 2013), the X-51A vehicle reached Mach 5.1 traveling more than 230 nm in over six minutes. It was the longest of four X-51A test flights and the longest air-breathing hypersonic flight powered by hydrocarbon-fueled scramjet (Mutzman and Murphy 2011; Rondeau and Jorris 2013).

13.4 X-43A: Air-lifted, Rocket-boosted Approach

The X-43A, formally the Hyper-X Research Vehicle (HXRV), was designed to be accelerated to its operational altitude and Mach number using a rocket-propelled launch vehicle, designated the HXLV. The HXLV was derived from a modified Pegasus launch vehicle stage one (Orion 50S) configuration. The HXRV was attached to the HXLV via the HXRV adapter, which provided services to maintain the desired HXRV environmental conditions during mated flight, and to separate the HXRV from the HXRV adapter for scramjet operation. This

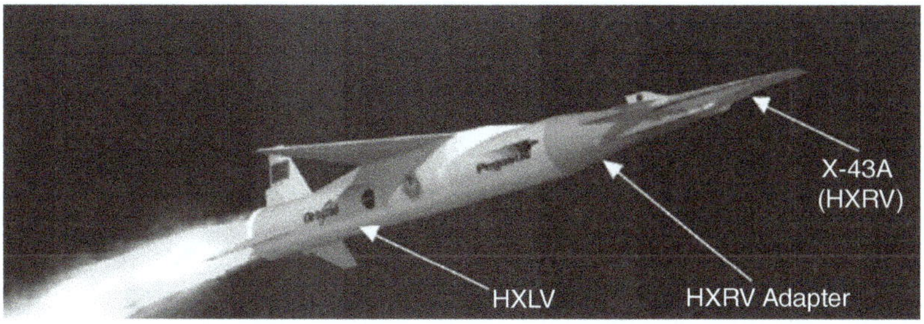

Figure 13.3 The NASA X-43A stack. Source: From Cotting (2004), NASA.

combination of the HXLV, the HXRV adapter and the HXRV was designated the X-43A stack (Figure 13.3).

The X-43A stack was integrated to the wing of the B-52 carrier aircraft and was flown to the launch area for deployment. Figure 13.4 shows a photograph of the X-43A mated HXRV-HXLV stack mounted under the wing of the B-52B airplane. The X43A HXRV was 12 ft (3.6576 m) long, 5 ft (1.524 m) wide, 2 ft (0.6096 m) high, and weighed about 3000 pounds (1360.777 kg). It was powered by a single hydrogen-fueled, dual-mode, airframe-integrated scramjet propulsion system.

Figure 13.5 illustrates the nominal flight test trajectory. The X-43A mission consisted of four flight phases: (i) Air-lifted X-43A stack by subsonic B-52B carrier aircraft, (ii) HXLV Boost, (iii) X-43A HXRV Separation, and (iv) X-43A HXRV Free Flight on scramjet power. Both the engine test and vehicle descent occurred during the HXRV Free Flight phase. As

Figure 13.4 The NASA X-43A stack mounted under the wing of a B-52B airplane, ready to be carried to test altitude. Source: Photograph from the Dryden Flight Research Center Collection/NASA/Public Domain.

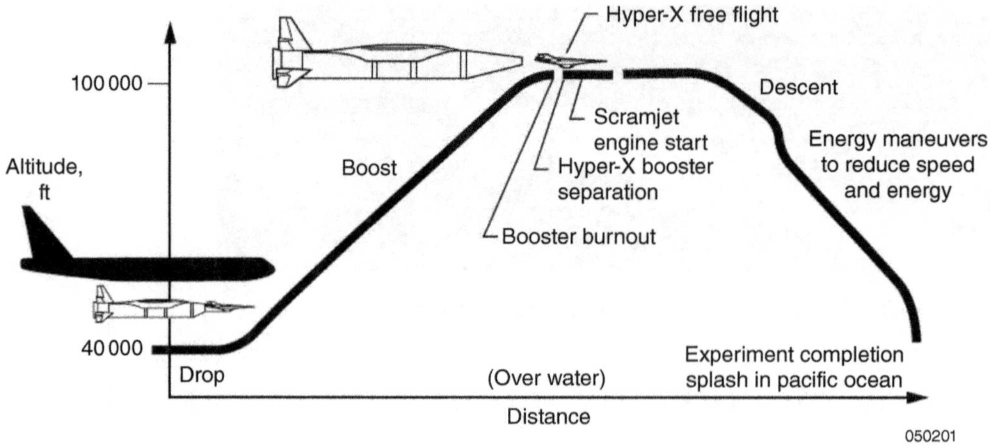

Figure 13.5 Nominal trajectory of X-43A flight tests. Source: Lux and Burkes (2008)/Public Domain.

shown in Figure 13.5, for all scramjet test conditions, the B-52B carried the Hyper-X stack to the designated launch point, Mach 0.8 at an altitude of 40 000 ft (12 192 m), approximately 50 nmi (92.6 km) off the southern California coast where the stack was released. The Pegasus rocket was then ignited to boost the X-43A vehicle to around 95 000 ft (~28.96 km) altitude. At that altitude, the rocket and vehicle were separated, and the scramjet engine was ignited, allowing the vehicle to cruise under its own power.

The first X-43A flight was attempted on 2 June 2001. The X-43A stack was carried under the wing of the B-52B. After the B-52 dropped the stack, the HXLV was to accelerate the HXRV to the required Mach number and operational altitude to obtain scramjet propulsion data. Unfortunately, the first attempt to test the scramjet in flight was unsuccessful due to a failure during the boost stage caused by a deficient control system design (McClinton et al. 2005). The boost was attempted at a lower altitude (22 244 ft) (higher dynamic pressure, 650 psf) than a typical Pegasus booster was designed for (see Section 13.10). A mishap investigation followed. Once the issue was resolved, two successful flight tests were carried out three years later.

Figure 13.6 illustrates the X-43A HXRV separation event, which occurred at an altitude of about 28.96 km. The artist's image shows the rocket booster after completing its task of carrying the research vehicle to the test altitude and speed, and the research vehicle has separated from the booster prior to scramjet ignition to demonstrate hypersonic flight with air-breathing propulsion.

In 2004, two X-43A vehicles successfully accomplished their hypersonic test flight over the Pacific Ocean from the Western Aeronautical Test Range (WATR), NASA DFRC. The flight profiles were similar for the Mach 7 and Mach 10 missions, except that the higher Mach number, higher altitude flight resulted in a longer descent trajectory. Also, the second vehicle reached a maximum speed of Mach 6.946, traveled about 400 nmi (740.8 km) downrange from the launch point, while the third vehicle, with a maximum speed of Mach 9.7, traveled more than twice that distance, to approximately 850 nmi (1574.2 km)

Figure 13.6 An artist concept of NASA's X-43A vehicle stage separation. The booster carried the research vehicle to the test altitude and speed, and the X-43A has separated from the booster prior to scramjet ignition. Source: Photograph from the Dryden Flight Research Center Collection/NASA/Public Domain.

downrange, which presented greater challenges for data acquisition and tracking (Lux and Burkes 2008).

13.4.1 Mach 7 Flight Test

Flying at an altitude of 12.2 km (40 000 ft) and Mach 0.8, the B-52 aircraft dropped the X-43A stack, and after a five-second free fall to about 12.0 km (39 500 ft), the rocket booster ignited. The HXLV executed a 1.9 g pull-up, followed by a 0.7 g pushover to achieve nearly level flight to the designated separation conditions. After booster burnout, the X-43A separated from the RV adapter by four explosive bolts and two pyrotechnically pressurized pistons to begin the free-flight phase and scramjet engine experiment. Approximately 2.5 seconds after separation, the engine cowl door opened to obtain five seconds of fuel-off performance baseline data. Then the scramjet ignited and operated for approximately 11 seconds, reaching Mach 6.946 when the dynamic pressure was 1024 psf (49 kPa) at an altitude of 94 049 ft (28.67 km) (Marshall et al. 2005b).

The scramjet cowl door was closed throughout the boost phase to block the inlet entrance and protect the internal engine components from high heat loads. After stabilization, the cowl door was opened to establish air flow through the engine. Following a few seconds of unpowered operation, hydrogen fuel was introduced and the powered portion of the flight began with the scramjet achieving Mach 6.83 for about 10 seconds, at a dynamic pressure of 0.463 atm (980 psf). This event was followed by five seconds of unpowered steady tare, followed by 10 seconds of Parameter IDentification (PID) maneuvers, (McClinton 2008). These PID maneuvers were conducted to quantify the vehicle's aerodynamic stability and control parameters, and to allow more accurate estimation of the scramjet engine thrust. After the cowl door closed again, the vehicle began a controlled descent over 560 km (300 nmi) to "splash-down" in the Pacific Ocean.

During descent, a series of PID maneuvers were performed as the vehicle slowed down starting at Mach 5 and at every integer Mach number down to Mach 2 (Morelli et al. 2005; Baumann et al. 2007). These maneuvers consisted of a series of step inputs, frequency sweeps, and an α sweep (Bahm et al. 2005). In 2010, Bauman et al. reported on the flight data from the boost ascent and unpowered post-experiment descent portions of the flight with the cowl door closed.

The flight results were compared with pre-flight predictions (Jentink and Ferlemann 2005; Meyer 2005). Marshall et al. (2005b) reported that the flight conditions were very close to the predicted separation conditions. Target values for this flight were Mach 7.075, altitude 93 932 ft (28.63 km), and dynamic pressure 1066 psf (0.503 atm). During powered flight, the maximum Mach number achieved was 6.83, and the X-43A flight controls and all separation conditions were within an acceptable tolerance.

13.4.2 Mach 10 Flight Test

The flight conditions for the B-52 aircraft carrier were the same as before (Mach 0.8 at an altitude of 40 000 ft or 12.2 km). The HXLV launch vehicle executed a 2.5 g pull-up to a flight path angle of over 30 degrees, followed by a 0.5 g push over to achieve nearly level flight. After booster burnout, the X-43A vehicle separated at Mach 9.702 at an altitude of 109 653 ft (33.4 km) (Karlgaard et al. 2005). After vehicle stabilization, the engine cowl was opened for about 20 seconds: three seconds of fuel-off tare, 11 seconds of powered flight and another six-seconds of unpowered steady tare. The pre-experiment tare was successfully performed and the engine fueling sequence was initiated approximately five seconds after separation. Both the ignitor and the fuel were delivered to the engine as predicted. The scramjet was fueled for approximately 10 seconds and provided the predicted thrust. Reported separation conditions reported by Marshall et al. (2005a) indicate the X-43A achieved Mach 9.736 at an altitude of 109 440 ft (33.35 km) and a dynamic pressure of 959 psf (0.4532 atm).

The scramjet-powered flight experiment concluded when the cowl door closed 22 seconds after separation. The powered portion of flight included five seconds of silane-piloted operation at two fuel equivalence ratio settings, followed by hydrogen-only fueled operation at two settings, then ramp down. This conservative approach assured good data before trying to operate the small engine without piloted fuel (Voland et al. 2005). The X-43A vehicle flew a controlled descent over 1600 km (800 nmi) to a "splash-down" in the Pacific Ocean.

During the descent portion of the mission, a series of PID maneuvers were performed as the vehicle slowed down, starting at Mach 7 and at every integer Mach number down to Mach 2 (Morelli et al. 2005; Baumann et al. 2007). The PID maneuvers consisted of two sets of frequency sweeps. The publicly released flight data are from the boosted ascent and unpowered post-experiment descent portions of these flights with the cowl door closed (Bauman et al. 2010).

13.4.3 NAWC-WD Sea Range That Supported the X-43A Flight Tests

The flight tests of the X-43A vehicle required multiple telemetry ground stations to provide continuous coverage of the captive carry, launch, boost, experiment, and descent phases of these missions. Lux and Burkes (2008) provided a view of the Naval Air Warfare Center Weapons Division (NAWC-WD) Sea Range that supported the flight tests. They also reported on the flight test operations, giving an overview of vehicle telemetry and distributed assets that supported telemetry acquisition, best source selection, radar tracking, video tracking, flight termination systems, and voice communications.

A comprehensive Infrastructure Readiness program was carried out to support these hypersonic flight tests. For example, an aircraft designated to take flight data was required to capture data for the entire X-43A flight (Mach 7 or Mach 10). The Navy NP-3D Orion aircraft (P-3) was selected for this task. Since the Mach 10 vehicle would travel further downrange in flight than the Mach 7 vehicle, two such aircraft were needed to capture data for the longer flight. Due to maintenance schedule, only one P-3 was available to support the Mach 10 flight. Thus, the team decided to position the aircraft such that it would capture the primary mission (boost through cowl closed) and capture as much data prior to splashdown as possible (Marshall et al. 2005b).

Lux and Burkes (2008) provide an overview of the extensive coordination required between the WATR and NAWC-WD to provide continuous coverage of the flight data, position information, video sources, and voice communications. Radar tracking, video tracking and transmission, flight termination systems, and voice communications for the X-43A program all performed according to plan for both Mach 7 and Mach 10 flights. Real-time data provided by the WATR and NAWC-WD assets were successfully transmitted into the NASA DFRC mission control centers to facilitate all safety and mission success decisions prior to launch (Lux and Burkes 2008).

13.5 Australia/USA Flight Experiments with Sounding Rockets

The Hypersonic International Flight Research and Experimentation (HIFiRE) program consisted of a series of flight experiments by US Air Force Research Laboratory (AFRL), the Australian Defense Science and Technology (DST), the University of Queensland, Boeing, NASA, and their contractors. HIFiRE used low-cost, sounding-rocket-based approach to hypersonic experimentation, based on experience gained through the HyShot and HYCAUSE programs carried out in Australia.

Because flight data involving supersonic combustion did not exist in the late 1990s, the University of Queensland (UQ) proposed flight testing a scramjet combustor using sounding rockets. The HyShot was a flight experiment designed to develop correlation between pressure measurements made of supersonic combustion in the T4 shock tunnel at UQ and the pressure measured in flight. The HyShot program was the least expensive as it was a component (combustor) flight test. It used a two-stage Terrier-Orion Mk70 rocket to boost the payload to an apogee of approximately 330 km. The trajectory is designed so that on descent (reentry), between 35 and 23 km, the payload is traveling at Mach 7.6. This is the part of the trajectory where the measurements of supersonic combustion are made. The nominal HyShot mission profile is shown in Figure 13.7.

The HyShot mission begins with ignition of the Terrier first stage, followed by burnout. Upon separation of the spent Terrier rocket, the Orion second stage ignites and burns. The spent Orion rocket remains attached to the experimental payload in order to provide aerodynamic stability on re-entry to the atmosphere (Smart and Stalker 2010). The flight

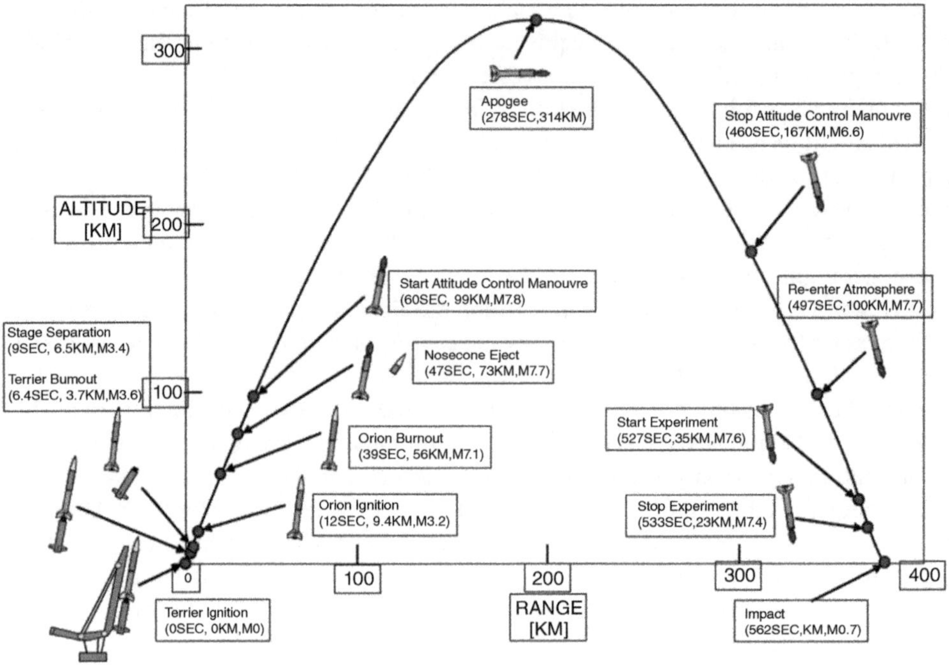

Figure 13.7 HyShot flight profile. Source: Smart and Stalker (2010).

vehicle then follows a parabolic trajectory, ascending to an altitude of 314 km and then reenters the atmosphere. While executing the descent maneuver, a control jet is activated to realign the flight vehicle so that it re-enters with the experimental model pointing downward. The steep descent trajectory allows the experiment to take place over a planned period of seven seconds as the vehicle descends from 35 to 23 km altitude (Paull et al. 2002; Smart et al. 2006). The last HyShot IV was launched on 30 March 2006.

The Hypersonic Collaborative Australian/United States Experiment (HyCAUSE) flight was a DARPA-sponsored scramjet technology program begun in 2004. It included ground tests and CFD analysis of two-dimensional and inward-turning scramjet engine concepts and a flight test aimed at obtaining engine data at Mach 10. In 2007, the HyCAUSE flight was conducted with a Talos-Castor launcher, a two-stage unguided booster launching the experiment payload on a ballistic trajectory with a target Mach 10. The test was unsuccessful.

The Hypersonic International Flight Research Experimentation (HIFiRE) Program was designed to demonstrate fundamental technologies critical to next-generation aerospace systems. HIFiRE consisted of extensive ground tests and computation focused on specific hypersonic flight technologies such as propulsion, propulsion–airframe integration, aerodynamics and aerothermodynamics, high-temperature materials and structures, thermal management strategies, guidance, navigation, and control, sensors, and system components (Dolvin 2009). Each technology program culminated in a flight test. HIFiRE objectives were to increase understanding in fundamental hypersonic phenomena and enable research and

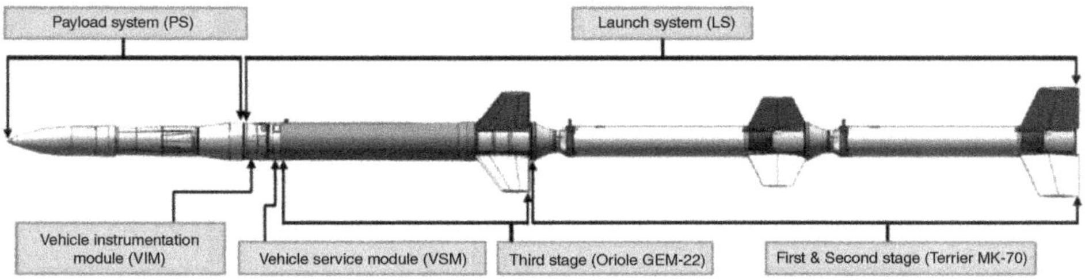

Figure 13.8 HIFiRE-2 payload mounted atop a three-stage rocket launch system. Source: Jackson et al. (2011)/Public Domain.

exploration in flight regimes that are expensive and difficult to model with CFD and test in ground test facilities. Various trajectories and various launch vehicles were employed to satisfy the requirements of the flight experiment. The flights were planned for launch at Woomera in South Australia, or an alternative range in Andoya, Norway. In 2009, the first (HIFiRE-0) trial successfully tested flight and mission control systems that were used in subsequent flight experiments.

Flights 1 and 5 of the HIFiRE Program were dedicated to aerothermodynamics. Their flight trajectory was parabolic (Adamczak et al. 2011), similar to that of HyShot depicted in Figure 13.8. The two-stage launch system for the HIFiRE-5 test consisted of an S-30 first stage and an improved Orion second-stage motor, with the payload mounted with a transition section atop (Kimmel et al. 2018).

13.5.1 HIFiRE-2

In May 2012, the HIFiRE team celebrated a successful test of a hydrocarbon-fueled scramjet research, on a flight from the Pacific Missile Range Facility on the island of Kauai, Hawaii. The HIFiRE 2 used a three-stage launch system (two Mk-70 Terrier boost motors and a GEM-22 Oriole sustainer motor) to test scramjet engine mode transition and operability. The flight vehicle configuration consisted of the launch system and payload, Figure 13.8. The hydrocarbon-fueled scramjet was boosted to 100 000 ft (30.480 km) altitude, accelerated from Mach 6 to Mach 8 (4567–6090 mph; 7350–9800 km/h), and operated for about 12 seconds. Figure 13.9 depicts the nominal trajectory.

The payload for HIFiRE 2 was a scramjet with an inward turning inlet, followed by a supersonic combustor and dual-exhaust nozzle. It was developed under a partnership between the AFRL and NASA, with contributions from the Navy's detachment at White Sands Missile Range, N.M., and Northrop Grumman Propulsion Test Complex (formerly GASL). This was the first time the HIFiRE team tested a hydrocarbon-fueled scramjet accelerating from Mach 6 to Mach 8. More than 700 instruments on board recorded and transmitted data to researchers on the ground. This flight test provided unique data on scramjets transitioning from subsonic to supersonic combustion, a complex process that wind tunnels cannot simulate. It also established the three-stage launch system as a new high-performance flight configuration to support future Air Force, Navy, and NASA flight research.

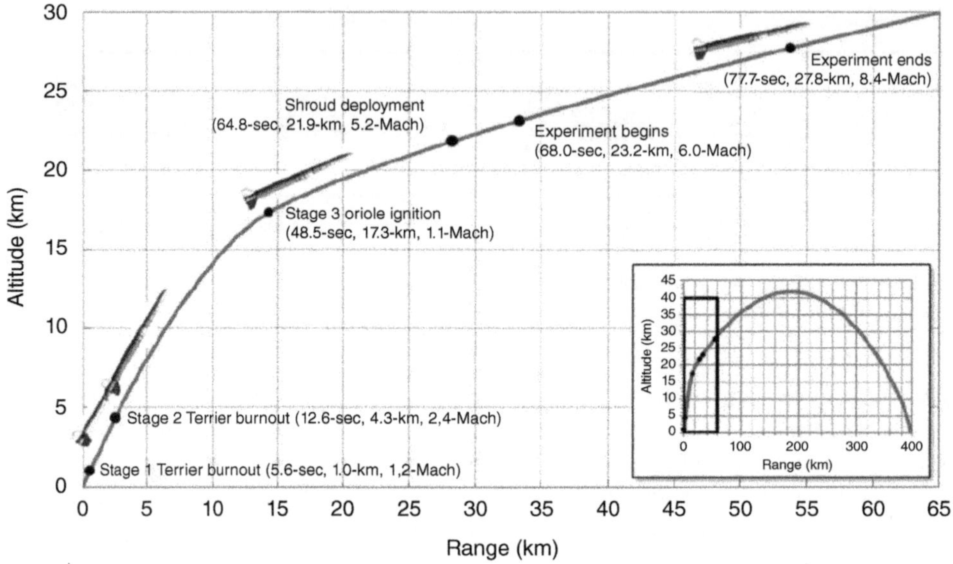

Figure 13.9 HIFIRE-2 trajectory profile. Source: Jackson et al. (2011)/Public Domain.

A follow-up flight to test a scramjet-powered waverider design was planned. It considered a depressed trajectory and the HyShot team planned to fly the scramjet vehicle horizontally for up to a minute at Mach 8. Unfortunately, the HIFiRE program ended before the waverider vehicle could be flown.

Technical details related to HIFiRE Program and its many accomplishments are found in Dolvin 2016; Jackson et al. 2011; Bowcutt et al. 2012; Kimmel et al. 2018, and references therein.

13.6 Russia CIAM and NASA Partnership for Scramjet Flight Testing

In the 1980s, the Russian Central Institute of Aviation Motors (CIAM) carried out the first successful flight tests of an unfueled two-dimensional scramjet model that had been tested in a free-jet test facility at Mach numbers 5 and 6. The cold scramjet was flown atop the SA5 surface-to-air rocket-powered missile, using "Kholod," Russia's HFL.

In the 1990s, CIAM designed and launched a subscale model of a Mach 6 dual mode, hydrogen-fueled scramjet engine. The first flight test took place in November 1991. The second flight test was carried out in November 1992. The trajectory segment over which the scramjet operated lasted for 23 seconds. Researchers reported stable combustion at hydrogen injection both from the subsonic and supersonic section of the fuel manifold at flight Mach number 5.3 (Sabel'nikov and Penzin 2000).

In November 1994, NASA contracted with CIAM to continue ground and flight research of the axisymmetric, dual-mode scramjet. The collaborative program incorporated ground and flight tests to explore the engine operating envelope from the ram/scram dual-mode below Mach 6 to the full supersonic combustion (scram) mode at Mach 6.5. This required a redesign of the combustor and active cooling system in order to cope with the higher heat loads it would experience at the higher Mach condition. At the same time, the SA-5 missile booster in the HFL required modifications to ensure the launcher would achieve the velocity and altitude specified for the new Mach 6.5 flight test. CIAM had to build four identical dual-mode scramjets to accomplish all required ground and flight tests: two engines for wind tunnel testing up to Mach 6-simulated flight conditions; one engine designated for flight test; and one backup flight test engine. The program required ground-to-flight comparisons for scramjet engine design tool methodology verification.

In 1998, CIAM performed a flight test of the hydrogen-cooled/fueled dual-mode scramjet engine over a Mach number range of 3.5–6.4. The flight test was conducted at the Sary Shagan test range in Kazakhstan (Russia). This rocket-boosted, captive-carry flight test of the axisymmetric scramjet reached the highest Mach number of any scramjet developed at that time and for the longest duration: it achieved 77 seconds of liquid-hydrogen fueled and regeneratively cooled engine data at Mach numbers ranging from 3.5 to 6.4 (Voland et al. 1999).

Post flight analysis revealed that, at the maximum flight Mach number condition, the engine operated as a dual-mode scramjet in a subsonic combustion mode. The NASA team concluded that this unacceptable result was probably caused by discrepancies between the as-designed and as-built inlet flowpath geometry, and also due to the engine control system not allowing fueling of the first stage injectors (which gave a false indication of inlet unstart). The most important lessons learned from this flight test activity emphasized the need to maintain the as-designed internal flowpath for predictable performance of scramjet engines and having incorporating fuel control laws with robust error detection and accommodation to avoid the loss of data due to unexpected flight events (Voland et al. 1999).

The flight and ground data generated during the NASA/CIAM joint test program were valuable for partial validation of scramjet engine design tools. However, the axisymmetric scramjet configuration tested was of little practical use for an operational hypersonic aircraft. The focus then shifted to developing fully integrated scramjet-powered cruise vehicles, which NASA demonstrated with the successful flight tests of the Hyper-X Program.

13.7 Hypersonic Flight Demonstration Program (HyFly)

The Hypersonic Flight Demonstration Program (HyFly) was a joint program between DARPA and the Office of Naval Research (ONR). It aimed to flight-test key technologies and demonstrate a missile-like vehicle with a range of 400 nm (740 km) and a maximum cruise speed of Mach 6. The HyFly program aim was to develop a hypersonic strike demonstrator vehicle, based on a hybrid propulsion concept called the dual combustor ramjet (DCR), developed at the Johns Hopkins University Applied Physics Laboratory. The DCR combines scramjet and ramjet flowpaths into an integrated engine to achieve performance over a wide speed range. It was designed to preburn liquid JP-10 fuel in a subsonic-combustion ramjet to generate a fuel-rich gas that is then burned in the scramjet.

The HyFly program included a subscale version of the DCR concept that was to be flight tested using the Freeflight Atmospheric Scramjet Test Technique (FASTT). As we noted before, this flight test approach was developed by GASL (now part of Northrup Grumman) as a low-cost technique to study subscale scramjet-powered vehicles in a true flight environment compared with full-scale flight testing.

The first flight test of HyFly was conducted on 26 January 2005 with an unpowered missile to demonstrate safe separation from a Boeing F-15E strike fighter jet, and to test vehicle guidance and control functions. The vehicle was launched from the NASA WFF at Wallops Island, Va. It was boosted by a two-stage, Terrier-Orion unguided solid-rocket system. Following separation from the booster, the scramjet engine ignited, propelling the vehicle to Mach 5.5 (speeds of 5300 ft/s), at an altitude of 63 000 ft (19.2 km).

On 26 August 2005, the Boeing Company, in partnership with DARPA and ONR, successfully demonstrated boost-phase performance of the hypersonic strike demonstrator vehicle. A Boeing F-15E launched the HyFly vehicle for the flight test over the US Navy's sea range at the Naval Air Weapons Center-Weapons Division (NAWC-WD). The solid rocket booster successfully ignited and accelerated the HyFly vehicle to its take-over speed.

The last flight test of HyFly was on January 2008. As before, the vehicle was launched from an F-15E fighter jet and boosted by rocket motor to the take-off speed where the dual-combustion ramjet was to ignite and accelerate the vehicle to Mach 6. However, the scramjet did not operate. After approximately 58 seconds of free, unpowered flight, the vehicle impacted the ocean. After three failures, the HyFly program was cancelled.

13.8 Phoenix Air-Launched Small Missile (ALSM)

To provide a quick turnaround and cost-effective hypersonic flight research capability, NASA Dryden proposed the Phoenix ALSM flight testbed. The ASLM would utilize existing assets to enable test and evaluation of hypersonic systems for flight Mach numbers ranging from 3 to 5 (Bui et al. 2007).

The Phoenix ALSM testbed is composed of the Navy Phoenix AIM-54 (Raytheon) long range, guided air-to-air missile and the NASA DFRC F-15B (Boeing) airplane. The Phoenix AIM-54 is a long-range, guided air-to-air missile designed to be carried by the US Navy F-14 Tomcat reconnaissance aircraft. Propelled by a solid-propellant rocket motor, the 13 ft (3.96 m) long missile has a diameter of 15 in (38.1 cm) and was designed to intercept air targets at a range in excess of 100 nmi. The AIM-54 was retired from fleet operation, thus presenting an opportunity for converting this flight asset into a new flight testbed.

In the proposed architecture for ALSM, a modified Phoenix missile (without warhead) would be used as a high supersonic–low hypersonic, precision-guided booster for flight research experiments. Analysis indicated that the modification would allow the missile to integrate a research payload volume of more than 7 cubic feet and an allowable weight of up to 250 pounds. By launching it from a high-performance 15B airplane, the performance of the standard Phoenix missile can be significantly increased. Preliminary studies indicated that, in this configuration, the guided Phoenix missile could boost research payloads to low hypersonic Mach numbers, enabling flight research in the supersonic-to-hypersonic transitional flight envelope (Bui et al. 2007).

13.9 Gun-Launched Scramjet Missile Testing

Bakos (2008) reported on gun-launched scramjet designs to demonstrate the feasibility of subscale gun launch as a means to provide low-cost scramjet free-flight data. The gun-launch approach was used to accelerate a projectile to scramjet take-over speeds and for flight in an atmospherically controlled ballistic range. The test vehicle was a 20-inch (50.8 cm) diameter missile operating in the range Mach 6-8.

In a gun launcher, a projectile is placed in a tube, then gas at high pressure is introduced behind it, and the projectile is driven down the tube by the expanding flow. When the driver is a gas of low molecular weight (hydrogen or helium), the device is called a light-gas gun. Light-gas guns produce higher projectile velocities than powder guns because accelerating the gas itself requires less energy (Gilbreath et al. 1988).

In the 1960s, the US Army Ballistic Research Laboratory and at McGill University in Canada began the development of gun-launched probes for upper-atmosphere studies under a program called the High-Altitude Research Project (HARP), a joint venture of the US Department of Defense and Canada's Department of National Defence. HARP was established to study ballistics of re-entry vehicles and collect upper atmospheric data for research. Unlike conventional space launching methods that rely on rockets, HARP instead used very large guns to fire projectiles into the atmosphere at extremely high speeds.

The HARP team designed a 100 kg scramjet vehicle to be launched from the HARP 16-inch gun in the island of Barbados. The scramjet's inlet design was based on a streamline traced of conical axisymmetric flow. Upon entering the scramjet, the freestream air flow was split into four equal streams that were compressed in the inlet and passed into four annular dump-type combustion chambers. The combustors burned three kg of triethylaluminum $[Al_2(C_2H_5)_6)]$ fuel, a compound that ignites on contact with air. The products of combustion were ejected through a 35-cm-diameter exhaust nozzle. The vehicle was designed to withstand acceleration loads up to 10 000 g. At launch, the vehicle became aerodynamically unstable because of structural damage suffered by the skin and stabilizing fins (Gilbreath et al. 1988).

Although a gun-launched air-breathing powered projectile is capable of very long range, packaging the projectile in gun-tube diameters and use of high-density fuels present unique challenges for testing ramjet–scramjet designs. The designs reported by Bakos (2008) proved difficult to optimize for both performance and strength for the gun-launch environment.

13.10 X-43A Flight Test Mishap

The first flight test attempt (2 June 2001) of the X-43A vehicle was unsuccessful. The Mach 7 mission failed during the boost phase when the launch vehicle HXLV departed from controlled flight. This failure resulted in a nine-month investigation to establish the root cause before the subsequent flight tests could be attempted. The X-43A Mishap Investigation Board (MIB) convened at NASA DFRC on 5 June 2001, and in the spring of 2002 the MIB attributed the mission failure to the launch vehicle. The following is a summary from their report (X-43A Mishap Investigation 2003).

One hour and 15 minutes after takeoff, the X-43A stack was released from a B-52 carrier aircraft. Just as planned, the HXLV solid rocket motor ignited 5.19 seconds later and began the pitch-up maneuver at eight seconds mission time. It was during this maneuver (at about 11.5 seconds) when the X-43A stack began to show a control anomaly, a diverging roll oscillation at a 2.5 Hz frequency. Two seconds later, the HXLV rudder electromechanical actuator stalled and ceased to respond to autopilot commands, leading to the loss of yaw control that caused the X-43A stack sideslip to diverge rapidly to over 8 degrees. Structural overload of the starboard elevon occurred at 13.5 seconds. This loss of control caused the X-43A stack to deviate significantly from its planned trajectory. Hence, the flight was terminated by range control 48.57 seconds after the vehicle was released from the B-52 aircraft.

The MIB concluded that the launch vehicle HXLV failed because the vehicle control system design was deficient for the trajectory flown due to inaccurate analytical models (Pegasus heritage and HXLV specific), which overestimated the system margins. In particular, the HXLV was launched and flown in an environment that was significantly different from previous Pegasus experience. At the time of failure (13.5 seconds mission time), the HXLV altitude was 22 244 ft (6.77 km), whereas a typical Pegasus altitude for the same flight duration would have been approximately 40 000 ft (12.19 km). Moreover, at Mach 1, which was near the failure point, the HXLV dynamic pressure was 650 psf (31 kPa), whereas a typical Pegasus dynamic pressure for the same Mach number would have been approximately 300 psf (16.75 kPa). This increase in dynamic pressure at transonic conditions was a major factor in the failure of the launch vehicle. Therefore, because the Pegasus departed from controlled flight, it could not boost the X-43A vehicle. Hence, it was not possible to carry out the first test of the Mach 7 scramjet in flight. Figure 13.10 depicts the notional dynamic pressure comparison of the flight 1 and flight 2 HXLV trajectories to a typical Pegasus first-stage boost. Note at the point of the mishap, the Pegasus dynamic pressure for the first flight was almost twice the dynamic pressure it should have at that Mach number.

Determining the cause of the X-43A flight mishap was a complex effort requiring a significant commitment of time and resources and the technical expertise of many individuals from NASA Centers and contractor organizations. After the MIB issued its findings, NASA updated test procedures based on flight 1 preparation experience and lessons learned. Where appropriate, additional testing was performed that provided the data for new models with improved fidelity that yielded a vehicle that was more robust in response to disturbances and better able to predict the results in flight (Marshall et al. 2005b).

Despite this mishap and failures in other programs, many achievements resulting from flight testing surpassed those setbacks and provided many lessons to guide future development efforts. In order to advance air-breathing hypersonic propulsion technology, we must be willing to accept technical risk and continue flight testing scramjet-integrated vehicles and their subsystems, as this is the only way to increase the probability of achieving useful scientific data and insight.

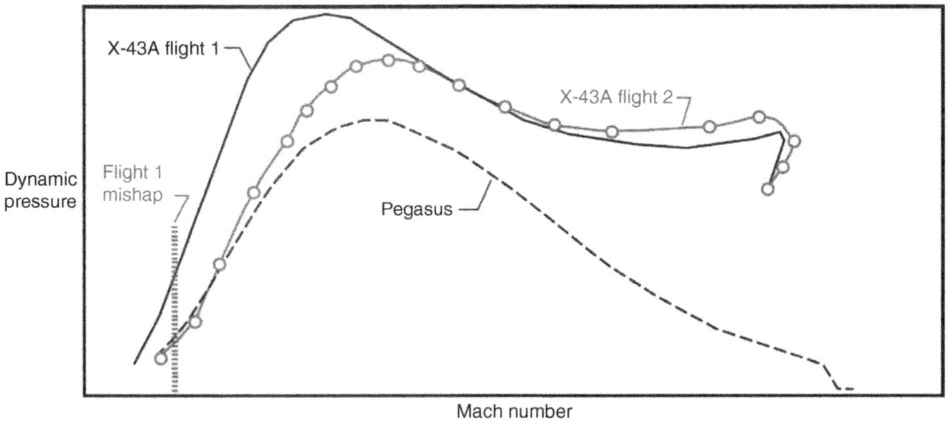

Figure 13.10 Comparison of boost trajectories for flights 1 and 2. Source: Adapted from X-43A Mishap Investigation Board (2003).

> *A flight program is the catalyst that drives technology development and synthesizes all of the efforts into a unified tool for development of the ultimate experiment, the flight of a hypersonic vehicle.* –Phil Drummond 2014.

13.11 Closing Remarks

We have reached the end of this book, which had the objective of providing a wide exposure to hypersonic air-breathing propulsion. It was intended to give the reader enough insight into the principles and the technologies that must be matured to achieve sustained hypersonic flight.

The NASA Hyper-X program culminated with the Mach 10 flight of the scramjet-powered X-43A vehicle, giving renewed confidence in the ability of design and analysis tools to effectively develop a hydrogen hypersonic engine with the desired performance. The U.S. Air Force SED-WR Program extended that capability with the demonstration flights of the hydrocarbon-fueled X-51A Waverider. These programs identified propulsion system components, integrations, and practical ground and flight test strategies to advance the technological maturity and accumulate in-flight operational experience. Although crucially important, these flights are only a first step on a long path to secure operational hypersonic air-breathing flight capability.

After successful flight demonstrations, it has become clear that the scramjet engine is a viable propulsion system, whether fueled with hydrogen or with hydrocarbons, and the question is no longer whether it is possible, but what technologies must be further matured to make hypersonic flight routine.

Although previous development programs have provided a wide range of analytical and testing methodology, much more work is required to advance high-temperature materials, TPS, instrumentation, and other technologies for the next generation of hypersonic air-breathing propulsion. The design of hypersonic vehicles will rely heavily on the use of computational fluid dynamics, because most of the experimental facilities are inadequate to simulate the full range of operating conditions (Mach number, Reynolds number, and enthalpy levels) to obtain the required data for design, analysis, and continued development of hypersonic technologies.

Flight control is a key enabling technology for flight demonstration of hypersonic air-breathing vehicle readiness. This topic is outside the scope of this book. However, based on what has been done to date, closed-loop flight control is essential to enable a successful stage separation (Blocker and Ruebush 2005), to achieve and maintain the design condition during the hypersonic air-breathing system test, and to provide a controlled descent of the test vehicle. Equally important is to incorporate flight testing with test article recovery to validate the design, verify analysis, and advance technologies of a new prototype reusable hypersonic vehicle.

I encourage you to study deeper into the basic sciences and engineering principles that provide the foundation for a true appreciation and understanding of hypersonic air-breathing propulsion. I hope you feel that with the material found in this book, you have acquired sufficient background to tackle the technical challenges that the field presents, whether you are a student undertaking a research project, or an engineer that was just recruited to support your organization's hypersonic program, or a manager needing to diversify your division's portfolio.

References

Adamczak, D., Kimmel, R.L., and DSTO AVD Brisbane Team (2011). HIFiRE-1 trajectory estimation and preliminary experimental results. *AIAA paper 2011-2358. 17th AIAA International Space Planes and Hypersonic Systems and Technologies Conference*, San Francisco, CA (11–14 April 2011).

Bahm, C., Baumann, E., Martin, J. et al. (2005). The X-43A hyper-X Mach 7 flight 2 guidance, navigation, and control overview and flight test results. *AIAA/CIRA International Space Planes and Hypersonic Systems and Technologies*, Capua, Italy (16–20 May 2005).

Bakos, R. (2008). Current hypersonic research in the USA. Advances on Propulsion Technology for High-Speed Aircraft. Educational Notes RTO-EN-AVT-150, Paper 10. Neuilly-sur-Seine, France: RTO. pp. 10-1–10-26. www.rto.nato.int.

Bauman, E., Pahle, J.W., Davis, M.C., and White, J.T. (2010). X-43A flush airdata sensing system flight-test results. *J. of Spacecraft and Rockets* 47 (1): 48–61.

Baumann, E., Bahm, C., Strovers, B., and Beck, R. (2007). The X-43A Mach 10 mission guidance and control updates, rationale, and flight test results. NASA TM-2007-214610, January 2007. NASA Dryden Flight Research Center Edwards, CA.

Berry, S.A., Kimmel, R., and Reshotko, E. (2011). Recommendations for hypersonic boundary layer transition flight testing. *AIAA Paper 2011-3415. 41st AIAA Fluid Dynamics Conference and Exhibit*, Honolulu, Hawaii (27–30 June 2011).

Blocker, W. and Ruebush, D. (2005) X-43A stage separation system - a flight data evaluation. *AIAA 2005-3335. AIAA/CIRA 13th International Space Planes and Hypersonics Systems and Technologies Conference*, Capua, Italy (16–20 May 2005).

Bowcutt, K., Paull, A., Dolvin, D., and Smart, M. (2012). HIFIRE: An international collaboration to advance the science and technology of hypersonic flight. *28th International Congress of the Aeronautical Sciences, ICAS 2012*, Brisbane, Australia (23–28 September 2012).

Bui, T., Lux, D.P., Stenger, M.T, et al. (2007). New Air-Launched Small Missile (ALSM) flight testbed for hypersonic systems. NASA TM 2007-214624.

Cotting, C. (2004). Seminar on HyperX (X-43), given at Virginia Tech, on 4 May, 2004. http://www.dept.aoe.vt.edu/~mason/Mason_f/ConfigAero.html.

Dolvin, D. (2009). Hypersonic international flight research and experimentation technology development and flight certification strategy. *AIAA 2009-7228. 16th AIAA/DLR/DGLR International Space Planes and Hypersonic Systems and Technologies Conference*, Bremen, Germany (19–22 October 2009).

Dolvin, D. (2016). High-speed flight research: insight briefing. *Basic Research Innovation and Collaboration Center (BRICC)*, Arlington, VA (27 June 2016).

Drummond, J.P. (2014). Methods for prediction of high-speed reacting flows in aerospace propulsion. *AIAA Journal* 52 (3): 465–485.

Foelsche, R.O., Beckel, S.A., Betti, A. et al. (2006). Flight results from a program to develop a freeflight atmospheric scramjet test technique. *AIAA 2006-8119. 14th AIAA/AHI Space Planes and Hypersonic Systems and Technologies Conference*, Canberra, Australia (6–9 November 2006).

Gilbreath, H.E., Fristrom, R.M., and Molder, S. (1988). The distributed-injection ballistic launcher. *Johns Hopkins APL Technical Digest* 9 (3): 299–309.

Hass, N.E., Smart, M.K., and Paull, A. (2005). Flight data analysis of hotShot 2. *13th AIAA/CIRA International Space Planes and Hypersonic Systems and Technologies Conference,* Capua, Italy (16–20 May 2005).

Jackson, K.R., Gruber, M.R., and Buccellato, S. (2011). HIFIRE Flight 2 Overview and Status Update 2011. *Report NF1676L-12610. 17th AIAA International Space Planes and Hypersonic Systems and Technologies Conference,* San Francisco, CA (11–14 April 2011).

Jentink, T.N. and Ferlemann, P.G. (2005) Mach 10 tip-to-tail preflight prediction. *28th JANNAF Airbreathing Propulsion Subcommittee Meeting,* Charleston, SC (13–17 June 2005).

Joyce, P.J., Pomroy, J.B., and Grindle, L. (2005). The Hyper-X launch vehicle: challenges and design considerations for hypersonic flight testing. *Paper AIAA 2005-3333. 13th International Space Planes and Hypersonic Systems and Technology Conference,* Capua, Italy (May 2005).

Karlgaard, C.D., Martin, J.G., Tartabini, P.V., and Thornblom, M.N. (2005). Hyper-X Mach 10 trajectory reconstruction. *AIAA Paper 2005-5920. AIAA Atmospheric Flight Mechanics Conference and Exhibit*, San Francisco, CA (15–18 August 2005).

Kimmel, R.L., Adamczak, D., Hartley, D. et al. (2018). Hypersonic international flight research experimentation-5b flight overview. *J. of Spacecraft and Rickets* 55 (6): 1303–1314.

Lux, J. and Burkes, D.A. (2008). Hyper-X (X-43A) flight test range operations overview. NASA TM-2008-214626. January 2008. NASA Dryden Flight Research Center, Edwards, California.

Marshall, L., Bahm, C., Corpening, G., and Sherrill, R. (2005a). Overview with results and lessons learned of the X-43A Mach 10 flight. *Paper AIAA 2005-3336. AIAA/CIRA 13th International Space Planes and Hypersonics Systems and Technologies Conference, Capua, Italy* (16–20 May 2005).

Marshall, L.A., Corpening, G.P., and Sherill, R. (2005b). A chief engineer's view of the NASA X-43A scramjet flight test. *Paper AIAA 2005-3332. AIAA/CIRA 13th International Space Planes and Hypersonics Systems and Technologies Conference*, Capua, Italy (16–20 May 2005).

McClinton, C.R. (2008). High speed/hypersonic aircraft propulsion technology development. Advances on Propulsion for High-Speed Aircraft. Educational Notes RTO-EN-AVT-150, Paper 1. Neuilly-sur-Seine, France. pp. 1-1–1-32. www.rto.nato.int.

McClinton, C.R., Rausch, V.L., Nguyen, L.T., and Sitz, J.R. (2005). Preliminary X-43 flight test results. *Acta Astronautica* 57: 266–276.

Meyer, B.M. (2005). Tip to tail preflight prediction. *28th JANNAF Airbreathing Propulsion Subcommittee Meeting*, Charleston, SC (13–17 June 2005).

Morelli, E.A., Derry, S.D., and Smith, M.S. (2005). Aerodynamic parameter estimation for the X-43A (Hyper-X) from flight data. *AIAA Paper 2005-5921. AIAA Atmospheric Flight Mechanics Conference and Exhibit*, San Francisco, CA (15–18 August 2005).

Mutzman, R. and Murphy, S. (2011). X-51 development: a chief engineer's perspective. *17th AIAA International Space Planes and Hypersonic Systems and Technologies Conference*, San Francisco, CA (11–14 April 2011).

Paull, A., Alesi, H., and Anderson, S. (2002). HyShot Flight Program and how it was developed. *AIAA 2002-4939. AIAA/AAAF 11th International Space Planes and Hypersonic Systems and Technologies Conference*, Orleans, France (29 September to 4 October 2002).

Rondeau, C.M. and Jorris, T.R. (2013). X-51A scramjet demonstrator program: waverider ground and flight test. *SFTE 44th International/SETP Southwest Flight Test Symposium*, Ft Worth, TX (28 October to 1 November 2013).

Roudakov, A.S., Schickhman, Y., Semenov, V. et al. (1993). Flight testing an axisymmetric scramjet: Russian recent advances. *44th Congress of the International Astronautical Federation*, Graz, Austria (16–22 October 1993).

Sabel'nikov, V.A. and Penzin, V.I. (2000). Scramjet research and development in Russia. In: *Scramjet Propulsion*, vol. 189 (ed. E.T. Curran and S.N.B. Murthy), 223–367. American Institute of Aeronautics and Astronautics (AIAA).

Smart, M. and Stalker, R. (2010). Scramjet combustion processes. *RTO-EN-AVT-185-11. Paper presented at the AVT-185 RTO AVT/VKI Lecture Series held at the von Karman Institute*, Rhode St. Genese, Belgium (13–16 September 2010).

Smart, M.K., Hass, N.E., and Paull, A. (2006). Flight data analysis of the hyShot 2 flight experiment. *AIAA Journal* 44 (10): 2366–2375.

Voland, R.T., Auslender, A.H., Smart, M.K. et al. (1999). CIAM/NASA Mach 6.5 scramjet flight and ground test. *Paper AIAA 99-4848. 9th International Space Planes and Hypersonic Systems and Technologies Conference*, Norfolk, VA (1–5 November 1999).

Voland, R.T., Huebner, L.D., and McClinton, C.R. (2005). X-43 hypersonic vehicle technology development. *IAC-05-D2.6.01. 56th International Astronautical Congress of the International Astronautical Federation, the International Academy of Astronautics, and the International Institute of Space Law*, Fukuoka, Japan (17–21 October 2005).

Walker, S., Rodgers, F., Paull, A., and Van Wie, D. (2008). HyCAUSE flight test program. *Paper AIAA 2008-2580 presented at 15th AIAA International Space Planes and Hypersonic Systems and Technologies Conference*, Dayton, Ohio (28 April to 1 May 2008).

X-43A Mishap Investigation Board (2003). *Report of Findings: X-43A Mishap*. Washington, DC: NASA.

Powering the Future of Transcontinental Flight and Access to Space

How close are we to changing transcontinental aeronautics and access to space with hypersonic air-breathing propulsion? A number of propulsion systems that ingest air from Earth's atmosphere are now developed to accelerate reusable vehicles from subsonic to hypersonic speeds to take us around the world within hours. R&D programs also include innovative space launch vehicles propelled by engines that combine synergistically scramjets with rockets. These space launchers are intended to place satellites in Earth orbit and spaceplanes to transport people, crossing the invisible Kármán line, which marks the edge of space.

Hypersonic Civil Transports

Hypersonic air-breathing propulsion is posed to revolutionize transcontinental commercial aviation. Developments in the past two decades hint at an exciting leap in technology that will transform how we travel across the globe. This is the field where substantial engineering effort is focused in the United States, Europe, Russia, Asia, and Australia.

The most promising Mach 5 airplane in the United States today is that being developed by Hermeus Corporation. Powered by turbine-based combined cycle (TBCC) propulsion, the beautiful hypersonic aircraft portrayed in Figure 1.15 will have a range of 7400 km (4600 miles) and is envisioned to shorten the flight across the Atlantic Ocean to less than two hours. In November 2021, Hermeus unveiled a full-scale prototype of their first aircraft identified as *Quarterhorse*. The autonomous, reusable aircraft will be propelled by a TBCC ramjet designed around a flight-proven GE J85-21 afterburning turbojet with a maximum thrust of 2850–3100 lbf (12.7–13.8 kN) (dry). Cruising at an altitude of 95 000 ft, Quarterhorse is intended to validate the TBCC ramjet.

Designed and manufactured by General Electric, the J85 turbojet was conceived as an engine for high-performance trainers and tactical aircraft. According to GE, with more than 75 million flight hours experience on military and commercial models, the J85 offers the highest thrust-to-weight ratio of any production engine in its class in the world.

In August 2021, Hermeus activated the hypersonic engine test site where the J85-21 engine is being tested. According to Hermeus, completing 50 engine runs within four weeks

Scramjet Propulsion: A Practical Introduction, First Edition. Dora Musielak.

is done to validate the predictive models for the Mach 5 TBCC propulsion system and compare them with real data.

Hermeus is an American venture-backed startup. In August 2020, the company received a $1.5 million contract from the US Air Force to develop their proposed aircraft. The objectives set forth are to design and build three prototypes by 2024 to demonstrate reusability and provide data for hypersonic wargaming fidelity and help USAF satisfy future payload integration. Some people have speculated that the Hermeus aircraft could be part of the American presidential fleet as a possible Air Force One. The unveiled prototype appears to lack landing gear or flight control surfaces. But those airframe features (which depend on the TBCC integration) must be part of the company's proprietary design, and at this stage of development they are closely guarded.

Let us focus on the TBCC ramjet that will propel Hermeus. The engine incorporates a precooler upstream of the turbojet proper. Somehow the engine flow will be directed to the afterburner, which will act as a ramjet once the flight velocity becomes high enough, burning an air/fuel mixture in a subsonic combustion process to produce the thrust required of high-speed flight. A question has been raised regarding the ability of this TBCC ramjet to accelerate the aircraft to truly reach hypersonic Mach 5 velocity. However, even if the engine can only propel the 40 ft (12.19 m) long Quarterhorse to Mach 4 or so, flying a reusable full-size aircraft at those speeds will surpass the record set forth by the SR-71, and it will allow propulsion engineers to gain flight data to continue pushing the operational limits of air-breathing TBCC propulsion.

The European Space Agency (ESA) is investing toward the development of an engine that could one day allow aircraft to fly anywhere in the world in just four hours. In England, Reaction Engines Ltd. (REL) is developing the Scimitar engine, a precooled air-turbo ramjet fueled by liquid hydrogen (LH_2) that can propel an aircraft from Mach 0 to Mach 5 speeds. Scimitar is a derivative of the Synergetic Air-Breathing Rocket Engine (SABRE), an air-breathing rocket concept that exploits the unique thermodynamic properties of liquid hydrogen by using lightweight heat exchangers. In addition to agreements with ESA and the UK Space Agency, REL has partnered with BAE Systems to support the first ground demonstrator Scimitar engine.

The European LAPCAT-II program continues advancing the development of a hypersonic civil transport. The program includes a Mach 5 airliner powered by a precooled air-turbo ramjet and a Mach 8 airplane powered by a ramjet engine, both fueled by LH_2. This four-year program is co-funded by the European Commission under the theme of air transportation. It involves 16 partners representing six European member states. LAPCAT II is a follow-up of the previous LAPCAT I program intended to reduce the duration of antipodal flights (flights between two diametrically opposite points on the Earth) to less than two to four hours.

The Mach 5 aircraft will be powered by a precooled turbo-ramjet developed under LAPCAT I. *Why precooling is needed for the air-breathing engine*? At Mach 5, the stagnation temperature of air approaches 1320 K; this condition is impractical at the inlet to compress the air due to the amount of work required and because the resulting compressor delivery temperature would be too high.

The design of the Scimitar engine incorporates a high bypass fan into the bypass duct, which encloses the core engine. This is needed to match the intake air capture flow to the engine demanded flow over the supersonic Mach number range. A hub turbine drives the bypass fan, using flow diverted from the core engine nozzle. The flow then discharges into the bypass and mixes with the bypass flow.

The most important technologies that must be matured for the Scimitar engine are the heat exchanger and the contrarotating turbine, which must operate reliably for extended periods of time, at all conditions from take-off to cruise and acceleration/deceleration. The Scimitar power cycle depends on maintaining a constant top-cycle helium temperature at around 1000 K. A preburner is required to heat the helium leaving the precooler during the low flight regime when the enthalpy of the captured air, and therefore the helium temperature, would be low. Advanced materials are thus required for the heat exchanger that can cope with the high temperature part of the engine cycle.

A few years ago, the LAPCAT team determined the feasibility of a Mach 8 aircraft propelled by a scramjet engine. The team concluded that the fuel consumption during acceleration would require a large fuel fraction, and this would adversely affect the aircraft's gross take-off weight. They also considered a first-stage rocket ejector concept, but predictions indicated that such propulsion system would give low range and large take-off mass. Clearly, for a civil transport, the integrated airframe-engine design should be one that optimizes range, cruise time, and fuel consumption.

Recent reports from the LAPCAT team addressed wind tunnel testing with a 1 : 120 scale model at Mach 8 speeds. The results were encouraging, as the test model could generate a positive thrust. The European team also reported on hydrogen fuel consumption measurements and aerodynamic heating. Again, acceleration and deceleration at hypersonic speeds are crucial for feasibility of civil transports. These two phases of flight are much more dominant in the propulsion optimization process due to the longer speed-up and slow-down phases and to the relatively shorter cruise phase. Hypersonic flight requires further optimization.

There are other issues related to hypersonic flight such as aerodynamic heating, environmental effects, and sonic boom that must be properly resolved. To deal with aerodynamic heating, for example, the LAPCAT aircraft will incorporate ceramic panels and other advance materials. The European team has also stated that hydrogen-fueled airliners would not emit greenhouse-increasing gases such as carbon dioxide, sulfur oxides, or soot like today's subsonic airplanes. However, water vapor produced by hydrogen combustion stays in the stratosphere for a long time and could be a contributing factor to global warming. This is an issue taken very seriously by the European researchers. The LAPCAT-II Program has the objective to develop the Mach 8 airliner to fly well above 33 km to minimize the environmental impact. The R&D program also considers the use of an alternative fuel such as supercold liquid methane. When stored as a liquid, methane needs far less storage space than gaseous methane, and thus it would reduce fuel tank volume, thus streamlining the design of the hypersonic aircraft.

Regarding sonic boom, the European hypersonic aircraft would fly over the North Pole and cross the Bering Strait, avoiding populated land. NASA is also working with Lockheed Martin and Boeing to design airplanes that break the sound barrier more quietly. I believe that with those improved designs, it is possible that future hypersonic airplanes could then exceed the sound barrier over populated land without causing major disturbances.

The Japanese Exploration Space Agency (JAXA) continues R&D on its experimental vehicle known as Hytex, a Mach 5 concept propelled by a precooled turbojet engine (PCTJ). The PCTJ, which has already been demonstrated with ground experiments and a flight experiment up to Mach 1.8, is designed to operate from take-off to Mach 5 and uses hydrogen both as coolant for the precooler and as fuel in the combustor.

Hypersonic Air-Breathing Propulsion for Military Applications

Today, the demand for hypersonic military assets for global reach to compete with increasing global threats among nations is growing. The research and development activities related to air-breathing hypersonic technologies is motivated by the need to produce hypersonic fighter planes that can execute rapid strikes, surveillance, reconnaissance, and other military activities.

The race continues to develop hypersonic weapons, drones, and missiles that can reach a target moving at more than five times the speed of sound. In the United States, most funding comes from the Hypersonic Air-breathing Weapon Concept (HAWC), a program managed by the Air Force Research Laboratory (AFRL) and the Defense Advanced Research Projects Agency (DARPA). HAWC is designed to develop the required technologies to achieve hypersonic speed in a cruise-missile-sized platform. It considers a JP-fueled scramjet engine to enable a medium-size rapid response intelligence and surveillance (ISR) vehicle with operationally relevant range capability.

Sponsored by the HAWC program, Raytheon Technologies built a hypersonic missile propelled by a scramjet engine designed by Northrop Grumman. In September 2021, the missile underwent a successful free flight test to verify vehicle integration and release sequence and demonstrate engine ignition and scramjet performance. Just like the very successful X-51A waverider, the HAWC concept uses a rocket booster to reach the required altitude and speed, and then the missile fires its scramjet engine to fly at Mach 5. *What will be the range of the new hypersonic missile*? Let us put this in perspective. In its final flight in May 2013, the X-51A demonstrator reached Mach 5.1 traveling more than 230 nautical miles in just over 6 minutes.

Raytheon competes with Lockheed on the HAWC program. The company is considering additive manufacturing for growing exotic components or the entire missile. Raytheon used 3D printing for components of some weapons already in production and now used by the Army. The Army's long range hypersonic missile, known as Dark Eagle, is expected to be operational by 2023.

In recent years, Lockheed Martin has stated publicly that the company is working on several innovative technologies to enable long-duration, maneuverable, hypersonic flight. Lockheed's breakthroughs include new thermal protection systems, innovative aerodynamic shapes, navigation, guidance, and control improvements.

A most intriguing idea is the SR-72 spy plane, a hypersonic concept unveiled by Lockheed Martin in 2013. Although shrouded in secrecy for obvious reasons, it appears that the SR-72 will be a 4000-mile/h reconnaissance drone with strike capability. As conceived, the futuristic hypersonic spy aircraft will evade assault, take photos from enemy's sites, and attack targets at speeds of up to Mach 6. That is twice as fast as the former SR-71 aircraft. The new spy plane could reach any location on any continent in an hour!

It is easy to speculate on the required air-breathing engines for the SR-72 spy plane. In order to be able to take off and reach up to Mach 6 speeds, the propulsion system may incorporate a turbojet that can take the SR-72 from runway to Mach 3 and a hypersonic ramjet/scramjet that can propel it to hypersonic speeds. The airframe design of such vehicle is also unique, as its aerothermodynamic performance cannot be decoupled from propulsion performance due to shared surfaces and complex flow field interactions.

At the time I wrote this section (December 2021), Pratt & Whitney (P&W) – developer of the J58 engine that powered the Blackbird – announced it has been working on a reusable, low-cost more advanced hypersonic engine called the *Metacomet*. To gain perspective on the P&W engine, remember that the J58 was an exceptional afterburning turbojet designed to transition from turbojet to turbo-ramjet engine in flight. Such capability was possible with a unique design that allowed the freestream air to bypass the turbomachinery and enter directly into the afterburner, mix with fuel, burning rapidly as it were a ramjet engine thus increasing the engine output thrust. In that mode, each J58 could generate up to 32 000 pounds of thrust, a spectacular feat for a propulsion system designed in the 1950s (without computers and without the technological advances available to today's air-breathing engines). Before the SR-71 was retired, it was the fastest aircraft in the United States. In March 1990, it flew from Los Angeles to Washington, D.C. in just over an hour, traveling at an average speed of 2144 miles/h.

This fact alone should convince us that with its new Metacomet program, Pratt & Whitney is positioned to deliver a new, more advanced turbo-ramjet designed with high-temperature materials and advanced thermal protection system, and thus it could generate the power to propel aircraft from Mach 0 to Mach 5 speeds. Why this flight range is critical? Because the new P&W turbo-ramjet could propel the Lockheed Martin SR-72 strategic reconnaissance aircraft, or the Mayhem demonstrator of US Air Force Research Lab, or even for a civil hypersonic cruiser that could compete with the Hermeus aircraft. The classified Mayhem is described as a "larger-scale expendable air-breathing hypersonic multi-mission flight demonstrator."

The airframe of future reusable hypersonic aircraft (civil or military) must include advanced materials to stay intact while subjected to high dynamic loads and to withstand the extreme aerodynamic heating of hypersonic flight, as air friction alone would melt conventional materials. Lightweight carbon composites used in experimental hypersonic vehicles can withstand temperatures of up to 2900 °F (1593 °C). But *can the current CC materials meet the reusability requirement of future hypersonic aircraft*?

Another issue of concern for a hypersonic aircraft is sonic boom. A sonic boom generates a 160-decibel noise that travels to the ground and – aside from the fact that it can permanently damage human ears – it would reveal the presence of the plane. A superboom will develop when the hypersonic airplane changes its speed, turns, or maneuvers. In a superboom, the ground noise is much louder. Of course, the hypersonic plane will fly at altitudes above 30 km, and so the shock waves are expected to spread out and produce a smaller shock wave on the ground. Nevertheless, I expect the design of the Mach 6 spy plane that will incorporate unique features to minimize its boom signature.

To develop a hypersonic spy plane requires extraordinary engineering and technology advances beyond propulsion and materials. Think about it: a plane moving at Mach 5 or Mach 6 will **require hundreds of miles to turn around and unique strategies to decelerate and land**. It will also require powerful and advanced guidance systems to line up targets, more than 30 km (~98 400 ft) below. And, *how would it inspect sites from that altitude, flying at hypervelocity*? No question about it, engineers involved in hypersonic flight project are developing truly groundbreaking hypersonic vehicles for the future.

Air-Breathing Propulsion for Access to Space

For space applications, the most prominent efforts are the development of reusable launch technology, including the NASA TSTO vehicle, the Skylon single stage to orbit spaceplane developed in England, and the three-stage launch system developed by Hypersonix Launch Systems Pty Ltd., a start-up company in Brisbane, Australia, co-founded by Michael Smart.

Funded by the British government and from the UK Space Agency, Reaction Engines, Ltd. (REL) continues developing the SABRE air-breathing rocket to power the Skylon spaceplane. The unique Sabre propulsion system is designed to allow Skylon to take off from a runway and accelerate to five times the speed of sound, before switching to rocket mode, propelling the spaceplane into orbit. REL's launch system has the potential to put satellites into orbit at a fraction of the current cost.

The air-breathing SABRE rocket is unique. Consider this: atmospheric air must be compressed to about 140 atm before injecting it into the rocket combustion chamber. Such compression raises the air temperature so high that it would melt the material walls. Sabre avoids this issue by first cooling the air to −150 °C (−238 °F, 123 K), using a precooler heat exchanger, until air is almost a liquid. Then a conventional turbo compressor compresses the air to the required pressure. The development effort in the past few years has concentrated in maturing the precooler heat exchanger.

In Australia, Hypersonix is developing a reusable three-stage launcher that incorporates air-breathing propulsion, intended to radically reduce the costs of placing satellites into Earth orbit. The second stage, known as Spartan, is designed around a powerful hydrogen-fueled scramjet engine that will operate in the Mach regime from 5 to 10. On the launch pad, the vehicle will look and launch like a conventional rocket, as the first stage will be a rocket booster. Once it reaches hypersonic speeds, however, the first stage will drop away and the scramjet-powered second stage will unfurl its wings to blast into the upper atmosphere. When it runs out of air, the scramjet second stage will separate and a small conventional rocket will carry the payload into space. Nice!

What makes the Australian launch concept more attractive is the reusability of the first two stages. Once completing their operation, both the scramjet stage and the rocket booster will fly back to a runway landing. Only the final third rocket stage will burn up in the atmosphere after releasing its payload into orbit.

In April 2021, Hypersonix began collaboration with Boeing Research and Technology (BR&T) to develop a "proof of concept" launch system with the integrated SPARTAN scramjet engine. A month later, the team reported that they had successfully tested the reusable engine in the shock tunnel at the University of Queensland, demonstrating that it can accelerate from Mach 5 to Mach 10 as designed.

Another interesting hypersonic technology development is the spaceplane concept by China Aerospace Science and Technology Corporation (CASTC). Although no details of the hypersonic propulsion system have been made public, I assume it will be a combined cycle engine concept since, according to public reports, the vehicle would take off from a runway like an aircraft, and for the hypersonic velocity range, it would use a scramjet engine to propel the vehicle to almost 100 km above sea level. Rocket boosters ultimately will provide the additional thrust, giving the Chinese spaceplane enough power to reach low Earth orbit.

With these developments and many more highly guarded industry-government programs related to scramjets and other high-speed propulsion technologies, I hope to convince the reader that the future of air-breathing propulsion for hypersonic flight is assured.

> *Hypersonic propulsion advances flight for us to soar above the clouds and reach the stars.*
>
> —Dora de Musielak

Glossary

Acceleration of gravity nominal gravitational acceleration of an object in a vacuum near the surface of the Earth, $g_0 \approx 9.81$ m/s^2.

Adiabatic wall temperature bounding- or zero-velocity streamline temperature that a wall would attain if it were perfectly insulated, also known as recovery temperature. See Chapter 9.

Actively Cooled Structure vehicle structure that uses a coolant fluid to flow through passages adjacent to the hot structure. The coolant is typically the fuel, especially when the vehicle carries a large volume of cryogenic fuel aboard. See Chapter 9.

Aerodynamics the study of the properties of moving air, especially focused on the interaction between the air and solid bodies such as aircraft. Aerodynamics studies the forces on a vehicle caused by it motion through the atmosphere. The rules of aerodynamics explain how an airplane is able to fly. All bodies that move through the air are subject to the principles of aerodynamics.

Aerodynamic heating the heat transfer from the hot boundary layer to the cooler vehicle surface – also called convective heating. This heating is aggravated when the shock-layer temperature is high enough that the thermal radiation emitted by the gas itself gives rise to a radiative flux to the surface – called radiative heating. For the slender hypersonic vehicles, aerodynamic heating is highest at leading edges, in the regions with shock interactions, in stagnation regions, and at locations where the flow transitions to turbulence. See Chapter 3.

Aerodynamic lift Lift force on an aircraft that overcomes the weight force. All aircraft utilize aerodynamic surfaces in order to generate the lift force. Spaceplanes can use a variety of wing shapes such as delta wings, straight wings, and lifting bodies. The lift force generated by these surfaces is many times that of the drag that they induce.

Aerospace Plane reusable exo-atmospheric vehicle that can take-off from the ground and reach low Earth orbit (LEO). It can be configured as SSTO or TSTO. See Space Plane.

Scramjet Propulsion: A Practical Introduction, First Edition. Dora Musielak.

Aerothermodynamics a key technology for design and optimization of hypersonic vehicles. It provides necessary databases for choice of trajectory, for guidance, navigation, and for thermal protection and propulsion systems. Computational aerothermodynamics is a powerful tool for improving our understanding of physical phenomena in hypersonics. See Chapter 3.

Aftbody the afterpart of a vehicle, which plays a huge role in the design and performance of an airframe integrated scramjet system. See Figure 2.1.

Afterburner duct section placed downstream of the engine turbine where an additional amount of fuel is combusted with the turbine discharge flow before the core flow expands in the exhaust nozzle. Afterburner is added as a mechanism to augment the thrust. See Chapter 10.

Air-breathing propulsion Propulsion system designed to ingest the air from the atmosphere and use the oxygen as oxidizer to mix and combust the fuel. Such propulsion can be based on gas turbines, or can operate on physical principles that do not require turbomachinery such as ramjets and scramjets. A rocket engine can also be designed to ingest air to reduce the amount of onboard oxidizer it carries. See Chapter 10.

Air-breathing Rocket a special kind of rocket engine designed with an inlet to take the oxidizer for combustion from the surrounding atmosphere rather than carrying it onboard as traditional rockets do. See SABRE, Chapter 10.

Aircraft a vehicle designed to fly within the Earth's atmosphere.

Airframe-integrated scramjet a scramjet propulsion system that uses the airframe or fuselage (forebody and afterbody) as external compression and expansion surfaces to extend the engine's inlet and nozzle, respectively. See Chapter 2.

Atmosphere the layer of gases surrounding a planet. Earth's atmosphere contains enough oxygen to supply the oxidizer required by air-breathing propulsion systems. See Chapter 2.

Atmospheric entry or reentry interface altitude that separates the atmosphere and outer space. For Earth returning spacecraft, the entry interface is located at an altitude 120 km, which is at the border where atmospheric effects on spacecraft become important.

Attitude the direction a vehicle points into the wind.

Augmenter afterburner on low-bypass turbofan engines. Core airflow and bypass (fan) airflow are mixed aft of the turbines in the exhaust. Fuel nozzles supply atomized fuel into the airflow and an igniter ignites the fuel–air mixture. Augmenters are used on low-bypass turbofan engines to increase thrust for short periods during takeoff, climb, and combat flight.

Boundary layer a layer of flow formed adjacent to a surface or wall wherein the dissipative mechanisms of viscosity, thermal conduction, and diffusion are strong. The viscous boundary layer is characterized by its thickness, denoted δ and defined as that distance above the wall where $V = 0.99\ V_e$; here, V_e is the velocity at the outer edge of the boundary layer. Hypersonic boundary layers are thicker than those developed at low speed, growing as the thickness is directly proportional to the square of the Mach number. See Chapter 3.

Boundary layer transition (BLT) Boundary layers can be laminar or turbulent, or may transition from laminar to turbulent. BLT refers to the region of transitional flow that occurs when the velocity profile over a surface begins to change as a result of instabilities and the boundary layer becomes enlarged. If the disturbances continue, the flow over the plate will become fully turbulent. See Chapter 3.

Borghi diagram a representation of combustion regimes on a diagram that qualitatively indicates the turbulence intensity versus turbulence length scale of a combustion system. Borghi diagrams define regimes of premixed turbulent combustion in terms of velocity and length scale ratios. See Chapter 12.

Breguet Equation formula to estimate the range or distance flown by an aircraft in a given time, not including takeoff, climb, approach, and landing distances. See Chapter 6.

Burner also known as combustor or combustion chamber, this is the part of the engine where the fuel is injected, mixed with the air, ignited, and burned continuously. See Chapter 5.

Bypass Ratio (BPR) ratio between the mass flow rate of air drawn through the fan disk that bypasses a turbofan engine core (uncombusted air) to the mass flow rate passing through the turbofan engine core. For example, a turbofan engine with a 10:1 bypass ratio means that 10 kg of air passes through the fan for every 1 kg of air passing through the core of the engine. A low bypass ratio is required for high flight Mach number operation.

Chemical kinetics branch of physical chemistry concerned with understanding the rates of chemical reactions. Since in supersonic combustion there is insufficient time available for the exothermic combustion reactions to reach equilibrium, the scramjet combustor analysis requires to consider the rate at which fuel/air chemical reactions proceed. See Chapter 5.

Choked Flow a condition in gases and vapors when the fluid velocity reaches sonic values at any point in the flowpath. See thermal choking.

Cold Structure vehicle structure that uses an external metallic heat shield and insulation material to protect the primary structure from aerodynamic heating. See Chapter 9.

Combined Cycle Propulsion a propulsion concept that integrates multiple engine cycles into a single system capable of operating in multiple modes (cycles) to extend the flight regime. These different modes allow a propulsion system to be extremely versatile and efficient while transitioning through varied flight conditions. Two common combined cycle configurations are turbine-based combined cycle (TBCC) and rocket-based combined cycle (RBCC). See Chapter 10.

Combustion a chemical process in which a fuel reacts rapidly with oxygen (oxidizer) and gives off heat. The fuel for a propulsion system can be a solid, liquid, or gas. For aircraft engines, the fuel is usually injected in liquid form. See Chapters 5 and 6.

Combustion chamber part of an engine where fuel and oxidizer are injected, mixed, ignited, and combustion takes place. See Chapter 5.

Compressor turbomachinery component of a jet engine where the atmospheric air is compressed. The compressor is driven by the turbine. It rotates at high speed, adding energy to the airflow and at the same time squeezing (compressing) it into a smaller space. Compressing the air increases its pressure and temperature.

Cowl it is the streamlined housing for the scramjet engine proper, a cover specifically designed to form part of the inlet and is attached to the airframe. See Figure 2.1.

Cracked JP-7 fuel Cracking is a chemical reaction in which larger saturated hydrocarbon molecules are broken down into smaller, more useful hydrocarbon molecules. Cracked JP-7 is therefore a JP-7 fuel that is heated prior to combustion and the result is a fuel with different composition, depending on the cracking method utilized. See Chapter 6.

Damköhler number parameter used to relate chemical reaction timescale to other phenomena occurring in a system, denoted Da. In analysis and simulation of scramjet combustion flowfields, the Da number is used for determining whether diffusion rates or reaction rates are more important for defining a steady-state chemical distribution over the length and time scales of interest. See Chapter 12.

DNS stands for direct numerical simulation. A method in which turbulent flow is directly simulated by numerically solving the full N-S equations, without any form of time or length averaging, i.e., both mean flow and all turbulent fluctuations (eddies) are simulated. DNS is not suitable for hypersonic flows. See Chapter 12.

Drag the force that opposes motion. Air-breathing engines require to ingest enough airflow in order to generate the thrust F to accelerate the vehicle and to overcome the drag forces on it.

Dry Thrust nonaugmented thrust. The maximum thrust produced by jet engines with no afterburner is also called military thrust.

Dual-mode Scramjet a scramjet engine that operates with combustion transitioning from subsonic to supersonic. It requires an isolator, a duct section separating combustor from engine inlet. Dual-mode scramjet must operate seamlessly over ramjet and scramjet speed Mach range, $3.0 < M < 15.0$. See Chapter 7.

Dynamic pressure of the atmosphere the component of the airflow pressure that represents fluid kinetic energy (i.e., motion). See Chapter 2.

Endo-atmospheric vehicle a vehicle that flies within the Earth's atmosphere, at altitudes below 100 km (328 999 ft).

Endothermic Fuel a fuel characterized by an additional heat sink capability over that of conventional fuels. The endothermic state is obtained by undergoing chemical reactions through heat absorption, reforming, and thermal cracking prior to injecting the fuel into the combustor. Endothermic processes such as cracking yield a substance that is much lighter than the heavy hydrocarbon fuel. Called cracking because the original long-chain hydrocarbon molecules are broken into lighter molecules that absorb heat (an endothermic process). See Chapter 6.

Entry an altitude when a space vehicle enters the atmosphere of a planet.

Equivalence Ratio combustor parameter defined as the actual fuel/air ratio divided by the fuel/air ratio required for complete combustion (stoichiometric fuel/air ratio). See Chapter 5.

Exo-atmospheric vehicle a space access vehicle with speeds up to Mach 25 (orbital).

External Burning method considered for thrust augmentation in which an additional amount of oxidant gas is injected into the expanding exhaust gas in the nozzle that still contains a significant amount of unburned fuel. See Chapter 8.

Flameholder a combustor device which generates turbulence, shock waves, and maintains a recirculation region through geometric effects in the flowpath. For example, fuel injectors can be integrated with wall cavities or steps in the scramjet flowpath. See Chapter 5.

Flamelet numerical method used to model turbulent reactions, assuming that chemical time scales are shorter than turbulent time scales, such that flame in a combustor can be approximated as an ensemble of laminar flamelets. See Chapter 12.

Forebody the part of the vehicle forward of the largest cross section and which plays a huge role in the performance of the airframe-integrated scramjet engine. The forebody is designed to provide the scramjet with a uniform flow at a prescribed back pressure.

A long forebody yields the largest air capture area. Its aerodynamic configuration is derived under a set of cruise conditions: flight Mach number, prescribed planar shock wave shape, and flight altitude. Data generate the flowfield from which the forebody design is derived. The forebody shape is derived from one planar shock wave that intersects with the cowl lip for effective engine mass flow capture. The forebody compression surfaces are designed by tracing stream-surfaces emanating from the prescribed shock waves. See Fig. 2.1.

Fuel the fluid injected in a combustor with the primary purpose of providing propulsive energy to the engine. In high-speed applications, the fuel can be hydrogen, or special type of hydrocarbons. It must burn rapidly and generate the thermal energy required for thrust production. The choice in fuels is of crucial importance and must be made with careful consideration to the mission and the vehicle type. See Chapter 6.

Fuel/Air Ratio the proportion (on a mass basis) of fuel to air in a combustor. See Chapters 1 and 5.

Gas turbine a combustion engine consisting of an upstream rotating compressor coupled to a downstream turbine and a combustion chamber in between.

Gravity turn motion by a launch vehicle on its ascent to orbit, when the upper or last stage gradually transitions from vertical to horizontal flight, an effect caused by the force of gravity as the last stage reaches its orbital location.

Heat Shield a protective cover designed for a hypersonic vehicle to survive very high aerodynamic heating.

Hot Structure a vehicle structure that does not employ a thermal protection system. The external surface of a hot structure acts as the primary load-bearing vehicle structure, and the external surface absorbs all the heat, operating efficiently up to 3300 K without active cooling or insulation. See Chapter 9.

Hypermixer a device used to enhance the mixing of the injected fuel with the main combustor flow. Examples of hypermixers considered for the scramjet combustor include ramps, struts, and pylons. Struts allow a deeper penetration because the fuel is directly injected into the airflow, above the boundary layer. Injecting the fuel downstream from a rearward facing step or ramp is also done to enhance mixing. See Chapter 5.

Hypersonics the field of study related to motion at high speeds greater than about five times the local speed of sound. The term *hypersonic* covers a wide range of speeds, from about Mach 5 to Mach 25, which is approximately the speed of atmospheric re-entry.

Hypersonic Air-Breathing Vehicle craft powered by an air-breathing engine capable of hypersonic speed ($M > 5.0$). Such hypersonic vehicle must operate at relatively low altitudes to maintain the high dynamic pressure required for maximum engine performance.

Hypersonic Cruiser Vehicle that can take-off from the ground, climb to a high altitude, and cruise through the atmosphere at hypersonic speeds to its destination, and land conventionally like any other airplane. It can be crewed or automatic, capable of carrying passengers and/or cargo. Capable of repetitive missions with minimum turn-around time and ground support infrastructure. Major loads on the cruise phase are thermal, aerodynamic, and thrust. See Chapter 10.

Hypersonic flight refers to a vehicle flight that can occur strictly within the atmosphere, called endo-atmospheric hypersonic flight, or outside the atmosphere, called exo-atmospheric hypersonic flight. Examples of vehicles in hypersonic flight: hypersonic cruise airplanes powered by air-breathing propulsion, rocket-powered space launch vehicles, and unpowered gliders.

Hypersonic Velocity High-speed flow regime where energy transfer between flow and thermodynamic and chemical processes become significant. Mach numbers at which supersonic linear theory fails. Flow regime where the specific heat ratio (γ) is no longer constant, and temperature effects on fluid properties must be considered. Flight regime with Mach number greater than 5 for streamlined aircraft. Blunt bodies cause large flow disturbances at Mach 3 so hypersonic regime could begin at that velocity. Flight at speeds where aerodynamic heating dominates physics of flight. See Chapter 1.

Hypersonic Vehicle a missile, aircraft or spacecraft that travels at least 5 times faster than the local speed-of-sound, or velocities $V > 1.5$ km/s, which is $M > 5.0$ in Earth's atmosphere.

Inlet also known as air intake, inlet is the engine duct designed to ensure smooth airflow into the engine despite air approaching the inlet from directions other than straight ahead. See Chapter 4.

Inlet Mode Transition In a TBCC engine with over-under dual flowpath configuration, mode transition refers to redirecting airflow between inlet flowpaths. See Chapter 10.

Inlet Speeding also known as "shock-on-lip" condition or scramjet design point. Condition when the inlet is most efficient as it captures all compressed freestream airflow. See Chapter 3.

Inlet Starting critical process required by the scramjet engine in which the air flow within the inlet or compression system is arranged to change from subsonic to supersonic as the vehicle accelerates,

	ensuring that efficient compression is obtained over the entire range of freestream Mach numbers. See Chapter 4.
Isolator	the duct in the scramjet engine, a section added to maintain stable operation at low hypersonic speeds. The isolator diffuser contains the inlet shock train. See Chapter 4.
Jet engine	a reaction engine discharging a fast-moving jet that generates thrust by jet propulsion. Turbojets, turbofans, rocket engines, ramjets, and pulse jets are jet combustion engines. See Chapter 1.
Karlovitz number	nondimensional parameter defined as the ratio of chemical time scale to the smallest turbulent time scale, used to characterize a combustion flow field. See Chapter 12.
Kármán line	A boundary commonly used to define outer space; it is the end of Earth's atmosphere at an altitude of 100 km (62 miles). In this region, where the gas density is very thin ($\rho = 0.6 \times 10^{-6}$ kg/m^3), the flow transitions from continuum to a rarified, free molecular regime, i.e., when the molecular mean free path is greater than a characteristic spacecraft dimension.
Launch to Orbit	a process that uses a vehicle to launch payloads into space or low Earth Orbit. Conventional launch systems use rockets. With hypersonic air-breathing propulsion, launch to orbit can be accomplished with one stage, two stage, or three stage vehicles.
LES	stands for Large Eddy Simulation, a computational method that solves the spatially averaged Navier–Stokes equations. LES reduces the computational cost by ignoring the smallest length scales, which are the most computationally expensive to resolve, via low-pass filtering of the Navier–Stokes equations. See Chapter 12.
Low Bypass Turbofan	Engine with a Bypass ratio (BPR) of less than 1.0. Military combat aircraft use engines with low bypass ratios to compromise between fuel economy and the requirements of combat: high power-to-weight ratios, supersonic performance, and the ability to use afterburners, all of which are more compatible with low bypass engines.
Mach number	the speed of a vehicle relative to the local speed-of-sound.
Mechanics	the study of how objects respond to forces.
MECO	stands for Main Engine Cut Off It refers to the moment when the main rocket stage has depleted the propellant or completes a burn and cuts off.
Nose Tip-to-Tail Analysis	methodology developed by NASA researchers to predict the performance of the fully integrated X-43A vehicles, including simulation of propulsion effects. The tip-to-tail analysis yields the integration of aerodynamic and propulsion flow fields, using diverse analytical tools that examine flow-field characteristics

and scramjet flowpath component performance and the external flowfields, combined to obtain preflight predictions of the X-43A longitudinal performance increments. See Figure 12.1.

Nozzle the exhaust duct of the air-breathing engine used to expand and accelerate the combustion gases produced by burning fuel and air. See Chapter 8.

Orbit the path an object follows around another object subject to gravitational force.

Orbiter Hypersonic vehicle that can take off and climb to a high altitude and then accelerate hypersonically. At an appropriate speed and altitude, a rocket engine helps to boost the vehicle to low Earth orbit. This vehicle can re-enter the atmosphere, descend and land. See Chapter 12.

Prandtl Number dimensionless parameter defined as the ratio of momentum diffusivity to thermal diffusivity, and is denoted Pr. For most gases over a wide range of temperature and pressure, this number is approximately constant, e.g., Pr~0.71 for air. Therefore, this number can be used to determine the thermal conductivity of gases at high temperatures, where it is difficult to measure experimentally.

Pre-cooled Air Breathing Propulsion an engine that utilizes methods to reduce the high temperature of air entering the inlet system. The precooled air can then be compressed to combustion chamber operating pressures with reduced power requirements. This approach can be used in turbo engines, or in air-breathing rockets (e.g., SABRE).

Propellant a chemical substance, usually a mixture of fuel and oxidizer, for propelling a rocket engine. Propellant can also be a pressurized gas or plasma that is used to create movement of a fluid or to generate propulsion of a vehicle or projectile.

Propulsion Engineer a person educated to use and apply the laws of physics, perform scientific experiments and analysis, and who utilizes analytical skills, engineering principles and mathematics to design, test, and develop propulsion systems and related technologies.

Ramjet The simplest air-breathing jet engine operating without turbomachinery. Ramjet engines work on the principle of ram compression, using the engine's forward motion to compress the incoming air. The ram effect occurs when a volume of air is forced into small space at high enough speeds, compressing the air to a higher pressure. Ramjet engines do not require any mechanical compression, that is, the compressor is not needed in the ramjet. A ramjet cannot produce thrust for takeoff as it requires ram compression, which develops until the flight Mach number is $M_0 > 1$. At the same time, the ramjet ceases to produce thrust at high flight velocity, $M_0 > 5$, when it experiences very high ram compression. See Chapter 10.

Ram drag engine inlet momentum term representing an opposing drag force that contributes negatively to the thrust generated. See Chapter 2.

Range the distance a vehicle can fly with a given load of fuel. The maximum distance that the aircraft can fly is equal to the ground speed times the maximum time that the plane can stay aloft. See Chapters 1 and 10.

RANS stands for Reynolds averaged Navier Stokes, and it refers to the numerical method that solves ensemble-averaged (or time-averaged) Navier–Stokes (NS) equations. The RANS include a form of Navier–Stokes equations in which additional terms (Reynolds stresses) are included to account for time averaged effects of turbulence. With this approach, all turbulent length scales are modeled. RANS is the most widely used approach for computing hypersonic flows. See Chapter 12.

RAS stands for Reynolds-averaged simulations in CFD. See Chapter 12.

Recovery factor a measure of how much of the kinetic energy of the fluid is converted to thermal energy on the surface. See Chapter 9.

Re-entry returning to Earth from space; entering the atmosphere of Earth from space.

Regenerative Cooling an active cooling technique that utilizes a liquid coolant, generally the fuel, circulated through specially designed channels or tubes over a hot structure (or blanket over the combustor) to remove some of the thermal load. The heated gaseous fuel is then injected directly into the engine combustion chamber. Regenerative cooling is the most commonly used cooling method in liquid propellant rocket engines. See Chapter 9.

Relaminarization reverse transition of turbulent flow, the result of a strong flow acceleration on a turbulent boundary layer, transitioning to a boundary layer which contributes to a transport of heat. This process is also called reverse transition of turbulent flow. See Chapter 8.

Reusable Multi-Stage Launch System a reusable launch system that is divided into two or more vehicles (stages), each having their own propulsion system, stacked one on top of the other or matted. The first vehicle (booster stage) carries the other stages to a predetermined altitude and velocity, and after separating from the upper stage flies back like an airplane, returning to the launching pad. The upper stages (or stage) continue their powered ascent mission until the last stage (orbiter) reaches outer space to deliver the payload to a predetermined Earth orbit. Examples of reusable multistage launch systems are SSTO, TSTO, and three-stage to orbit configurations. See Chapter 10.

Reynolds Number defined as the ratio of inertial forces to viscous forces within a fluid which is subjected to relative internal movement due to different fluid velocities. Denoted Re, the Reynolds number helps predict flow patterns in different fluid flow situations. At low Reynolds numbers, flows tend to be dominated by laminar flow, while at high Reynolds numbers flows tend to be turbulent. See Chapter 3.

Rocket a propulsion system that is a self-contained device that operates on the action and reaction principle, designed to expel a small amount of mass of hot gas out of the rear at high velocity (i.e., momentum).

Rocket-Based Combined Cycle (RBCC) a propulsion system that integrates an air-augmented rocket and a dual-mode scramjet. See Chapter 10.

SABRE stands for **S**ynergistic **A**ir-**B**reathing **R**ocket **E**ngine, the hybrid propulsion system that will propel the 84-m long Skylon spaceplane designed by UK company Reaction Engines Ltd. SABRE is a unique engine that will use atmospheric air in the early part of the flight before switching to rocket mode for the final ascent to orbit. The Skylon-SABRE concept paves the way for true spaceplanes – lighter, reusable, and able to fly from conventional runways.

SBAO stands for Surrogate-Based Analysis and Optimization. SBAO is a set of computational tools that can reduce the number of expensive, high-fidelity simulations by using a succession of approximate (surrogate) models, which are created from data or the results of submodels. See Chapter 12.

Schmidt number dimensionless number defined as the ratio of momentum diffusivity (kinematic viscosity) and mass diffusivity. Denoted Sc, this parameter is used to characterize fluid flows in which there are simultaneous momentum and mass diffusion convection processes.

Scramjet a hypersonic air-breathing engine where the airflow through the engine always exceeds the speed-of-sound, even when it mixes with the fuel in its combustion chamber.

SERN see Single Expansion Ramp Nozzle.

Shock-on-lip condition where the shock waves generated by the vehicle forebody are focused (impinge) on the leading edge of the engine cowl. This is the design point of the scramjet. The engine cowl leading edge is designed for maximum air capture to maximize thrust. It requires that the vehicle is configured and operated such that the compression shocks produced by the forebody ramps intersect the cowl leading edge. See Chapter 3.

Shock Train system of oblique or normal shocks developing in the scramjet isolator section, where pressure rises and Mach number decreases. The increase of static pressure occurs within the initial duct region, after which the boundary layer grows as the flow moves downstream. See Chapter 4.

Shock wave disturbance wave that occurs in air when a vehicle is flying faster than the speed-of-sound.

Shock-wave/ boundary-layer interactions (SWBLI) when a shock impinges and reflects from the boundary layer, the boundary layer thickens and, for higher shock strengths (equivalent to high wedge angles), it will separate. Examples of shock-wave/boundary-layer interactions are: (i) shock impingement, (ii) ramp flow, and (iii) transonic flow over a bump. See Chapter 3.

Single Expansion Ramp Nozzle (SERN) typical exhaust nozzle used by asymmetric vehicle-integrated scramjet; this is an asymmetric nozzle, with only one solid surface upon which the combustion hot gases expand. See Chapter 8.

Single Stage to Orbit (SSTO) a launch system characterized by a vehicle with horizontal take-off and landing capability, able to utilize conventional airfields, accelerate to hypersonic speeds, and achieve low Earth orbit (LEO) in a single stage. SSTO examples: NASP, Skylon. See Chapter 10.

Skylon a single-stage-to-orbit spaceplane developed by British company Reaction Engines, Ltd. Powered by two SABRE air-breathing rockets, Skylon is designed to do the same job as traditional rocket launchers while operating like an aeroplane, potentially revolutionizing access to space. See Chapters 1 and 10.

Space Launch Vehicle (SLV) vehicle propelled by a rocket or combined cycle propulsion system with the purpose of carrying a payload (cargo or crew) from Earth's surface into outer space.

Spacecraft a vehicle that can move in space, typically designed for space travel or exploration of interplanetary space such as a space probe, space telescope, or a planet orbiter.

Spaceplane a winged space vehicle designed to take off horizontally, fly into orbit, carry out a space mission, and return to Earth, landing on runways like airplanes at an airport.

Scramjet Powered Accelerator for Reusable Technology AdvaNcement (SPARTAN) SPARTAN is a scramjet engine developed as a second stage of a reusable launch vehicle that will be used to deliver satellites into Low Earth Orbit (LEO). The scramjet engine is made from CMC composites, reusable, scalable, robust, and reliable. It generates thrust to travel at speeds from Mach 5 to Mach 12. See Chapter 10. Visit https://hypersonix.com.au/

Specific Impulse performance parameter denoted Isp that indicates how much thrust the engine produces per every unit mass of propellant (fuel plus oxidizer) it uses per second. Since the air-breathing engine does not need to carry oxidizer on board, its specific impulse is much higher than that of the rocket. With the highest (fuel)-specific impulse, dual-mode scramjets are the most efficient airbreathing engines in the flight regime $3 < M < 12$. See Chapters 1, 2, and 10.

Stanton Number dimensionless heat transfer coefficient, a quantity that must be most carefully selected on the basis of data and/or experience to give valid results for the convective heat transfer situation under consideration. See Chapter 9.

Stoichiometric Fuel/Air Ratio this is a unique proportion of fuel to air for combustion that results in neither excess oxygen nor any excess fuel. See Chapter 5.

Surrogate-Based Analysis and Optimization (SBAO) approach to reduce computational burden by approximating and replacing an expensive simulation model with a cheaper-to-run surrogate model. SBAO tools have demonstrated to be effective for the design of computationally expensive models. SBAO tools can reduce the number of expensive, high-fidelity simulations by using a succession of approximate (surrogate) models, which are created from data or the results of submodels. See Chapter 12.

Takeno Flame Index (TFI) a metric indicative of flame activity in the combustor, which can help to identify regions of mixed and premixed combustion. See Chapters 5 and 12.

Take-over speed the speed at which a ramjet or scramjet engine can begin operation.

TET stands for turbine entry temperature and represents the maximum engine cycle temperature limited by the ability of turbine blades to remain intact. Current jet engines may have a nominal TET of about 2000 K (cooled) or higher. See Chapter 10.

Thermal Choking process where the flow approaches Mach 1 after absorbing a critical amount of heat. Heat addition will cause both supersonic and subsonic flow to approach Mach 1, resulting in choked flow. See Chapter 7.

Thermal barrier coating a special high-temperature resistant coating applied to engine components to ensure they can withstand the combination of mechanical, thermal, and chemical loads. See Chapter 9.

Thermal Protection System (TPS) specially designed engineering system incorporated into the hypersonic vehicle with the objective of minimizing heat conduction into the vehicle proper to maximize performance without adding unnecessary mass. A reusable TPS consists of materials that are mechanically or chemically unchanged by flight missions and can be safely flown a number of times (with or without servicing). See Chapter 9.

Thrust Augmentation techniques for generating additional thrust beyond that derived from the simple ramjet or scramjet engine added thrust is available to use at critical points of the mission (e.g., take-off, climb acceleration, transonic point). Thrust can be augmented by incorporating ejectors (e.g., ejector ramjet), with an afterburner (e.g., afterburning turbojet), by fuel and oxidizer enrichment, or via external burning. See Chapter 10.

Thrust Potential parameter used to more objectively quantify the potential performance of a combustor flowpath. The thrust potential determines the momentum portion of ideal potential net thrust that could be obtained when a flowpath is truncated at a given streamwise location and coupled at that location to an ideal thrust nozzle. See Chapter 5.

Transonic thrust pinch refers to the flight region near Mach 1 where the excess thrust of the engine is significantly reduced, that is, where the available thrust and the required thrust curves are "pinched" closer together. See Chapter 10.

Transonic Speed the flight condition in which a range of velocities of airflow exist surrounding and flowing past an air vehicle or an airfoil that are concurrently below, at, and above the speed of sound in the range of Mach 0.8–1.0, i.e., 965–1236 km/h (600–768 mph) at sea level.

Thrust the force that propels a vehicle. The time rate of change of linear momentum is equal to a force. The reaction is an opposite force called thrust that propels a rocket or jet engine forward.

Turbine a rotary mechanical device that extracts energy from a fluid flow and converts it into useful work. In the jet engine, the work produced by the turbine is used for producing thrust. A turbine is a turbomachine with one moving part called a rotor assembly, which is a shaft or drum with blades attached. Moving fluid acts on the blades so that they move and impart rotational energy to the rotor.

Turbine-Based Combined Cycle (TBCC) a propulsion system that integrates a turbojet or turbofan with a dual mode scramjet engine. See Chapter 10.

Turbofan turbojet engine that incorporates a ducted fan at the front of the engine designed to ingest additional air to augment the thrust and improve propulsive efficiency.

Turbojet an air-breathing jet engine consisting of a gas turbine with a propelling nozzle. The gas turbine has an air inlet, a compressor, a combustion chamber, and a turbine (that drives the compressor). An afterburning turbojet is fitted with a uniquely-designed duct downstream of the turbine, a second combustion chamber where additional fuel is burned to augment the turbojet engine thrust. See Chapter 10.

Turboramjet a propulsion system that integrates a turbojet and a ramjet in a wraparound configuration using a single flowpath. This concept is used to extend the operational range of either engine. See Chapter 10.

Turbulence model a mathematical model used to describe flow turbulence, based on the concept of scales (velocity, length, and time). The objective of turbulence modeling is to prompt equations to anticipate the time-averaged velocity, pressure, and temperature fields. Hence, a number of turbulence models have been constructed and used; these include, Spalart-Allmaras, k-ε, k-ω, low Reynolds number k-ε, SST, and v2-f turbulence models. The Reynolds Stress Model is the most complete turbulence model with regard to representing turbulent flow. See Chapter 12.

Turbulent Prandtl number nondimensional parameter defined as the ratio between the momentum eddy diffusivity and the heat transfer eddy diffusivity. Denoted Pr_t, this ratio is useful for solving the heat transfer problem of turbulent boundary layer flows. The simplest model for Pr_t is the Reynolds analogy, which yields a turbulent Prandtl number of 1. From experimental data, Pr_t has an average value of 0.85, but ranges from 0.7 to 0.9 depending on the Prandtl number of the fluid in question. See Chapter 12.

Turbulent Schmidt number important parameter used in a turbulence model, defined as the ratio between coefficients of turbulent (eddy) viscosity and turbulent (eddy) diffusion of scalar property. Denoted Sc_t, this number determines the diffusion of a mixture (gaseous) in the mixing process. Many CFD studies assume $Sc_t = 0.7$ since this value agrees with some experiments. However, as the Schmidt number is decreased, the mixing rate is found to increase and increases the diffusion process in the flow. However, no universal value of Sc_t has been established and empirical values are typically used. See Chapter 12.

Two-Stage-to-Orbit (TSTO) a launch system consisting of an air-breathing-powered first-stage hypersonic cruise vehicle (CV) matted to a rocket-powered-second-stage ascent and reentry vehicle (ARV). Staging or separation of the two vehicles occurs at a given altitude and predetermined Mach number when the CV turns around to return to landing site, while the ARV continues its ascent to orbit powered by conventional rocket engines. Aerothermodynamic phenomena critical to the design of TSTO differ for the two stages. These differences are due to differences in the trajectories and in the different configurations used in the two vehicles. See Chapter 10.

VULCAN-CFD stands for Viscous Upwind aLgorithm for Complex flow Analysis, it is an advanced analysis code developed at NASA Langley

Research Center for high-speed flows that has become a standard for simulating scramjet external and internal flows to the present day. See Chapter 12. For details on the code, visit https://vulcan-cfd.larc.nasa.gov/

Wet Thrust augmented engine thrust, typically generated by the use of an afterburner. An engine producing maximum wet thrust is operating at maximum power.

Nomenclature

Latin	Definition
a	Local speed of sound
A	Cross-sectional area
C_D	Coefficient of drag
C_L	Coefficient of lift
c_p	Specific heat at constant pressure
c_v	Specific heat at constant volume
e	Specific internal energy
F	Thrust force (uninstalled)
f	Fuel/air ratio (mass basis)
g_0	Gravitational acceleration on the surface of the Earth
H	Flight altitude over S.L.; enthalpy
h	Specific enthalpy
h_t	Specific total (stagnation) enthalpy
h_{PR}	Ideal fuel energy density (fuel heat of reaction)
I_{sp}	Specific impulse
k	Thermal conductivity
LH_2	Liquid hydrogen fuel
LHC	Liquid hydrocarbon fuel
M	Mach number
m	Mass
$\dot{m}$	Mass flow rate
Pr	Prandtl number
p	Static pressure
p_t	Total (stagnation) pressure
$\dot{Q}$	Total heat load

(Continued)

Scramjet Propulsion: A Practical Introduction, First Edition. Dora Musielak.

Latin	Definition
$\dot{q}$	Specific (local) heat flux
q	Dynamic pressure
R	Cruise range ; Universal Gas constant
Re	Reynolds number
r	Radius
St	Stanton number
S	Entropy
s	Specific entropy
T	Static temperature
T_t	Total (stagnation) temperature
t	Time
V	Velocity
W	Vehicle weight
x	Axial direction
Greek	
α	Angle of attack
β	Oblique shock wave angle
γ	Ratio of specific heats
δ	Boundary-layer thickness
μ	Fluid viscosity
η_m	Fuel/air mixing efficiency
η_o	Overall propulsion efficiency
η_p	Propulsive efficiency
η_r	Static pressure recovery
η_{th}	Thermodynamic efficiency
θ	Turning angle; boundary-layer momentum thickness
π	Total pressure ratio
ρ	Fluid density
τ	Total temperature ratio
φ	Fuel/air equivalence ratio
ψ	Engine cycle static temperature ratio
Subscripts	
1, 2, 3, etc.	Engine station numbers
0	Freestream
*	Choking or critical point

Latin	Definition
c	Compression process
e	Exit or exhaust
f	Final
i	Initial; ignition
r	Radial direction
SL	Sea-level reference
s	Isentropic process
t	Throat; tangential direction; total or stagnation condition; turbine
w	wall
y	Transverse direction

Index

Scramjet Propulsion: A Practical Introduction, First Edition. Dora Musielak.

d

e

u

v

w

x

y